A PRACTICAL UNDERSTANDING OF PRE- AND POSTSTACK MIGRATIONS

VOLUME 2
(Prestack)
2007 Edition

by

John C. Bancroft, Ph.D.
University of Calgary

Course Notes Series No. 13
Lawrence M. Gochioco, Series Editor

A PRACTICAL UNDERSTANDING OF PRE- AND POSTSTACK MIGRATIONS

Volume 2

(Prestack)

by

John C. Bancroft, Ph.D.

2007 Edition

Copyright February 2007

©

No figures or examples may be reproduced by any means without permission from the author.

Library of Congress Cataloging-in-Publication Data

Bancroft, John C.
 A practical understanding of pre- and poststack migrations / by John C. Bancroft. -- 2007 ed.
  p. cm. -- (Course notes series ; no. 13, 14)
 Includes bibliographical references.
 ISBN 1-56080-144-1 (v. 1) -- ISBN 1-56080-145-X (v. 2) -- ISBN 0-931830-48-6 (series)
 1. Seismic prospecting--Mathematical models. 2. Seismic waves--Mathematical models. I. Title.

TN269.8.B36 2007
622'.1592--dc22

2007027206

SOCIETY OF EXPLORATION GEOPHYSICISTS

The international society of applied geophysics
Tulsa, Oklahoma, U.S.A.

Contents

Preface to Volume II

Acknowledgements

Prestack Abbreviations and Symbols

Volume II (Prestack)

Chapter 7 Prestack Modelling

7.1	Introduction to Prestack Data	7.2
7.2	Modelling of Source Gathers (Shot Records)	7.18
7.3	Constant Offset Sections	7.42
7.4	Prestack Eikonal Equation Modelling	7.62
7.5	The Marmousi Model	7.64
7.6	MO Processing of Prestack Data	7.70
7.7	Cheops Pyramid for 2-D Data	7.72
7.8	Moveout (MO) Processing of Cheops Pyramid	7.76
7.9	Comparison of Prestack Summation Surfaces	7.78
7.10	Modelling Linear Reflectors with Scatterpoints	7.84
7.11	Prestack Modelling of 3-D Data	7.90
7.12	Summary of Points to Note in Chapter 7	7.91

Chapter 8 2-D Dip Moveout (DMO)

8.1	Introduction	8.2
8.2	DMO of Source Gathers	8.20
8.3	DMO of Constant Offset Sections	8.24
8.4	Dip Moveout (DMO) Algorithms	8.32
8.5	Pre-Processing for DMO (and Prestack Migration)	8.72
8.6	DMO Processing Loops	8.76
8.7	Real World DMO	8.80
8.8	DMO Examples on Modelled Data	8.83
8.9	Kinematics of DMO on Cheops pyramids	8.91
8.10	Summary of Points to Note in Chapter 8	8.94

Chapter 9 Prestack Migration (and 3-D DMO)

9.1 Introduction to Prestack Migration 9.2
9.2 DMO and Prestack Migration 9.6
9.3 DMO, Prestack time, IMO to stack, poststack depth migration 9.8
9.4 Direct Kirchhoff migration from prestack volume 9.10
9.5 Prestack Migration of Source Records... 9.12
9.6 Inversion Method of Imaging a Source Record 9.38
9.7 Source-Receiver (Shot-Geophone) Method 9.44
9.8 Prestack Migration of a Constant-Offset Section 9.48
9.9 Constant Angle and τ-p Migration 9.54
9.10 Stolt Prestack Migration of 2-D Data 9.56
9.11 Gardner's DMO-PSI 9.58
9.12 Equivalent Offset Migration (EOM) 9.68
9.13 Kinematic Comparison between DMO-PSI and EOM 9.74
9.14 Prestack Migration Aliasing and Irregular Geometry 9.74
9.15 Prestack Migration of 3-D Data Volumes 9.78
9.16 2-D Prestack Time Migration 9.80
9.17 3-D DMO 9.84
9.18 Cross Dip, Down Dip, and Azimuth in Acquisition Design 9.92
9.19 General comments 9.95
9.20 Summary of Points to Note in Chapter 9 .... 9.96

Chapter 10 Examples of DMO and Prestack Migrations

10.1 Description of Figures 10.2
10.2 Comparison of DMO and Prestack Migration Ellipses 10.8
10.3 3-D Model Test for DMO 10.9
10.4 Example of the DMO Operator 10.12
10.5 Testing and Evaluating DMO 10.16
10.6 DMO of an Aliased Dipping Event 10.20
10.7 DMO Processing of Real Data 10.24
10.8 DMO of a Source Record 10.28
10.9 Prestack Migration of a Source Record 10.30
10.10 Comparisons of DMO and Migrations on Real Data 10.32
10.11 Prestack Migration of a Structurally Complex Model 10.44

10.12 Examples of 3-D Salt Imaging 10.48
10.13 Summary of Summary of Points to note in Chapter 10 .... 10.52

Chapter 11 Equivalent Offset Migration (EOM)

11.1   Introduction 11.2
11.2   The Equivalent Offset h_e 11.18
11.3   Mapping an Input Trace to a CSP Gather 11.26
11.4   The CSP Gather 11.36
11.5   3-D Processing 11.48
11.6   Depth migration examples (including anisotropy) 11.52
11.7   Rugged Topography Processing 11.56
11.8   Residual Statics Before NMO 11.58
11.9   Converted Wave Processing 11.66
11.10 Other Features of EOM 11.70
11.11 Benefits of EOM Processing 11.73
11.12 Disadvantages of the Process 11.74
11.13 Summary of Points to Note in Chapter 11 .... 11.75

Chapter 12 Comparisons and Evaluations

12.1   Which Algorithm is Best for a Given Application? 12.2
12.2   Differences Between Time and Depth Migrations? 12.4
12.3   What Factors Are Considered when Choosing a Migration Algorithm? 12.5

Appendix 4 Kinematic Derivations for DMO-PSI and EOM Appen-4.1

Appendix 5 Stacking Chart for 2-D Data Appen-5.1

References (Full ID number sort)

ID number sort .... Reference 1
Author sort .... Reference 12

Preface to volume II

I have been told on many occasions that I should not include the old technology of DMO, especially when there are many efficient prestack migrations. In a similar thread of thought, I have also been told that we should not poststack migrate or use any time migrations.

Conventional DMO is based on constant velocity models, and in some places it may do more harm than good. There are however, areas where DMO does work. In addition, the fundamentals of DMO are the foundation of some prestack migrations, especially the DMO-PSI process.

The objective of the seismic process is to use surface measurements (and what ever other information is available) to estimate the geology of the subsurface in depth i.e. (*x, y, z*). <u>Standard processing</u> is the first stage in which noise is reduced, bandwidth recovered, near surface model resolved, static corrections estimated, and amplitudes balanced. The output should be a prestack time migrated section that has required minimal input from an interpreter.

<u>Advanced processing</u> requires extensive interaction with an interpreter (geologist) to aid in building a depth model for depth migration. Algorithms for accurate depth migration have been available for many years; however, estimation of the depth model remains a difficult task in areas with complex geology. The inclusion of anisotropy, mode converted waves, and/or the varied use of wavefront times (maximum energy, first arrival, etc.) continue to improve these algorithms and to help build better models.

The ultimate objective of "seismic processing" is to create an accurate depth model of the subsurface (that may also include rock parameters). However, in many areas, a prestack time migration may be adequate to identify a drilling location. In other areas, the geology may be so complex that it is not possible to define a depth model accurately enough for a depth migration and the best result remains the prestack time migration. Consequently, economics will continue to determine the level of processing that is required for a particular project. These economics must be continually evaluated in response to the costs of acquisition and processing.

It is the intent of these course notes to provide exposure to the basic prestack migration algorithms and thereby enable accurate and informed decisions.

Prestack migration should always produce a superior image in comparison to one produce by a poststack migration. The exceptions are usually caused by inferior implementations of the prestack algorithms, especially in areas with an uneven acquisition geometry. The balancing of energy for a poststack migration is accomplished in the stacking process; one that is very powerful. The balancing of energy in a prestack migration is more difficult, but must match the quality of that used in preparing the poststack data.

JCB, April 2005

These notes are in the middle of a major change. Figure numbering may not be complete, and page numbers may not correspond to those in the Table of Contents.

Prestack Abbreviations and Symbols

AAF	Antialiasing filter
MO	moveout *as in moveout correction* (dipping and horizontal)
NMO	normal moveout (historically used for horizontal layering)
	my term for moveout of horizontal layers
DD-MO	dip-dependent moveout,
	my term for moveout of dipping layers
DMO	dip moveout, prestack partial migration
	(historically used for moveout of dipping layers)
DMO-PSI	Gardner's method of prestack migration [322]
Kirchhoff MO	MO correction that combines antialiasing filters, scaling, and interpolation.

V_{rms} RMS velocity

1. <u>defined</u> from interval velocities of horizontal layers
   - defines the curvature at the apex of a diffraction in horizontal layers
   - moveout for short offsets will approximate an hyperbola
   - will equal stacking velocities if <u>short offsets, isotropic, and horizontal layers</u>
   - simplifies raypaths to be linear on the time section
   - used with Dix equation to estimate interval velocities

2. Usually <u>equated to stacking velocities in horizontal layers</u>
   - not equal to stacking velocities in the presence of anisotropy
   - usually used in violation of the requirement for horizontal reflectors and short offsets (should use the term Stacking velocities)
     1. will not give a good estimate of interval velocities
   - used with Kirchhoff time migrations to:
     1. simplify time section raypaths to be linear
     2. enable use of the double square-root (DSR) equation
   - velocities are defined along an image ray

V_{stk} dip dependent velocity for DD-MO of dipping data: $V_{stk} = V_{rms}/cos(dip)$
   - can vary with the length of the offset range

V local velocity of interest (often RMS)

Time variables are usually <u>two-way time</u>.

- T input time
- T_0 vertical zero-offset time to scatter point (image ray)
- T_n after NMO correction when using RMS velocities
- T_{dn} dip dependent NMO correction using $V_{stk} = V_{rms}/\cos(\beta)$
(same as zero-offset time from CMP for a dipping reflector)
- T_d tangential time after NMO-DMO
- T_g tangential time (relative to T_0) after Gardner's DMO
- T_c time at offset b in DMO unit circle.
- $t = 2h/V$ radius (in time) of DMO unit circle (x, t) from $-h$ to h.
- $rj\omega$ Root differential filter for 2-D Kirchhoff migrations, (45° phase shift).

Figures relating scatterpoint, reflectors, reflections, traveltimes, and raypaths.

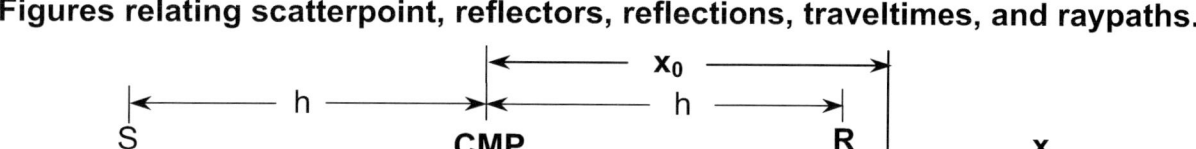

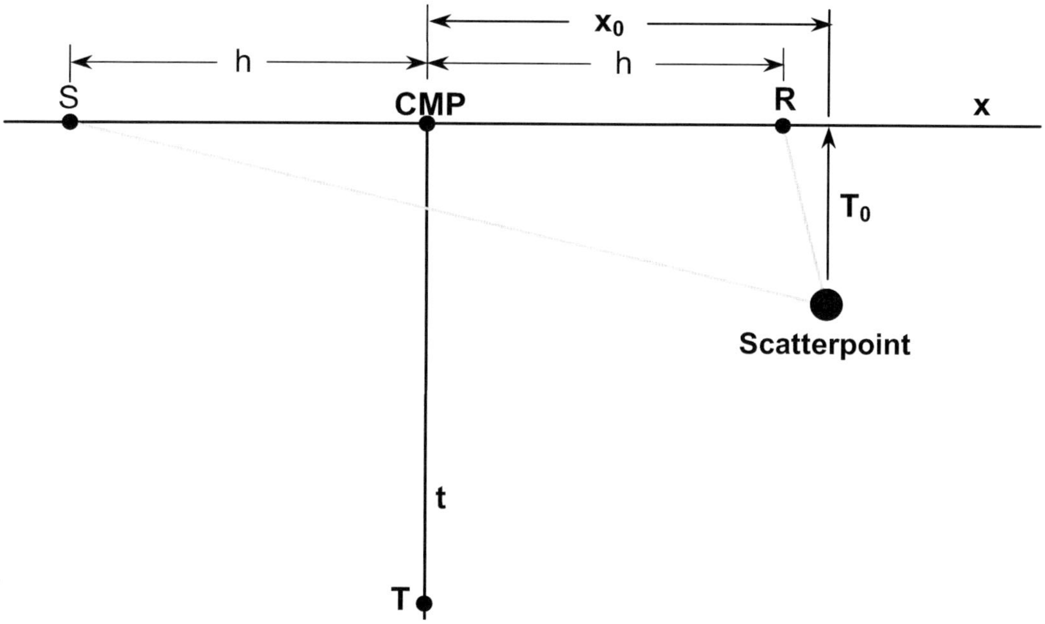

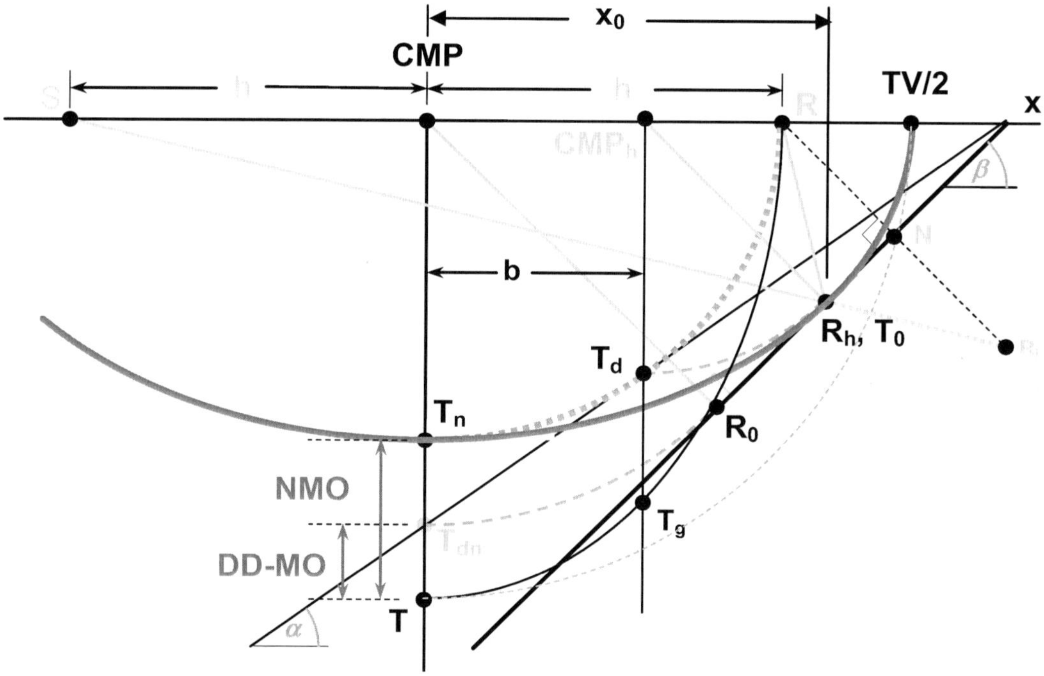

The velocities used in the above figures allow two-way travel times to equal one-way distances, i.e. V = 0.5.
- The distance CMP – R_0 is equal to the zero-offset travel time T_{dn}.
- Half the distance S – R_i or S – R_h - R is equal to the input time T.

Acknowledgments

Preparation of this second volume in its present form has received support from the CREWES project and a number of processing companies.

I especially thank:

Darren Foltinek	CREWES
Henry Bland	CREWES
Shaowu Wang	CREWES
Hugh Geiger	CREWES
Gary Margrave	CREWES
Xinxiang Li	CREWES
Yong Xu	CREWES
Saleh Al-Saleh	CREWES
Xiang Du	CREWES
Irene Kelly	PanCanadian
Carmine Militano	C and C Systems
David Ganley	Arcis (EXSSEL)
Rick Wallace	Ulterra Geoscience
Mike Marcoux	Veritas DGC
Olav Barkved	Amoco

Sponsors of the CREWES project

Development of the EOM method of prestack migration has involved a number of persons who have contributed theoretical development, software programs, and data processing examples. I especially recognize Hugh Geiger and Gary Margrave for many stimulating discussions and their numerous supporting publication.

Chapter Seven

Objectives

- Know that the incident and reflected rays follow different paths, but maintain equal angles of incidence and reflection.

- Define and become familiar with the various ways of displaying prestack data such as source gathers, constant offset sections, or CMP gathers in the prestack volume (x, h, t).

- Construct a source gather and a constant offset section.

- Know that conventional stacking is only valid for horizontal reflectors.

- Become aware that care must be taken when forming limited-offset stacks from a range of offsets to create a constant-offset section.

- Know that diffractions don't stack.

- A scatterpoint forms a surface in the prestack volume (x, h, t) which is referred to as Cheops pyramid.

7.1 Introduction to Prestack Data

7.1.1 -D prestack volume for 2-D data

The prestack traces from 2-D data may be sorted and displayed in a <u>3-D prestack volume</u> (x, h, t) as illustrated in Figure 7.1a, where x is the CMP position, h the half source-receiver offset, and t the two-way time.

When the <u>velocities (V) are constant</u> the vertical times t of the zero-offset plane (x, h = 0, t) may be converted to depth ($z = tV/2$), to allow the <u>superposition</u> of a coincident plane that contains the geometry of the 2-D line (x, z). This plane could contain the source and receiver locations for an input trace, along with <u>wavefronts or raypaths</u>. The surface location of a source and receiver is indicated in Figure 7.1a.

Figures 7.1.b-c illustrate <u>source records</u> within the prestack volume. The grid on the surface identifies the CMP locations and offset increments that correspond to the CMP interval. The station interval for receivers is two grid locations. Sources are spaced at two station intervals.

The top surface of the cube is similar to a <u>surface stacking chart</u> and may illustrate the location of source gathers, constant offset sections, CMP gathers, etc.

Figure (b) shows <u>left-sided</u> source records identified on the surface (t=0) by dots. One record contains vertical lines that represent the traces.

Figure (c) shows only the surface locations of <u>right-sided</u> sources.

All 2-D prestack traces may be mapped into the 3-D volume (x, h, t). Many of the traces in the volume are unique with respect to offset and CMP location, however duplications may exist. The duplications may be reduced or eliminated by using a signed value for the offset.

Sources, receivers, reflectors, scatterpoints, and raypaths are all located on the zero-offset plane.

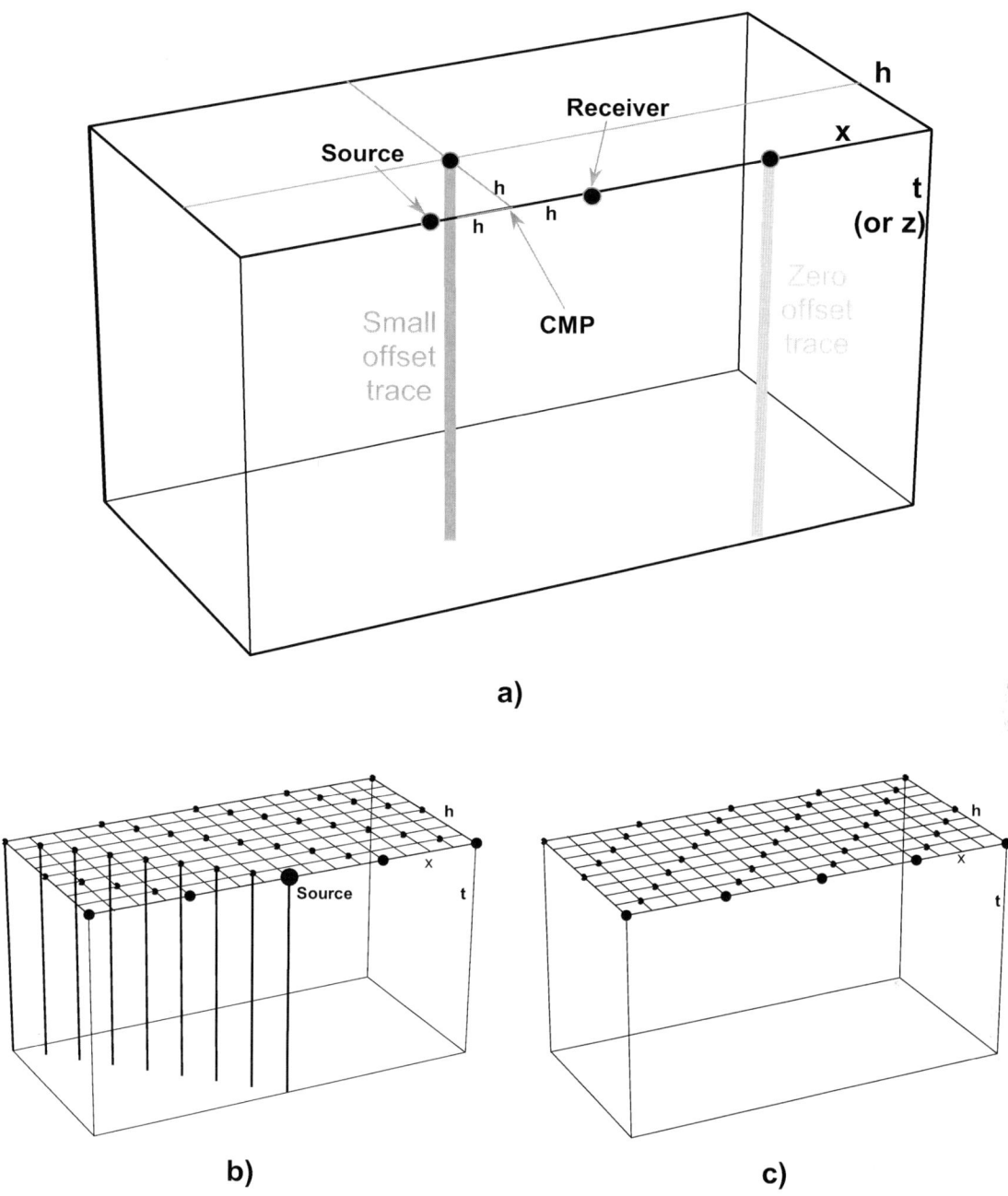

Figure 7.1 Prestack cube of data showing, a) traces at zero and mid-offset, b) the surface positions of left-sided source records, and c) a number of right-sided source records showing only the surface location.

3-D seismic data (x, y, t) would require a 4-D volume (x, y, *h*, t) to display a similar figure.

7.1.2 Source gathers

Source gathers are also referred to as <u>shot records</u>. The term "gather" is used for a collection of traces based on a parameter, which in this case is a common source. The term "<u>source</u>" is preferred to "shot" to generalize the traditional explosive source to include other types such as vibroseis or air guns.

The source gather is a <u>natural</u> arrangement of the data as it was acquired in this format and stored on tape.

The <u>memory requirements</u> of large 3-D projects may require leaving all the prestack data in source record format until the final stack.

2-D data in source record format may have <u>prestack processes</u> of DMO or prestack migration applied. In these cases the <u>number of traces</u> in the new source records may increase at the edges of the record. Sometimes these additional traces are <u>ignored</u> to preserve the original geometry of the line (a poor option).

Figure 7.2a illustrates one shot location and five receivers. The plan view illustrates locating the traces in offset space (x, h), that form a line that is at 45 degrees.

Source records may preserve <u>the sign of the half offset h</u>, however, data in CMP gathers and constant offset sections usually ignore the sign of h.

Most of the prestack volumes (x, h, t) in these notes will use the magnitude of h for simpler illustrations.

Figure 7.2b illustrates a collection of traces from a <u>two-sided source record</u> displayed in the prestack volume. Part (c) of this figure shows the vector form that maintains the sign of the offset h, while part (b) shows an absolute value of h.

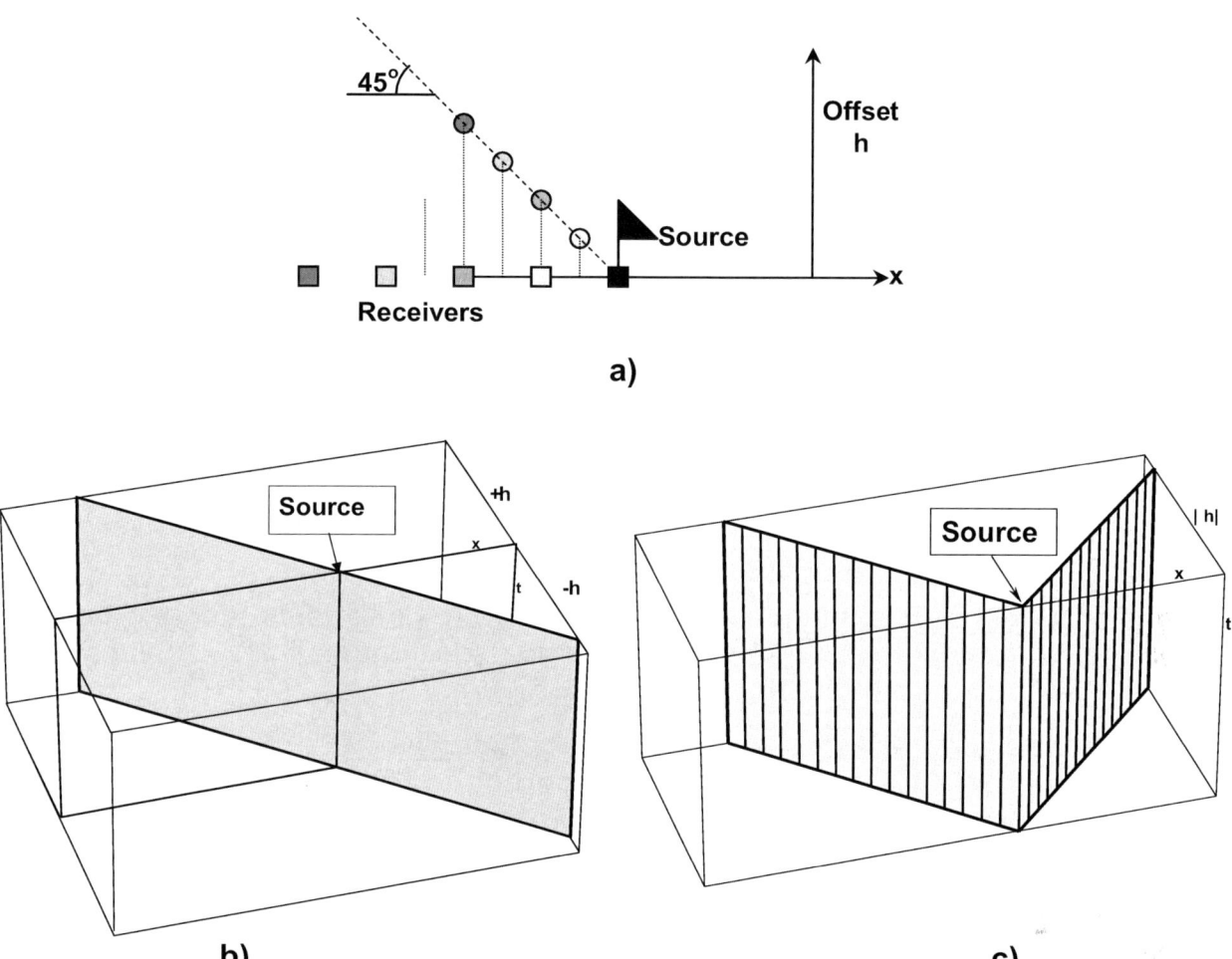

Figure 7.2 Source records with a) showing one shot and five receivers with the corresponding midpoint location, b) the vector location of traces in a two-sided source gather in the prestack volume (*x, h, t*) while b) shows the magnitude or scalar value off the offset.

7.1.3 Constant offset sections (gathers)

Data may be sorted into bins that have a limited range of source/receiver offset. These limited offset sections may then be referred to as a constant offset section that has a fixed offset h defined at the center of the offset range. Processes such as MO, DMO, prestack migration, or inverse MO will assume the fixed offset value.

Traces within each constant offset section are sorted by CMP position as illustrated in Error! Reference source not found.. Conventional 2-D acquisition has regular offset spacing with the number of offsets defined by the geometry of the spread. Typically, the number of offsets is determined by the number of receivers on one side of the source.

The data or traces in constant offset sections may be quite sparse, especially when sources are located at fourth station intervals. This lack of coverage may be used as an argument to increase the range of offsets used in creating the constant offset sections. (See the constant offset sections in Figure 7.1c where the sources are at every second station).

- Loss of temporal resolution may result if the range of offsets is too large. The change of moveout within the offset range, (see Figure 7.3a), causes a time smear that acts as a high cut filter.

- Some processing streams may use the actual true offset to apply MO (and sometimes DMO) to each trace before insertion to the constant offset section. This maintains resolution in gathers with large offset ranges. Inverse MO may then be used at the fixed offset.

- A partial MO correction, referred to as differential moveout correction, may also be applied to the traces within the range of offsets to align traveltimes at the center of the offset range.

DMO and prestack migration may be performed more economically on constant offset sections.

Some 3-D projects may also be sorted into constant offset volumes (*x, y, t* or *z*). It should be noted that some processes, (such as DMO), still require source-receiver orientation within the 3-D volume.

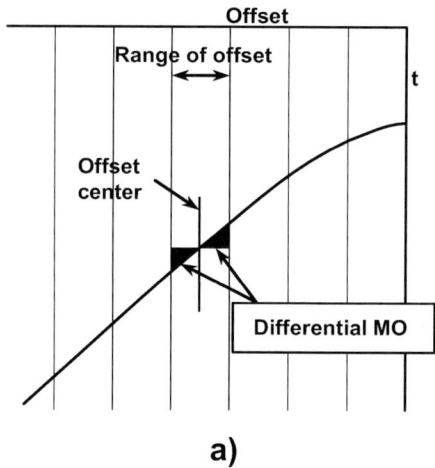

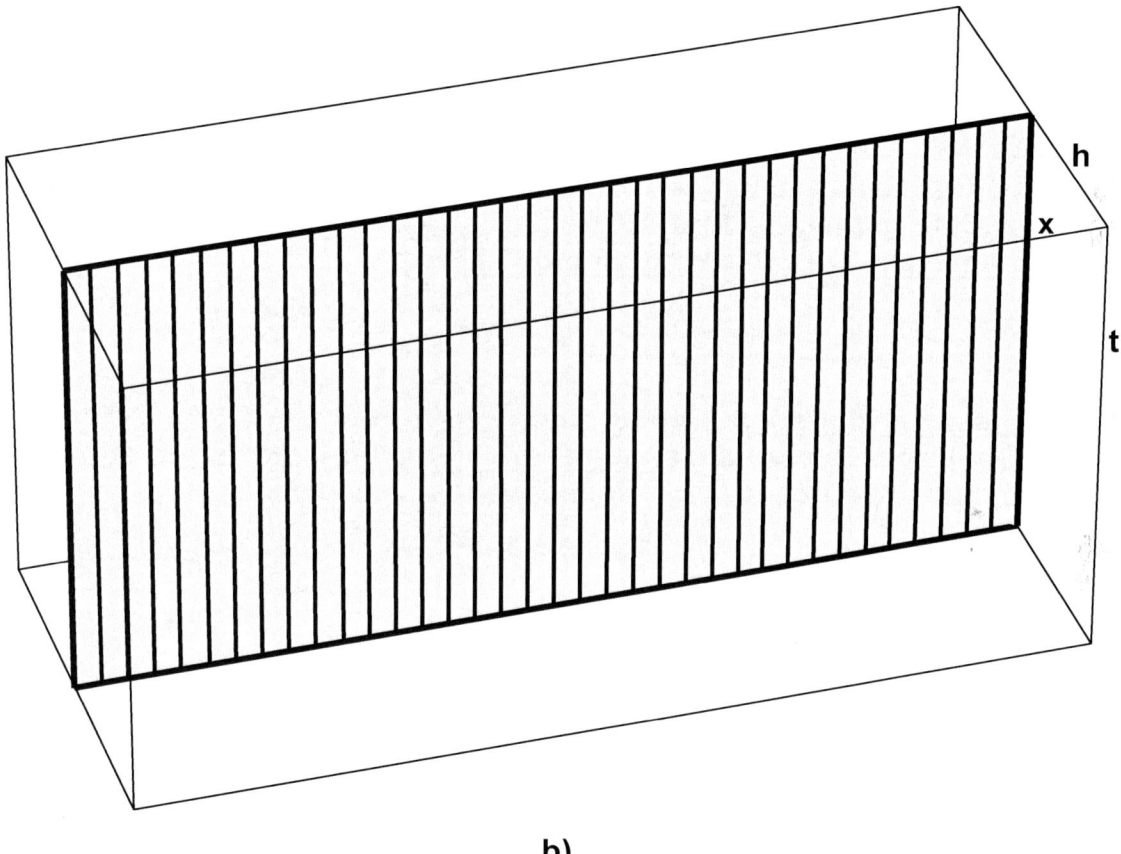

Figure 7.3 Formation of a constant offset section, a) showing residual MO correction on a CMP gather, and b) the constant offset section in the prestack volume (x, h, t).

7.1.4 Common midpoint (CMP) gathers

The common midpoint (CMP) is a <u>subsurface location</u>, halfway between the source and receiver and represents the reflection points in horizontally layered geology.

Traces with the same CMP location are <u>sorted by offset</u> to form a CMP gather as illustrated in Error! Reference source not found.**a**.

After MO processing, all the traces in a CMP gather should be similar to the zero-offset trace and when summed form a <u>stacked trace</u>. (Differences may be due to MO stretch, multiples, coherent noise, etc.)

The <u>effects of dip</u> are reduced in CMP gathers making them suitable for evaluating velocities.

The <u>number of traces in a CMP gather</u> is usually much less than the number of original offsets. This is again due to a lack of surface coverage. See Error! Reference source not found.**c** where the sources are at every second station.

<u>Trim statics</u> are usually estimated and applied to the traces in the CMP gather prior to stacking.

The lack of offset coverage in one CMP gather may be supplemented by summing a number of CMP gathers to form COFF's or <u>super CMP gathers</u>. These gathers <u>aid velocity analysis</u>, but will <u>smear dipping energy</u>.

<u>DMO and prestack migration</u> spread energy from one trace in a CMP gather to neighbouring traces along a <u>constant offset section</u>, (or source record), as illustrated by the gray area in Error! Reference source not found.**b**.

Therefore, DMO or prestack migrations cannot be applied <u>directly to the CMP gathers</u>.

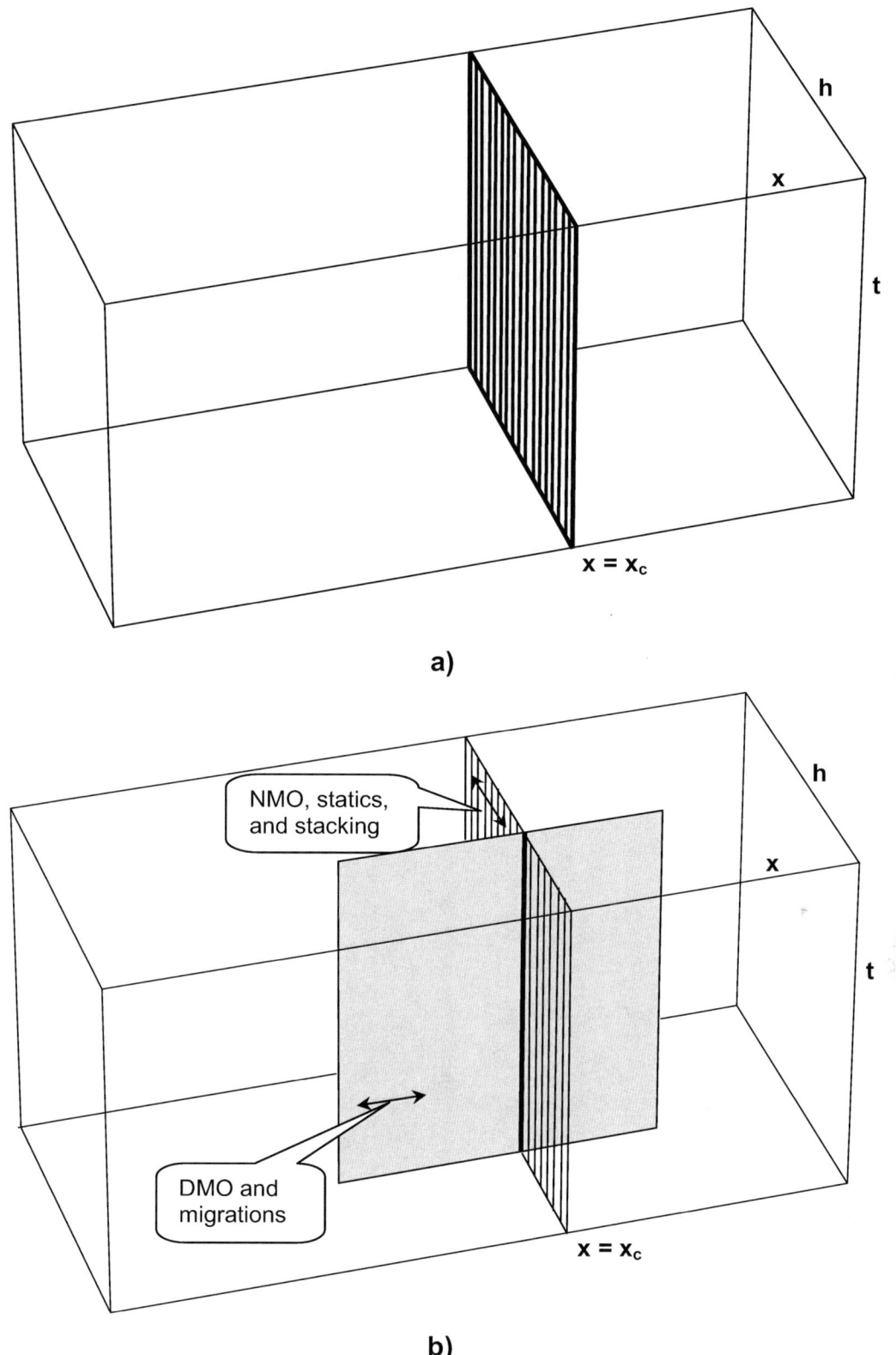

Figure 7.4 Prestack volume illustrating traces in a CMP gather (x = xc, h, t).

7.1.5 Common scatterpoint (CSP) gathers

A common scatterpoint (CSP) gather is a <u>prestack migration gather</u> that collects all input traces that contain energy from a scatterpoint.

- Offsets in a CSP gather are defined by the distance from the scatterpoint to the source and receiver,
- …not the source-receiver offset.

A CSP gather is similar in appearance to a CMP gather as they both define the <u>subsurface location</u>, and the traces are <u>sorted by offsets</u>.

All traces within the <u>prestack migration aperture</u>, regardless of the source or receiver position, may be used to form a CSP gather. Traces within the CSP gather are sorted by offset. The <u>choice of offset</u> is a topic of considerable interest. (Try half the CMP interval.)

Figure 7.5 illustrates the movement of five input traces to the CSP gather by curved arrows. The arrows follow circular paths defined by the offset h_{csp} defined by

$$h_{csp}^2 = x^2 + h^2 \tag{0.1}$$

where h is half the source-receiver offset, and x is the distance between the CMP and CSP gathers. The final CSP gather would include all input traces, including the five zero-offset traces and the three traces in the CMP gather.

CSP gathers can have:

- a very <u>high fold</u> at each offset,
- much <u>larger offsets</u> than a CMP gather.

Additional information on the CSP gather is presented in Chapters 10 and 11.

Other prestack migration gathers may be formed by CMP processes that apply MO, DMO and zero-offset migration on constant offset sections or source gathers. The data is then sorted back to CMP gathers for velocity analysis. These gathers may also be referred to as CRP gathers to indicate the more correct positioning of the reflection points. This type of processing is only accurate when the velocities are accurate; velocity analysis iterations are required.

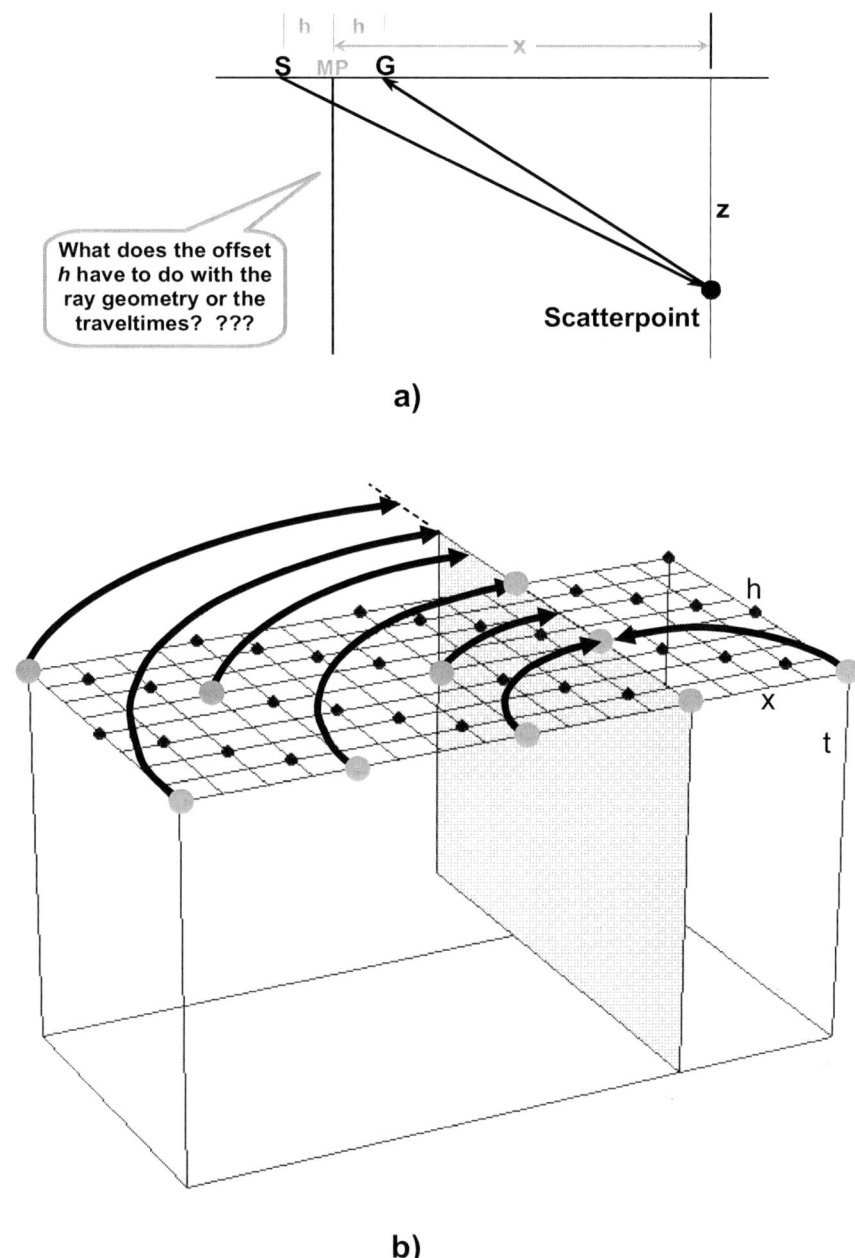

Figure 7.5 Formation of a CSP gather, a) illustrating on offset ray paths where the offset h has little to do with the traveltime, and b) the prestack volume illustrating the inclusion of six input traces to one CSP gather.

Any input trace can contribute energy to any output trace providing the recording time is long enough. Why?

7.1.6 Common receiver gathers

The data may be sorted into common receiver gathers as illustrated in Figure 7.6. Traces in these gathers all have <u>the same receiver location</u>, but different sources. A plot of common receivers would appear similar to the source gather of Error! Reference source not found.b, but would usually contain fewer traces that are spaced proportional to the receiver interval.

Receiver gathers are used for surface consistent processing (such as amplitude or statics), and for the s-g (source-geophone) method of prestack migration.

Use of the term common <u>geophone</u> gather is discouraged, as in land acquisition it is normal practice to have many geophones strung together and connected to one receiver position, while in marine acquisition, hydrophones are used.

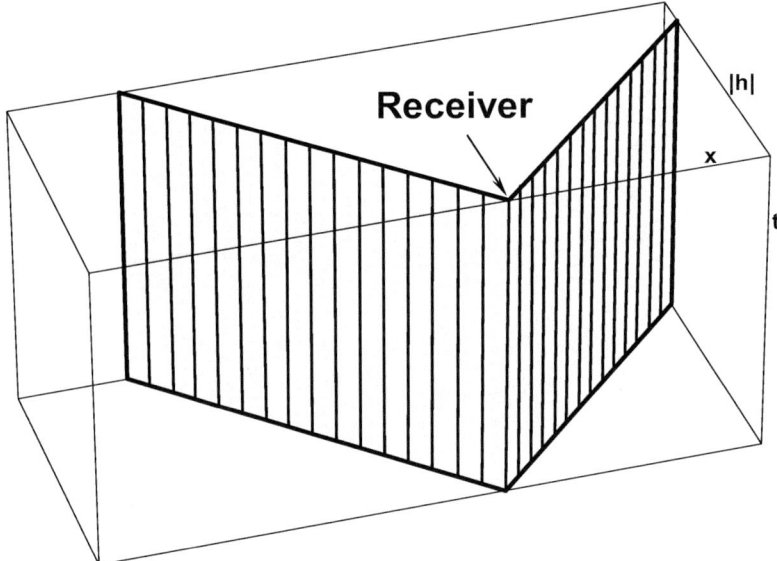

Figure 7.6 Location of traces in a two-sided receiver gather in the prestack volume (x, h, t) with the magnitude or scalar value off the offset.

Removal of noise from seismic data may require processing both the source and receiver gathers, as described by Vermeer [546].

7.1.7 Single trace

It is important in prestack analysis to recognize the contribution of a <u>single input prestack trace</u>.

- **Each input trace contains reflection energy from all neighboring scatterpoints.**
- **Each input trace will migrate energy to all neighboring migrated traces.**

In 2-D processing, <u>DMO and prestack migration is typically applied to each input trace</u>. The resulting traces are stacked or added into either a source gather, or a constant offset section.

Single trace processing becomes a necessity for 3-D DMO as the DMO'd traces lie in a vertical plane between the source and receiver. The resulting traces are stacked into bins of the 3-D volume.

The <u>stacks</u> resulting from DMO or prestack migration of <u>source</u> or <u>constant offset</u> gathers should be identical.

- **All these traces are assumed to have <u>zero-offset</u>.**
- **Traces are usually left in the source or constant offset gather for velocity analysis.**

7.1.8 Reflection point construction

When the source and receiver are at different locations, the problems of ray tracing become more complicated.

- **Consider the source S, and the receiver R in Figure 7.7.**
- **Construct the ray path from S to R, making sure that the incident and reflected angles are equal.**
- **Practice on (a), and save (b) for the exact solution.**

> **Refer to the last page of this section if you require with this construction.**

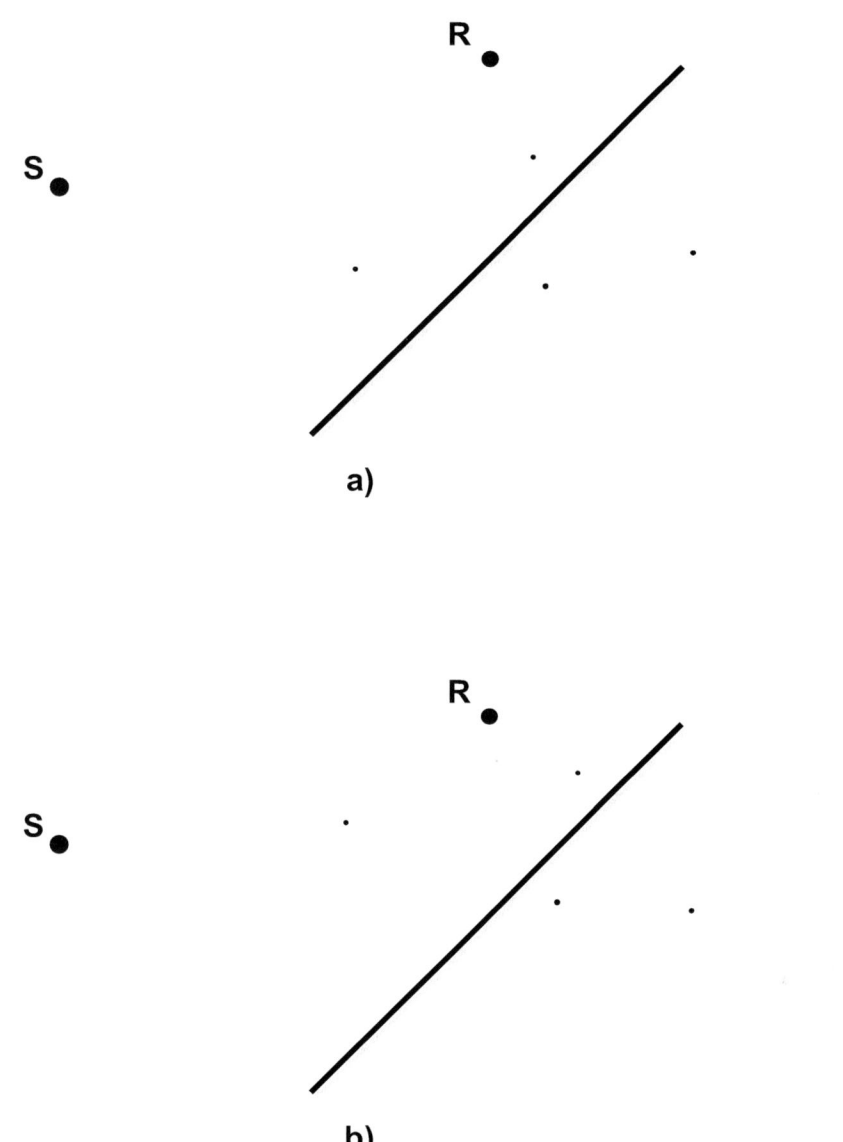

Figure 7.7 Construction for incident and reflected ray paths, a) for an initial attempt, and b) for the exact solution.

7.1.9 DD-MO and velocities for dipping events

Theory for conventional MO with RMS velocities is based on horizontal events. The following section defines a velocity modification by Levin [115] that allows the stacking of dipping events. This theory is based on constant velocities.

Figure 7.8a contains an event with dip β with a zero-offset raypath from a CMP location (*CMP*) to a reflection point R0 in gray. An offset (*2h*) raypath with the same mid point (*CMP*) begins at the source S, reflects at R_h, and ends at receiver R. Scaling allows the simultaneous plotting of vertical two-way times T and T_{dn}.

- The dipping zero-offset time T_{dn} is plotted below *CMP*.
- R_i defines the mirror image of the receiver about the dipping reflector.
- The distance from S to R_i defines the length of the offset raypath *VT*.
- The line (*CMP, N*) is half (*S, R_i*) and defines the time *VT/2*.
- Why is (*CMP, N*) is half (*S, Ri*)? _____

Figure 7.8b is similar to (a), but is re-drawn to emphasize the highlighted triangle (*CMP, N, R_0*) and the added dashed line (*R, Q*) with dip β.

The right-angle triangle (*CMP, R, Q*) defines the line (*R, Q*) to be,

$$(R, Q) = h \cos \beta = (N, R_0). \tag{7.2}$$

An equation for defining the dip-dependent moveout DD-MO correction may now be derived.

(*R, Q*) is equal in length to (*N, R0*). With (*CMP, N*) defined by *VT/2*, and (*CMP, R0*) defined by $VT_{dn}/2$, the Pythagorean theorem gives,

$$\left(\frac{VT}{2}\right)^2 = \left(\frac{VT_{dn}}{2}\right)^2 + h^2 \cos^2(\beta). \tag{7.3}$$

Shifting terms gives the familiar dip-dependent moveout (DD-MO) equation,

$$T^2 = T_{dn}^2 + \frac{4h^2 \cos^2(\beta)}{V^2} = T_{dn}^2 + \frac{4h^2}{V_{stk}^2} \tag{7.4}$$

where, for convenience, the stacking velocity V_{stk} is defined as,

$$\boxed{V_{stk} = \frac{V_{rms}}{\cos(\beta)}}. \tag{7.5}$$

> Dipping events may be stacked coherently using equation (7.4). The reflection at R_h will be co-located with R_0, causing energy to smear along the dip.

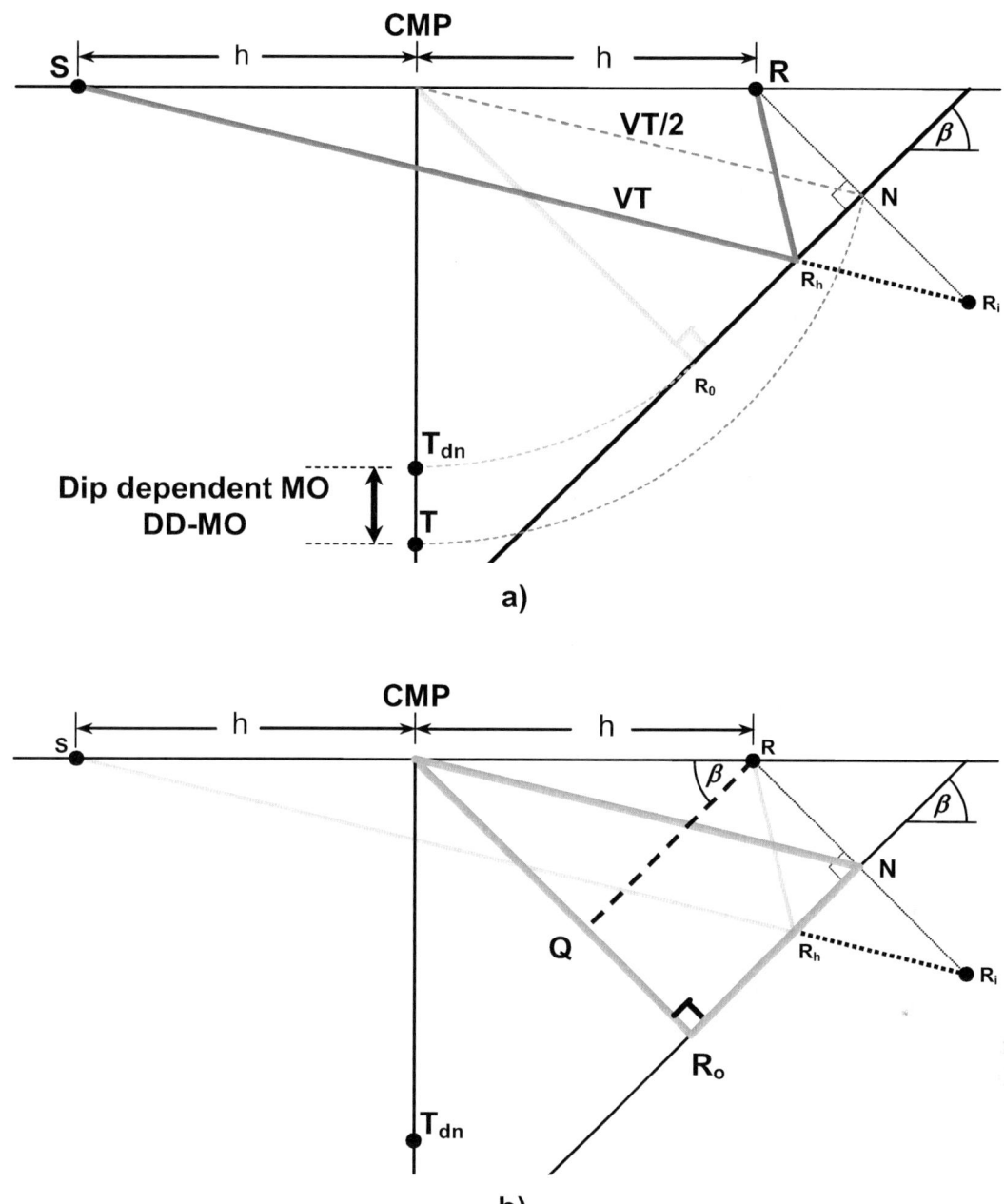

Figure 7.8 Geometry of dipping reflector for zero and fixed offsets with a) emphasizing the ray path, and b) the geometry for dip-dependent MO.

The dip β is geological, and not from a time section before migration.

The stacking velocities Vstk are picked automatically by processors.

This method will not resolve conflicting dips.

Reflection points will be smeared when stacked, i.e. Rh is assumed to be at R0.

7.2 Modelling of Source Gathers (Shot Records)

7.2.1 Exploding point model

A source record may be created similar to the exploding reflector model of Section 2.5. Now, however, an <u>explosion</u> occurs at a <u>point source</u> location.

Figure 7.9 shows "multiple exposures" of wavefronts emanating from a source. The reflected and transmitted <u>energy from a horizontal reflector</u> with increasing velocity is shown in Figure 7.9a. Energy reflected <u>upward</u> from a scatterpoint is shown in Figure 7.9b.

Energy arriving at the surfaces of Figures 7.8a-b may be used to create a source record ($x, t, z = 0$). Figure 7.10 through Figure 7.12 plot Figure 7.9 in a 3-D perspective view of the volume (x, z, t).

The energy radiating from a buried point source is shown in Figure 7.9 with only the direct arrival reaching the surface.

Where does the reflected energy in Figure 7.9a appear to be coming from?

Where does the energy from the scatterpoint in Figure 7.9b come from?

How is the shape of the scatterpoint energy affected by the source location?

The kinematics in Figure 7.9 are only shown for the acoustic P waves. <u>Amplitudes</u> and mode conversions may be computed using:

- the 3-D volume (x, z, t) wave-equation methods of Chapter 2,
- direct (x, t) diffraction modelling with hyperbolic assumptions,
- diffraction modelling with ray tracing which include transmission and reflection coefficients, or by
- including the reflection angle to compute more accurate amplitudes (and mode conversions) using Zoeppritz's equations.

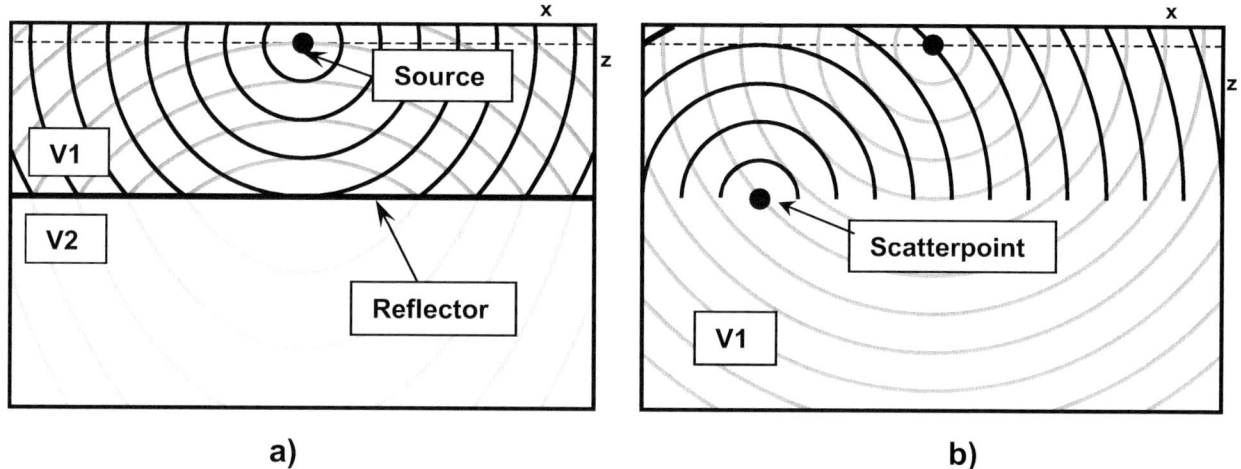

Figure 7.9 Multiple exposure of incremental wavefronts from a buried source in a 2-D plane (x, z) showing a) reflected energy from a horizontal reflector, and b) energy reflected from a scatterpoint (only upward propagating energy shown).

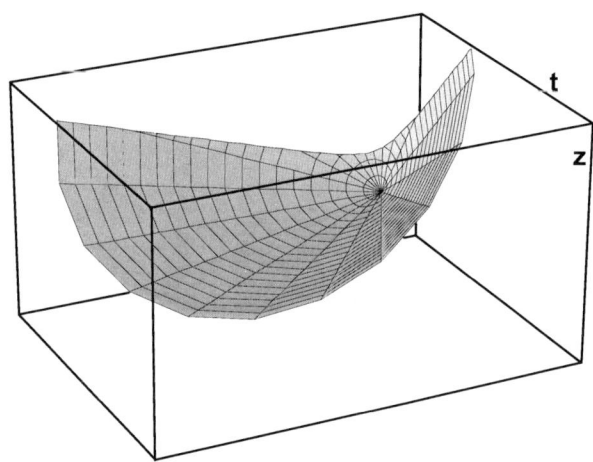

Figure 7.10 Wavefront from a buried shot with the added dimension of time, i.e. (x, z, t)

Source record of horizontal reflector

The reflection energy from a plane surface is illustrated in the (x, z, t) volumes of Figure 7.11. Part (a) shows source energy propagating to the reflecting surface, while (b) shows only the reflected energy. The source and reflected energies are shown plotted together with two different perspective views in parts (c) and (d).

It is important to realize the 2-D reflecting planes in Figure 7.10 are the 1-D line of Figure 7.9a; i.e. the geological geometry of the (x, z) plane is independent of time in the (x, z, t) volume.

The top surface ($z = 0$) in Figure 7.11 (c) and (d) shows a simplified source record with a first arrival and a hyperbolic reflection.

Source record of a scatterpoint

A source model for energy reflected from a scatterpoint is shown with two perspective views in Figure 7.12. For simplicity, the source energy is only plotted up to the time at which it arrives at the scatterpoint. It is emphasized that a scatterpoint is independent of t and may be considered a line when plotted in the (x, z, t) volume, and should be compared with Figure 7.9b.

Once again the energy on the top surface ($z = 0$) shows a simplified source record. Only one scatterpoint is shown. Actual reflectors would be composed from a number of scatterpoints as illustrated later in Section 7.1.3.

The source records of Figure 7.11 (c) and (d) form a basis for using normal moveout (MO) correction to flatten the reflection to represent a horizontal reflection.

In contrast, Figure 7.12 (a) and (b) form the basis of Kirchhoff prestack migration in which the scattered energy is relocated at the actual scatterpoint.

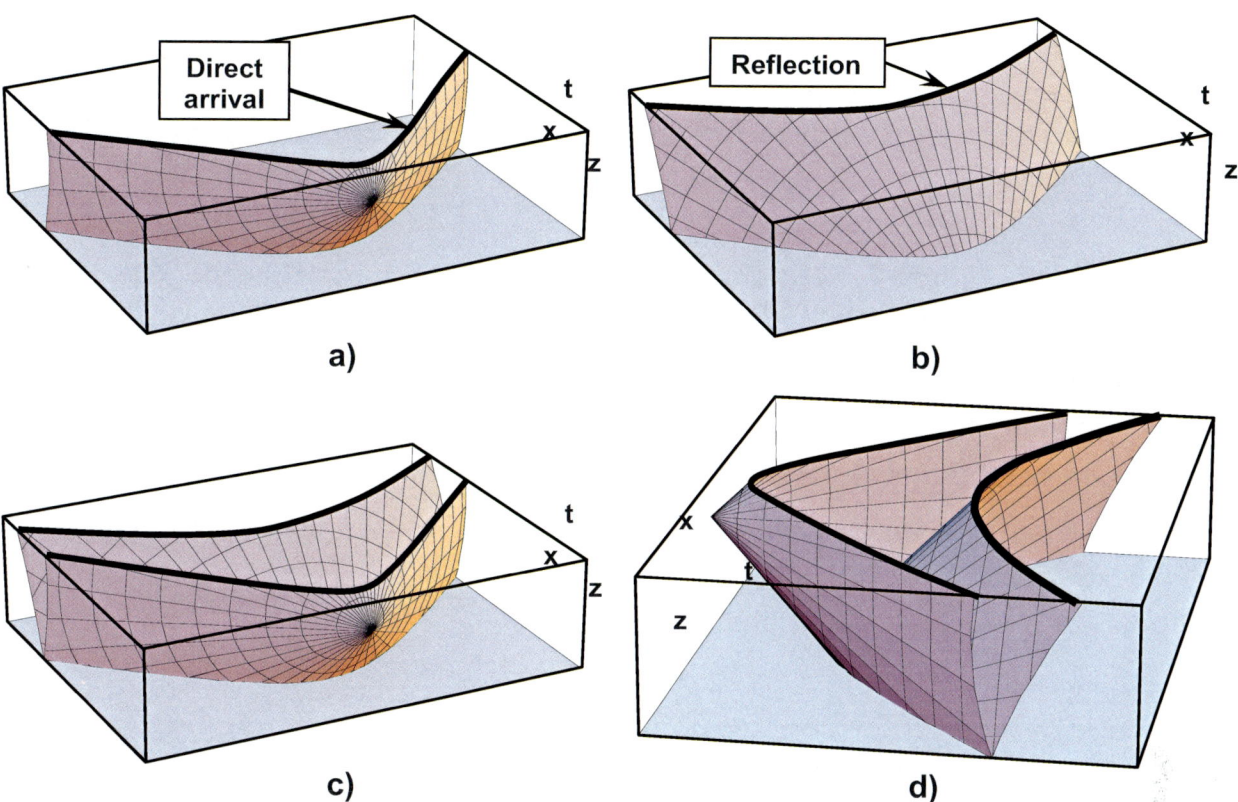

Figure 7.11 Wavefront model for a horizontal reflection with a) showing the wavefront emanating from the source, b) the reflected energy, c) the combined source and reflected energy, and d) an alternate perspective view of (c).

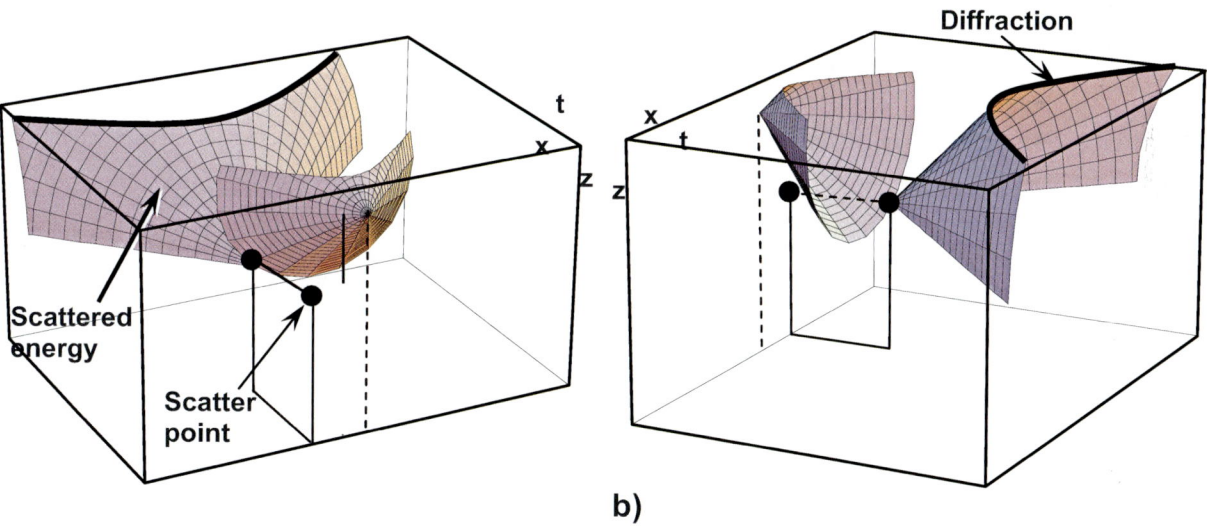

Figure 7.12 Two perspective views of a partial source wavefront and energy reflected upward from a scatterpoint.

7.2.2 Compass construction of source record

Source records may be modelled by estimating traveltimes from rays that travel from the source via a reflector to the receivers, with the traveltimes plotted below the receivers.

Use Figure 7.14 - Figure 7.16 to construct a kinematic source record. The source is located at the flag S, and all other surface locations represent receivers.

For the horizontal event on Figure 7.13:

- Locate the image of the source relative to the horizontal event.
- Using the source image, define the total raypath to a receiver ($t_s + t_r$).
- Plot the length of the ray path (i.e., proportional to two-way time), below the corresponding receiver or surface location, using V = 1.0.
- Repeat for all receiver positions.

Use the above method for constructing the reflection from the dipping event on Figure 7.15 of the following page.

Construct two scatterpoint diffractions using the total ray path on Figure 7.16:

- Measure the traveltime from the source to a scatterpoint. (The time will be independent of the receivers, but will vary for each scatterpoint.)
- Draw a horizontal line across the section at a time equal to the source-receiver traveltime.
- Measure the traveltime from the scatterpoint to a receiver.
- Add this receiver time to the line (source time) below the receiver.
- Repeat the last two steps for each receiver location.

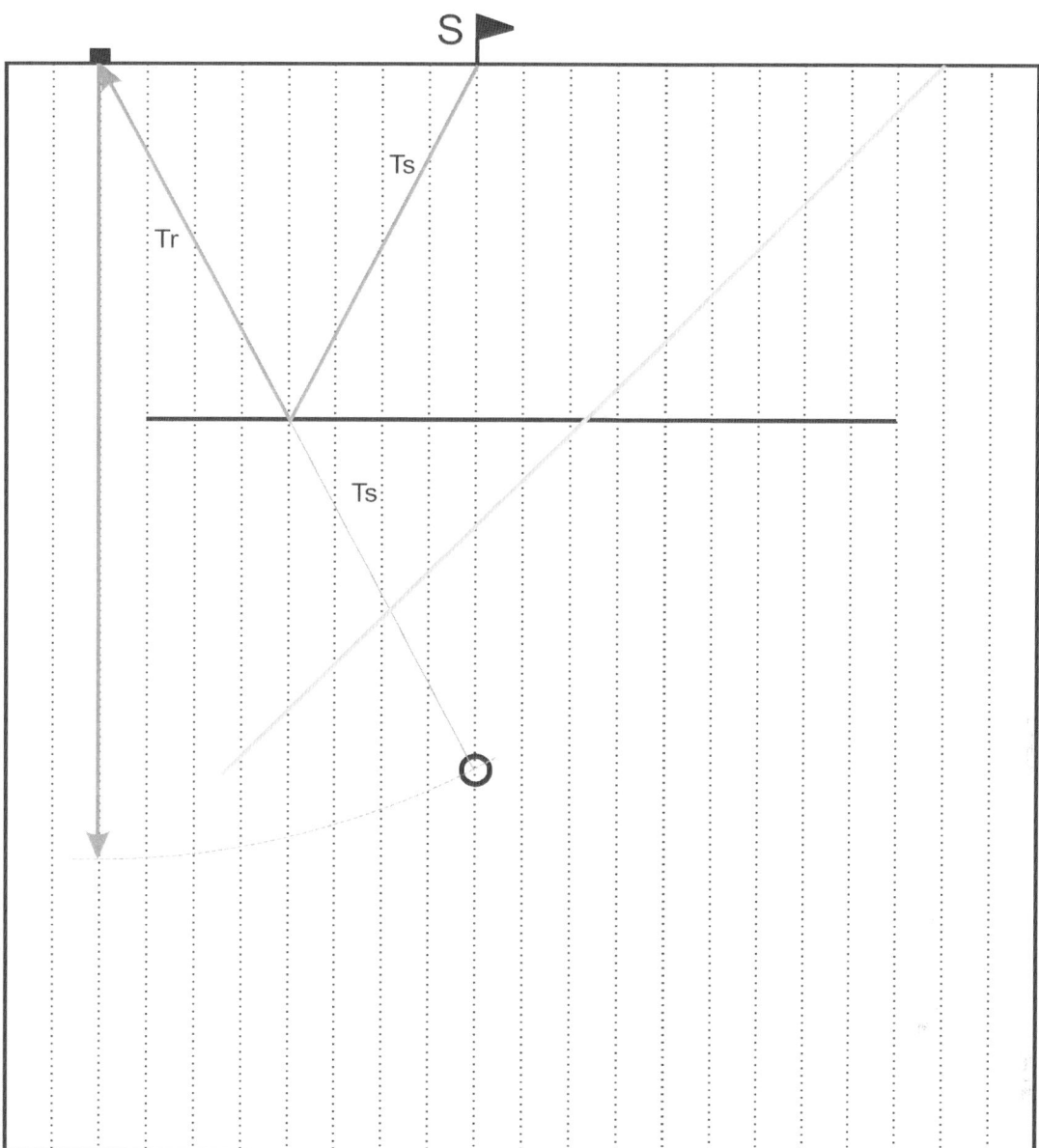

Figure 7.14 Construction for a source gather: horizontal reflector.

What is the shape of a <u>reflection</u> from a <u>horizontal</u> event in a source-gather?

Where do the asymptotes intersect?

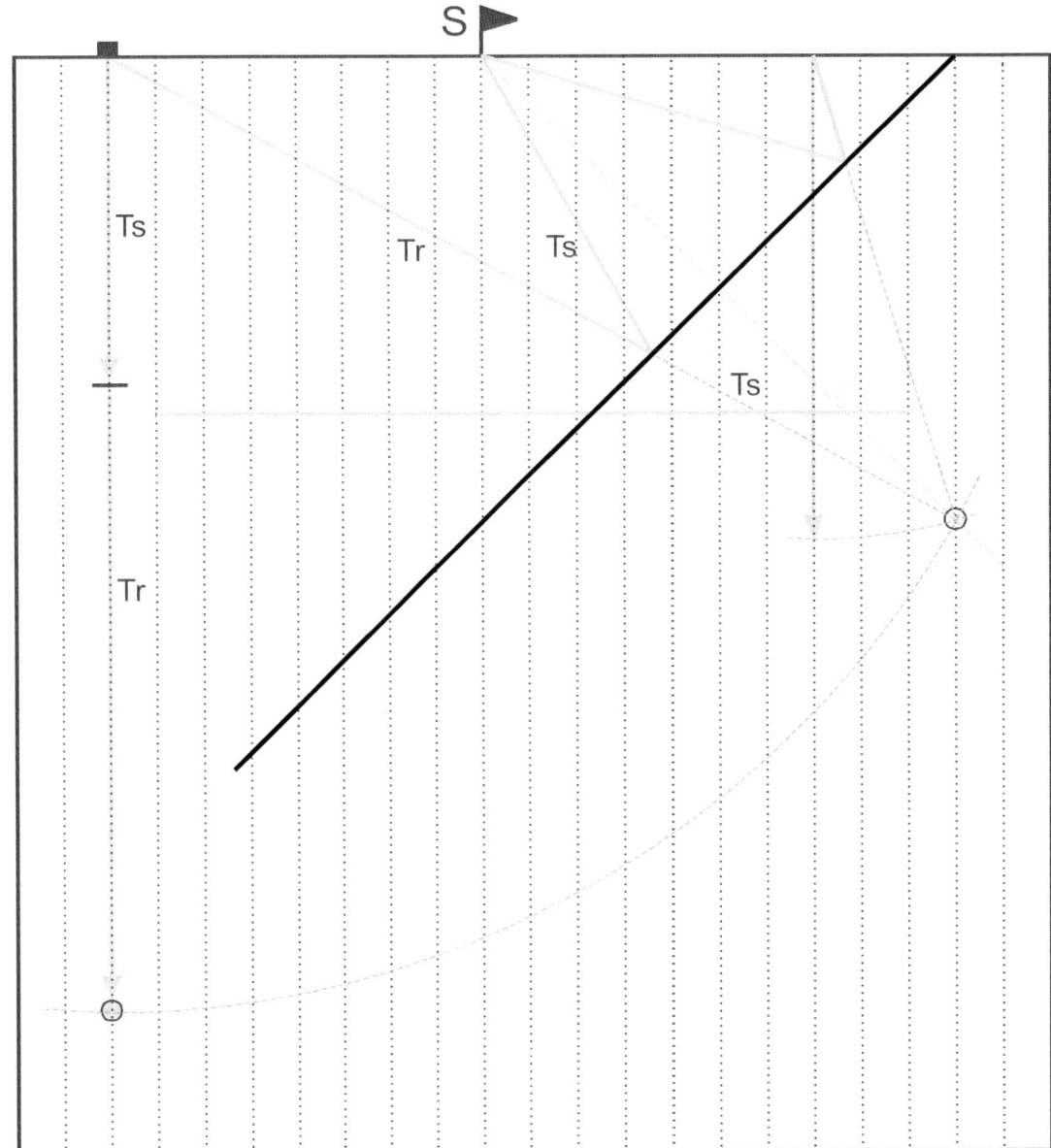

Figure 7.15 Construction for a source gather: dipping reflector.

What is the shape of a reflection from a <u>dipping</u> event in a source-gather?

Where do the asymptotes intersect?

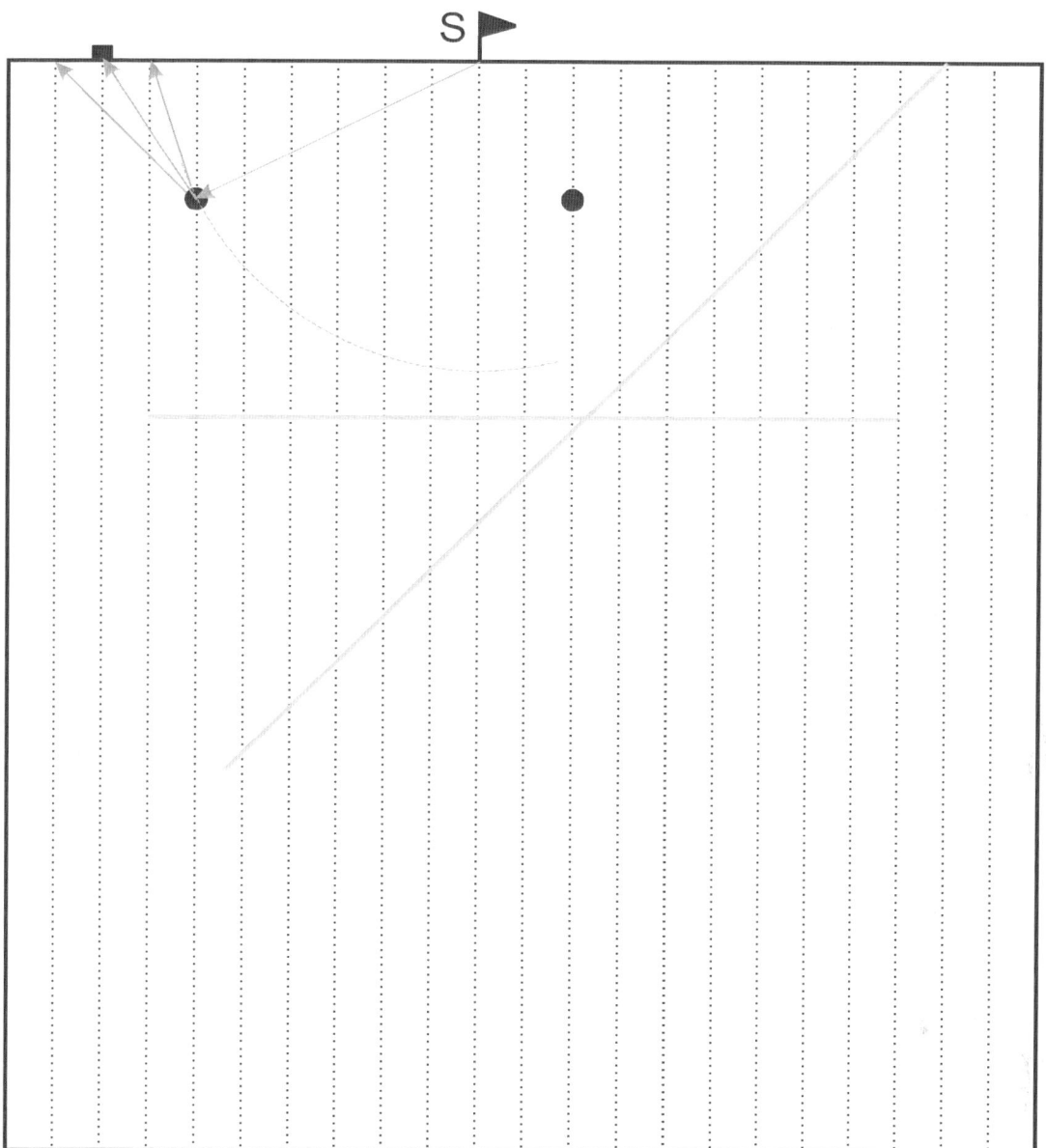

Figure 7.16 Construction for a source gather: scatterpoints.

What defines the shape of a source-gather <u>diffraction</u>?

Where do the asymptotes intersect?

7.2.3 Diffraction modelling of source gathers

A source gather may be modelled by diffractions in a manner similar to the post-stack case in section 2.2. However, in a source gather the shape of the <u>diffraction</u> not only <u>varies with depth</u>, but also with <u>offset</u>.

Figure 7.17a illustrates constant depth scatterpoints that produce diffractions with the same shape, but they occur at later times as the offset is increased. The times in the figure are two-way times, and the data is plotted below the receiver location (not at the CMP location).

The scatterpoints in Figure 7.17b illustrate diffractions with the same apex time. They are located at different depths and have different hyperbolic shapes.

Modelling is accomplished by:
- computing the traveltime from the source point to a scatterpoint,
- computing the traveltimes from the scatterpoint to each receiver location,
- adding the source time to the receiver times to define the diffraction shape,
- spreading the amplitude of the reflected energy along the diffraction,
- repeating the process for each scatterpoint in the input structure.

Where do the asymptotes of the source record diffractions intersect?

An interesting (and rather obvious) observation in Figure 7.17a is the envelope of the diffractions and how it forms the MO response of a flat reflector located at the same depth as the scattering points.

How do you migrate a source record with Kirchhoff migration?

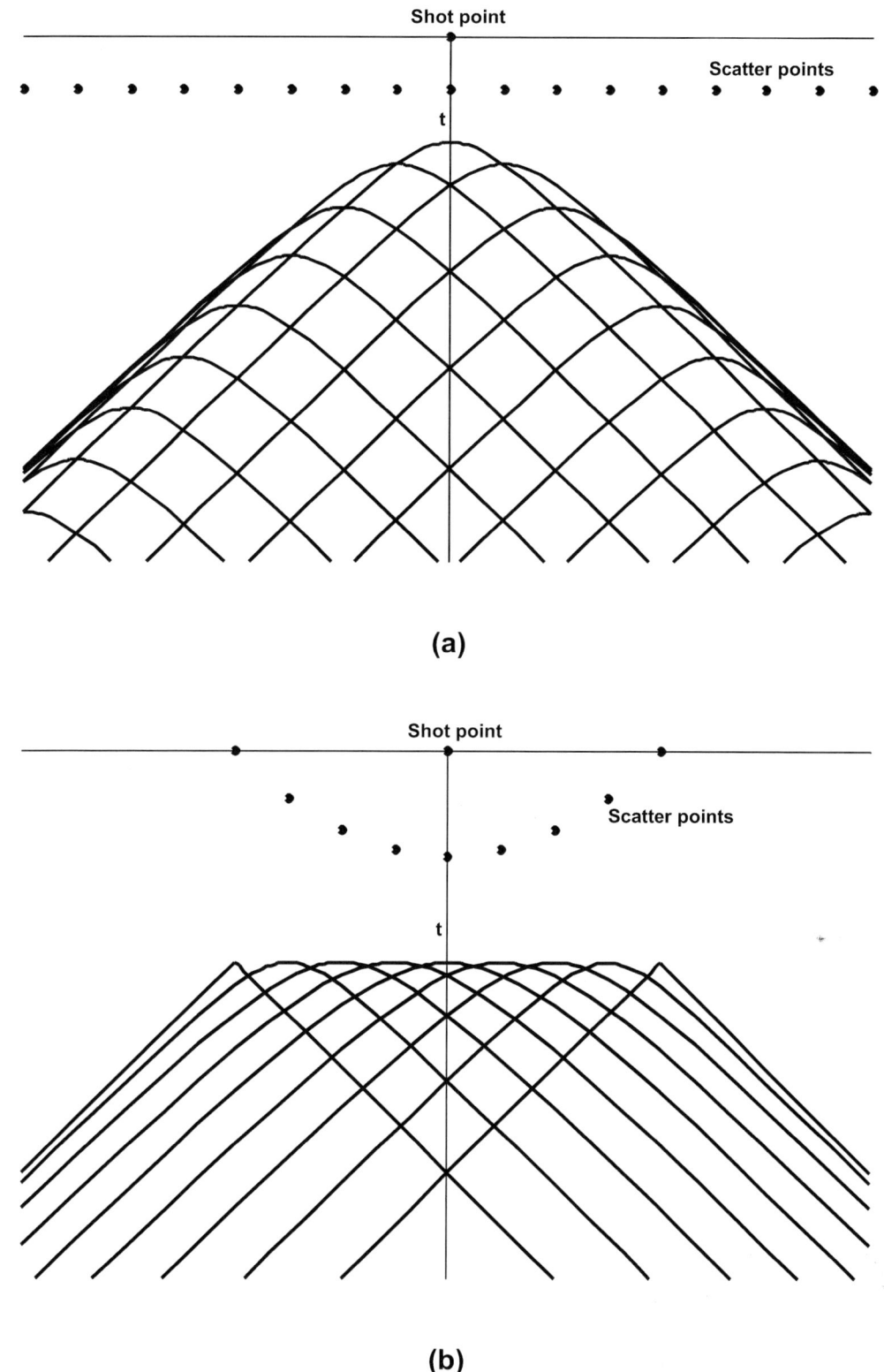

Figure 7.17 Source modelling by diffractions, a) at constant depth dispersion points, and b) diffraction with the same apex times.

Figure 7.18 Illustrates a number of source record that are recorded over a single scatterpoint.

Note:
- The diffraction in each record has the identical shape.
- The apex of each diffraction identifies the receiver that is directly above the scatterpoint.

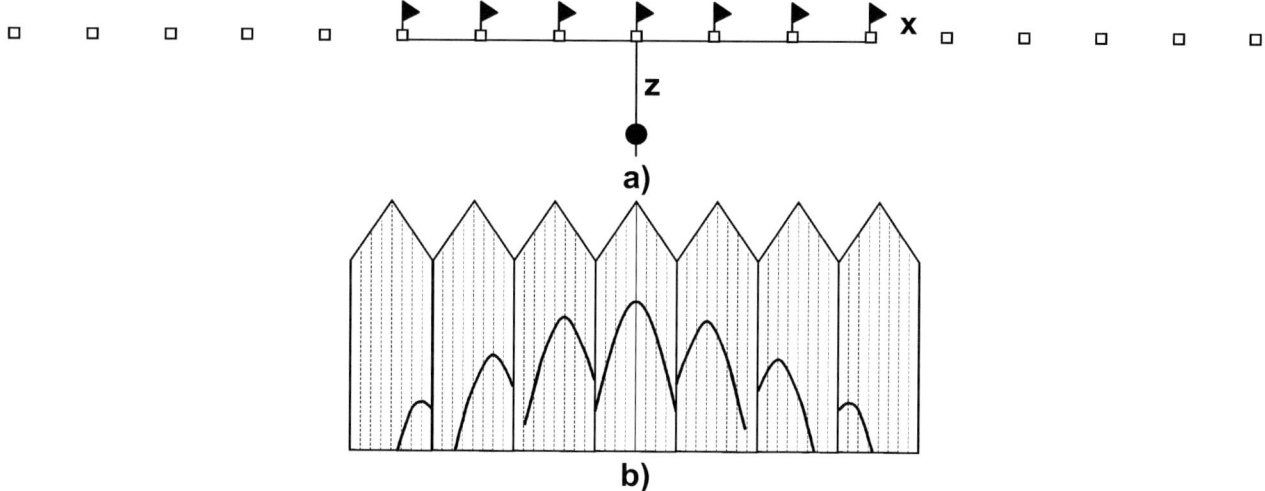

Figure 7.18 A number of shots at various locations above a scatterpoint.

7.2.4 Modelling of source records using an exploding point source

The exploding reflector algorithm for modelling may be used to produce a source model. The exploding reflector becomes a point located at the source position. Energy arriving at the surface is recorded as the source gather.

A model is shown below in Figure 7.19 that is used to create the images in Figure 7.20 and Figure 7.21. In these figures, there is no reflection from the earth's surface. The reflection from the earth's surface is included in Figure 7.22 to produce a rather complex image from a simple geological structure.

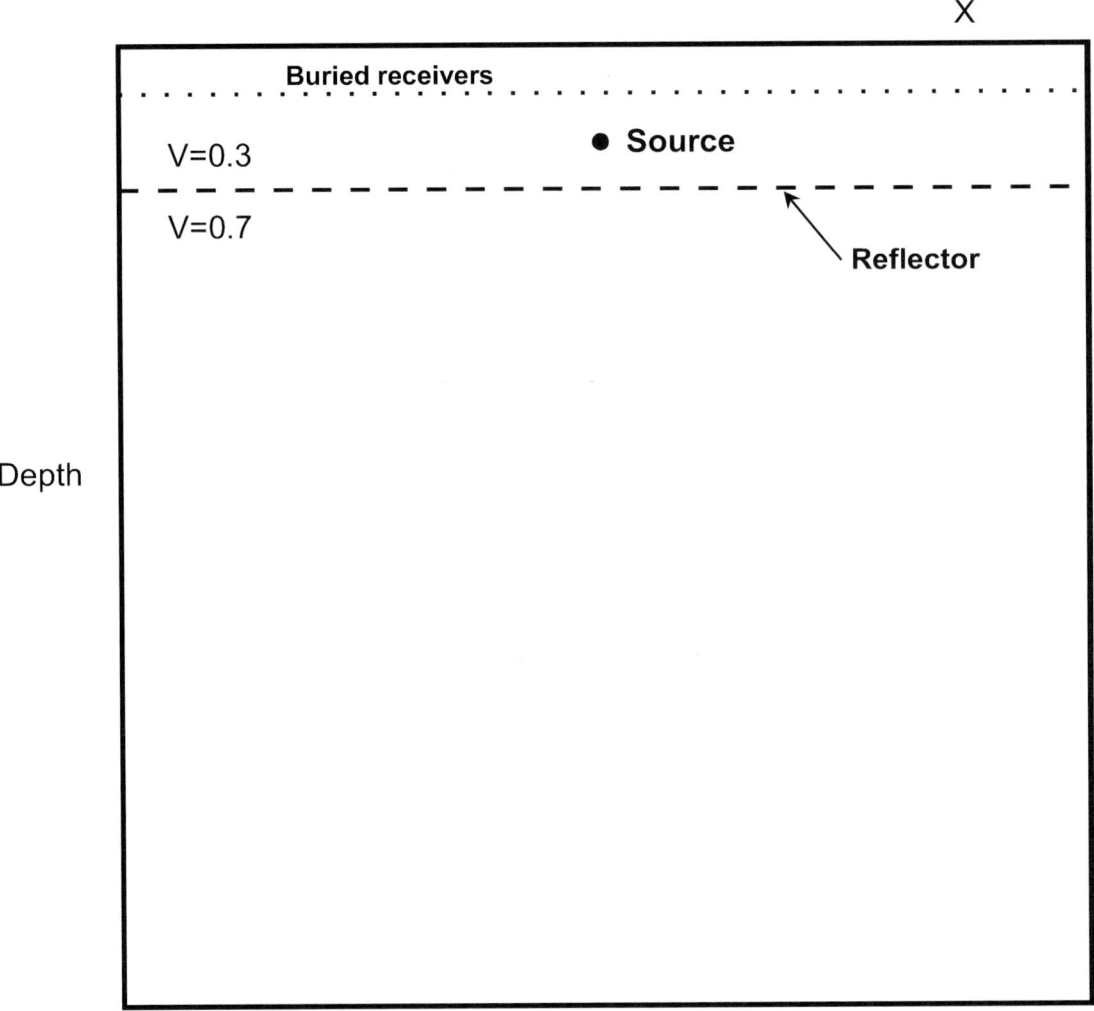

Figure 7.19 Structure for the exploding source model.

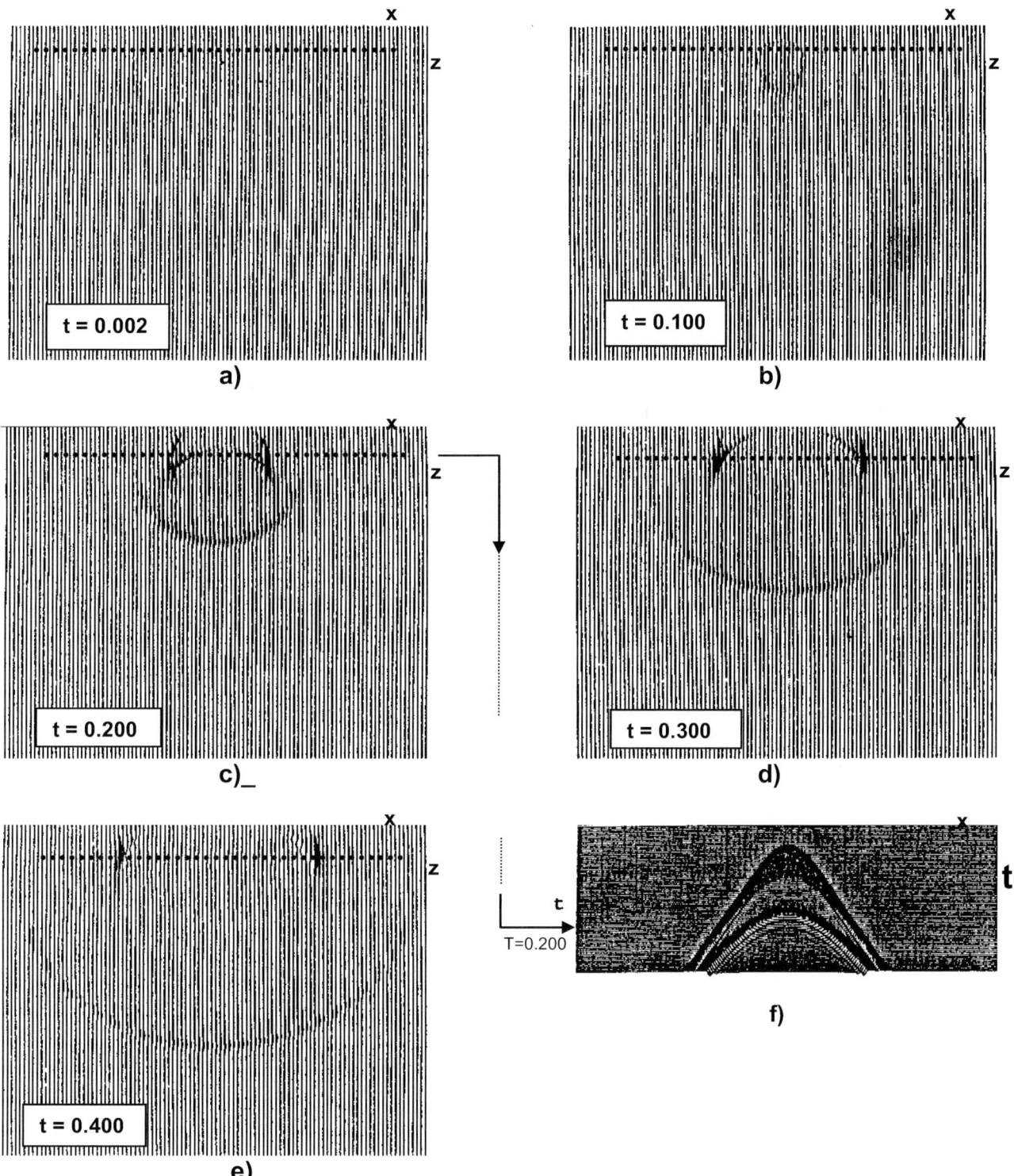

Figure 7.20 Time frames of exploding source model in (x, z) at times a) 0.002, b) 0.100, c) 0.200, d) 0.300, and e) 0.400 seconds. Energy at the receivers of (a) through (e) contribute horizontal lines to the (x, t) time response f).

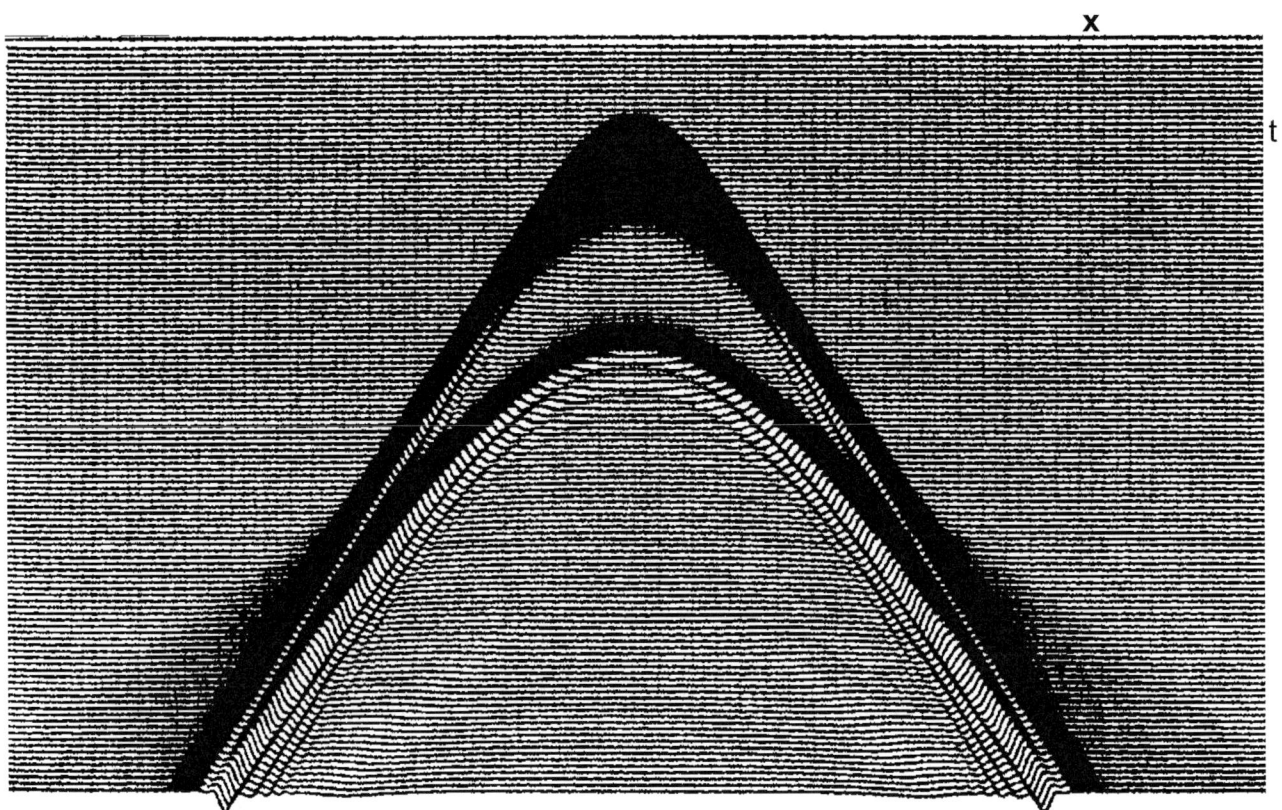

Figure 7.21 The time response (x, t) of an exploding source after 600 time units. Note the direct wave, reflected wave, and the head wave energy.

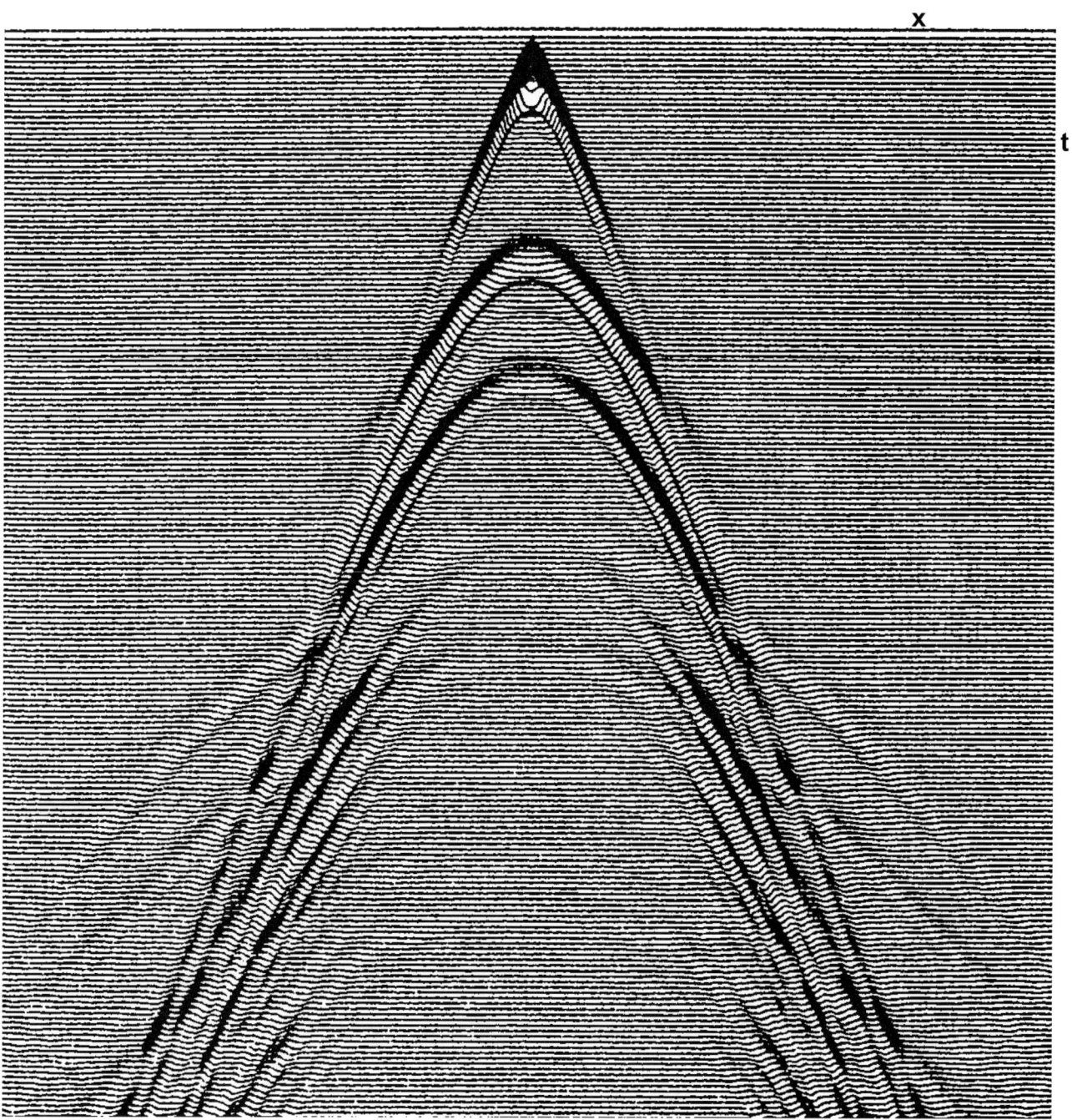

Figure 7.22 Source explosion model (x, t) in which surface reflections are permitted. Note the ghost reflection.

7.2.5 Prestack numerical model

Figure 7.23 shows a simple numerical model that contains short horizontal reflector, a dipping event with a small gap, and a scatterpoint. Because of its shape, it is referred to as the "hockey stick model".

The model contains:

- 101 sources with 96 split spread receivers,
- velocity 10,000 fps,
- CMP spacing of 100 ft,
- maximum offset at 10,000 ft,
- recording time of 2.0 seconds at 4ms sample rate.

Rays were traced from each source to the corresponding reflectors and then to the appropriate receivers. Figure (a) illustrates one shot location in the center of the line, but the ray density is so dense that they cover the rays to the scatterpoint. Consequently, only the rays to the fifth receiver are shown in (b).

Note that the plotting in these figures is from right to left, or that the origin is at the upper right (third quadrant). Some later figures may plot the data from left to right.

The modelling method used the image of the source with both reflectors to define the raypaths, and those image points are also shown. Only the specula energy from each reflector was used. There was no diffraction energy computed at the edges of the reflectors or gap. Diffraction energy was represented by the single scatterpoint.

The traveltimes to the reflectors was used to insert energy on the appropriate trace. The energy was linearly weighted between time samples. The same amplitude was used for all reflections, including the reflected energy from the scatterpoint.

Shot gathers were created, but the data was also sorted into CMP gathers, and into constant offset and limited offset gathers.

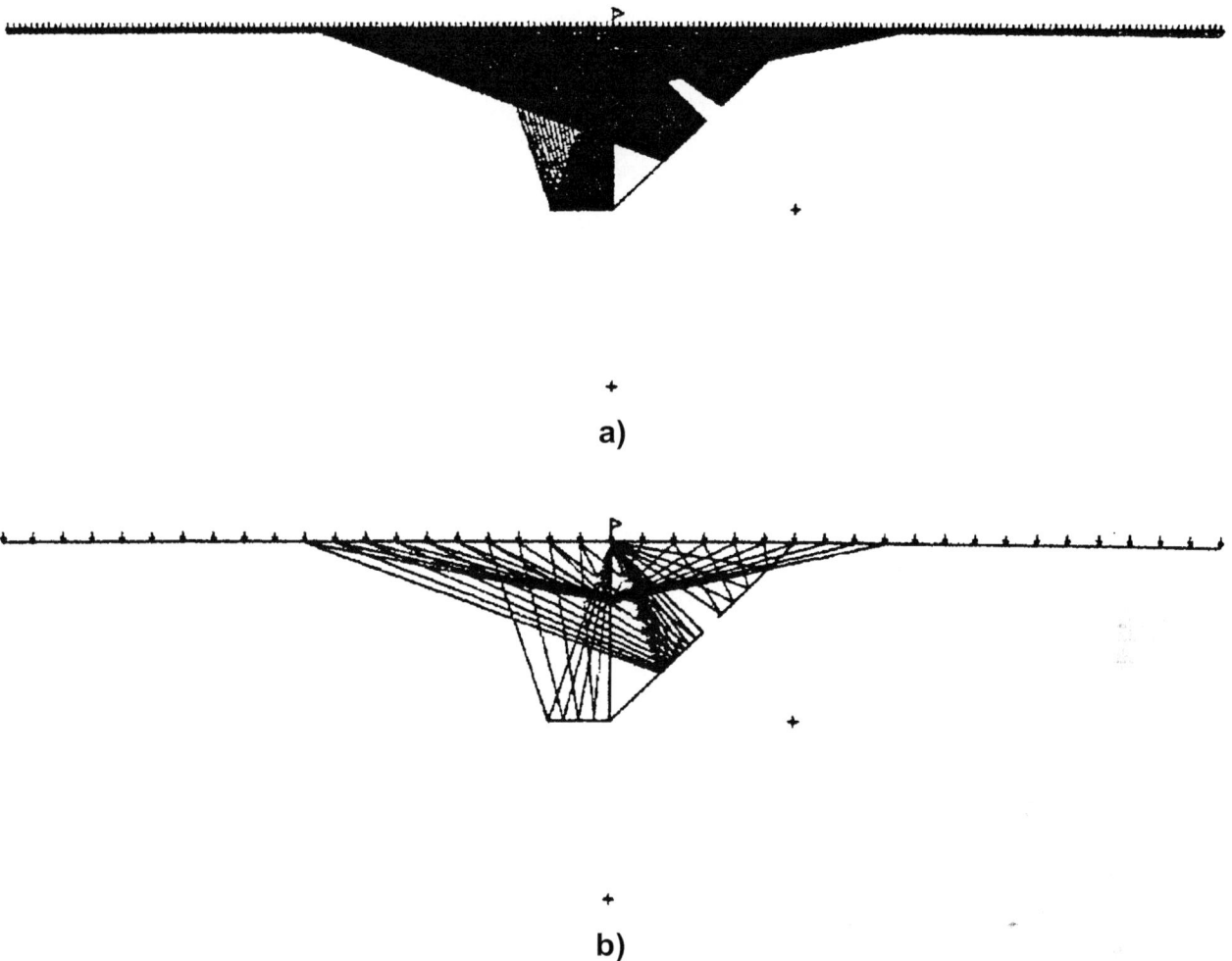

Figure 7.23 Full ray tracing of source model, a) with all the rays, and b) with every fifth ray to allow viewing of the scatterpoint rays..

The figures on the following pages show the same three source records, that were selected at equal spacing across the line.

Figure 7.24 shows the ray diagrams (every fifth ray).

Figure 7.25 shows the three source records as "recorded" in (a). These same source records are repeated in (b) with NMO using the model velocity, and again in (c) with dip-dependent moveout (DD-MO) using a velocity computed for the dipping event.

The same three shot record were then stack as shown in Figure 7.26a. Parts (b) and (c) of this figure contain the NMO'ed and DD-MO corrected sections.

Note the misalignment of energy in all three records of Figure 7.26, especially when the model is constant velocity.

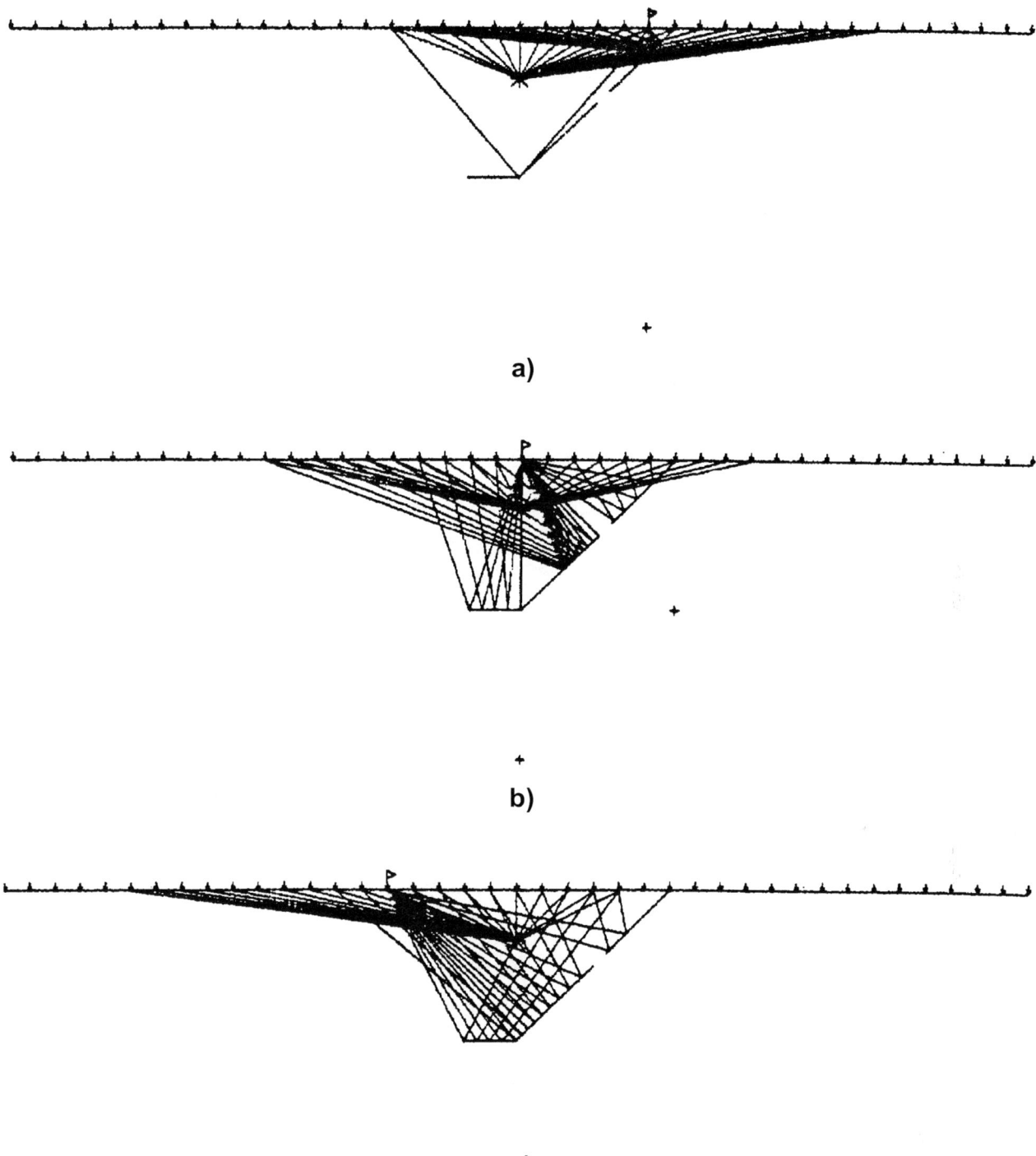

Figure 7.24 Three models showing every fifth ray path with sources located at stations a) 151, b) 201, and c) 251.

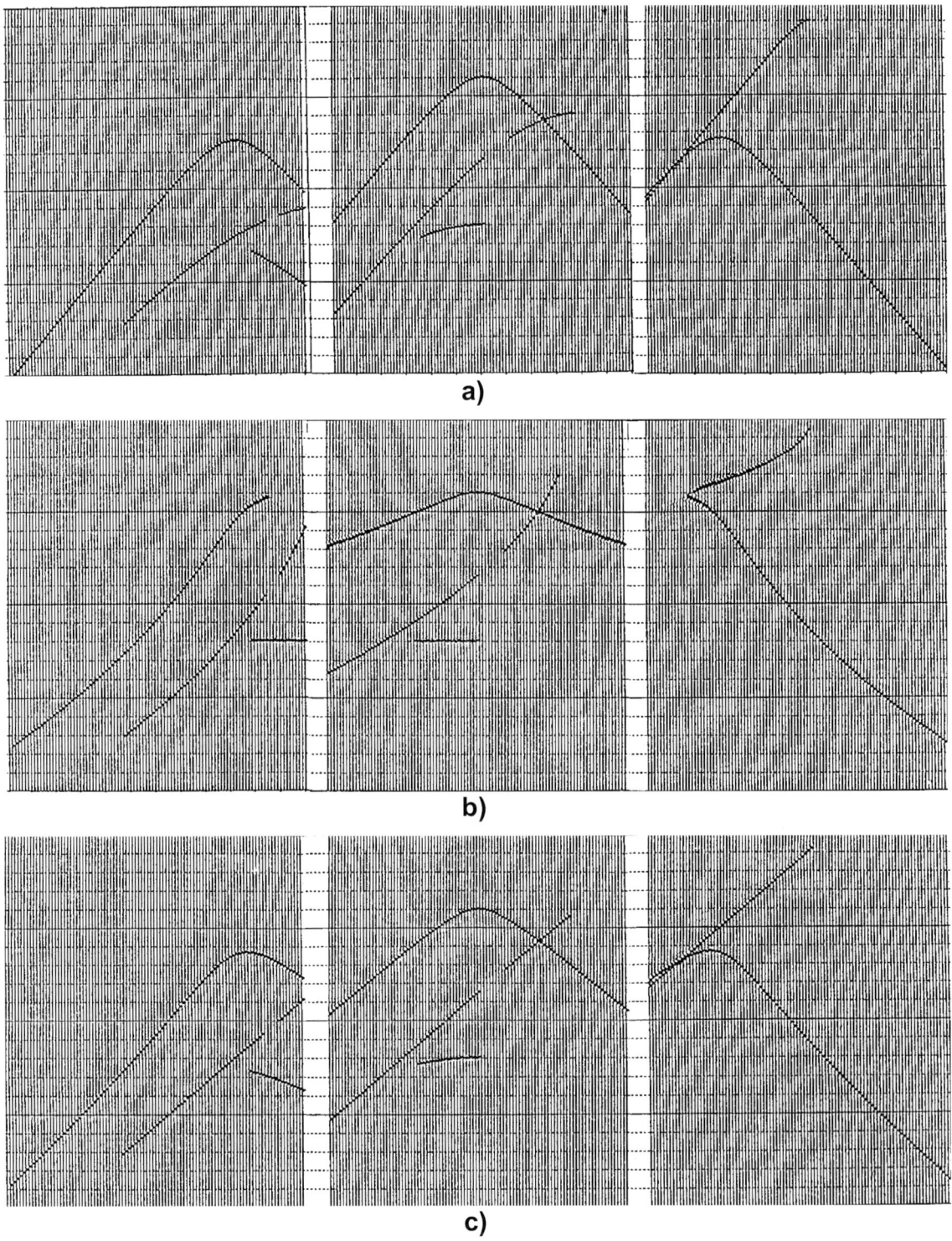

Figure 7.25 Three source model records (x, t): a) the time response, b) NMO applied for flat reflector, and c) DD-MO applied for dipping reflector.

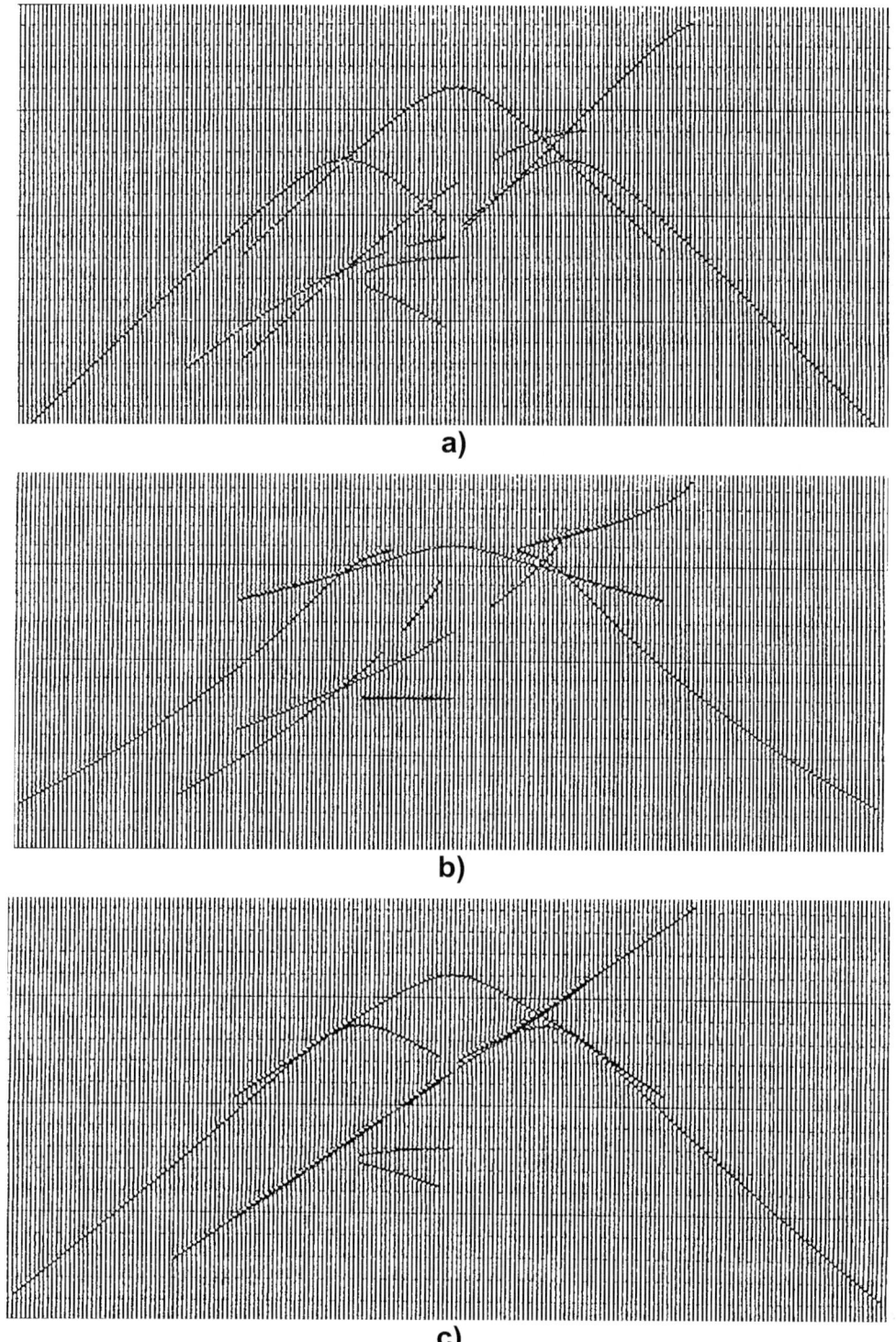

Figure 7.26 Stack of three source records (x, t) with a) no MO correction, b) with NMO correction for flat reflector, and c) with DD-MO correction for the dipping reflector. <u>Note the miss alignment of data in each stack</u>.

7.2.6 Comments and conclusions:

- **Diffractions don't stack.**
- The concept for stacking CMP gathers is only valid for <u>horizontal reflectors</u>. Even in this case, diffraction energy from the edges of horizontal reflectors will be miss-stacked.
- DD-MO will position a dipping reflection at the zero-offset traveltimes, but the reflection point are <u>smeared along the dip</u>.
- … and this is a constant velocity model.

7.3 Constant Offset Sections

7.3.1 Compass construction of constant offset section

The construction of traveltimes on a constant offset section may be accomplished by defining the ray paths from the source-receiver positions as they move across the top of the structure. Traveltimes are estimated from the length of the ray paths, with the reflections plotted below the CMP positions. Two-way times are convenient for the construction; however, the one-way times (half the two-way time) located at the midpoint (CMP) are a closer match to the geology.

Figure 7.27 to Figure 7.31 contain the construction of a constant offset section with a horizontal event, a dipping event and one scatterpoint. The source-receiver offset is defined by the line from *A to E*. Other source receiver pairs *are* *B – F, C – G*, ..., and *G - K*. The image positions of *A, B, C, ..., J*, relative to the dipping event are at *a, b, c, ..., j*.

- The construction first estimates the two-way travel times at the midpoint location.
- The two-way times are converted to one-way travel times for comparison to zero offset data.

> When modelling source records, surface locations and two-way times were used to preserve the geometrical shapes.
>
> When modelling constant offset sections, the CMP or subsurface locations are used with one-way times to preserve the geometrical shapes.

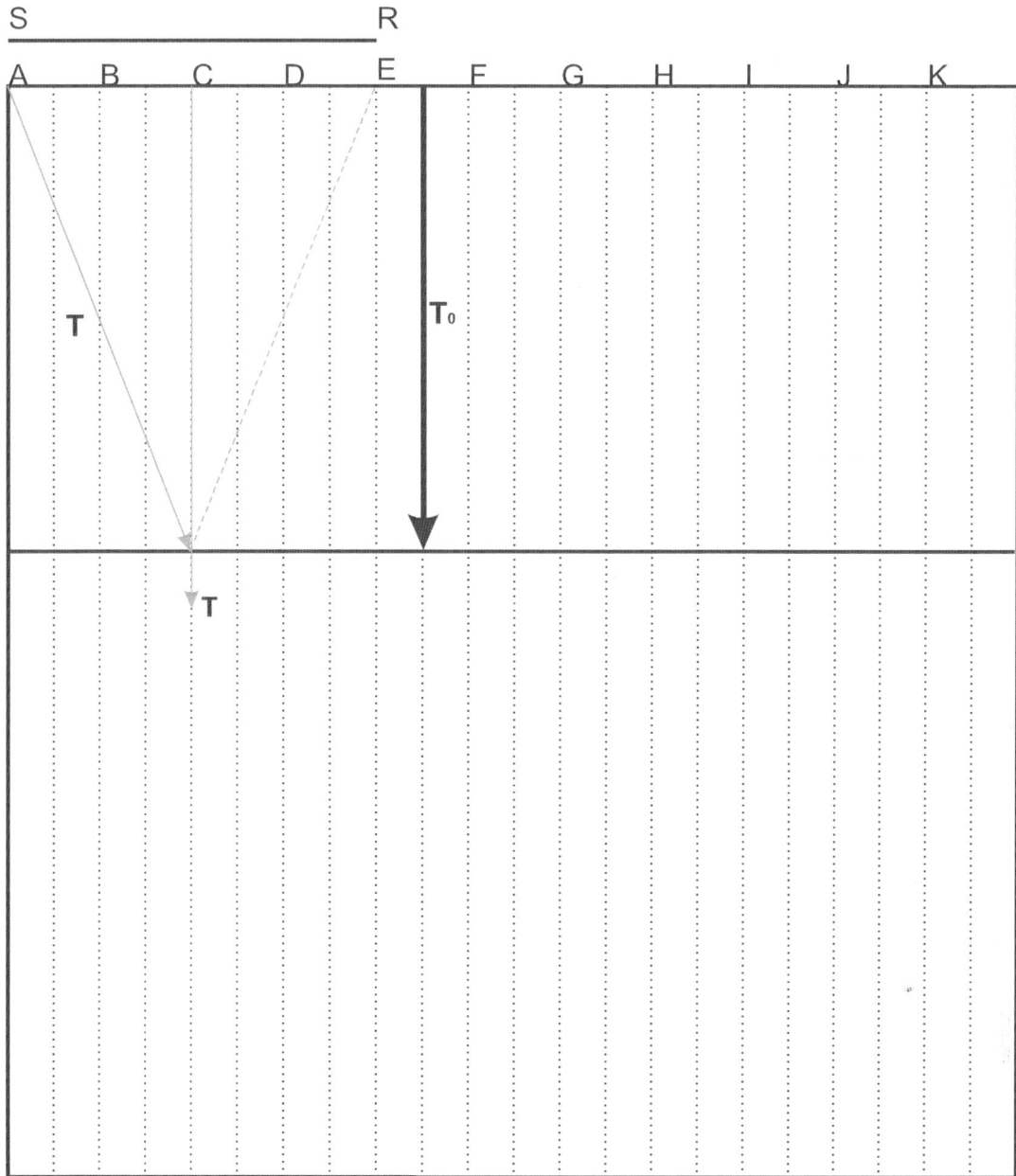

Figure 7.27 Constant offset construction for a horizontal reflector.

On Figure 7.27 construct:
- **the offset reflection from the flat reflector,**
- **become familiar with NMO correction,**
- **apply NMO correction to the flat reflection.**

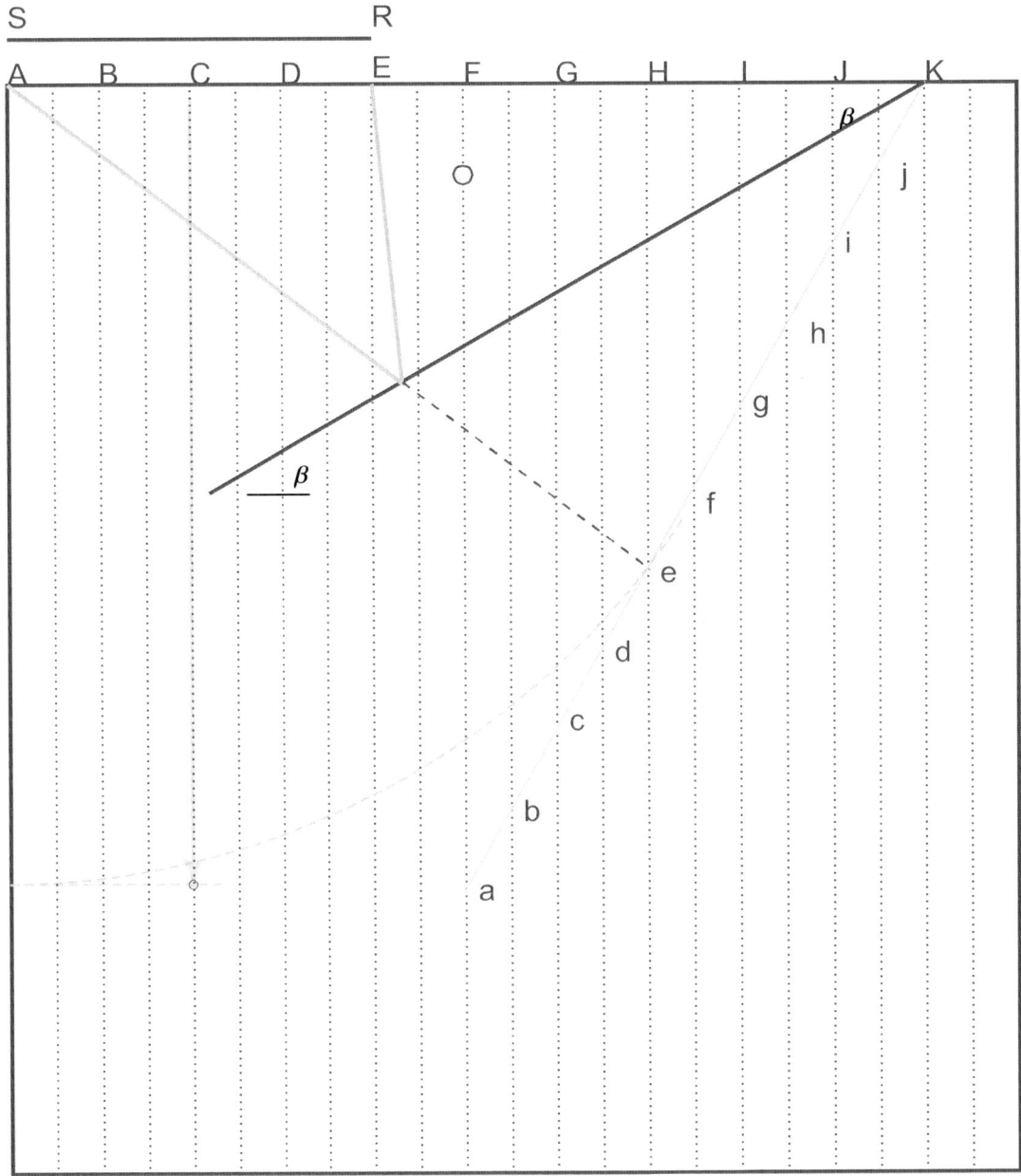

Figure 7.28 Constant offset construction of dipping reflector.

Construct the dipping event by:

- **Define the two-way traveltime by the distance of the ray path from A to e. (Source to image of receiver).**
- **Plot the two-way traveltime below the CMP position at C.**
- **Repeat the procedure for all source-receiver combinations, B-F, C-G, etc.**

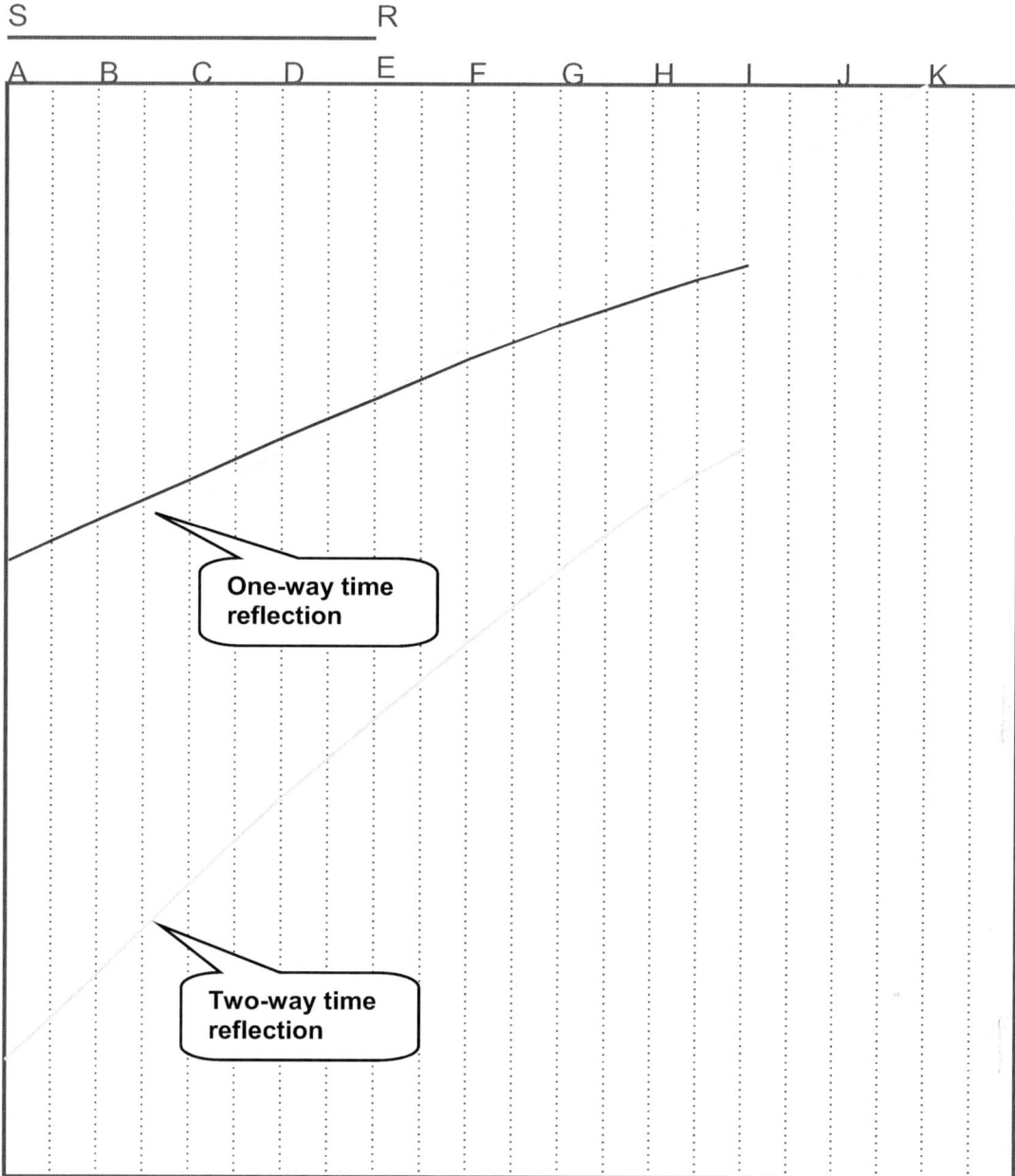

Figure 7.29 Constant-offset construction of dipping reflector.

The two-way times from Figure 7.28 are converted to one-way times above.

- Add the zero-offset reflection (one-way times).
- Apply dip-dependent moveout DD-MO to the offset reflection using Vstk for the dipping reflector.
- Apply NMO correction to the offset reflection using V.
- Where does the reflection at location "I" come from?

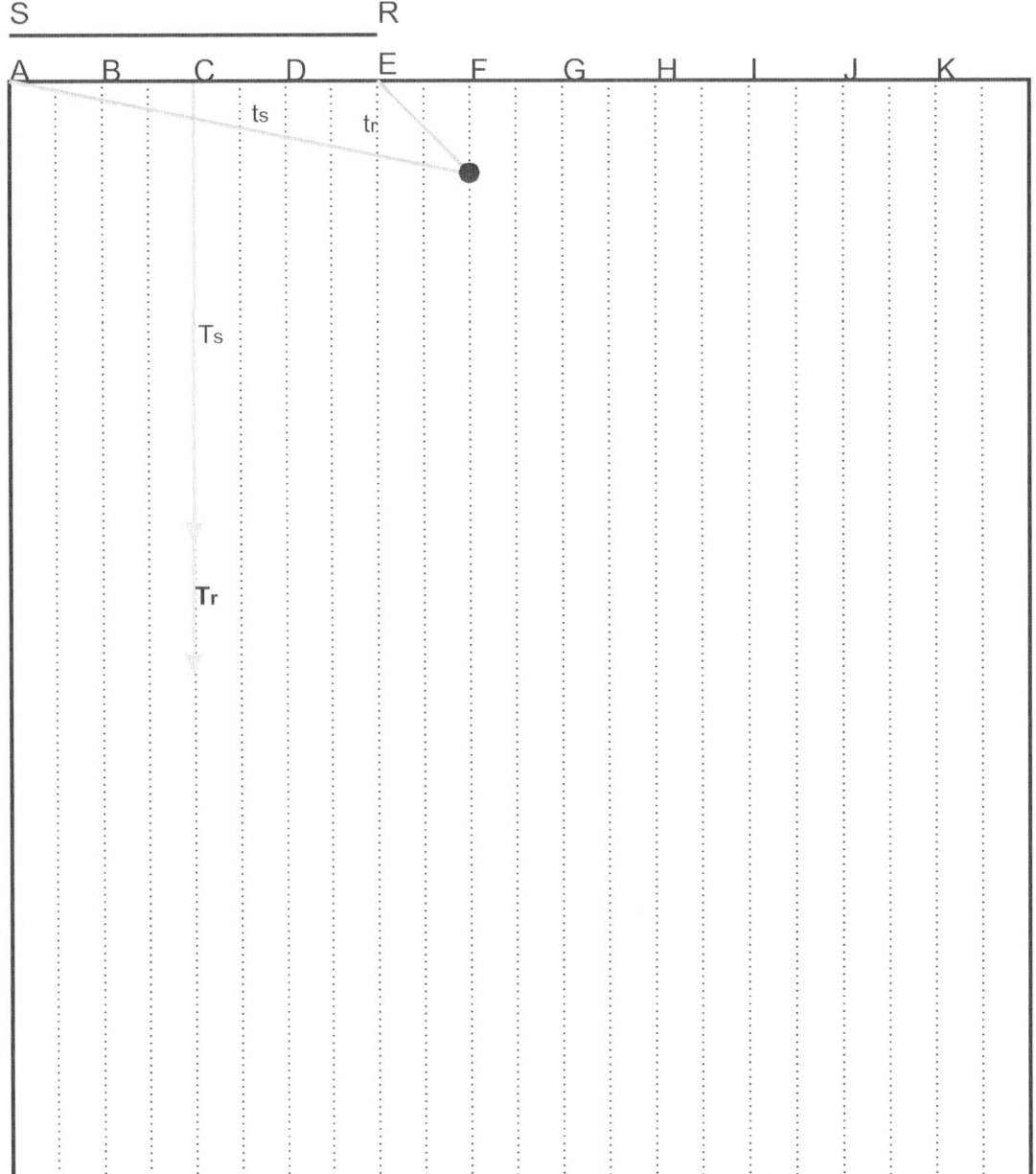

Figure 7.30 Constant offset construction of diffraction.

Construction of the <u>scatterpoint diffraction</u> by:
- **Measure the ray path from the source at A to the scatterpoint.**
- **Plot this time below the mid-point C.**
- **Measure the ray path from the scatterpoint to the receiver E.**
- **Add this time to the source raypath <u>below the CMP</u> at C.**
- **Repeat this procedure for all source-receiver pairs, B-F, C-G, etc.**

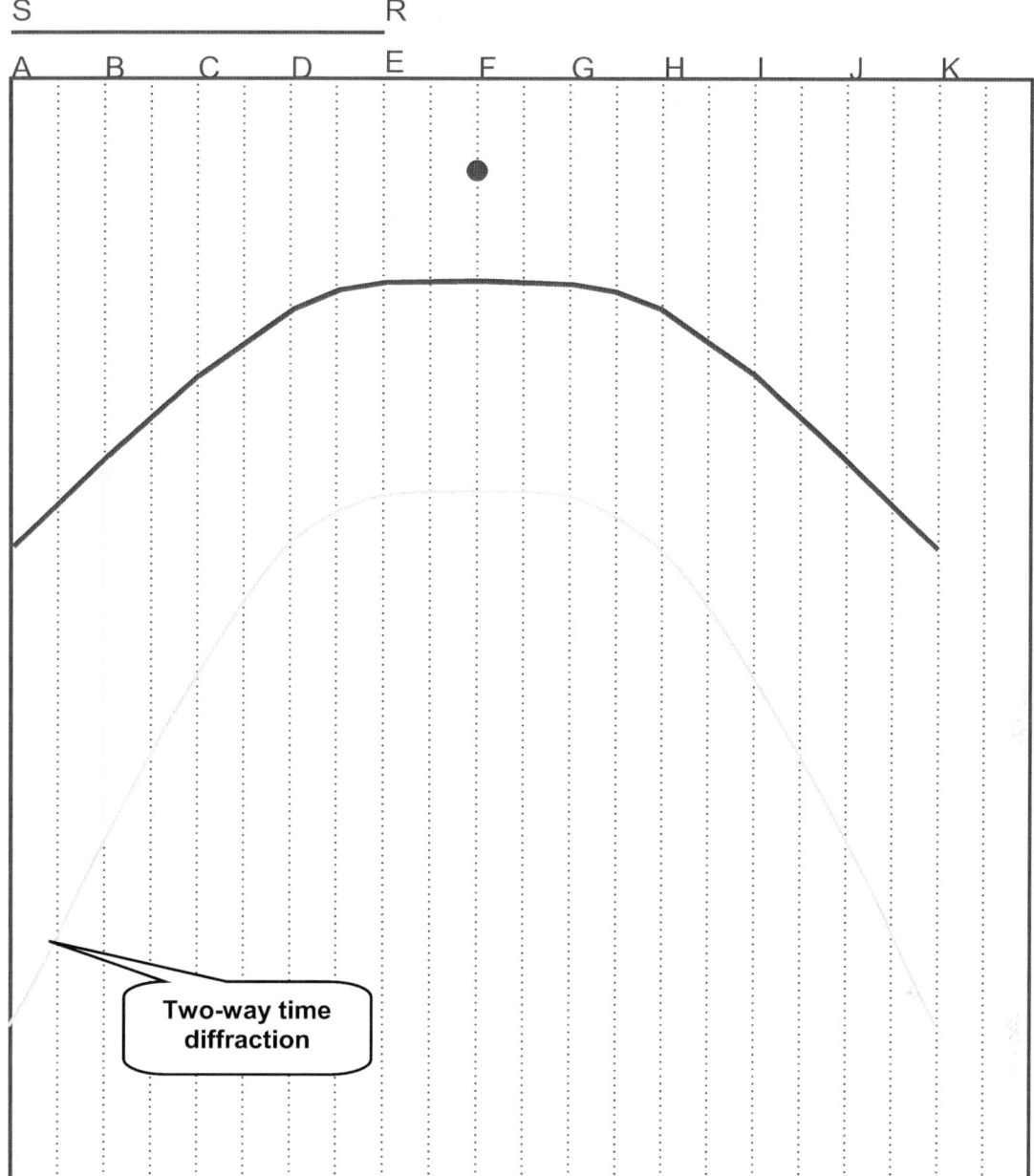

Figure 7.31 Constant offset construction of diffraction.

The two-way times from Figure 7.30 are plotted above as one-way times.
- **Add the zero offset diffraction to the above figure.**
- **Add NMO correction to the offset diffraction**
- **Compare with the offset diffraction.**

7.3.2 Scatterpoints and diffractions of constant offset sections

Examples of scatterpoints and their constant offset diffractions are illustrated in Figure 7.32 by the black curves. The source-receiver offset is indicated by S-R. For comparison, the zero-offset diffractions are shown in gray.

Note:

- The <u>apex</u> of the zero-offset diffraction passes through the scatterpoint.
- The offset diffractions fall below the scatterpoint.
- The offset diffractions are <u>bound</u> by the zero-offset <u>asymptotes</u>.
- The <u>apex of the offset diffraction is flatter</u> than the apex of the zero-offset diffraction.

Verify the offset diffraction time T is defined by the double square root (DSR) equation

$$T = \left(\frac{T_0^2}{4} + \frac{(x+h)^2}{V^2} \right)^{1/2} + \left(\frac{T_0^2}{4} + \frac{(x-h)^2}{V^2} \right)^{1/2}, \qquad (7.6)$$

where T_0 is the vertical two-way time from a scatter point, x is the horizontal distance from the scatterpoint to the CMP, h is the half offset, and V is the RMS velocity.

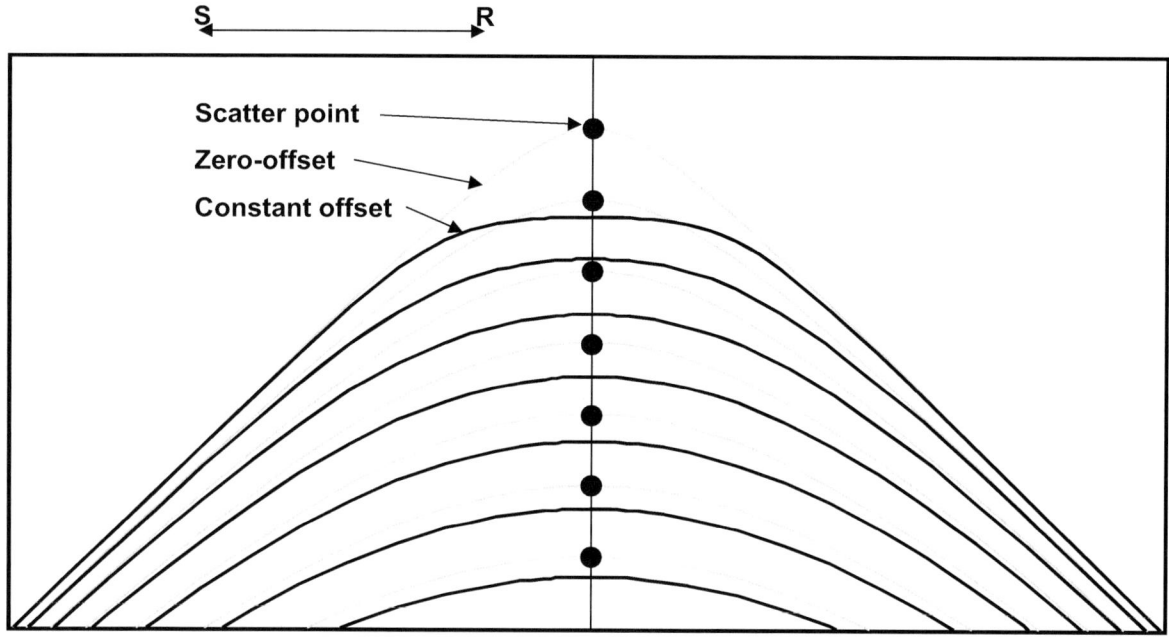

Figure 7.32 Diffractions from scatterpoints with zero-offset in gray, and constant offsets in black. The source-receiver offset is indicated by S-R.

7.3.3 Modelling of Constant Offset Sections

From the three source records in Figure 7.25 and Figure 7.26, we are aware the MO and stacking process does not locate dipping data in the zero-offset location; even in the simple case with constant velocity. We emphasize that:

STACKING SMEARS DATA

We will continue to evaluate CMP stacking concept using constant offset sections from the Hockey-stick model.

Figure 7.33 represents a <u>zero-offset section</u> and is the <u>standard</u> by which the following offset sections should be compared.

Figure 7.34 to Figure 7.41 show various stacks with:

- NMO for horizontal events,
- DD-MO for the dipping event,
- and various offset ranges.

Figure 7.38 to Figure 7.41 illustrate dip-dependent moveout (DD-MO) correction will align energy along a linear dipping reflection. Note, however, the gap moves down dip with increased offset. A stack of the various offsets will result in a smear of reflection points along the dipping reflection.

Some figures include <u>large offsets</u> that accentuate the smearing of energy. These large offsets at shallow times would be <u>muted</u> in conventional processing.

The apparent noise on the upper portion of the diffraction in Figure 7.41 results from gathering data over a range of offsets before MO.

What is the time smear that results from the offset range, and how will it affect our data?

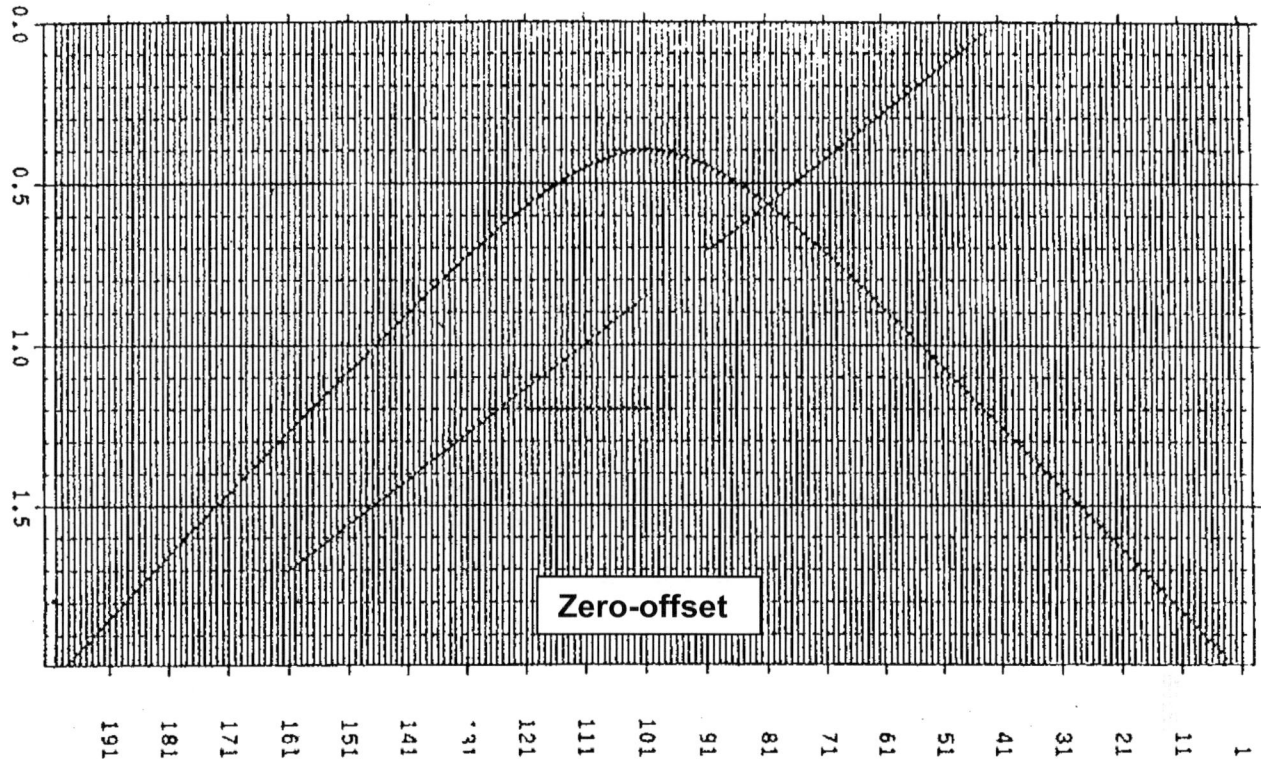

Figure 7.33 Zero-offset stack (*x*, *t*) representing the desired result of stacking all the data.

Question: Is it possible to correct a constant offset section to be equivalent to a zero-offset section?

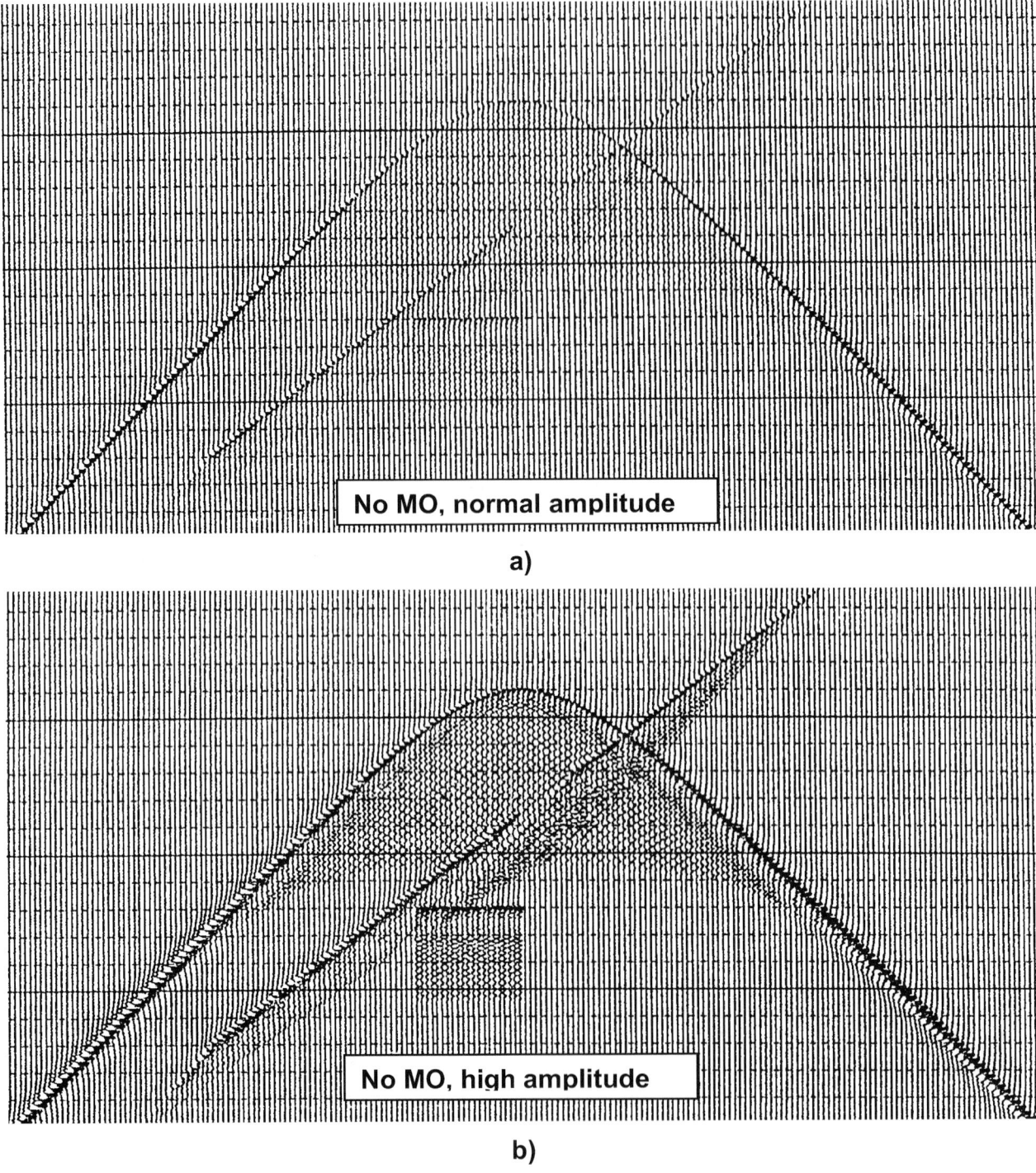

Figure 7.34 Stack (*x, t*) with <u>no moveout correction</u> at a) normal amplitudes and b) high amplitudes. The steeper dips of the diffraction tend to stack quite well, which illustrates the use of infinite velocity stacks for steeply structured data.

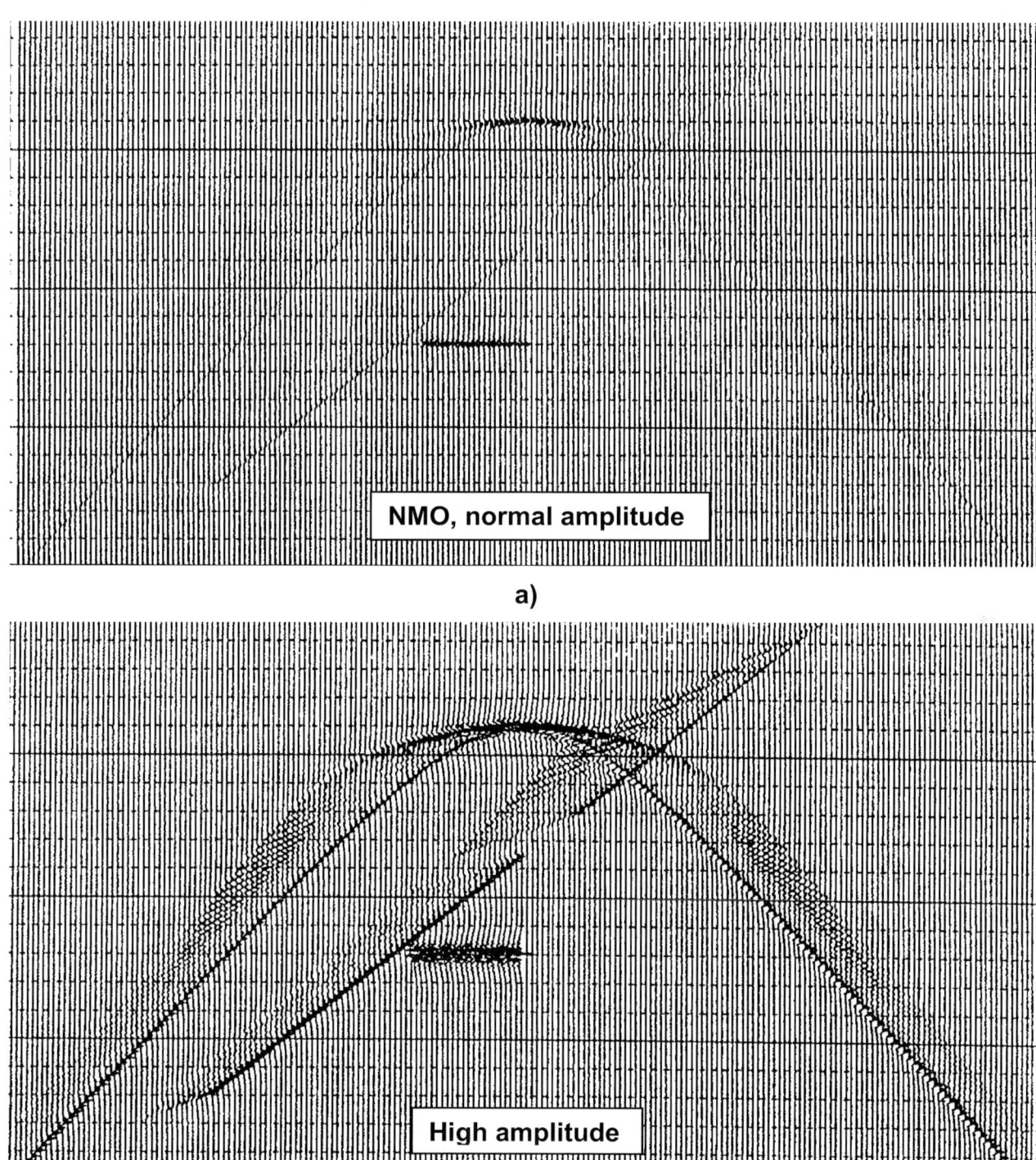

Figure 7.35 Normal stack (*x, t*) with NMO correction for horizontal event plotted at a) normal amplitude, and b) high amplitude. Note the energy at the flat portion of the diffraction is coherent while dipping energy is dispersed.

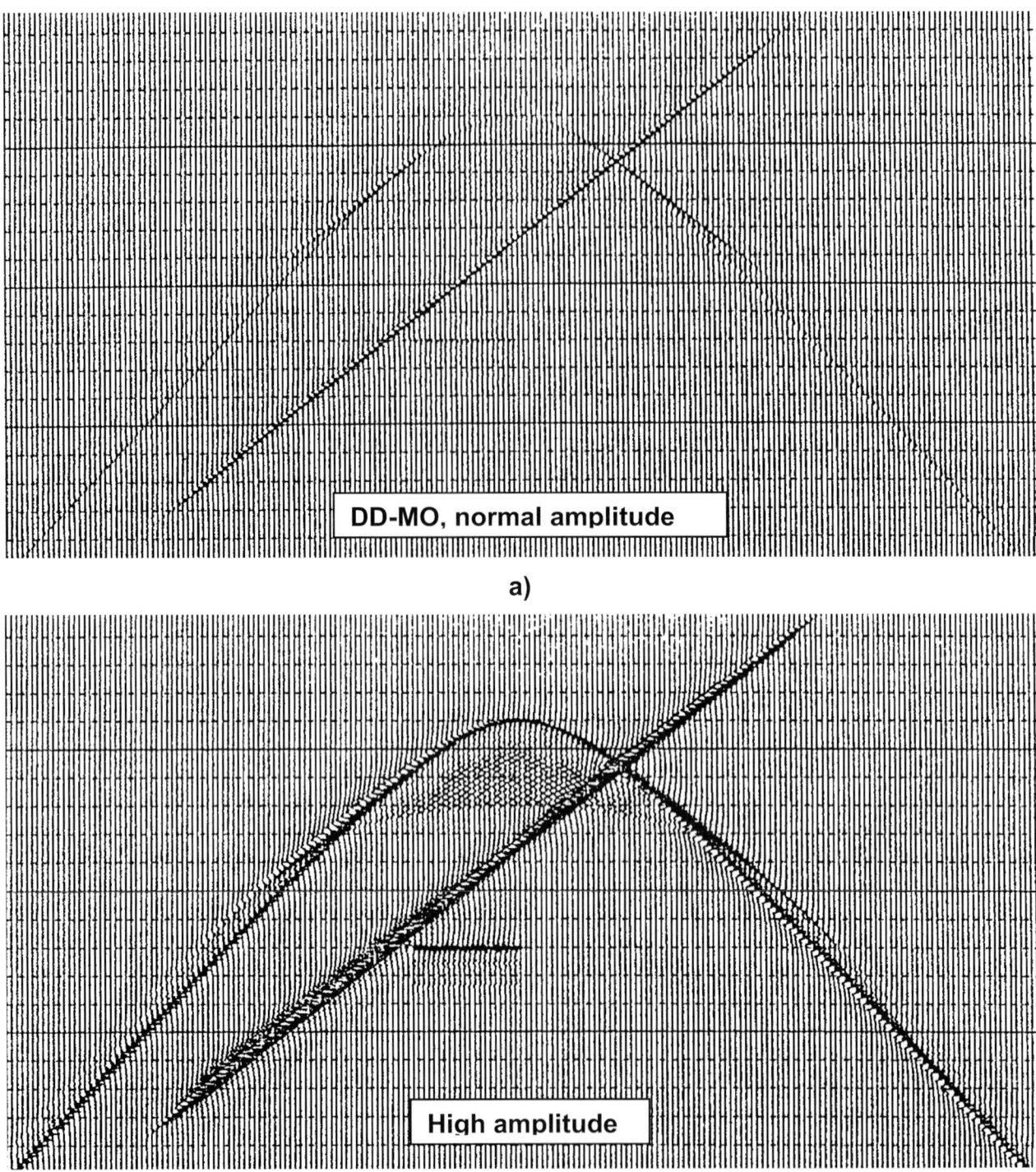

Figure 7.36 Stack (*x, t*) with dip-dependent moveout (DD-MO) a) normal, and b) high amplitudes. Note the flat event is dispersed, while the dipping event and the parallel dips on the diffraction are coherent. The gap is smeared along the dip.

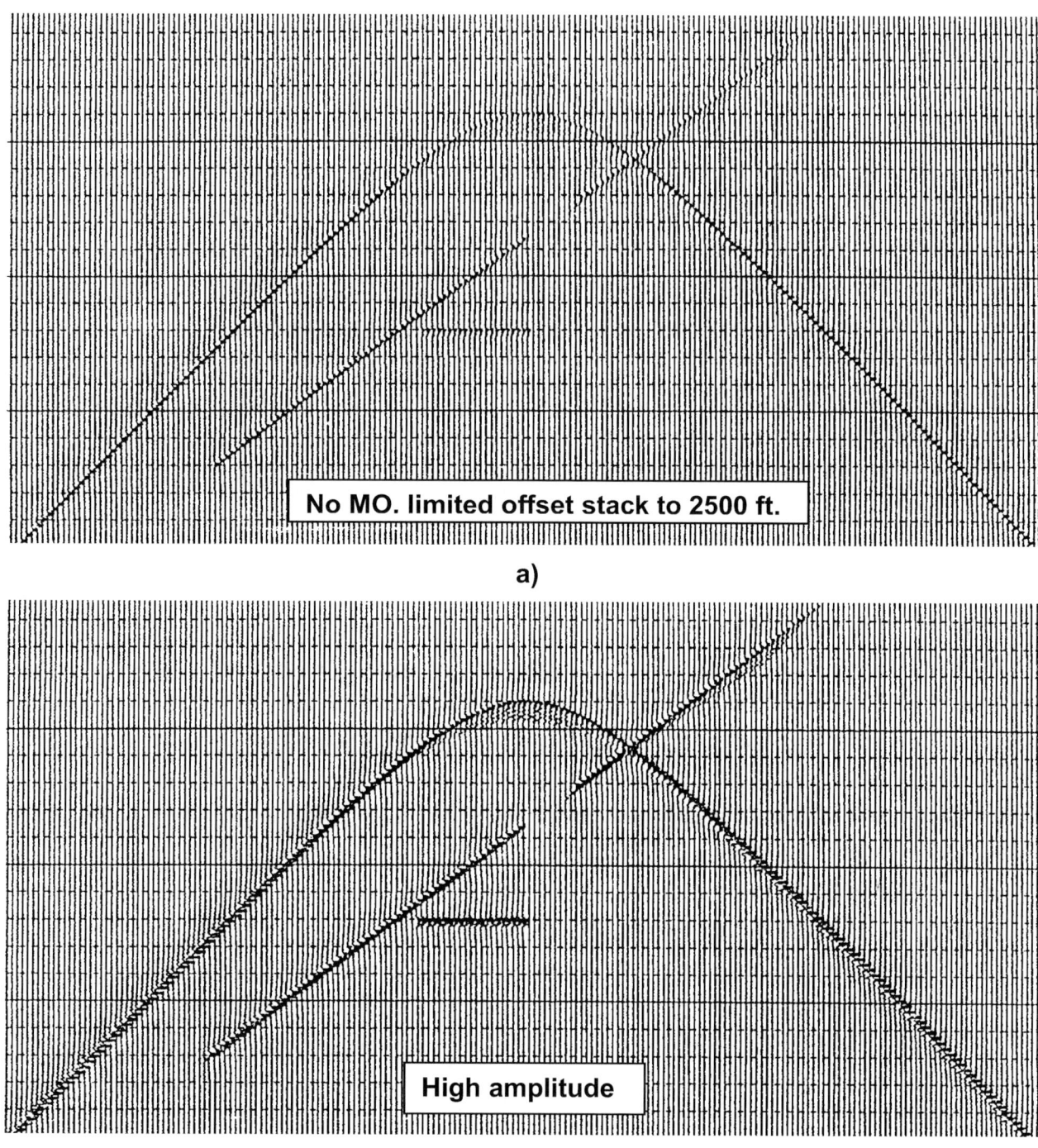

Figure 7.37 Limited offset stack (*x, t*) to 2,500 ft at a) normal amplitudes, and b) at high amplitudes.

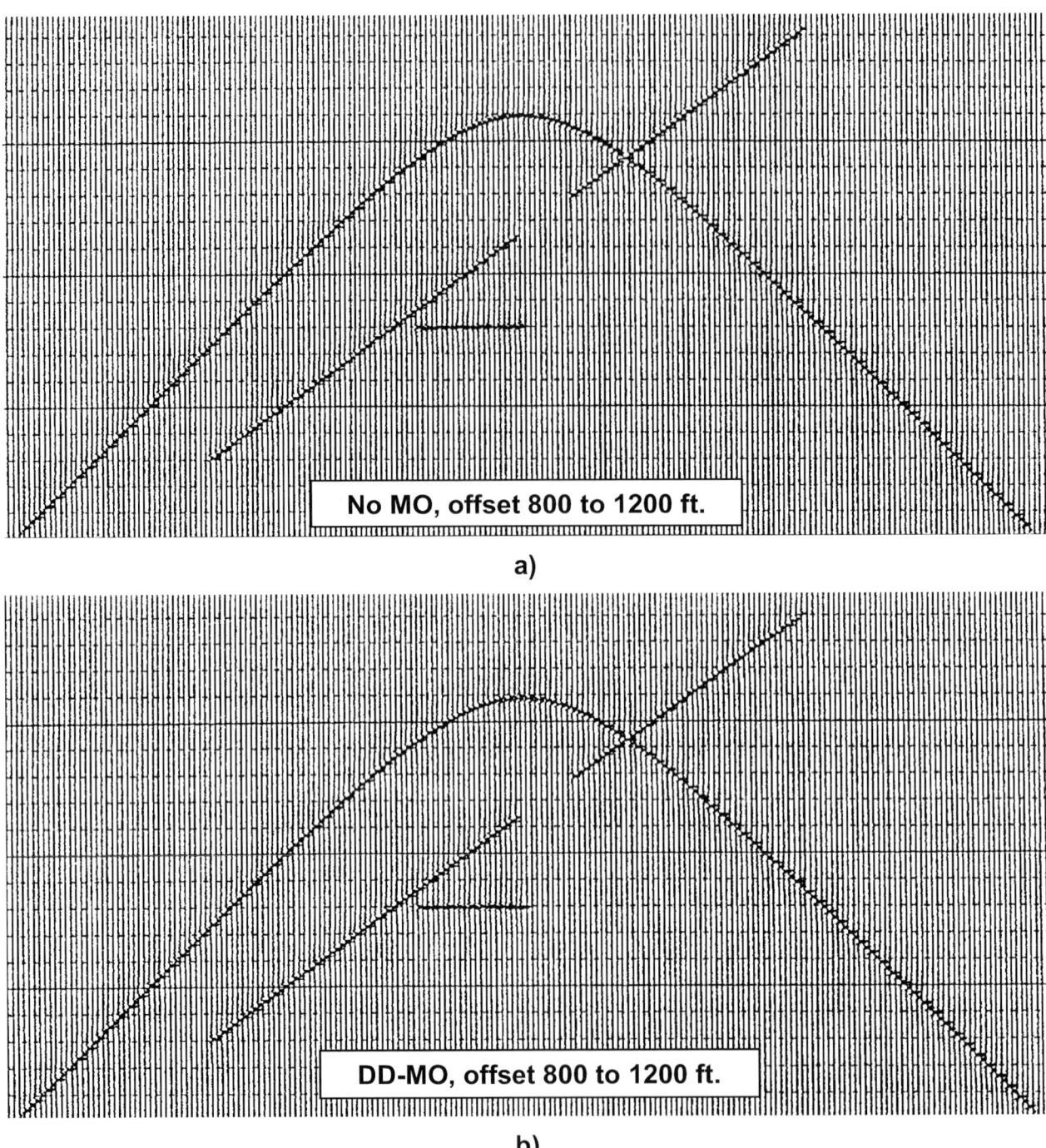

Figure 7.38 Limited offset stacks (*x, t*) from 800 to 1200 ft. with a) no MO, and b) DD-MO correction for the dipping event. Note the location of the gap is mainly to the right of the horizontal event.

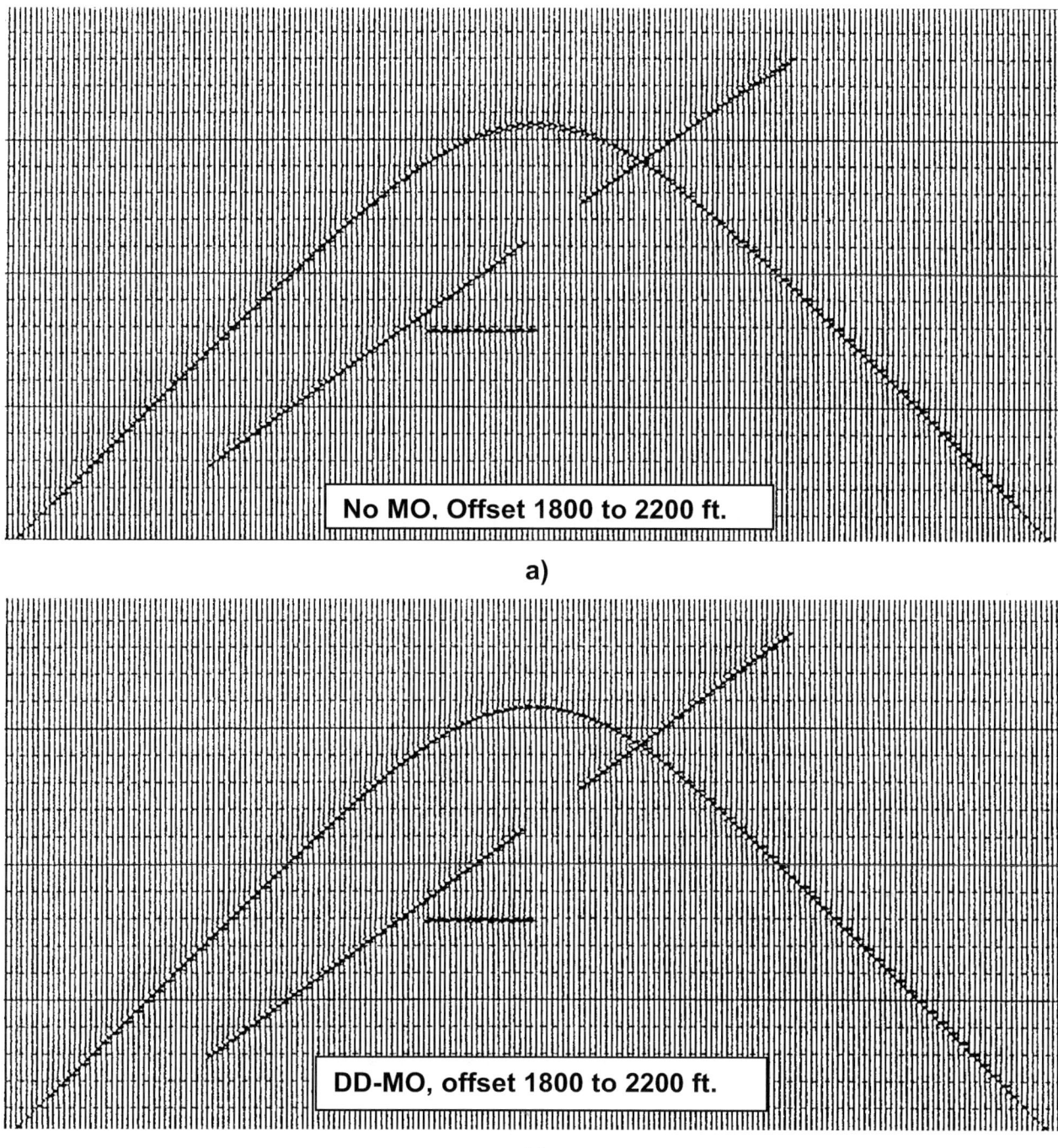

Figure 7.39 Limited offset stacks (x, t) from 1800 to 2200 ft. with a) no MO and b) DD-MO correction for the dipping event.

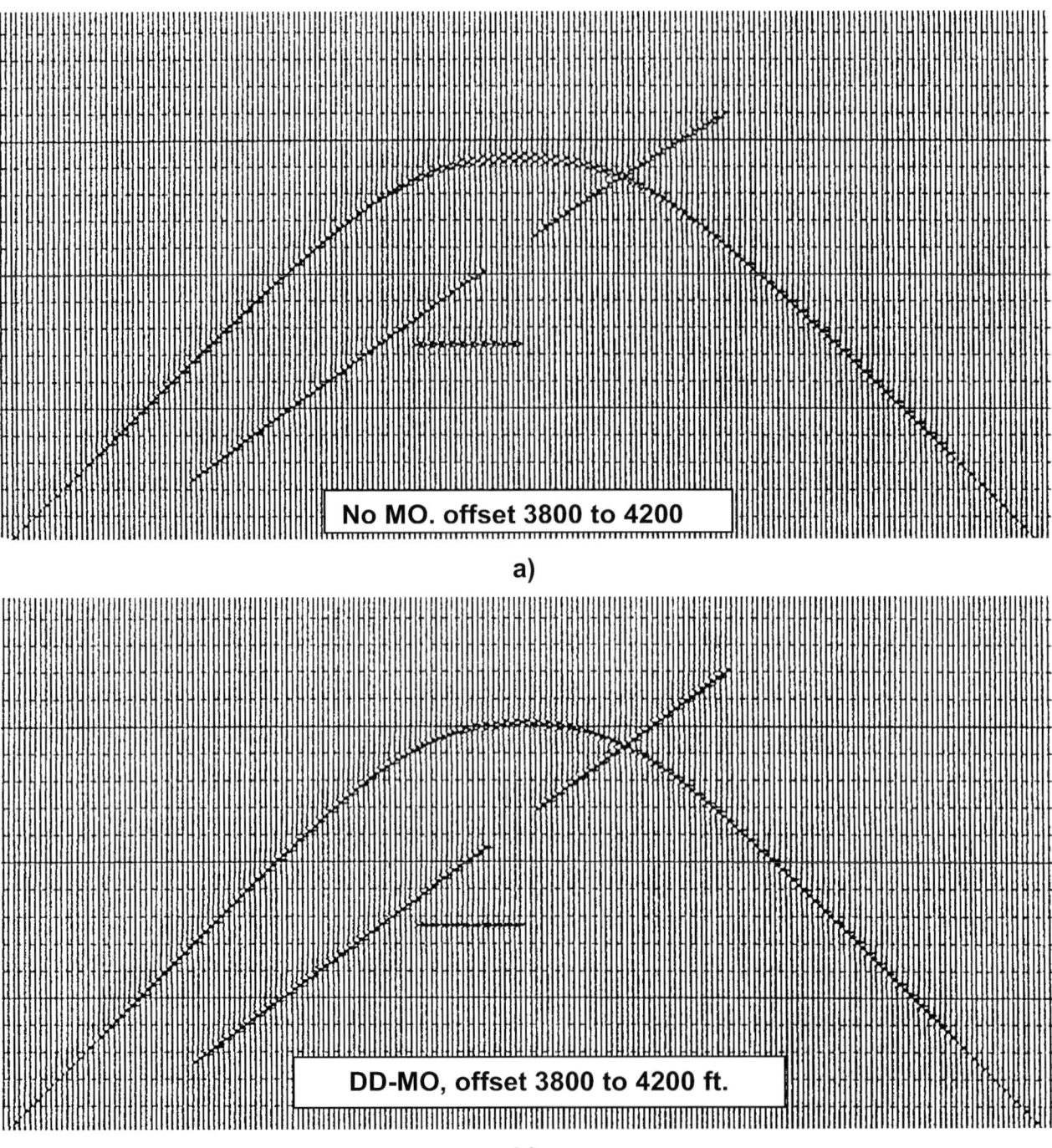

Figure 7.40 Limited offset stacks (*x, t*) from 3800 to 4200 ft. with a) no MO and b) with DD-MO. Note the location of the gap is moving down dip with increased offset range.

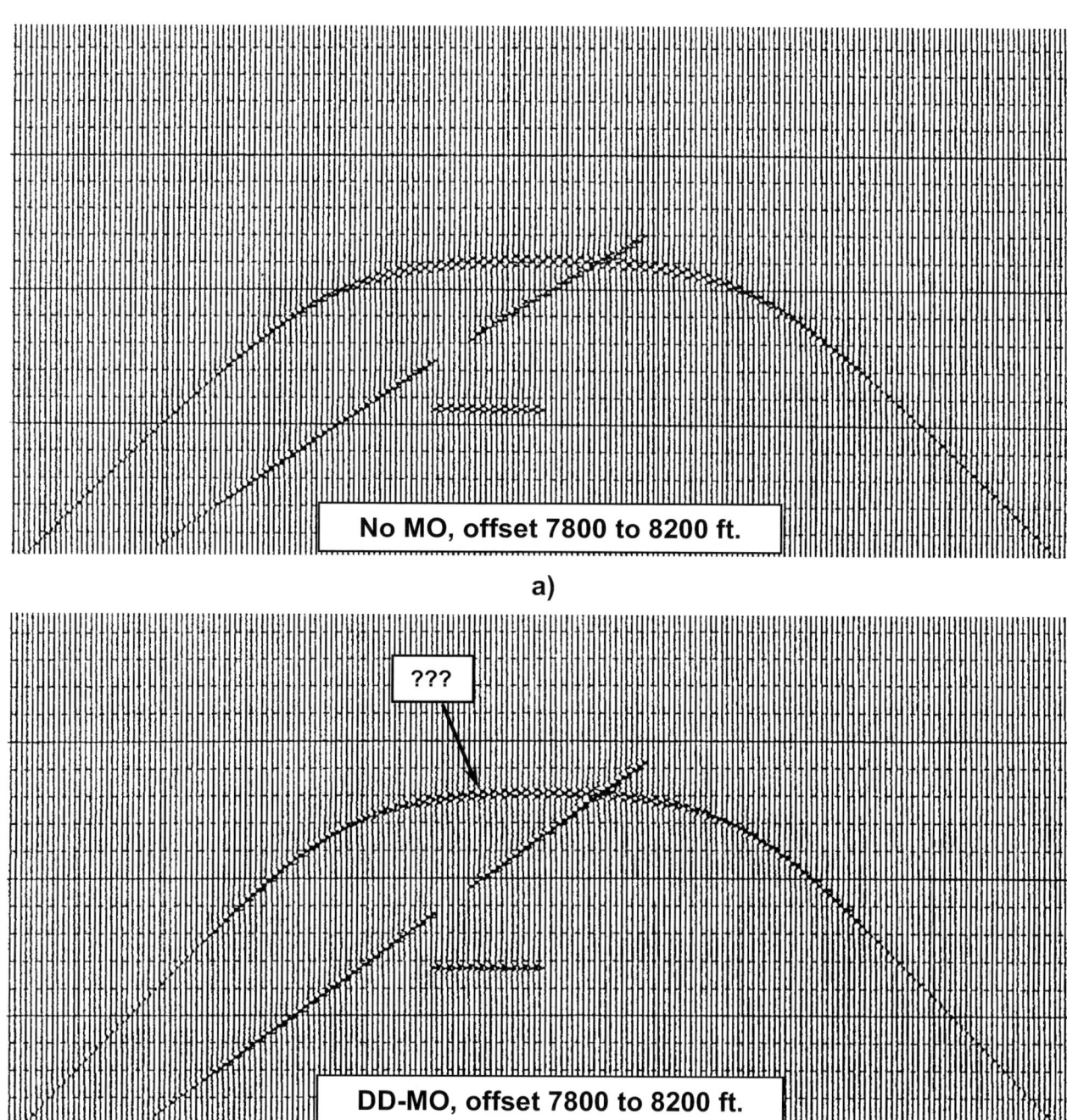

Figure 7.41 Limited offset stacks (*x, t*) from 7800 to 8200 ft. with a) no MO, and b) with DD-MO. These offsets are very large relative to the reflection times (0 to 2.0 sec's). Note the location of the gap has moved down dip to be above the <u>left edge</u> of the horizontal event.

The apparent noise near the top of the diffraction results from various MO times from different offsets within the offset range of 7800 to 8200 ft. or ± 2.5%.

7.3.4 Wavefield modelling of constant offset sections

Both a zero offset section and source record can be modelled using wavefield extrapolation.

- The exploding reflector computes the wavefield as it propagates from the reflector to the surface, where the recorded wavefield represents the zero offset section.
- A source record initiates the wavefield at a known source location then propagates the wave field through the depth space to the receivers.

That is not the case for a constant offset section.

At this point, we can't propagate a wavefield from a scatterpoint directly to a constant offset section.

It is of course possible to model the all source records then extract the constant offset section, but that is not computing it directly.

Constant offset sections, when modelled directly, use the diffraction technique.

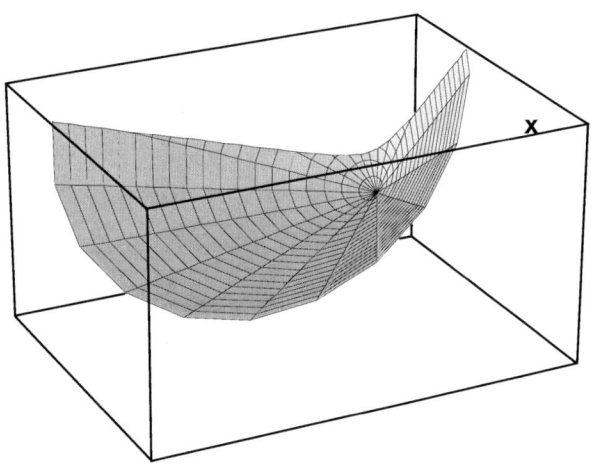

Figure 7.42 Wavefront propagating from a scatterpoint in <u>zero offset</u> (*x, z, t*) space.

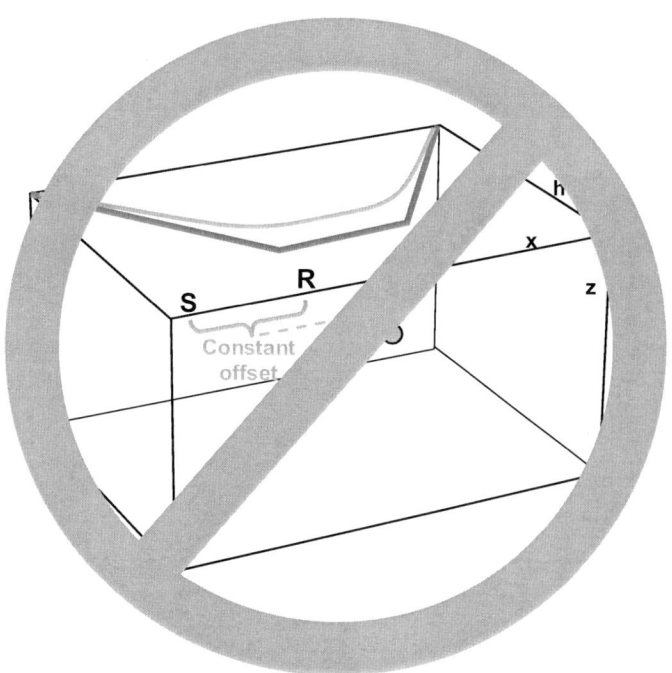

Figure 7.43 Wavefield propagation can't be used to propagate energy from a scatterpoint directly to the <u>offset diffraction</u>.

7.4 Prestack Eikonal Equation Modelling

The Eikonal method of modelling computes a <u>traveltime map</u> of wave fronts emanating from a defined location. An example of four traveltime maps with various surface locations is illustrated in Figure 7.44a. Traveltime contours (isochrons) represent the wave front at equal time intervals. Raypaths will propagate from the defined location and will be normal to the wavefronts.

Section 2.7 introduced the concept of Eikonal equation modelling for the zero-offset case. Figure 2.26 illustrated computing the diffraction <u>traveltimes for the scatter point</u>. This method requires computing the traveltime maps for each sample in the geological structure and is very inefficient.

A more <u>practical method</u> computes the traveltime maps from the surface at each <u>source and/or receiver location</u>. The <u>diffraction traveltimes</u> for one prestack subsurface point are found by defining the source and receiver for each input trace, and combining their traveltimes ($T_s + T_r$) from their respective traveltime maps. (See the raytracing method described by Gray in 1986 in [64].)

Various schemes for organizing data allow prestack modelling or migration by either summation (Kirchhoff) or dispersive (Hagedoorn) methods.

Exercise:

- Figure 7.44b shows two overlapping traveltime maps for one seismic trace. The velocity is constant and the isochron interval is 100 ms.
- What is the total traveltime T at the indicated point? $T = \ldots\ldots$
- Sketch the aplanatic surface or all points that have this same travel time.
- Repeat this procedure for a number of traveltimes.

These aplanatic surfaces define:

- A Hagedoorn type prestack depth migration where energy from an input trace is spread along the surface.
- Prestack modelling where the average reflection coefficient from all locations on the aplanatic surface is inserted into the seismic trace at time T.

Complex structures with arbitrary velocity distributions may have <u>problems</u> with this method as multiple ray paths may arrive at the same depth point (e.g. reflection and refraction). Only one time at each location is saved, and that is usually the <u>first arrival time</u> or the ray path with <u>maximum energy</u>.

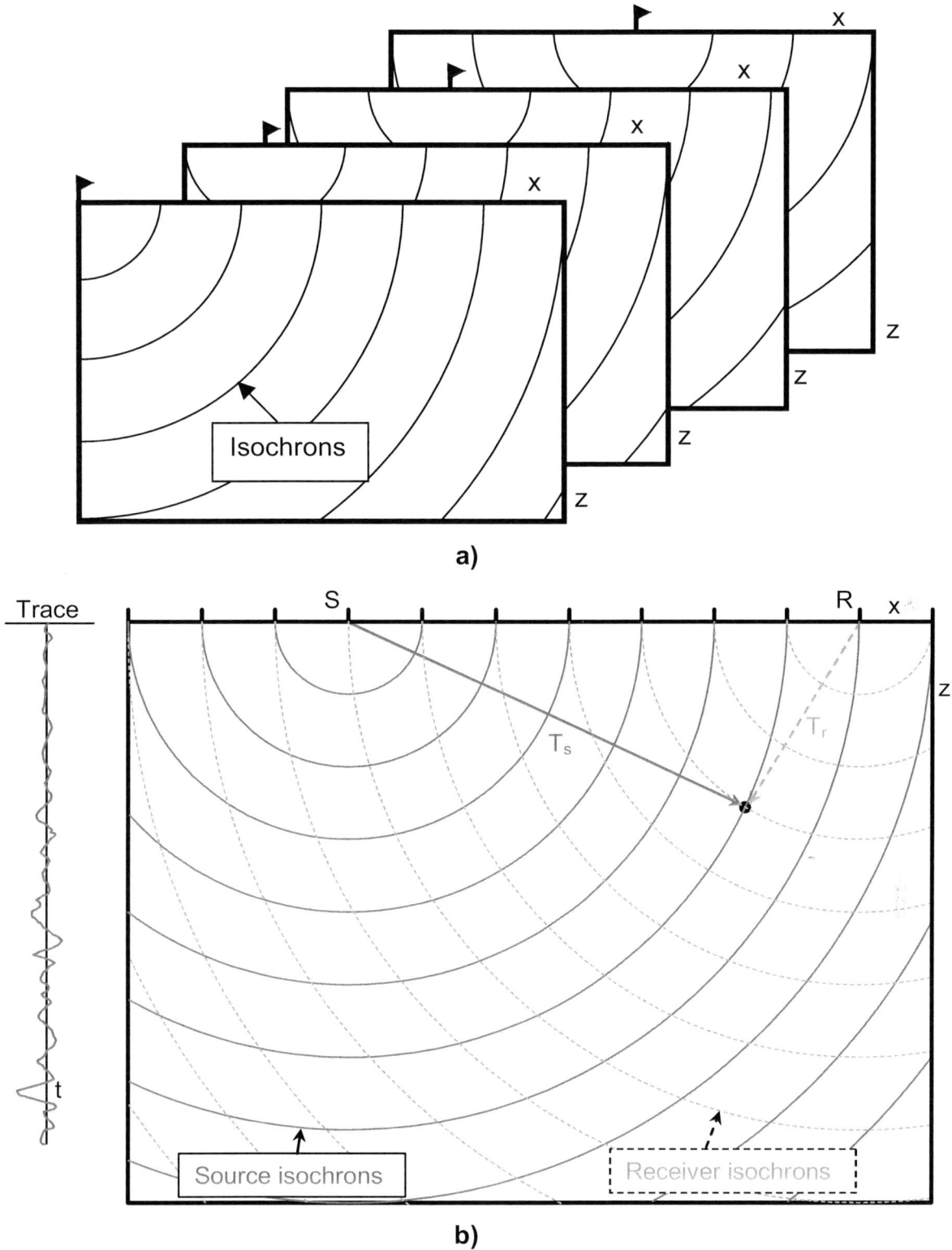

Figure 7.44 Traveltime maps, with a) illustrating four traveltime maps for different surface locations, and b) an example of estimating a raypath traveltime from two maps at the source and receiver location.

7.5 The Marmousi Model

In 1988 the Institute Francais du Petrole (IFP) created a complex 2-D geological model and generated synthetic seismic data. This model was distributed to numerous institutions and companies for a <u>blind test</u>. The velocity model was withheld from the participants until May 1990, when it was released at the EAEG workshop on "Practical aspects of seismic data inversion".

<u>Thirty-one invitations</u> resulted in <u>eight participants</u> who were willing to compare their blind tests. The proceeding of the conference [151] contains their results.

The model is complex and was based on a profile across the North Quenguela trough in the Cuanza basin in Angola. Both density and velocity models with high horizontal and vertical velocity gradients were used. The length of the model was 9200 meters with a depth of 3000 meters. The grid size was 4 by 4 meters.

The seismic data simulated a marine acquisition from 3000 meters to 8975 meters across the model, and consisted of <u>240 sources, each with 96 traces</u>.

The results of the <u>blind test show considerable differences</u> between the final velocity models, illustrating the problems in the depth migration process. One interesting result, however, was the <u>average of the models was superior</u> to the individual models. (See Versteeg's Doctoral thesis [152]).

Figure 7.47 on the opposite page shows the velocity model, a zero-offset section, and a post-stack depth migration (by Xiang Du).

The data has become a <u>standard 2-D test set</u> that is used to compare or illustrate the success of various pre or post-stack, time or depth migrations.

> The 1990 EAEG workshop proceedings [151] and Doctoral thesis by Versteeg [152] are an excellent source of material on the Marmousi experiment. (Note that the title is in French, but the text is in English.)

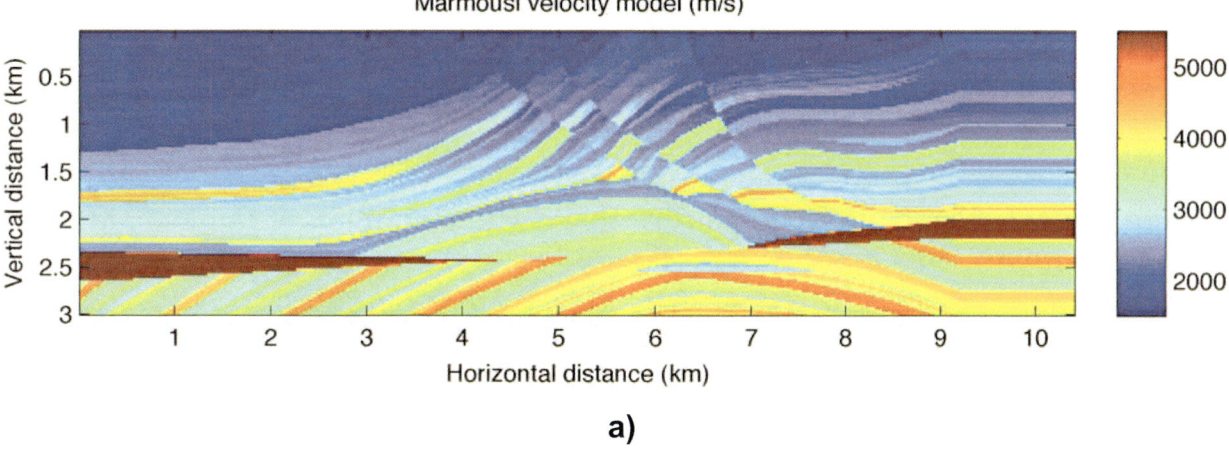

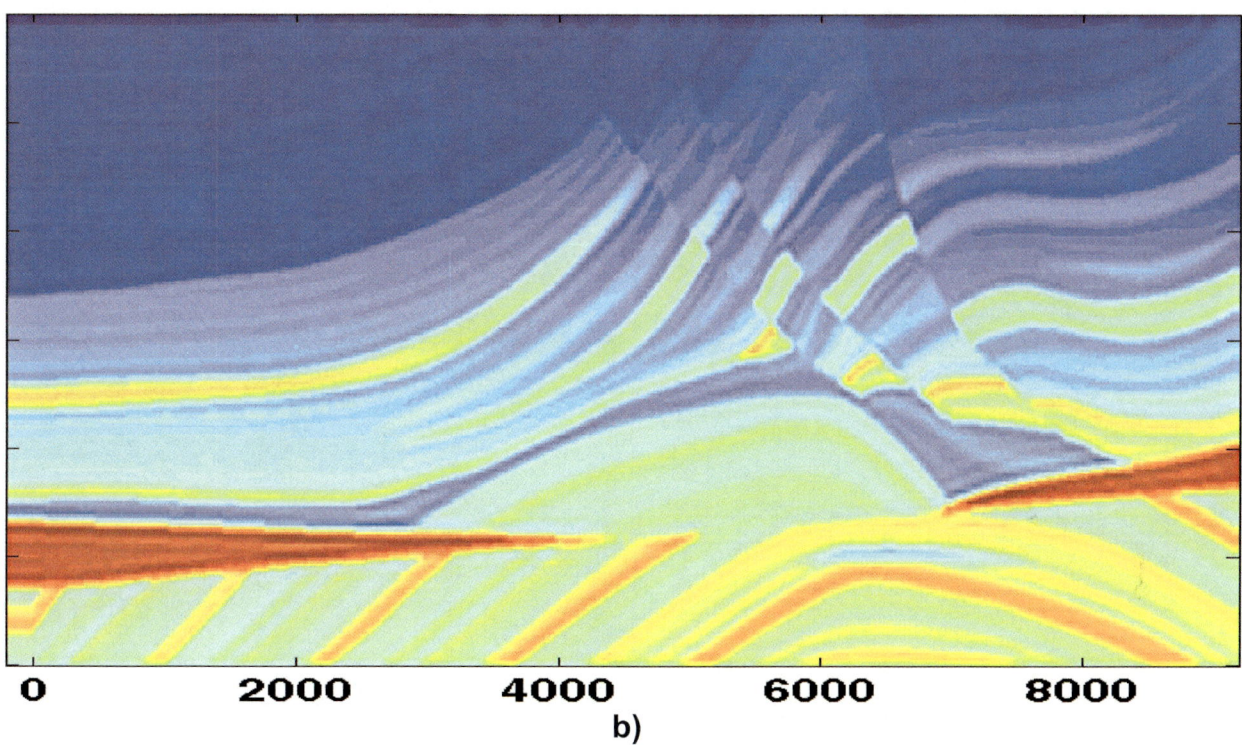

Figure 7.45 The Marmousi model a) with equal horizontal and depth scales and a velocity legend, and b) an enlargement showing the velocity details.

7.5.1 Waveform modelling of a source in the Marmousi model

The following images were captured at various times from a movie created by Gary Margrave to demonstrate the full P-waveform propagation from a single source point in the Marmousi model. (A nine point Laplacian was used with a 30 Hz minimum-phase wavelet.)

Incident, reflected, and multiple energy is evident. Note that <u>surface seismic</u> only records the energy that reaches the surface.

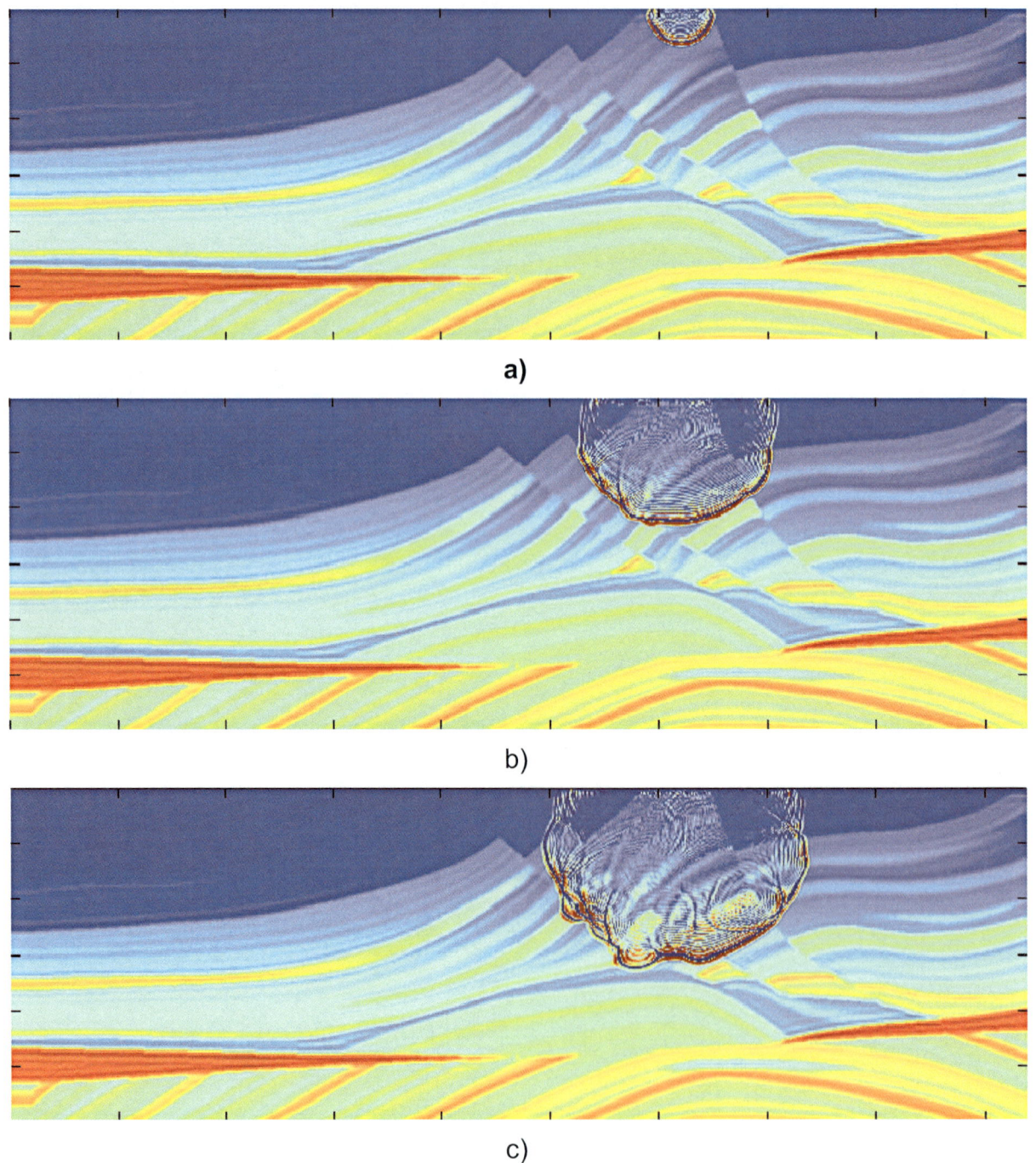

a)

b)

c)

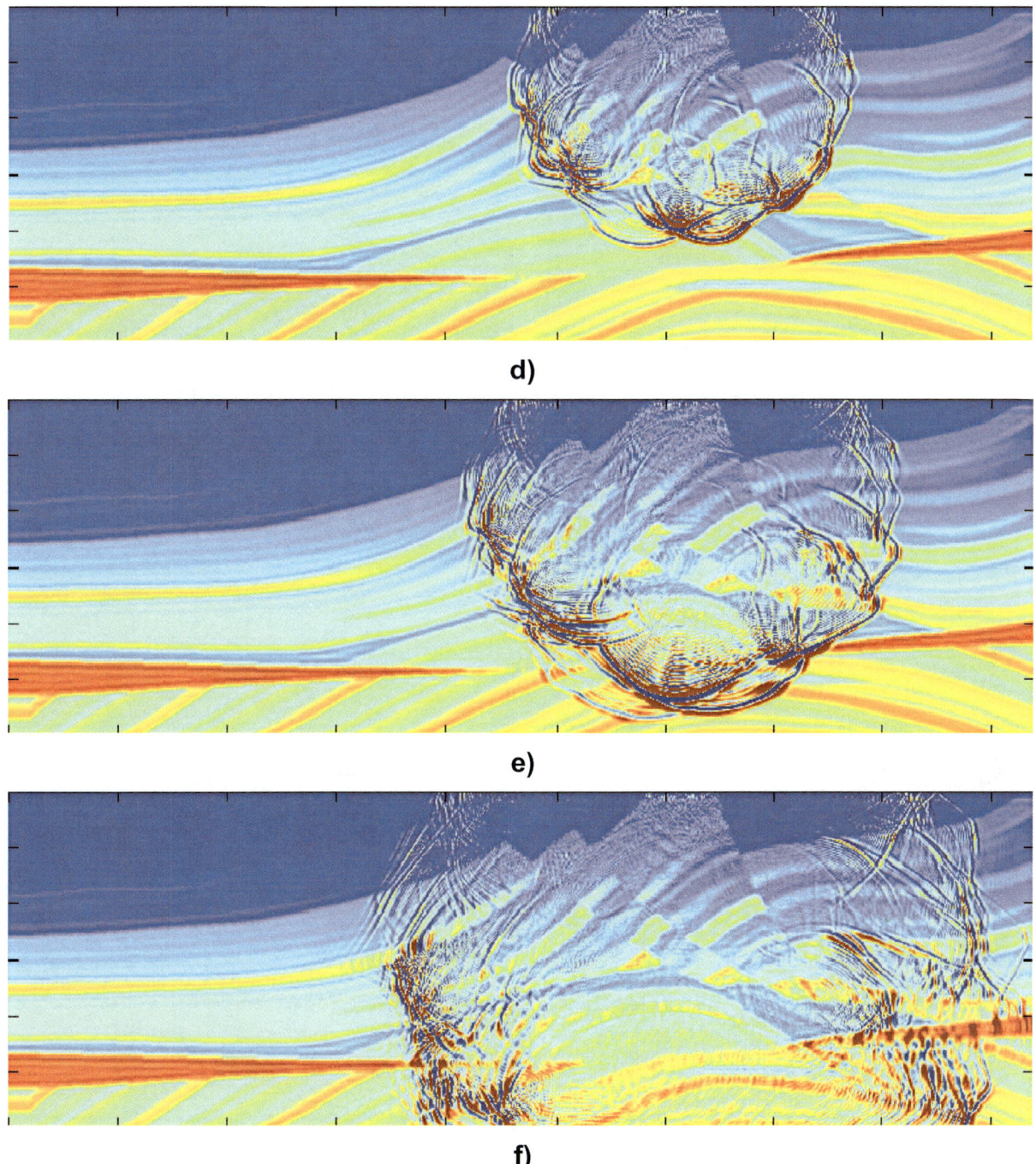

Figure 7.46 P-wave energy from a single source point propagating in the Marmousi model. Each panel shows the energy at different times. (Margrave 2006)

Features to note: complexity of the wavefront, head waves from higher velocity layers, multiple arrivals, reflection energy arriving at the surface, etc.

7.5.2 Examples of processing Marmousi data

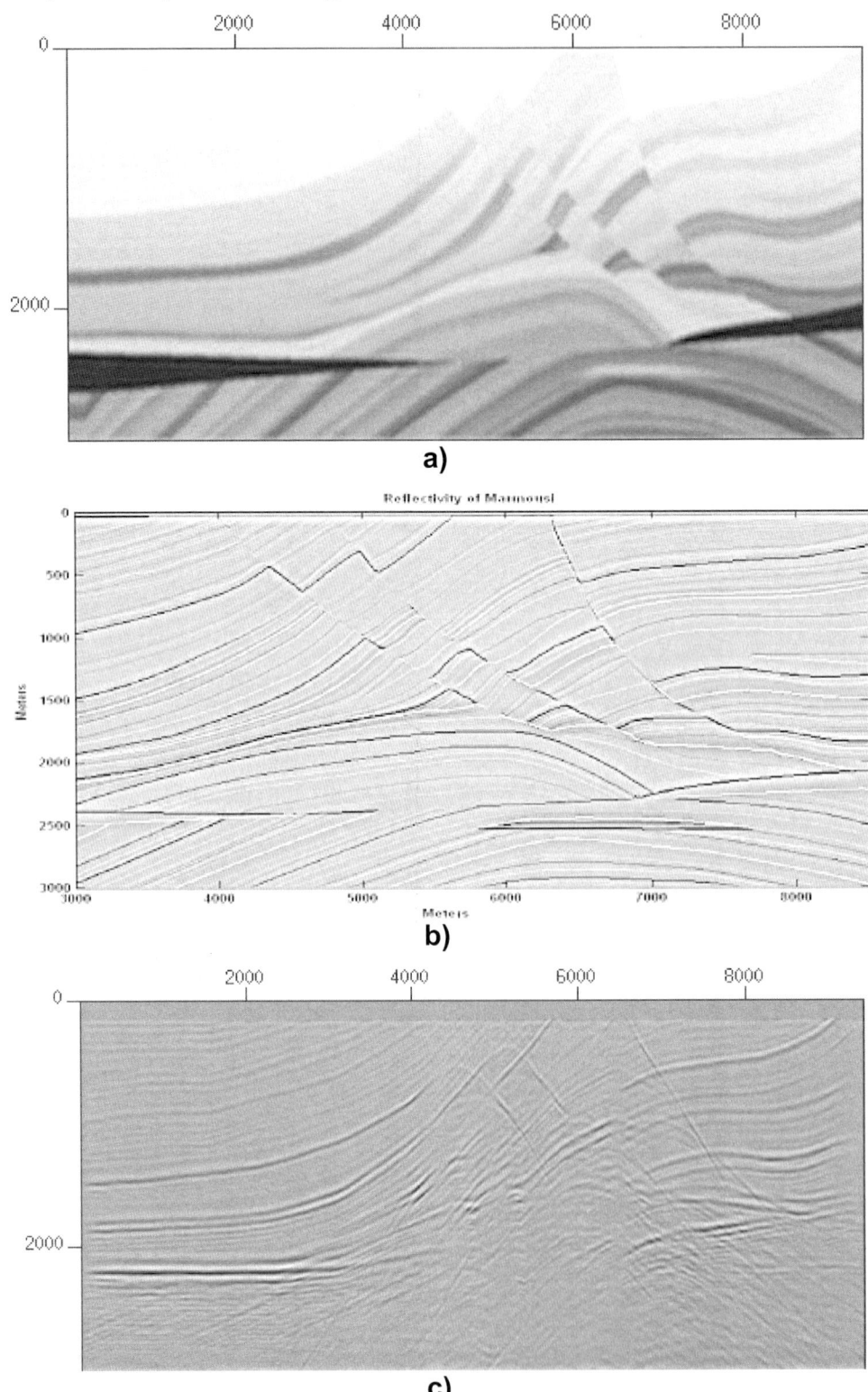

Figure 7.47 The Marmousi model with a) showing a grayscale of the velocities, b) a reflectivity model, and c) a stacked section.

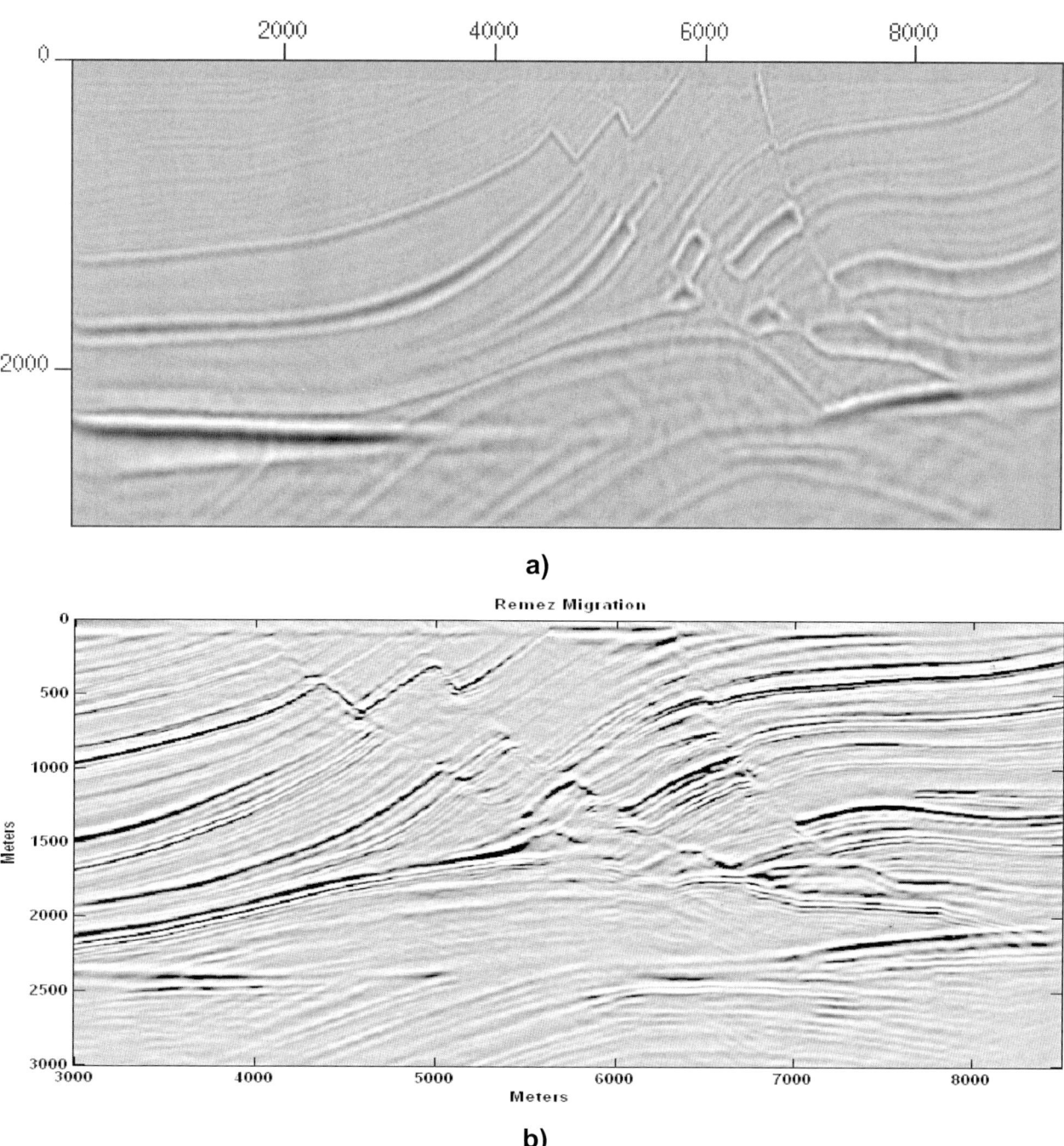

Figure 7.48 Migration examples, a) a poststack reverse time finite element-finite difference (FE-FD) depth migration (by Xiang Du), and b) a prestack migration using the Remez construction for an ωX migration that is based on the phase shift method. (by Saleh Al-Saleh)

7.6 Moveout (MO) Processing of Prestack Data

Previous examples have shown the application of a single velocity to stack horizontal and dipping events.

This section also shows moveout (MO) processing on a constant offset section with a constant propagation velocity, however, the stacking velocity is allowed to vary with the zero-offset dip of the data.

Figure 7.49 is a constant offset section (offset 2h) that shows a number of diffractions from evenly spaced scatterpoints. Zero-offset diffractions are shown in gray. Each figure shows:

a) Diffractions.

b) NMO correction using the constant velocity of the medium.

c) Dip-dependent moveout (DD-MO), where the stacking velocity was optimized for each dip α on the zero-offset diffraction. The stacking velocity was defined by

$$V_{stk} = \frac{V_{rms}}{\cos\beta} = \frac{V_{rms}}{(1-\sin^2\beta)^{1/2}} = \frac{V_{rms}}{(1-\tan^2\alpha)^{1/2}}. \qquad (7.7)$$

The results in (c) may appear more distorted than in (b), however, note the time errors in (c) are in general much less than in (b), especially at larger offsets.

These figures show only one offset (x, t) for a constant h, i.e. (x, h = hc, t). A more comprehensive view of the processed data may be gained by viewing the data in a 3-D volume that shows an added dimension of offset, i.e. (x, h, t).

These volumes are shown in the following sections.

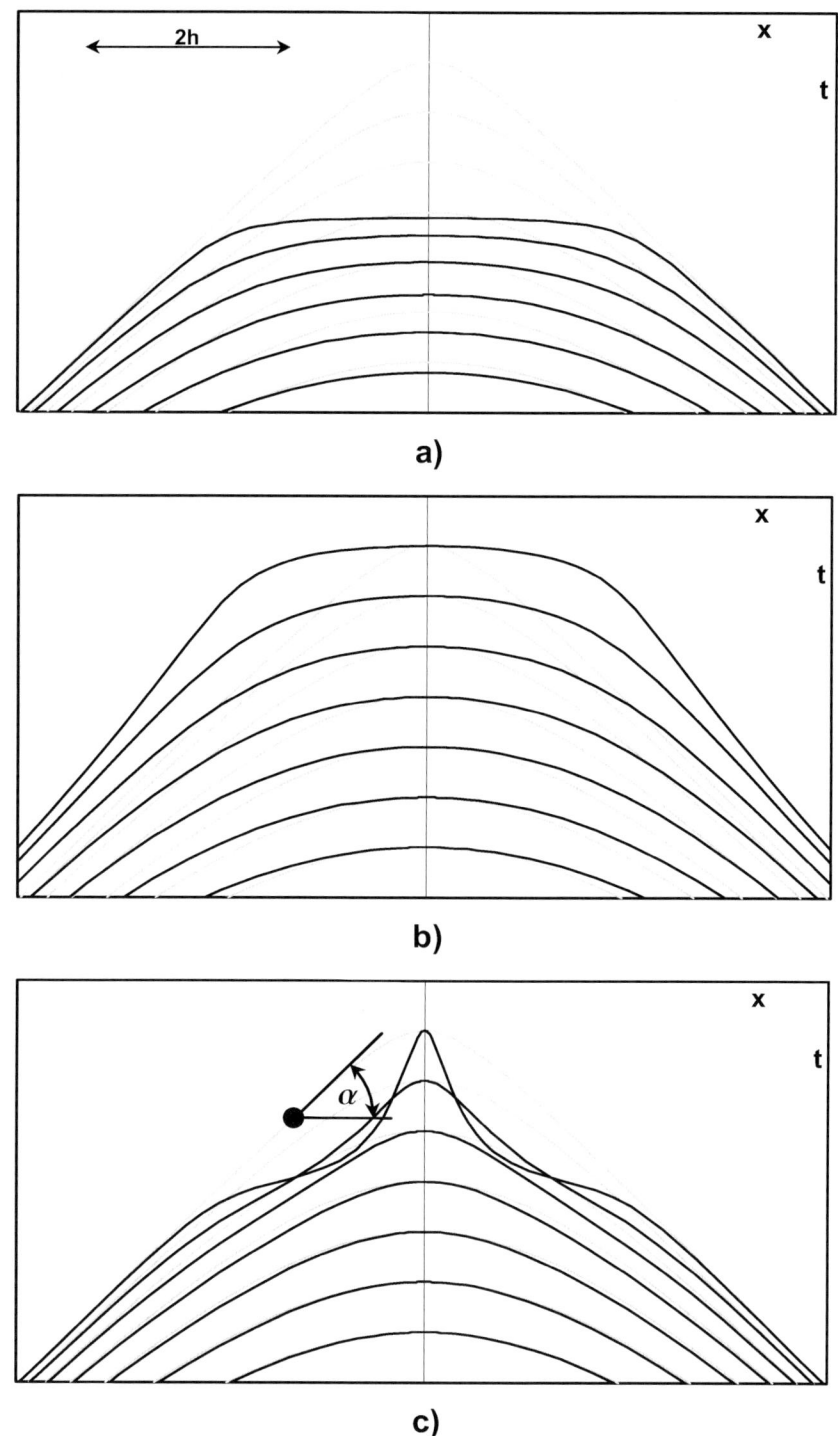

Figure 7.49 Constant offset sections illustrating a) the diffractions from a number of evenly spaced scatterpoints, b) after constant velocity NMO correction, and c) with DD-MO that varies with the zero-offset dip α. Each figure shows zero-offset diffractions in gray.

7.7 Cheops Pyramid for 2-D Data

The travel times of energy from a scatter point are defined by the double square root (DSR) equation (7.6) which is repeated for convenience, (scatterpoint located at $x = 0$),

$$T = \left(\frac{T_0^2}{4} + \frac{(x+h)^2}{V^2}\right)^{\frac{1}{2}} + \left(\frac{T_0^2}{4} + \frac{(x-h)^2}{V^2}\right)^{\frac{1}{2}}.$$

The diffraction shapes for eight different offset from one scatterpoint is shown in Figure 7.50. The first or zero offset is shown in a lighter grey and is hyperbolic. The finite offset diffraction shapes are shown in a darker gray with the tops of the curves flattening with increased offset.

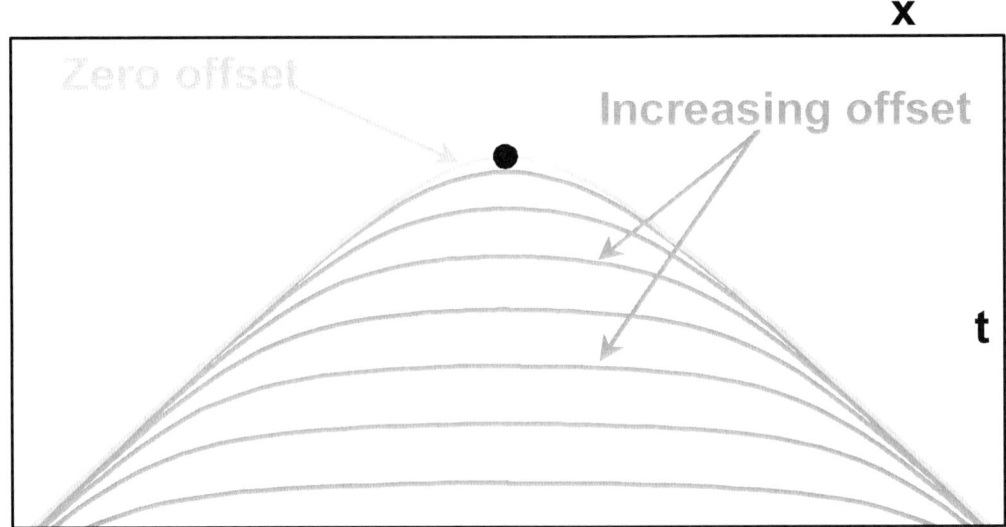

Figure 7.50 Diffraction energy from one scatterpoint at eight constant offsets.

A continuous range of the midpoint x and offset h form a surface in the prestack volume (x, h, t) illustrated in Figure 7.51. This surface is referred to as Cheops pyramid (Ottolini [373] and Claerbout [294]) as it resembles the eroded shape of Cheops pyramid on the Gaza plateau in Egypt.

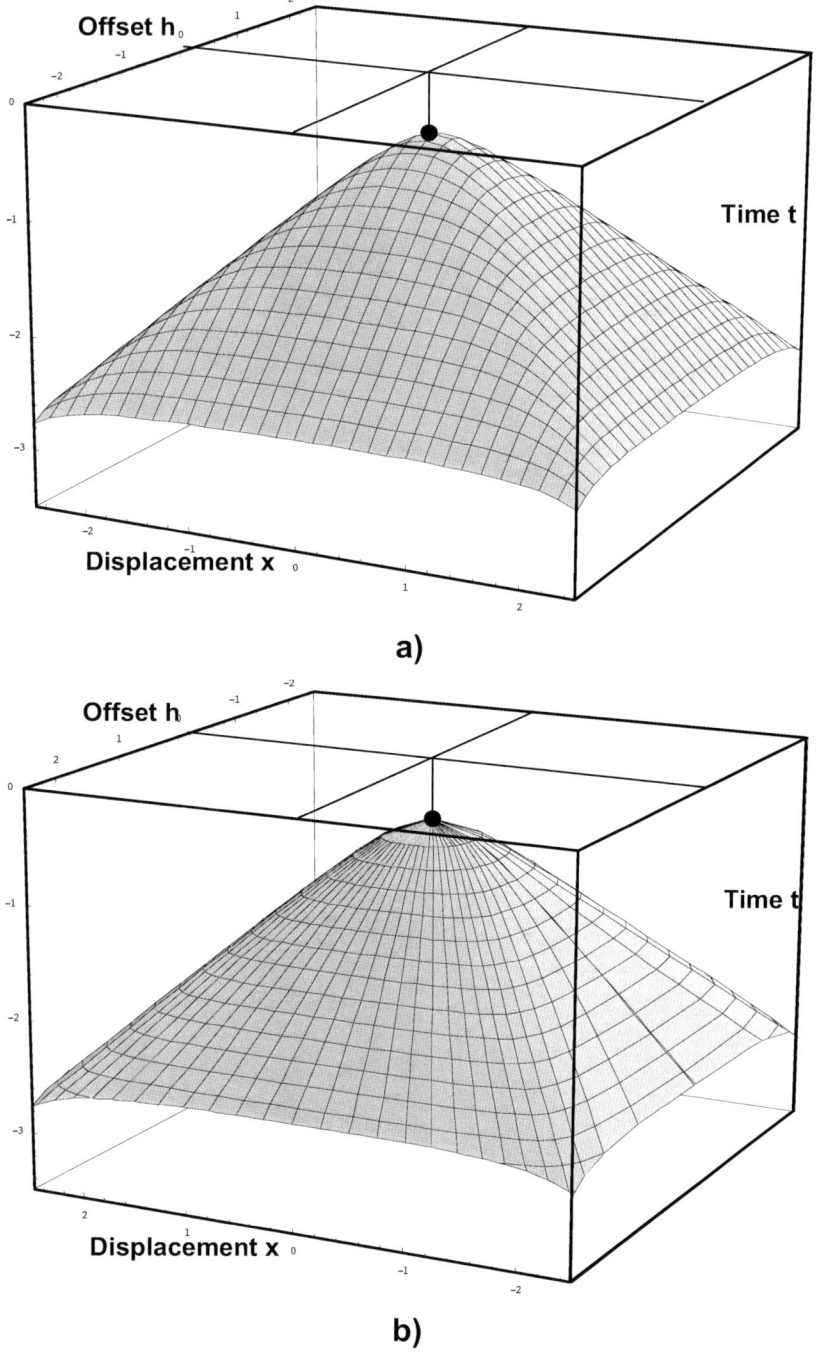

Figure 7.51 Cheops pyramid for continuous range of midpoints and offsets from one scatterpoint is referred to a Cheops pyramid with a) showing a grid in x and h, and b) showing contours at equal times.

Hyperbolic intersections are formed when a number of **planar surfaces** intersect Cheops pyramid and are shown as **black** curves in Figure 7.52a – d:

- The single CMP gather through the apex of the pyramid, $x = 0$ in equation (7.6).
- The zero-offset section, $h = 0$ in equation (7.6).
- Dipping planes that rotate through the $t=0$, $h=0$ edge of the volume as illustrated in Figure 7.52c.
- All source or receiver gathers, as evident in Figure 7.52d.

Non-hyperbolic intersections are shown as white curves in Figure 7.52b and are:

- Constant offset sections (except non-zero-offset),
- CMP gathers (except through the apex or scatterpoint).

Interesting points:

- The **velocity at the scatterpoint** defines the shape of Cheops pyramid.
- Structured seismic may be modelled by **many Cheops pyramids**.
- Structured energy in a CMP gather **will not be corrected** with hyperbolic MO.
- Small offset energy in a CMP gather tends to be **horizontal** when the dip is large (i. e. at a large displacement from the scatterpoint). Consequently, stacked **sections with no MO will tend to stack** the steeply dipping events.

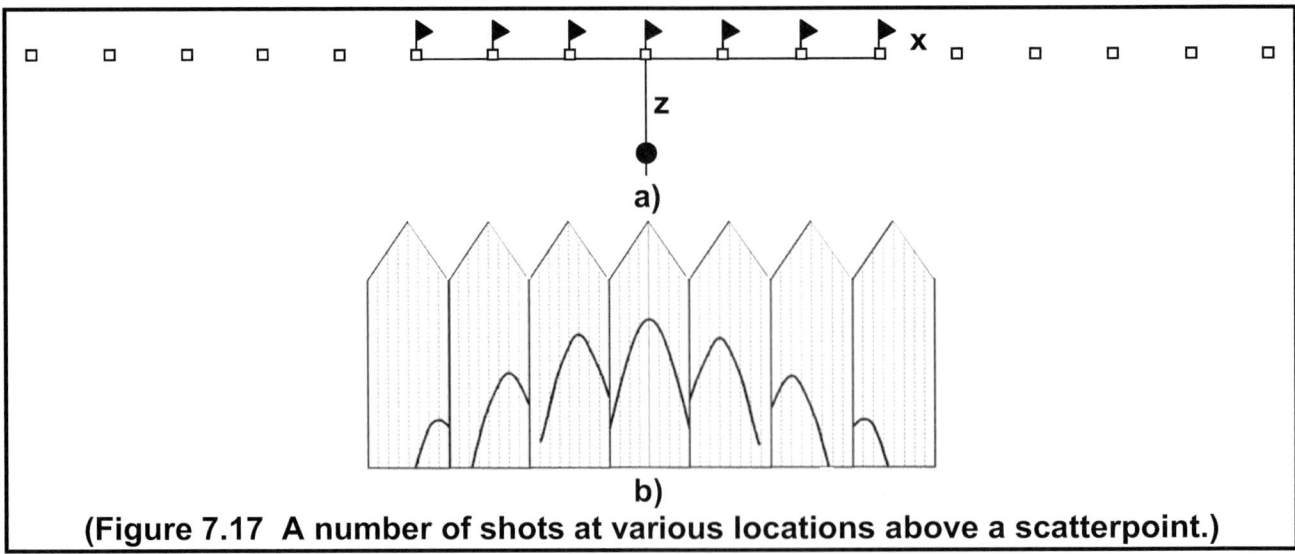

(Figure 7.17 A number of shots at various locations above a scatterpoint.)

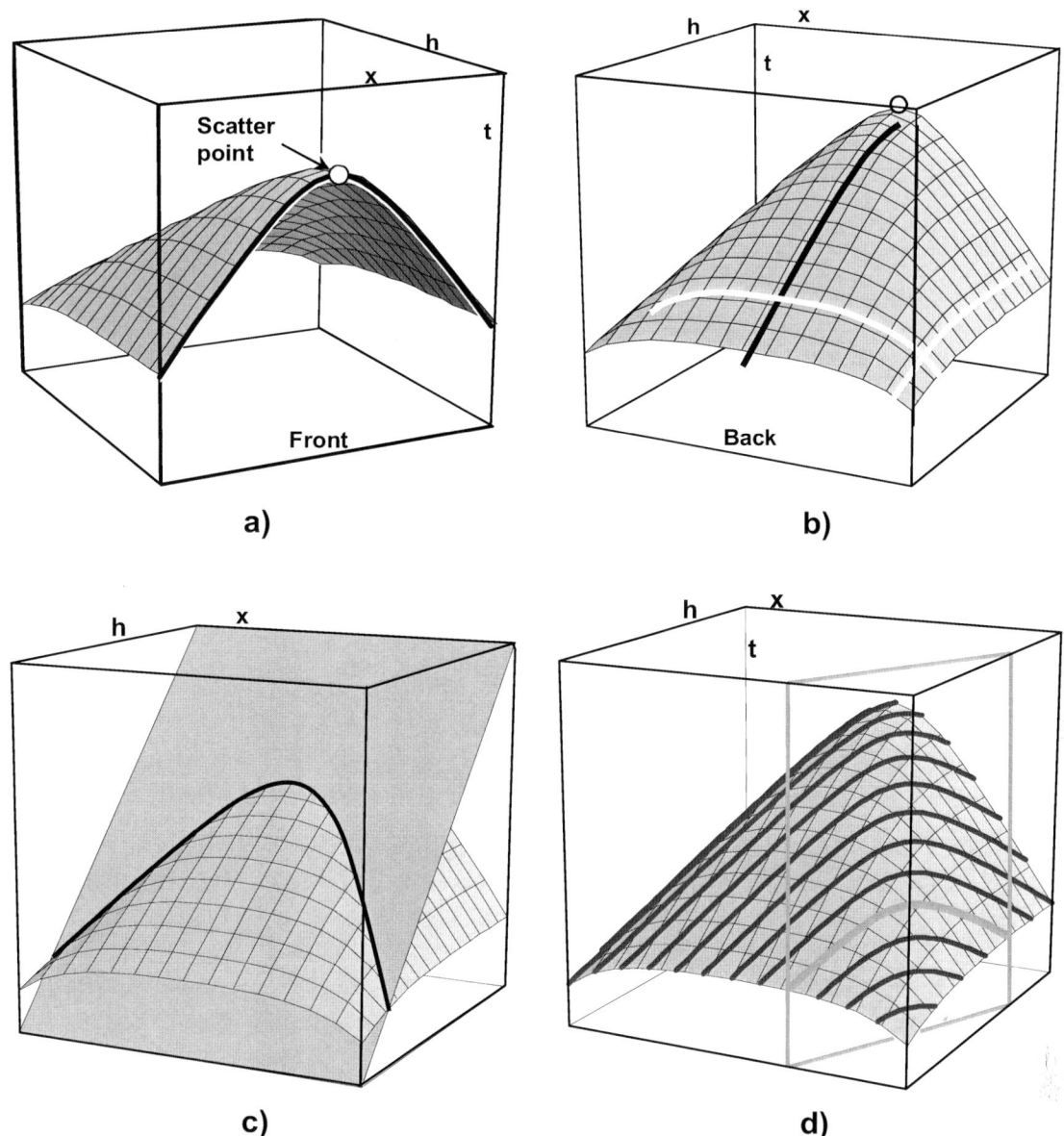

Figure 7.52 Cheops pyramid in the prestack volume (x, h, t), showing hyperbolic and non-hyperbolic intersections. a) shows the zero-offset hyperbola, b) the CMP hyperbola at the scatter point, while other CMP gathers and constant offset sections are non-hyperbolic (in white), c) a dipping plane, and d) one-sided source or receiver gathers.

> Describe the shape of the diffractions in each source (or receiver) gather of Figure 7.38d

7.8 Moveout (MO) Processing of Cheops Pyramid

Conventional processing of the diffraction energy for one scatterpoint, in a constant velocity medium, is shown in Figure 7.53. <u>Each figure:</u>

a) <u>shows the diffracted energy with the Cheops</u> pyramid shape.

b) contains a <u>hyperbolic cylinder</u> that is the desired shape after some form of normal moveout correction. Stacking would sum all the energy to the hyperbolas

c) the result of constant-velocity <u>NMO correction</u>.

d) the result of <u>DD-MO correction</u>, where the stacking velocity varied according to the zero-offset dip α.

Note the area of the surfaces in (c) and (d) that minimize the difference in (b).

Once again, the DD-MO correction will provide a better stack, especially with smaller offsets.

The dip-dependent stacking velocity on the flank of (d) is defined by its zero-offset dip α and RMS velocity of the scatterpoint or apex of the pyramid.

In a variable velocity medium, the scatterpoint velocity will not match the RMS velocity of the flank, which then leads to the problem of stacking conflicting dips.

This section shows the consequence of NMO processing. The following section is similar, but takes a different approach by comparing the summation surfaces that attempt to gather the scattered energy and place it back at the scatter point; the main concept of prestack migration.

What kind of processing would create the surface shown in Figure 7.39b?

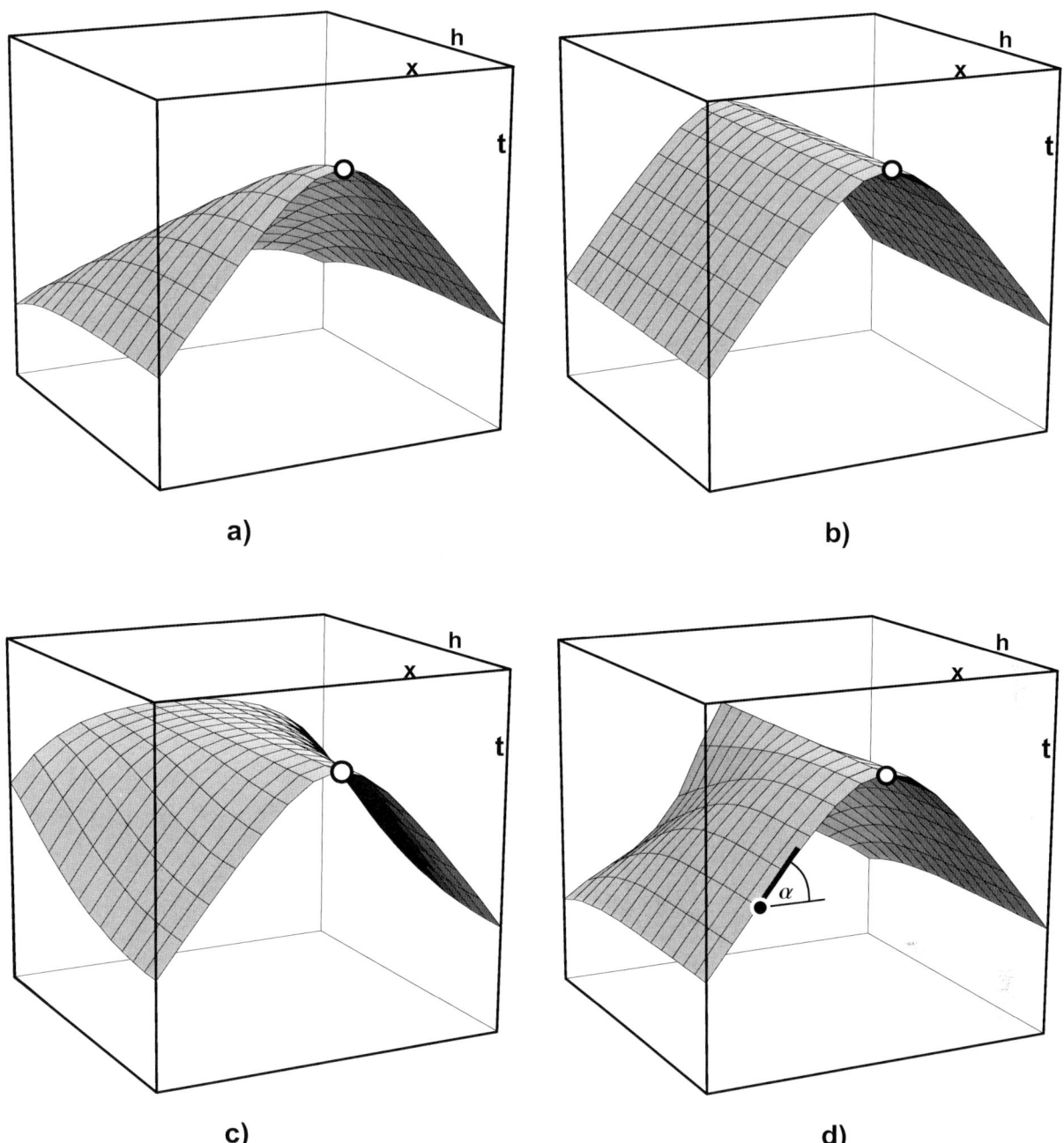

Figure 7.53 MO processing of scatterpoint energy with a) showing the input energy distributed on Cheops pyramid, b) the ideal moveout correction, c) conventional constant velocity NMO correction, and d) DD-MO correction (dip α is defined at zero-offset).

7.9 Comparison of Prestack Summation Surfaces

7.9.1 Summation surface in prestack volume

In contrast to section 7.7 that processed scattered energy, this section defines summation surfaces that collect scattered energy to place it back at the scatter point, the fundamental principle of prestack migration. Front and rear views of the three prestack migration surfaces shown in Figure 7.54 are:

a) Cheops pyramid or the DSR equation (the desired shape for time migrations).

b) NMO $T^2 = T_n^2 + \dfrac{4h^2}{V^2}$ and post-stack migration $T_n^2 = T_0^2 + \dfrac{4x^2}{V^2}$.

c) $T^2 = T_n^2 + \dfrac{4h^2 \cos^2(\beta)}{V^2}$ and post-stack migration.

The differences between (b) and (c) with the Cheops pyramid (a) indicate their ineffectiveness in gathering the prestack data.

<u>Traveltimes on the surface are defined by:</u>

a) The surface of Cheops pyramid is usually computed using equation (7.6). However, for comparison with (b) and (c), it is written as (see App. 4)

        Error! Objects cannot be created from editing field codes..   (7.8)

b) A combination of the NMO and post-stack migration equations give the equation of a hyperboloid in Figure 7.40b. i.e.,

        Error! Objects cannot be created from editing field codes..   (7.9)

c) DD-MO is combined with post-stack migration to give a more complex shape shown in Figure 7.40c, i.e.,

        Error! Objects cannot be created from editing field codes..   (7.10)

Differences are identified by the "offset" terms (…) in the above equations.

The "dip" β of a scatter point at displacement x, is found from the dip α on the zero-offset diffraction and the $tan(\alpha) = sin(\beta)$ relationship.

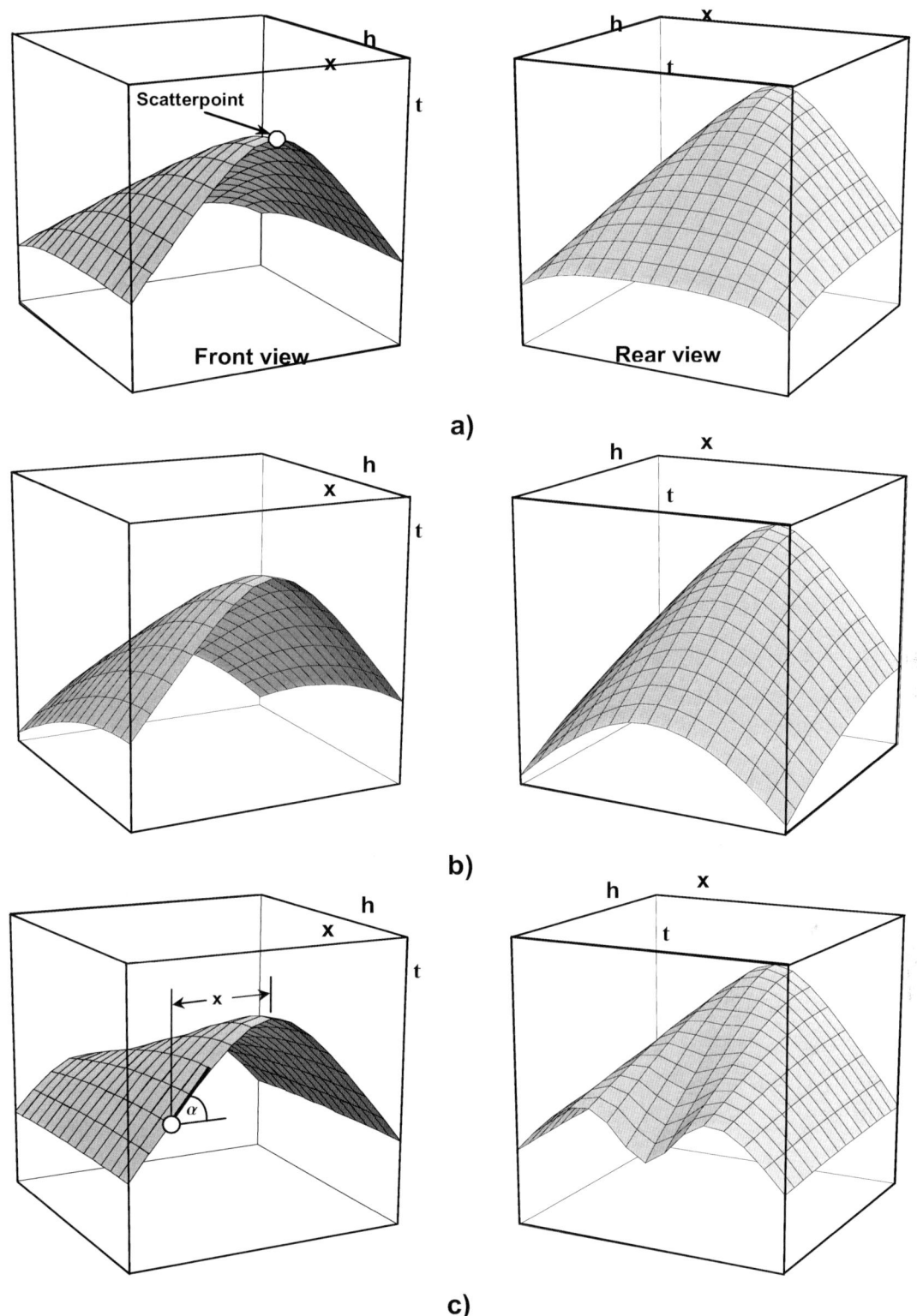

Figure 7.54 Front and rear views of summation surfaces for prestack time migration, a) Cheops pyramid, b) post-stack migration hyperboloid, and c), DD-MO and post-stack migration.

7.9.2 NMO and DD-MO summation surface

The shape of the moveout from the previous images can be compared more effectively from a side projection, or looking at a combination of CMP gathers at equal intervals of x. These side views in (h, t) are shown in Figure 7.55.

The curves in (a), (b), and (c) have the <u>same times at zero-offset</u> and match the zero-offset diffraction or hyperbola of Figure 7.40a.

Figure 7.55 shows:

a) Cheops moveout for a number of CMP gathers.

b) Hyperbolic NMO using velocity V, provides a poor match to Cheops moveout.

c) Hyperbolic DD-MO using velocity V_{stk}, provides a better match to Cheops moveout to larger offsets.

d) Superposition of the three previous images to make the above comparisons easier.

Note:

- The poor fit of NMO correction using V (the dashed lines).
- Reasonable fit to medium offsets when using DD-MO correction with V_{stk}.
- The <u>curvature</u> of Cheops pyramid at zero-offset matches the curvature of the hyperbolic MO using V_{stk}. i.e. P(x, h=0, t) = P(x=0, h, t).
- At increased displacements x, (or increase in zero-offset time), the moveout in (a) tends to be more horizontal. Infinite velocity stacks, (no MO correction) with limited offsets will produce a reasonable stack of steeply dipping energy.

At zero-offset, $sin(\beta)=2x/T_{MO}V$, and $T = T_{MO}$, which may be substituted into equation (7.10) to produce an equation similar to (7.8). This relationship has led many to believe <u>incorrectly</u> that equation (7.10) is identical to equation (7.8).

A similar argument has led some geophysicists to believe that DD-MO (with the cosine term), is the same as Hale's DMO: it is only equivalent when multiple stacks and special dip limiting filters are incorporated into the process.

The above discussion has assumed the zero-offset velocities are defined at the scatter point. What effect will real data have on the shapes of (a), (b), and (c)?

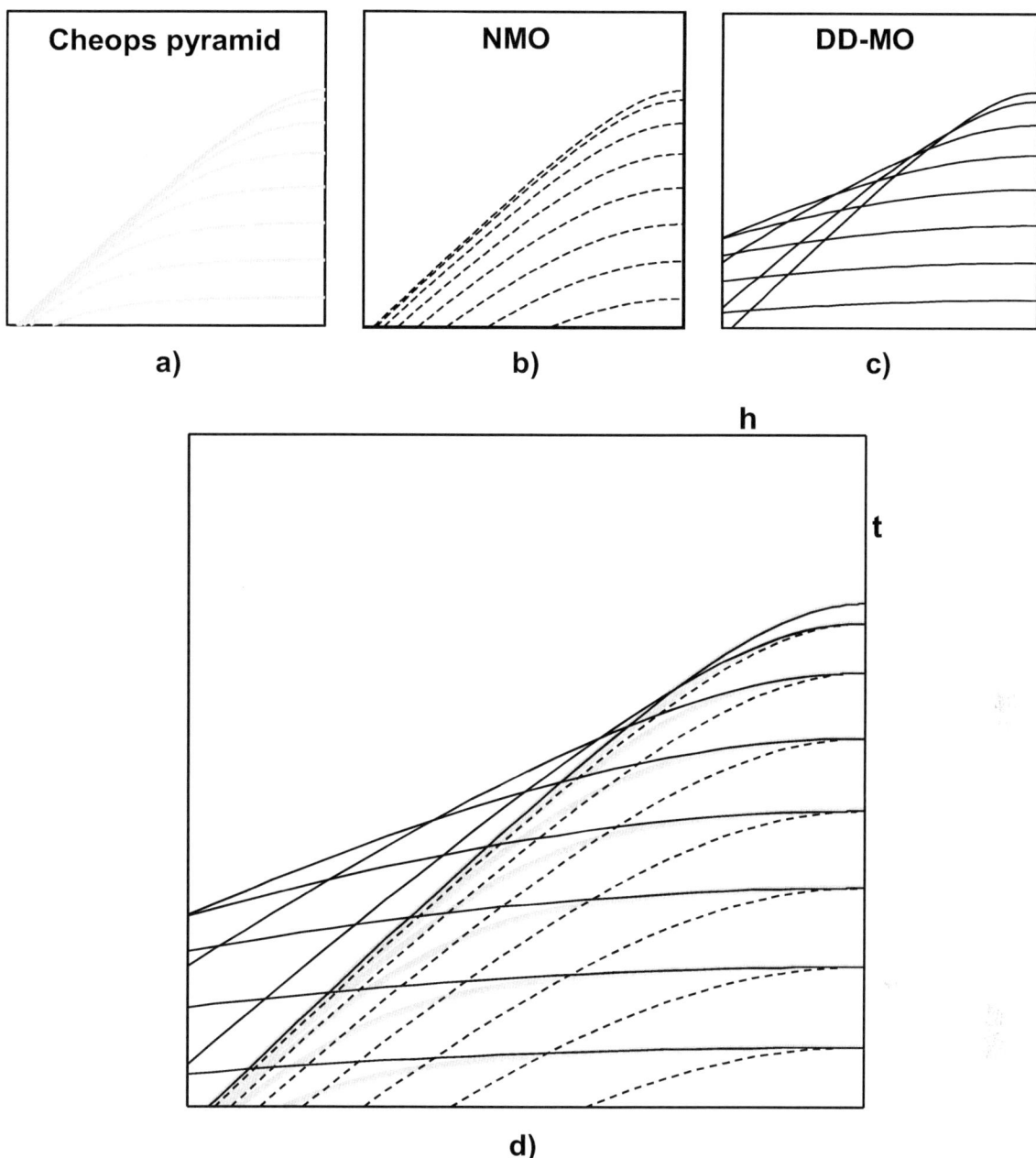

Figure 7.55 Combined MP gathers (h, t) illustrating MO correction for a) Cheops pyramid, b) conventional NMO, c) for DD-MO, and d) a superposition of (a), (b), and (c).

Muting the DD-MO to small offset provides a good match to Cheops times. However, the stacking velocities are defined relative to the velocity at the scatterpoint, not at the times of the offset curves.

Prove the curvatures of Figure 7.55a and 7.41c are equal at zero-offset.

7.9.3 Conclusions and summary of prestack surfaces

- The DSR equation defines the time of all energy reflected from a scatterpoint.
- The velocity of scattered energy is defined at the scatter point; (prestack time migration).
- The accurate gathering of prestack data requires the velocity of the scatter point.
- NMO, stack, and post stack migration only moves a portion of the prestack energy to the scatter point location.
- For small offsets, DD-MO provides a good match to Cheops pyramid; their curvatures are equal at zero-offset.
- Conventional processing with real data will only consider the local velocity and not the velocity of a scatter point.

The complete gathering of prestack energy can only be accomplished by processes that:

- Sum over Cheops pyramid for a prestack time migration.
- Compute the actual scatterpoint traveltimes for a depth migration.

7.10 Modelling Linear Reflectors with Scatterpoints

7.10.1 Introduction

<u>Horizontal reflector</u>

The moveout energy from a flat reflector is often displayed in a CMP gather (h, t) and assume to lie on a hyperbolic curve. The hyperbolic curve is defined by the normal moveout (NMO) equation with variables half-offset h, and the RMS velocity V_{RMS}

$$T^2 = T_{dn}^2 + \frac{4h^2}{V_{RMS}^2} \quad . \tag{7.11}$$

In the prestack volume (x, h, t), a horizontal reflector will produce a hyperbolic surface. When the RMS velocity is constant with x, the hyperbolic moveout is identical a each CMP location and the surface is referred to a hyperbolic cylinder as illustrated in Figure 7.42a. This surface represents the location of specula energy from a flat reflector in the prestack volume.

<u>Dipping reflector</u>

When the reflector is dipping, the energy in a CMP gather is still assumed to be hyperbolic as derived in Section 7.0.9 for a constant velocity. The dip-dependent moveout (DD-MO) is similar in form to equation 7.11, but now uses the stacking velocities V_{stk}, that modifies the velocity by the cosine of the dip β. In practice, the constant velocity is replaced by the RMS velocity in equation 7.12

$$T^2 = T_{dn}^2 + \frac{4h^2 \cos^2(\beta)}{V_{RMS}^2} = T_{dn}^2 + \frac{4h^2}{V_{stk}^2} . \tag{7.12}$$

Even for a constant velocity, this hyperbolic surface varies at each CMP location as illustrated in Figure 7.56b. Note the dip of the reflector is β and that the zero-offset dip on the reflection is α.

> **Question:** What reflection points are assumed in a CMP gather of dipping energy?

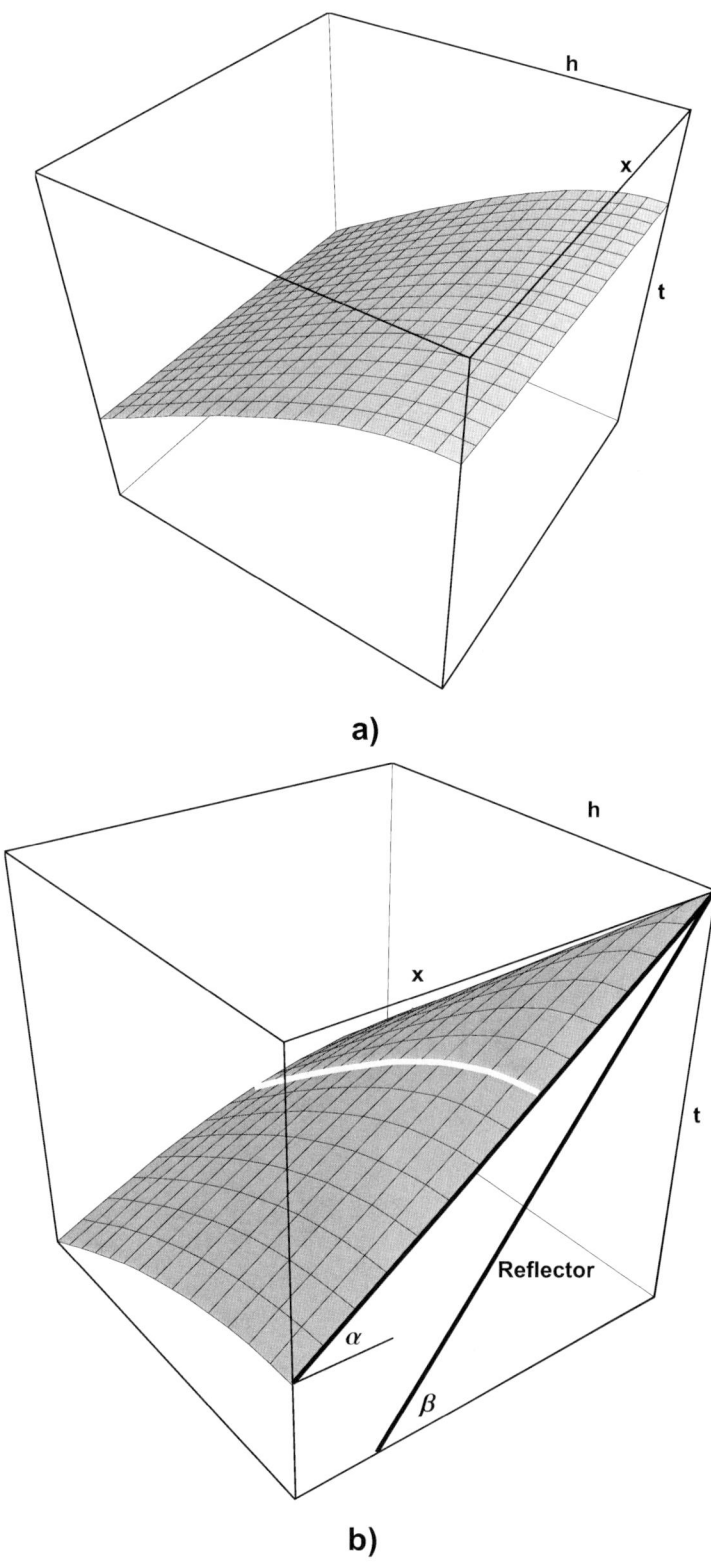

Figure 7.56 Hyperbolic prestack reflection surfaces from a) a horizontal reflector, and b) a dipping reflector.

7.10.2 Modelling a horizontal reflector.

<u>Zero-offset data</u> (*x, t*) may be constructed from a series of scatterpoints located along an event and inserting diffractions at each scatterpoint as previously illustrated in Section 2.2.

<u>Offset data</u> (*x, h, t*) my be modelled in a similar manner by placing energy on the surfaces of Cheops pyramids as illustrated in Figure 7.57. This figure shows the Cheops pyramids from seven horizontal scatterpoints. The upper surfaces of these pyramids tend to form the prestack hyperbolic surface of a flat reflector. When a sufficient number of scatterpoints are used, the energy will construct the hyperbolic surface, while the energy below this surface will deconstruct and sum to zero.

The images in Figure 7.58 shows the <u>hyperbolic cylinder</u> that represents the hyperbolic moveout from a horizontal reflector in a constant velocity medium. Also shown in these images is one Cheops pyramid from a scatter point that lies slightly above the horizontal reflector.

If the scatter point was located on the reflector, the Cheops pyramid would be tangent to the hyperbolic cylinder. However, by placing the scatterpoint slightly above the reflector, the area of tangency for finite bandwidth energy between these two surfaces my be identified. This area represents the <u>specula energy</u> from a horizontal reflector that will be summed in a Kirchhoff prestack migration, i.e. when summing energy over a Cheops pyramid.

When the shift between the hyperbolic cylinder and Cheops pyramid is a half-wavelength of the wavelet, then the intersecting area also defines the Fresnel zone.

A prestack surface (*x, h, t*) may be modelled from a 2-D structure of scatterpoints by placing energy on the surfaces defined by the corresponding Cheops pyramids. The shape of the Cheops pyramids is defined by the RMS velocity at the scatter point.

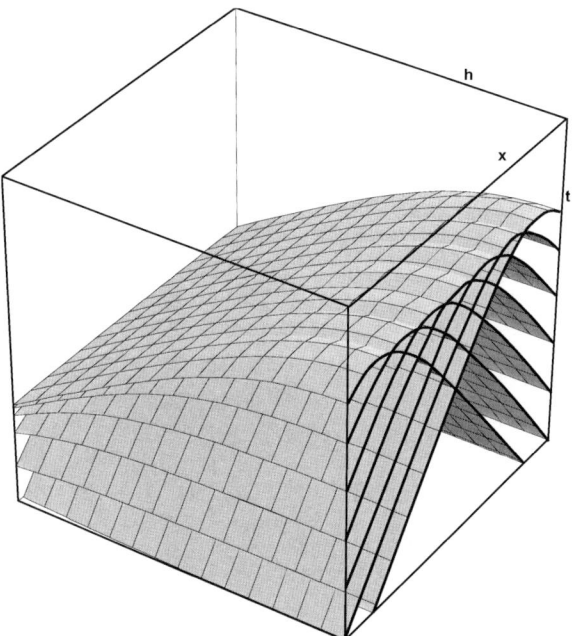

Figure 7.57 Cheops pyramids from seven horizontal scatterpoints that tend to form the prestack hyperbolic surface of a flat reflector. When a sufficient number of scatterpoint are used, the energy will reconstruct the hyperbolic surface, while the energy below this surface will sum to zero.

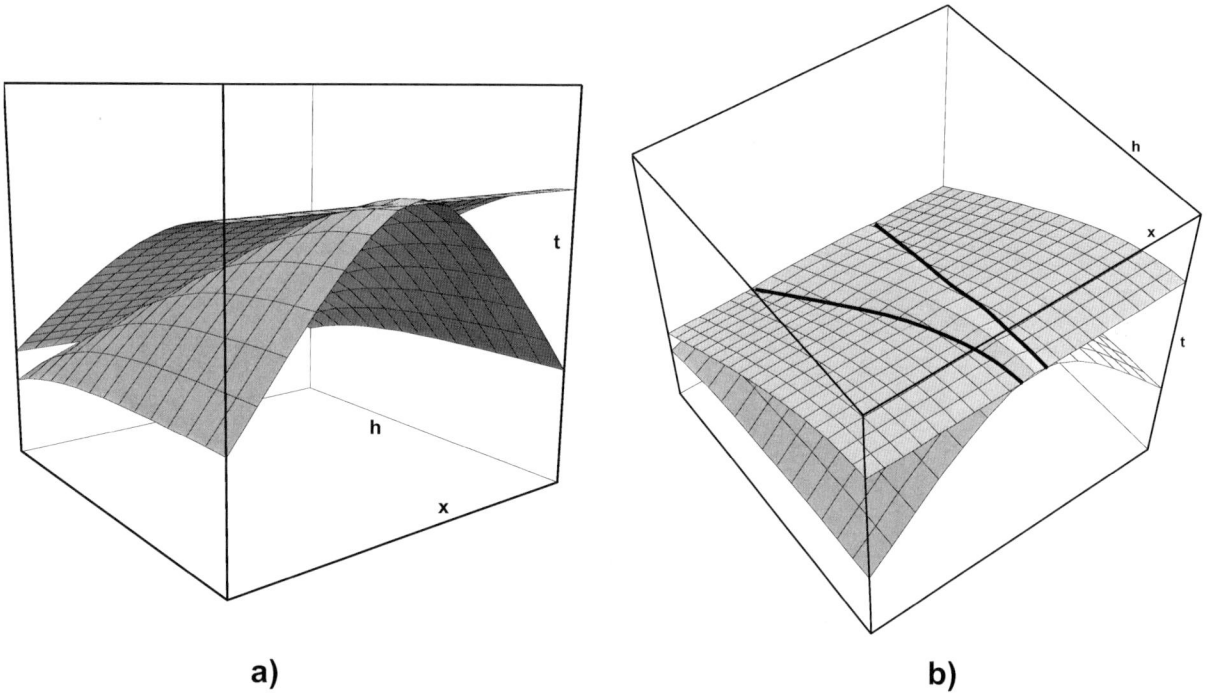

a) b)

Figure 7.58 One Cheops pyramid and the hyperbolic surface from a flat reflector from a) a low view and (b) from a higher perspective view point to identify the area of tangency between the two surfaces.

7.10.3 Modelling a dipping event

A 2-D dipping event, illustrated in the ray tracings of Figure 7.59, may be modelled in the prestack volume (x, h, t) to produce the surface in Figure 7.60a. We showed in section 3.0.9 that this surface is exactly <u>hyperbolic in each CMP gather</u>. Dip-dependent moveout (DD-MO) corrects offset time in CMP gathers to a zero-offset time; however the actual offset reflection points move updip from the CMP gather as re-illustrated in (Figure 7.60a).

In a manner similar to 2D diffraction modelling, the prestack hyperbolic surface of Figure 7.59a may also <u>be modelled by a reconstruction of Cheops pyramids</u>, formed from a series of scatterpoints that are located along the dipping event, as illustrated by three scatterpoints in Figure 7.60b.

Figure 7.59b below, shows various offset raypaths for a single reflecting point, or dipping element, on the dipping event. Their midpoints are identified by the black dots, and as the offset is increased, the midpoints move in the down dip direction. The location of these midpoints that vary with offset are illustrated by the red line in Figure 7.60c.

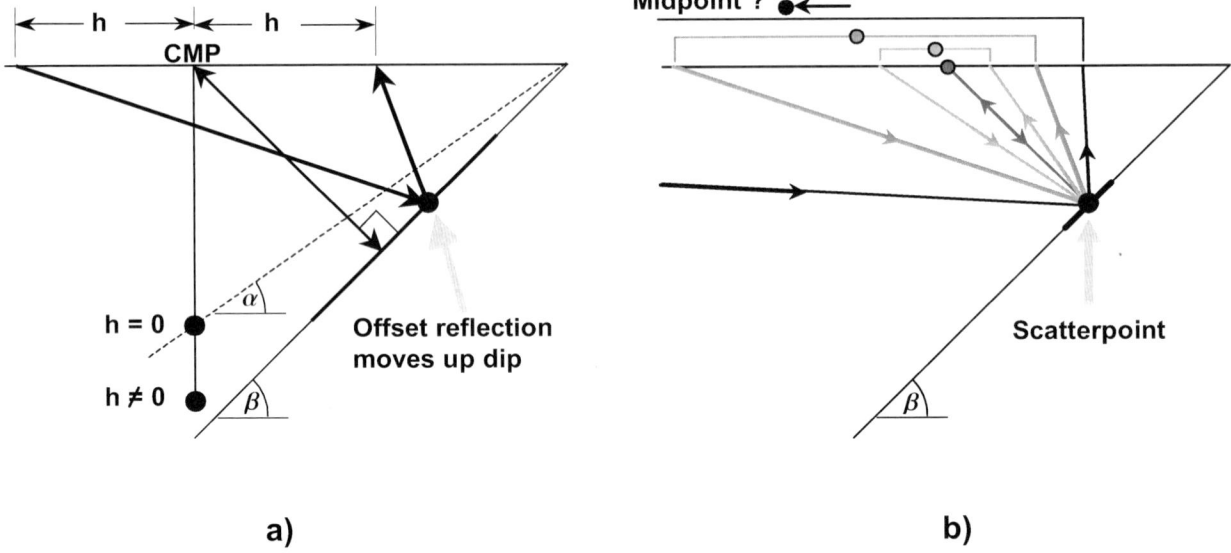

Figure 7.59 Dipping reflections illustrating in a) reflection times in a CMP gather, and b) the offset raypaths for a single reflection point.

The surface in Figures 7.46a and one Cheops pyramid from (b) are combined in (c) to illustrate the area of tangency between the dipping surface and Cheops pyramid. The black line defines the theoretical tangency location while the gray band was formed by slightly reducing the time of the Cheops pyramid. As the offset is increased, the area of tangency curves down dip corresponding to that shown in Figure 7.46c. It is the specula energy in this tangency band that is summed to the reflection point during a prestack migration, (that sums energy over a Cheops surface).

Figure 7.46d illustrates the area of tangency for a shallower dip. When the shift between the specula energy and Cheops pyramid is a half wavelength, then the gray area corresponds to the area of the Fresnel zone.

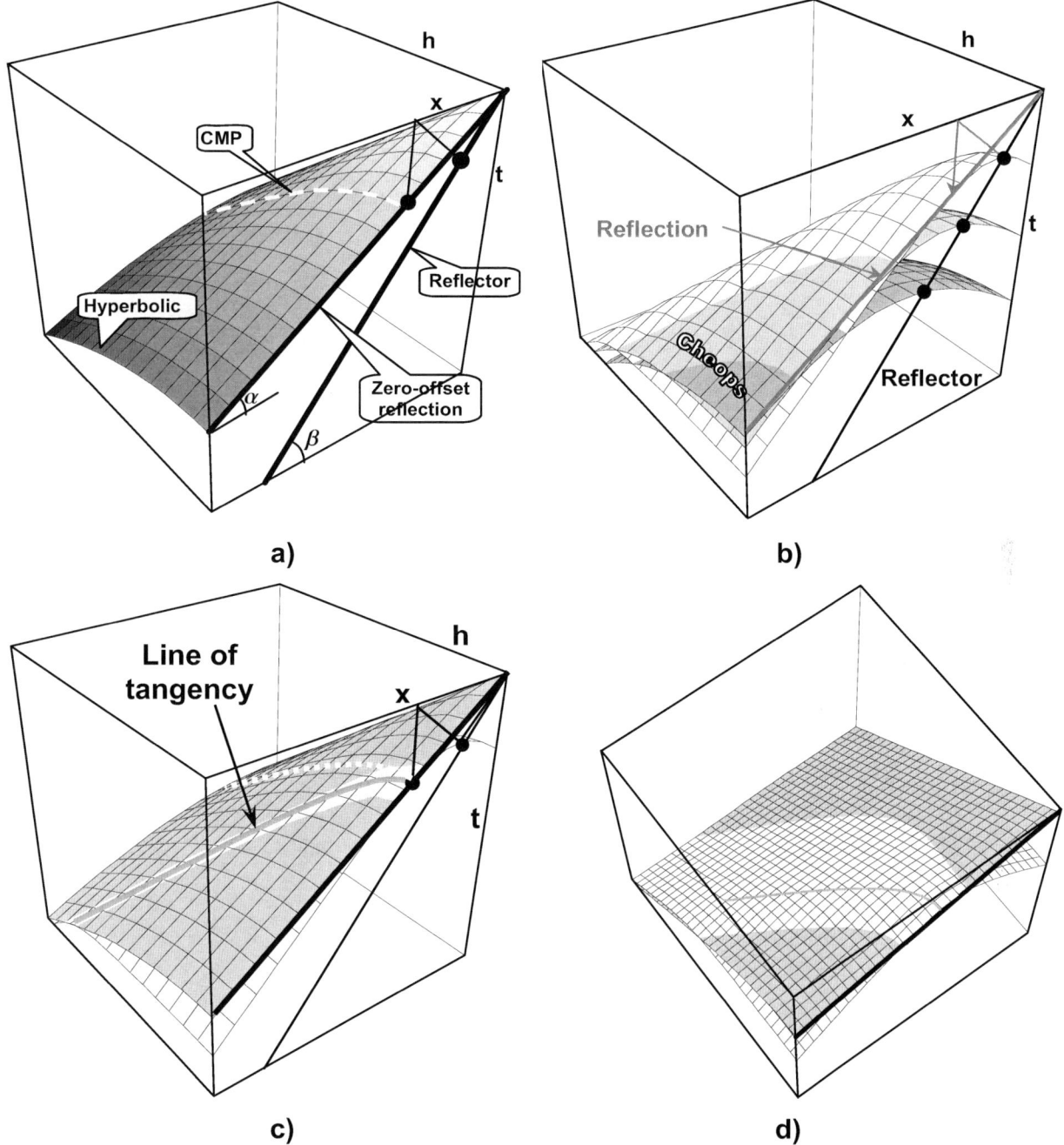

Figure 7.60 Dipping prestack hyperbolic surface a) formed in b) from three scatterpoints on the reflector and the three corresponding Cheops pyramids. Part c) is a combination the hyperbolic surface and one Cheops pyramid illustrating the line (red) and area of tangency. Part d) is similar to (c) but contains a shallower dip and a lower frequency with a resulting increase in size of the Fresnel zone.

7.11 Prestack Modelling of 3-D Data

Modelling <u>post stack</u> 3-D data can be expensive; modelling <u>prestack</u> 3-D data is even more expensive and time consuming. Techniques include wave equation methods, ray tracing methods, and hyperboloid approximations using the RMS velocity assumptions.

As with 2-D models, the effects of stacking and velocity analysis can be evaluated. In 3-D models, effects such as azimuth may be evaluated.

Industry standard 3-D models have been created by Aminzadeh, Brac, and Kunz and are available to the public through the SEG or EAGE. The data is on two CD's and sold as "SEG/EAGE 3-D Modeling Series No. 1: #-D Salt and Overthrust Models" [638].

All the modelling techniques discussed in these course notes have been numerical models that are very predictable and are constrained by the design criteria. It should be noted that modelling is also accomplished by building a physical model scaled from a geological structure. Physical modelled seismic data is acquired from high frequency sensors attached to or included in the models. These physical models mimic seismic data by including energy from multiple reflections, along with various signals that may result from mode conversion, (i.e. conversion between P and S waves).

7.12 Summary of Points to Note in Chapter 7

- Prestack modelling is <u>more complex</u> than the zero-offset case.
- NMO correction and stacking is only valid for <u>flat reflectors</u>, and then only using the velocity for flat reflectors.
- DD-MO correction for dipping events will help stack the energy along the dip, but will still <u>smear</u> the structure along the dip.
- Diffractions do not stack. (Unless the velocity happens to increase at a rate that matches the dip on the diffraction???)
- Care must be taken when forming <u>limited offset sections</u>.
- Reflectors may be composed of many <u>scatter points</u>.
- The traveltimes from a scatter point are computed using the <u>DSR equation</u>.
- A scatter point will create a 3-D surface in (x, h, t) that is referred to as <u>Cheops pyramid</u>.
- The intersections of many planes with Cheops pyramid are <u>hyperbolic</u>.
- The CMP intersections with Cheops pyramid are <u>non-hyperbolic</u>.
- The <u>complete collection of scattered energy</u> requires a process that gathers the energy on Cheops pyramid (i.e. for a time migration; depth migrations require a similar process that uses some form of raypath computation).
- The <u>area on Cheops pyramid that contains specula energy</u> may be identified by <u>offset varying Fresnel zones</u>.

Comment:

<u>Zero-offset</u> or post-stack modelling always assumes

- <u>orthogonal reflections</u> with amplitudes proportional to the reflection coefficients.
-

<u>Prestack</u> modelling allows

- the <u>amplitude</u> of the reflection to vary with incident, reflected, and transmitted angle using Zoeppritz's equations,
- <u>converted waves</u> (with incident P-wave energy and reflected S-wave energy) may also be modelled.

Construction help for page 7.15:

Recall the principle of mirror images. Define the image of the receiver by:

- Construct the <u>normal</u> from the receiver to the reflector (and onto the image side of the reflector).
- Measure the normal distance from the receiver to the reflector.
- Locate the image point at the normal distance on the image side of the normal line.
- The reflection point is the <u>intersection</u> of the reflector with a line drawn from the source to the image of the receiver.

Note the length of the reflected raypath is identical to the length of the image raypath.

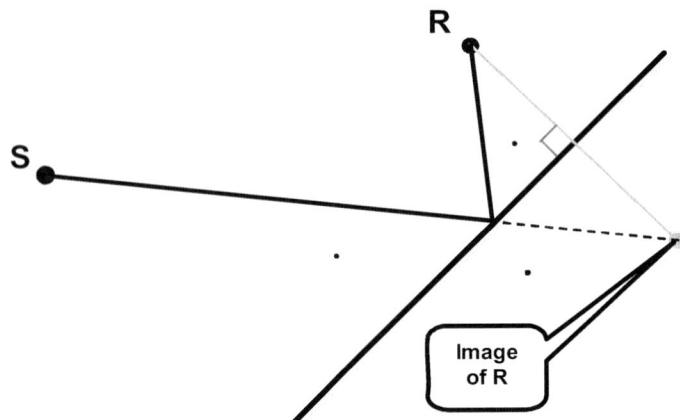

Chapter Eight

2-D Dip Moveout (DMO)

Objectives

- Know that prestack migration is equivalent to DMO and poststack migration for constant velocities.
- Know that DMO (constant velocity) enables all dips to be MO'd with the same (RMS) velocity i.e. NMO.
- Know that DMO (constant velocity) eliminates dip smear.
- Know that constant velocity assumption may be extended to areas with smoothly varying velocities.
- Understand the basic DMO algorithms.
- Become aware that some DMO algorithms are available for variable velocities.
- Be able to identify the different types of processes referred to as DMO.

8.1 Introduction

8.1.1 Why the need for Prestack Migration or Dip Moveout?

- Present <u>stacking</u> methods are only valid for flat data.
- <u>Diffractions</u> don't stack.
- <u>Energy smear</u> along dipping events.
- MO stacking velocities <u>vary with dip</u>.
- Can only process one event when there are <u>conflicting dips</u>.
- DMO and poststack migration is faster than prestack migration.
  (That was the case but is now questionable. Some prestack migrations may be as fast as DMO.)
- After DMO, a number of poststack migrations may be run more economically than a number of prestack migrations.

List other reasons for the need for prestack migration or DMO.

8.1.2 Present stacking methods only valid for flat data

The purpose of applying MO correction to offset traces is to modify the time scale to be equivalent to zero-offset traces, enabling them to be stacked. It is only <u>valid for flat events</u>, as illustrated in Figure 8.1a.

The dip-dependent stacking velocities V_{stk} become $V_{rms}/cos(dip)$ to form a better stack as illustrated in Figure 8.1b. However, there is still <u>smear</u> of reflections along the dipping event.

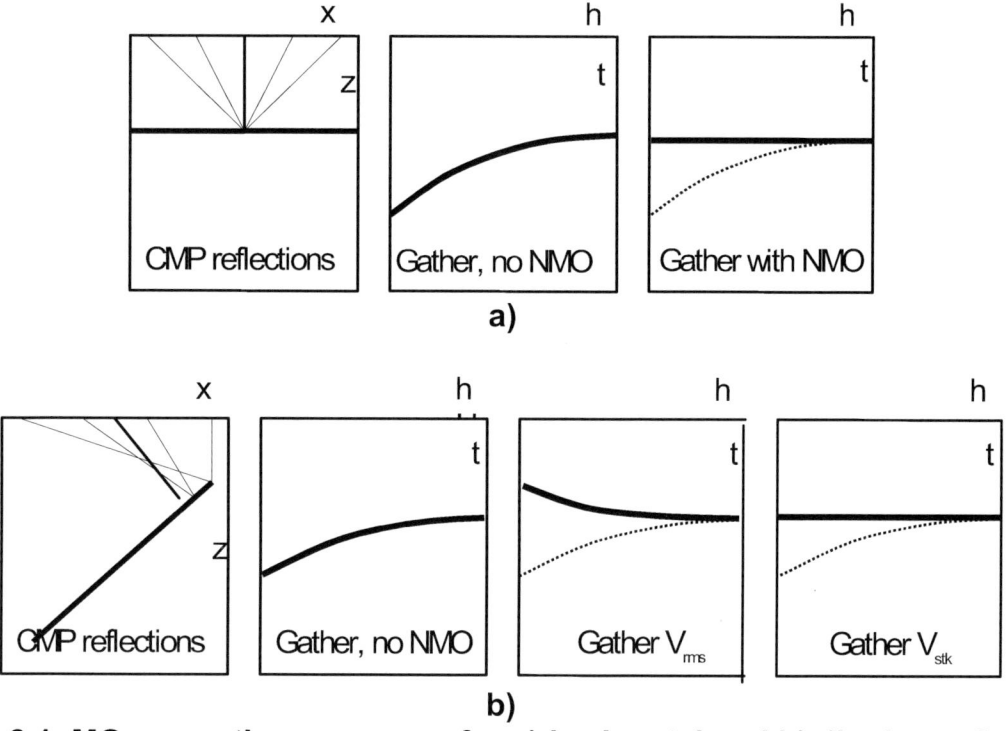

Figure 8.1 MO correction sequence for a) horizontal and b) dipping reflectors.

8.1.3 Diffractions do not stack

Scatter point diffractions form Cheops surface in (x, h, t), and its intersection with a CMP gather produces non-hyperbolic moveout. The non-hyperbolic energy will not stack correctly with conventional MO. Consequently MO and stack will disperse the dipping energy of the diffraction. When the MO velocity is increased, only a part of the diffraction will appear coherent as illustrated in Figure 8.2 .

There are possible conditions in which the velocity increases with depth at a rate that may help the diffraction stack.

Illustrations representing the poor stacking of diffractions may be seen in Figures 7.16, 7.20 and 7.21.

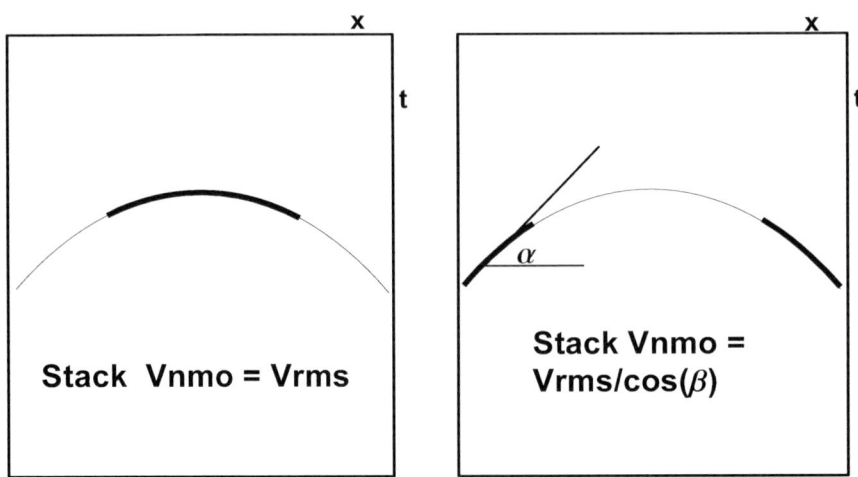

Figure 8.2 Stacking of diffractions.

8.1.4 Problems of MO and stacking conflicting dips

Geological structures rarely contain conflicting dips. The resulting <u>time sections</u> may contain areas in which the reflections from dipping events overlap as illustrated in Figure 8.3.

A seismic processor must choose an MO velocity for one of the events. Typically the shallower dipping event is chosen for the best stacking image.

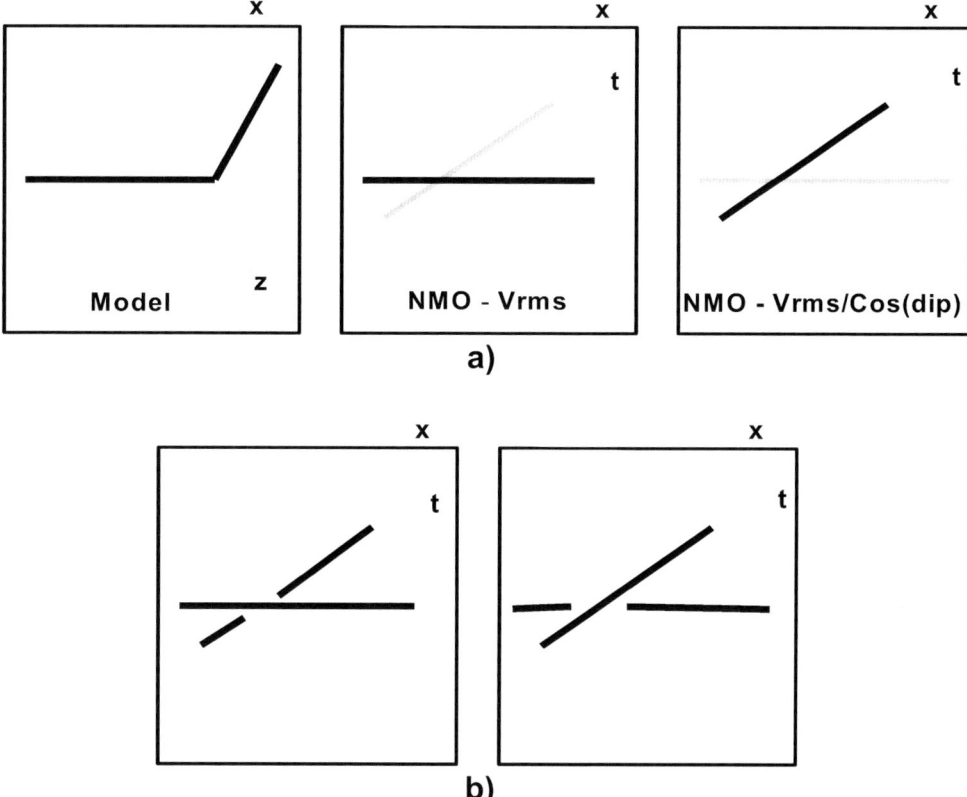

Figure 8.3 Illustration of conflicting dips in processing, a) shows a geological model with two stacked sections, and b) shows the result of two best processing sequences to emphasize either the horizontal or dipping event.

8.1.5 Offset reflection times for a point on a dipping reflector.

Dipping reflection energy, in a CMP gather, will contain <u>reflection points that move up dip</u> as the source-receiver offset is increased. Dip-dependent moveout (DD-MO) correction will allow the reflection points to be stacked at the <u>zero-offset location</u>, smearing energy along a dipping event. Consider the geometry in Figure 8.4a that shows of a dipping event (*A, B*) and the zero-offset reflection (*A, C*). The <u>offset</u> raypath has a reflection point at R_h, reflection time T, DD-MO corrected time T_{dn} (which is equal to the zero-offset time at the CMP), and moveout (MO) corrected time T_n. The times T, T_{dn}, and T_n are plotted below the CMP location and scaled to one-way distances. The dashed curve illustrates the zero-offset time on the reflection at R_0.

The input time T may be converted to the zero-offset time T_{dn} by using DD-MO, developed in Section 7.1.9, as

$$T^2 = T_{dn}^2 + \frac{4h^2 \cos^2 \beta}{V^2}. \tag{8.1}$$

<u>Construction</u> Figure 8.4a;

- Use a compass to verify the location of the input time T, (See section 7.10).
- Verify the time T_n represents <u>conventional NMO</u> for horizontal reflectors.

<u>Construction</u> Figure 8.4b;

- Define the zero-offset surface location (*CMP$_h$*) for R_h by constructing a line from R_h to the surface that is normal to the reflector.
- Construct the zero-offset reflection time T_d for R_h immediately below *CMP$_h$*.
- Verify the time T_d is on the zero-offset reflection (*A - C*).

Comment
- Dip-dependent moveout moves the offset reflection time T to T_{dn}.
- The above construction defines the true position to which the energy should be moved.

- DD-MO and poststack migration will relocate energy at T to _____ .
- NMO and DMO will relocate energy at T to _____ .
- Prestack migration will relocate energy at T to _____ .

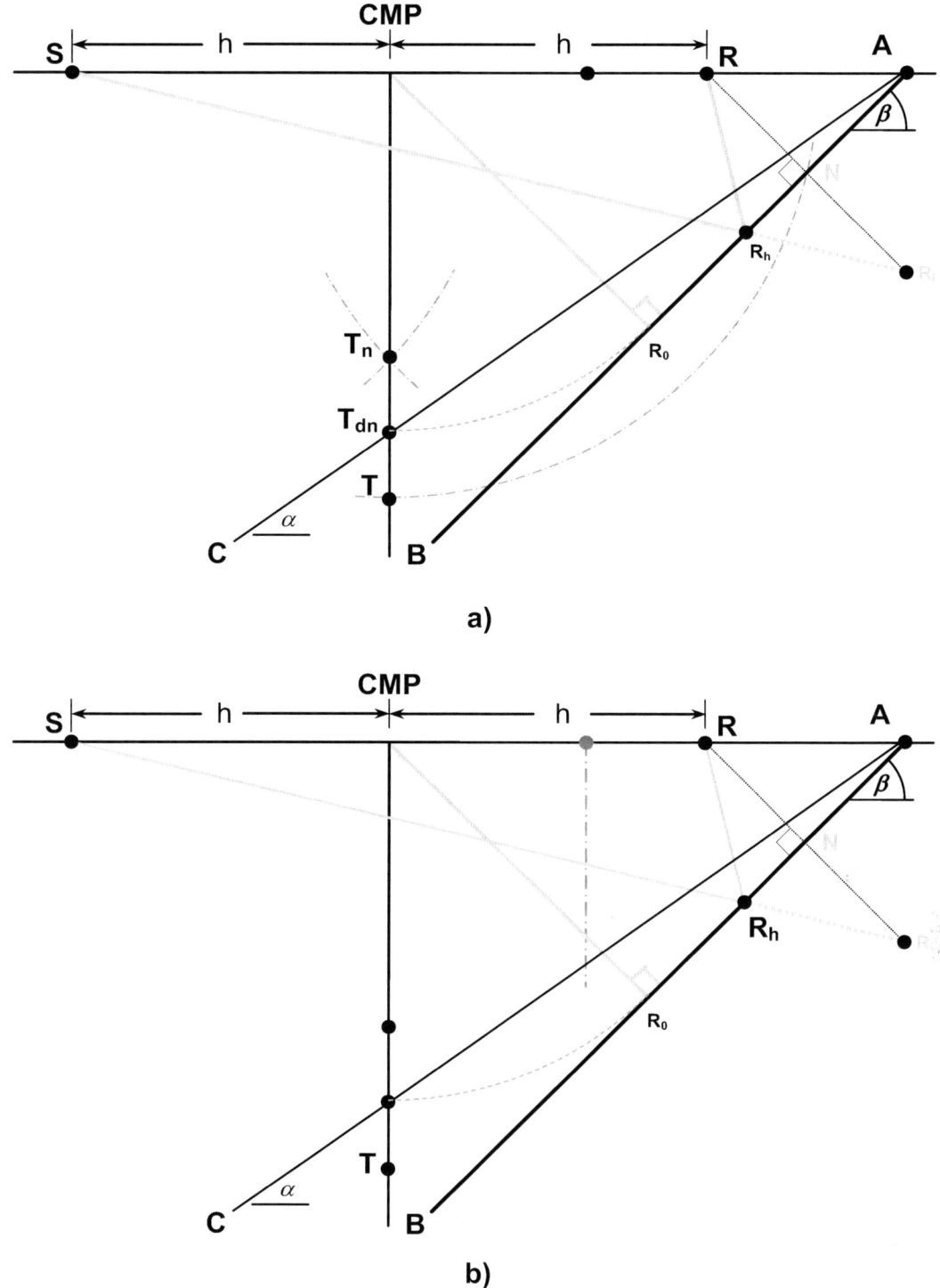

Figure 8.4 Construction of reflection energy for a dipping event

8.1.6 Relocation of offset energy to its zero-offset position

The previous construction showed the zero-offset location of the offset reflection point should be moved to a new spatial location. It is repeated in Figure 8.5 at time T_d.

The **objective of DMO** is to move energy from T to T_d at a new spatial location CMP_h as illustrated in Figure 8.5a. This may be accomplished by a number of techniques such as:

- **Direct movement of energy** at the input sample T to T_d.

- Using a **wave equation approach** by spreading energy along the dashed curve from T_n to T_d. Energy from other input traces will reinforce or cancel along the reflection.

Note that **NMO correction prior to DMO** uses RMS velocities to move the input sample at T to time T_n, i.e.

$$T_n^2 = T^2 - \frac{4h^2}{V^2}. \tag{8.2}$$

The **objective of prestack migration** is to move the energy from T to R_h as illustrated in Figure 8.5b. This also may be accomplished by the two techniques discussed above. In comparison, the prestack migration ellipse is bounded by T_n and $VT/2$.

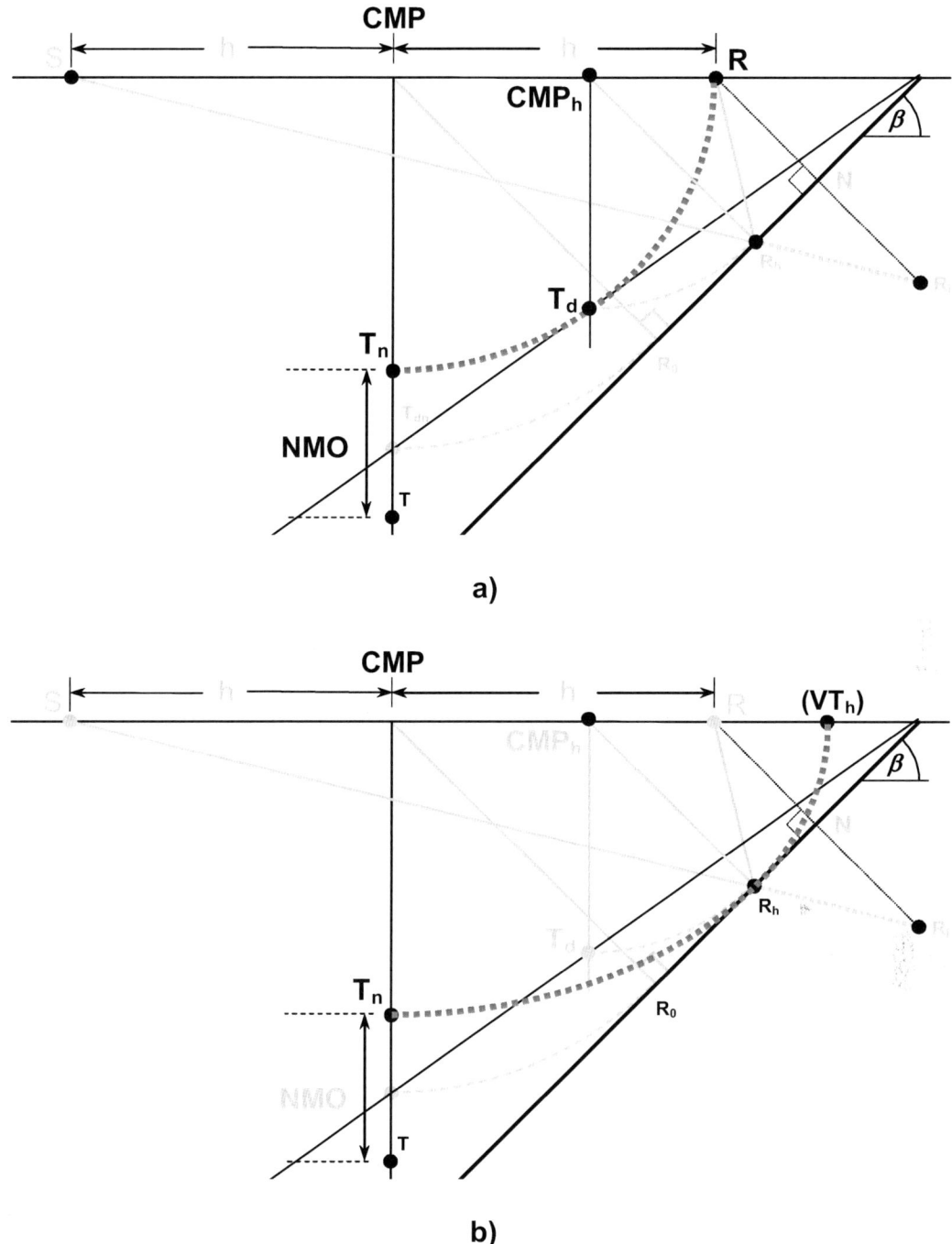

Figure 8.5 Energy movement for a) DMO and b) prestack migration.

8.1.7 Prestack migration and the prestack migration ellipse

Prestack migration is a process that locates reflection energy back to the scatter point by <u>wave front reconstruction</u>.

Prestack migration may be in <u>time or in depth</u>.

Prestack migration may be required in a <u>structurally complex area where DMO fails</u> to completely focus the data.

Some prestack migrations use <u>DMO and post-stack migration</u>.

One approach to prestack migration is similar to the Hagedoorn method. Energy from a reflection point in the time section is spread along all possible reflector points.

For the constant velocity case, the <u>energy is spread along an ellipse</u> Rockwell 1971 [789], as illustrated in Figure 8.6. It shows the source and receiver at the foci of the ellipse. The boundaries of the ellipse are defined by the MO time T_n (equation (8.2)) and $\pm TV/2$, where V is the velocity, and T the total traveltime. The equation of this ellipse is:

$$\frac{4x^2}{T^2V^2} + \frac{t^2}{T_n^2} = 1. \tag{8.3}$$

All reflected raypaths from S to R (off the ellipse) have the same total time T. The length of the raypath is the distance VT; the width of the ellipse.

The prestack migration ellipse is only applicable for constant velocities, similar to the semicircle of poststack migration [300]. In a geological model with linearly <u>increasing velocities in depth</u>, wavefronts are semicircular, but the center of the circle moves down the z-axis as the traveltimes are increased. A time section (x, t) shows considerable distortion from a semicircle (page 162 of [300]).

In contrast, diffractions approximate hyperbolic shapes for geological dips approaching 65 degrees and beyond. (See also section 4.3.5).

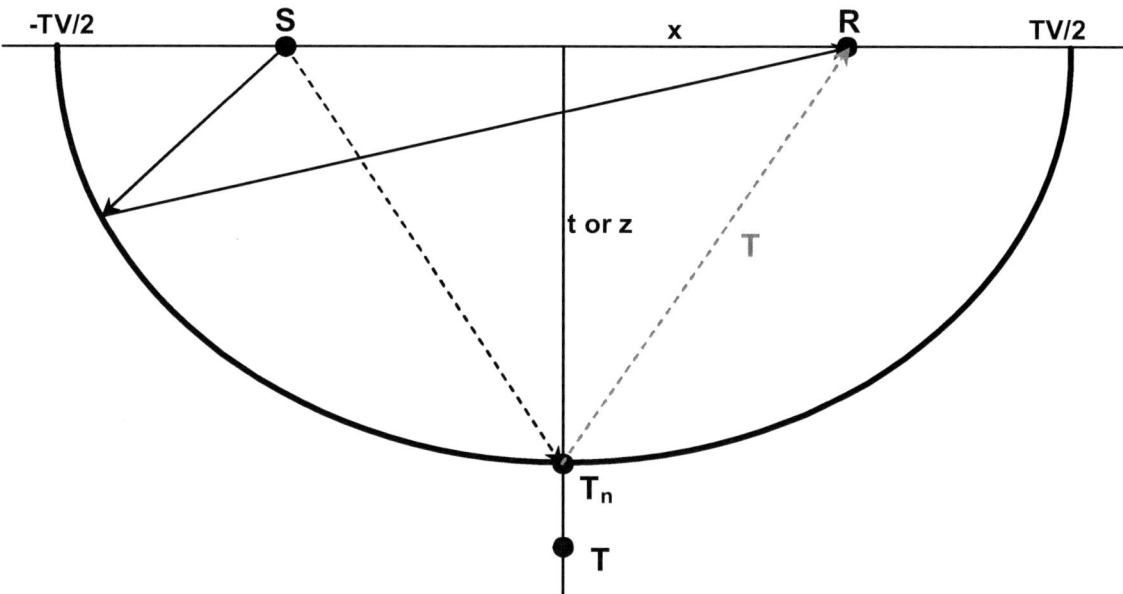

Figure 8.6 Prestack migration ellipse.

8.1.8 Dip Moveout (DMO)

Dip moveout, or DMO, is a term that is used to define a process that is more accurately referred to as "**Prestack Partial Migration**" (PSPM).

- The "partial" term implies the migration is not complete.
- After "PSPM", **post**stack migration is still required to complete a prestack migration.
- Use of the term **DMO now implies PSPM**, however, in the past, DMO was used to refer to any process that applied MO to dipping events.
- Current use of the term **DMO also implies the wave equation solution** in which energy is partially spread along the DMO ellipse.
- I use **DD-MO** to replace the historical term DMO.

DMO is a process that **follows NMO** and will focus the data to be equivalent to **zero-offset** data.

The processing steps of NMO, DMO, and poststack migration are equivalent to **prestack migration** for constant velocities.

Being a partial migration, **DMO moves energy to neighboring traces** similar to poststack migration. However, the DMO energy is only spread along a portion of an ellipse.

A seismic section, with DMO applied, is still limited to dips less than **45 degrees**; recall that post stack migration is still to follow. Consequently energy on the DMO ellipse is limited to dips less than 45 degrees.

Deregowski [76] listed 10 advantages of DMO in an ideal environment:

1. Each trace is migrated to zero-offset.

2. Reflector point dispersal is removed for dipping events.

3. 2-D cross-line ties improved as direction of offset is effectively removed.

4. An excellent interpolator.

5. Removes noise with its dip filtering capabilities.

6. Improves signal-to-noise ratio.

7. Velocities become independent of dip, ~~eliminating conflicting dips~~. (Only for constant velocities)

8. Velocity analysis tends to RMS velocities for improved migration and inversions.

9. Diffractions preserved for improved definition of discontinuities.

10. DMO and post-stack migration is less expensive than prestack migration.

Item 7 is still only valid for constant velocities. **Conflicting dips** with different velocities may still occur when each dip has a different RMS velocity (such as a shallow diffraction crossing a horizontal reflector). This problem is resolved with true **prestack migration**.

8.1.9 The DMO ellipse for wave equation solutions

Energy after DMO is partially spread along an elliptical curve which is similar to the prestack migration ellipse; however, the <u>horizontal boundaries</u> now become the <u>source S and receiver R</u>, (with half offset *h*) as illustrated in Figure 8.7. The DMO kinematic equation is therefore;

$$\frac{x^2}{h^2} + \frac{t^2}{T_n^2} = 1.$$ (8.4)

The energy in a DMO'd section is restricted to <u>dips less than 45 degrees</u> as illustrated by the thicker portion at the bottom of the ellipse. Energy beyond this limit may contribute <u>aliased energy</u> to the poststack migration.

8.1.10 Construction of prestack migration ellipse from DMO ellipse

Semicircular post stack migration of the DMO ellipse will reconstruct on the prestack migration ellipse. <u>Verify</u> with your compass on Figure 8.8.

- At each vertical line, draw the migration semicircles with centers at the surface, and radius defined by the intersection with the DMO curve.
- Each semicircle should become tangent to the prestack migration ellipse.
- Note that a 45 degree dip on the DMO ellipse migrates to 90 degrees on the prestack migration ellipse.

What would happen to energy on the DMO curve that extends <u>beyond</u> the 45 degree dip limit?

This exercise is intended to verify that <u>prestack migration</u> may be achieved by combining the steps of NMO, DMO and poststack migration.

Chapter 8 2-D Dip Moveout (DMO)

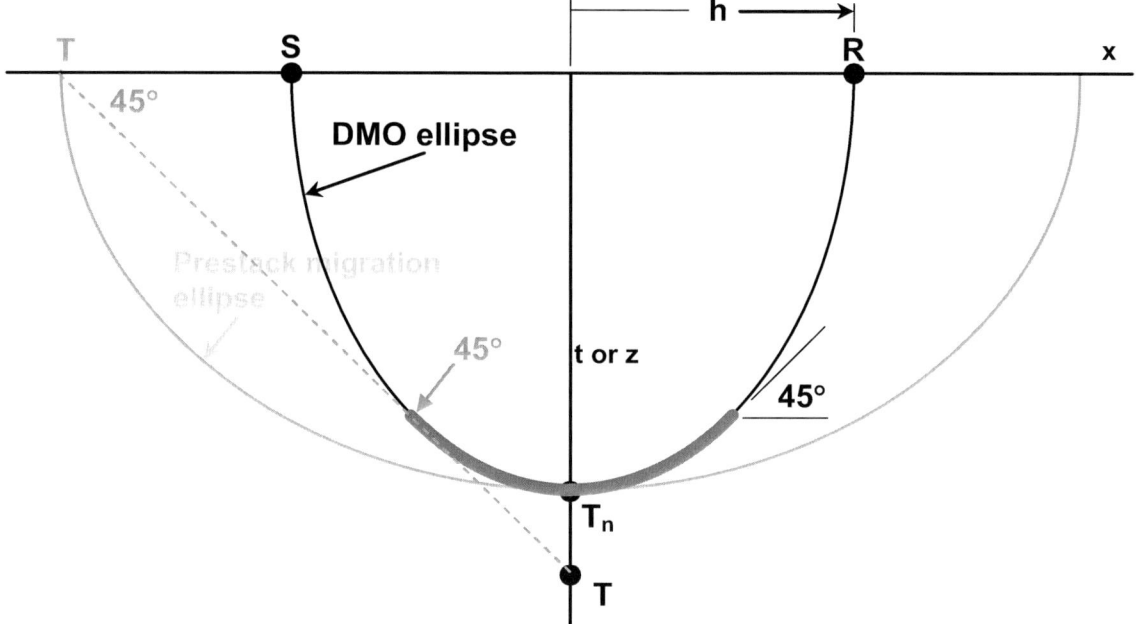

Figure 8.7 The DMO ellipse.

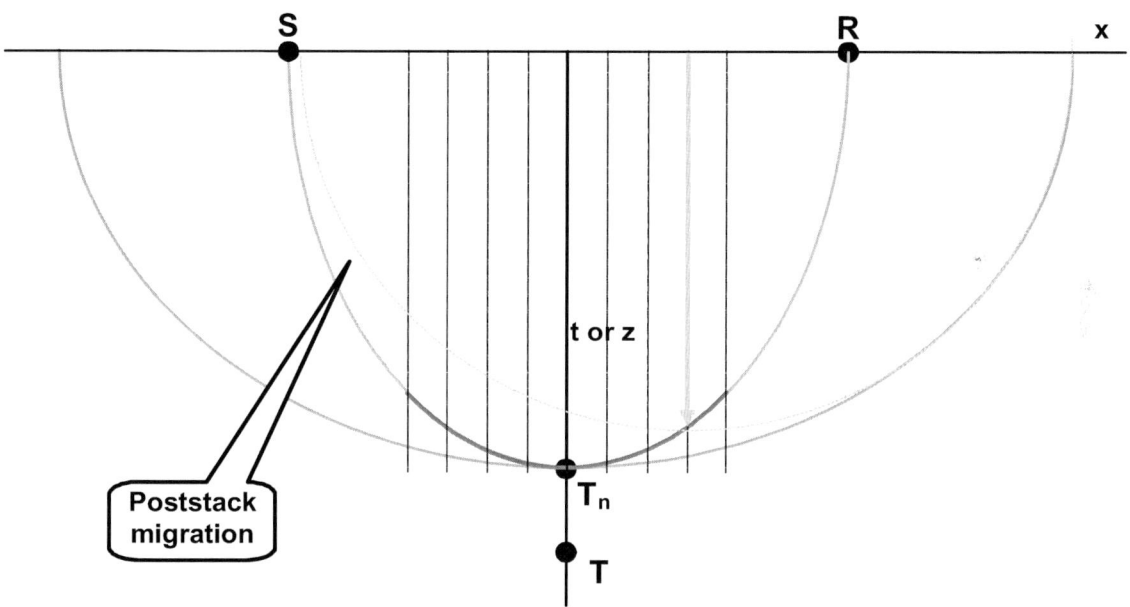

Figure 8.8 Construction of prestack migration ellipse from DMO ellipse.

8.1.11 Formation of DMO ellipse from the prestack migration ellipse

Zero-offset sections may be constructed from hyperbolas as described in Section 2.2. When the geological structure has the shape of a prestack migration ellipse, zero-offset modelling with hyperbolas will construct the DMO ellipse.

DMO modelling is illustrated in Figure 8.9 where the hyperbolas reconstruct the DMO ellipse from the prestack migration ellipse. (This is simply the inverse procedure of creating the prestack migration ellipse from the DMO ellipse).

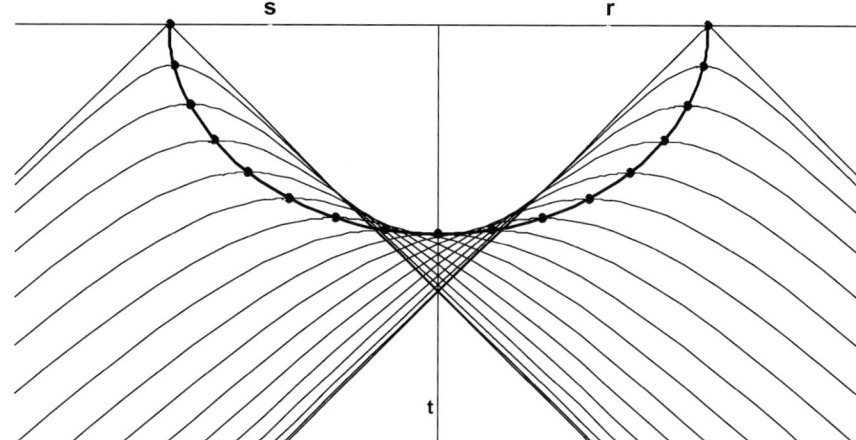

Figure 8.9 Modelling the DMO ellipse from the prestack migration ellipse.

Modelling the DMO operator from the prestack migration operator is a valuable tool. In the constant velocity case, the curves are ellipses and hyperbolas.

In areas of complex geology where the RMS assumptions are valid, the <u>*elliptical assumption*</u> of DMO and prestack migration may (will) fail.

<u>Modelling with the hyperbola, however, is still valid</u>. In these cases, the prestack migration curve may still be modelled to create the DMO curve.

In cases where the <u>prestack migration curve deviates from an ellipse</u>, modelling indicates the DMO curve will have a <u>larger distortion</u> from the DMO ellipse.

In some moderate velocity areas, the modelled DMO curve is severely distorted indicating that use of DMO may not be beneficial.

8.1.12 DMO of a single trace

DMO of a single input trace <u>moves energy into the neighboring traces</u> as shown in Figure 8.10. Part (a) shows an input trace and (b) DMO of the input trace. Part (c) is a kinematic illustration of the DMO data in (b). Figure 8.11 is an enlarged portion of (b).

Note:

- <u>Energy in (b) is spread to neighboring traces</u> (as in poststack migration).
- Energy follows <u>elliptical paths</u> as illustrated in (c).
- All elliptical <u>paths meet</u> at either the source *s* or receiver *r*.
- Energy is limited to dips less than <u>45 degrees</u>.
- The zero phase shape of the <u>input trace wavelets</u>.
- The <u>phase change</u> on the DMO'd traces.
- The <u>amplitude</u> distribution of the DMO'd traces.

A single trace DMO algorithm <u>may</u> be used for:
- stacking into a limited offset section,
- stacking into source gathers,
- stacking into the final stacked section.

A single trace algorithm is <u>required for 3-D</u> (*x, y, t*) projects.

Some algorithms allow all traces in a 2-D constant offset section or a 2-D source record to be DMO'd together.

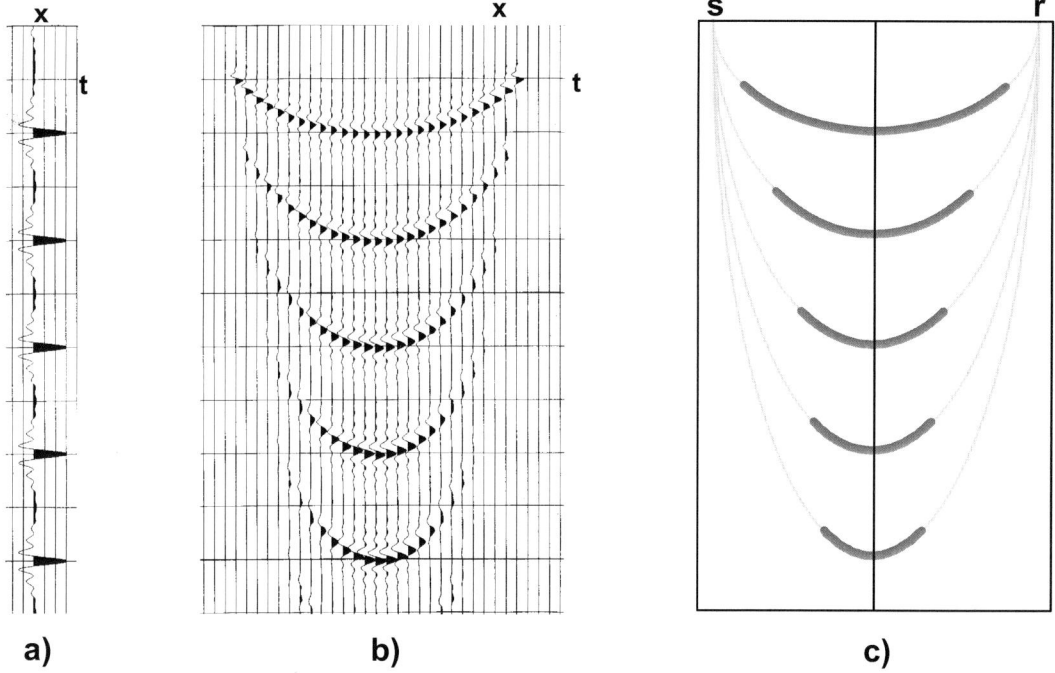

Figure 8.10 Example of DMO spreading energy to neighboring traces, a) input trace, b) the DMO'd traces, and c) a kinematic illustration of (b).

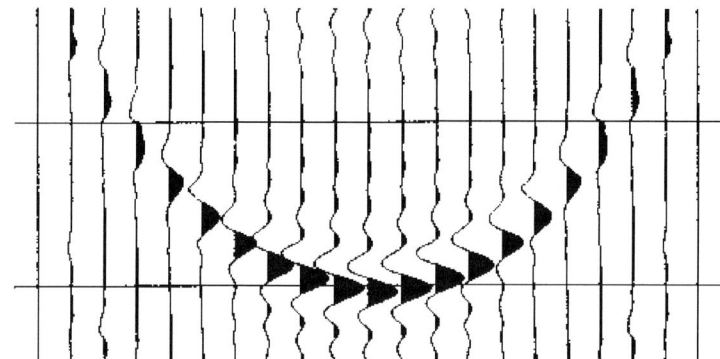

Figure 8.11 Close-up view of the DMO ellipse; note the phase of the wavelet is different from that of the input trace.

8.2 DMO of Source Gathers

Source gathers are a convenient form for prestack processing, as the data is <u>naturally acquired</u> in this format. The record is usually prestack processed with DMO, or prestack migration. The traces are then sorted to CMP gathers for velocity analysis, or stacked directly to the final section.

The advantages of applying DMO to source records are:
- the <u>geometry can</u> remain simple (except for additional traces on ends of each source record),
- all traces in the source gather are live and contiguous in offset.

> <u>In contrast</u>, the traces will be sparse when sorted to CMP gathers, or to constant offset sections.

Figure 8.12 illustrates DMO on a one-sided source record. Note:
- The DMO result is for <u>two</u> traces.
- One edge of each family of DMO ellipses is always at <u>zero-offset</u>.
- The traces after DMO are spread across <u>different offsets</u>.
- A <u>projection</u> of the DMO'd traces onto the <u>zero-offset plane</u> should be identical to the projected traces from the constant offset section.

> Many believe that no aliasing occurs in either source record DMO or source record prestack migration.
> - This concept is only partially true.
> - DMO or prestack migration of a particular <u>source record should not be aliased</u>.
> - However, the <u>summation of a number of source records</u> to create a stack <u>may introduce aliasing</u>, especially with the steeper dips on a prestack migration.

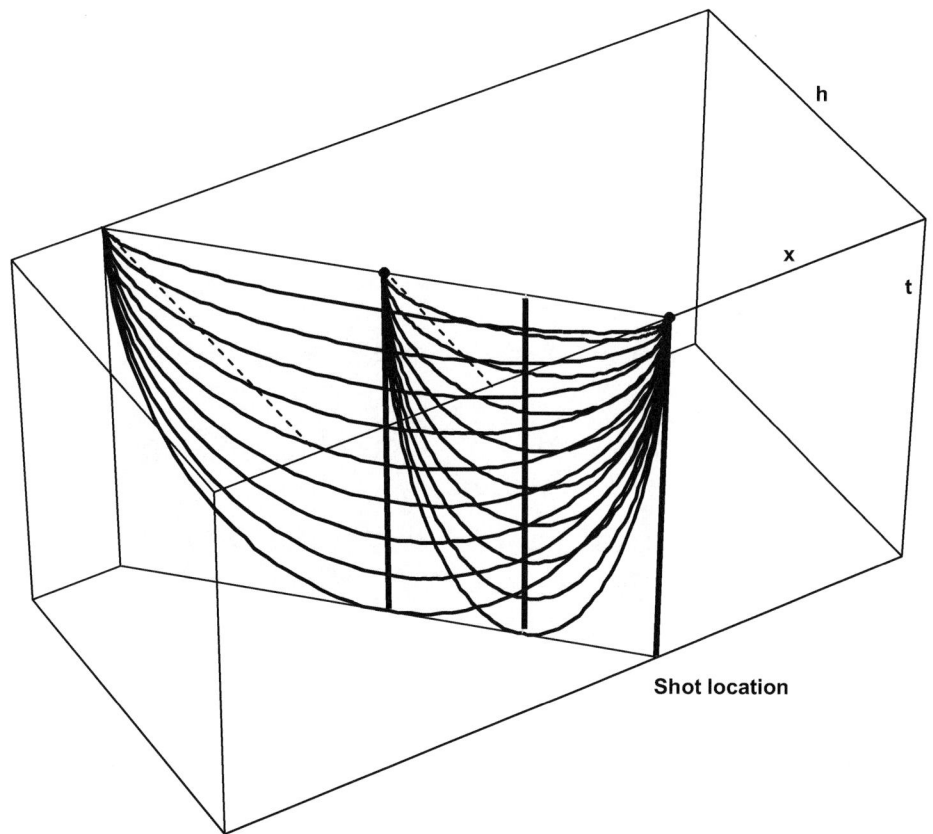

Figure 8.12 DMO on a one-sided source record.

DMO of source records may be accomplished by:
- DMOing a single trace, then adding the new traces to the source record, or
- using a log-log stretch method to DMO one side of a source record [287], (more to follow in section 8.4.17).

Points to note in source record processing:
- DMO adds a minimal number of traces to the edges of the source record that are <u>usually ignored</u>.
- Prestack migrated source records contain many <u>additional traces</u> at the edges of the record that should not be muted.
- Energy migrated <u>beyond</u> the range of the original source record may be <u>steeply dipping</u> and require a <u>high concentration of source records for wave front reconstruction.</u>
- Source record traces are now <u>equivalent to zero-offset</u> and could be stacked.
- The <u>assumed offsets are questionable</u> when inverse MO is applied to source records for velocity analysis.

Figure 8.13 illustrates an outline of energy distributed in (a) a source gather, (b) after DMO, and (c) after prestack migration. DMO adds little energy to a few traces beyond the original geometry of the source. In contrast, prestack migration adds a considerable number of traces to a source record.

Restricting the number of traces to the original source geometry may have little effect on DMO, but will have a considerable effect on prestack migration.

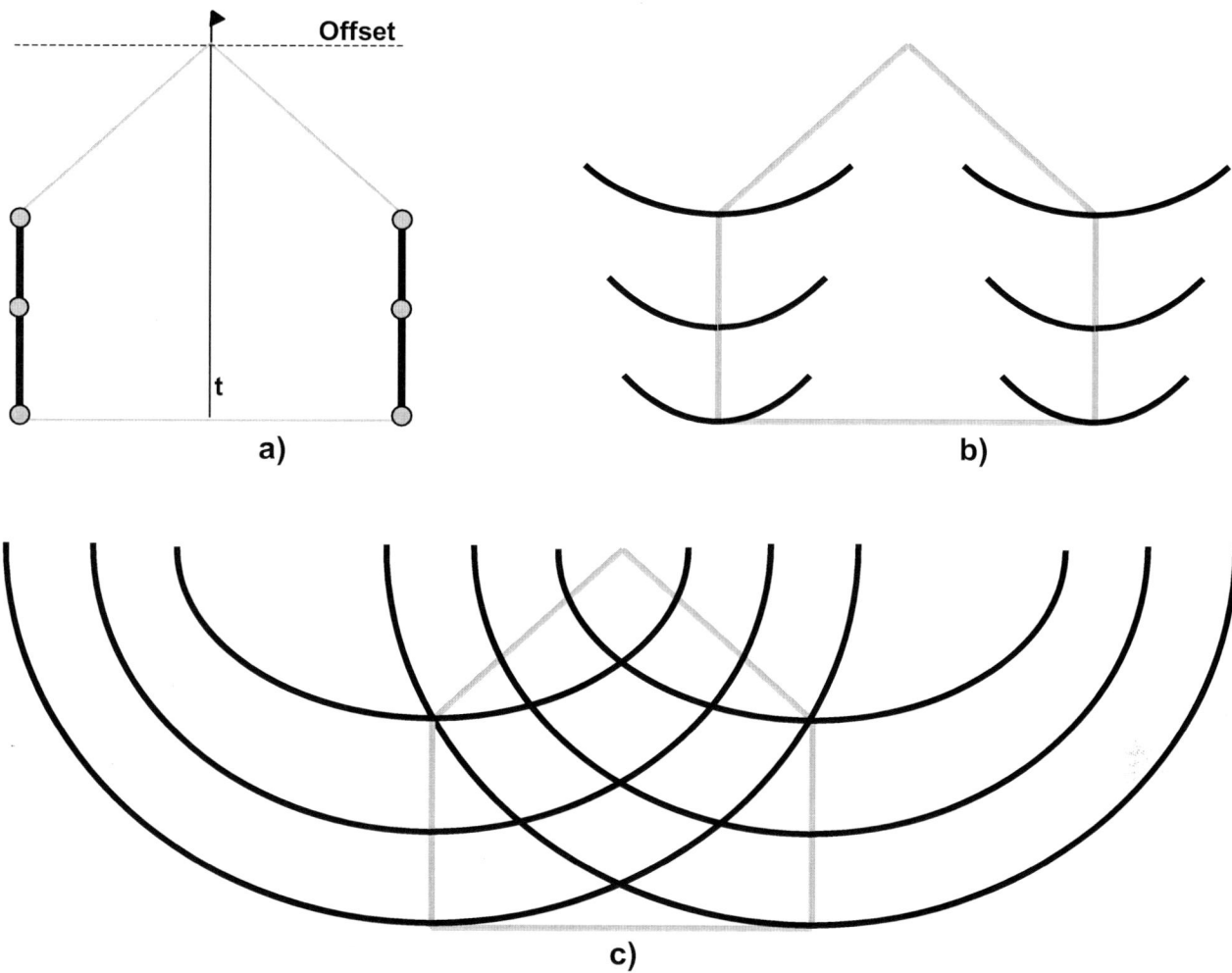

Figure 8.13 Illustrations of a) outline of a MO corrected source record (x, t), b) the DMO source record, and c) the prestack migrated source record.

8.3 DMO of Constant Offset Sections

8.3.1 Introduction

Constant offset or limited offset sections, are suitable for 2-D and some 3-D marine data. Algorithms may be written to take advantage of the data when the entire section may be <u>regarded as having one offset</u>.

The advantages of constant offset sections are:

- useful for evaluating prestack data,
- horizontal events are still horizontal, and
- fast computing of DMO on all traces at constant offset.

Figure 8.14a illustrates the DMO operator is the same for all spatial positions on a constant offset section. Algorithms take advantage of the constant offsets for efficient computation

The offset in each constant offset section varies from zero (at the front) to a maximum at the back. The DMO operators may be seen to vary with offset h in Figure 8.14b.

DMO algorithms are usually most <u>efficient</u> when using constant offset sections.

For 3-D data, the azimuths of DMO'ed energy are random and must be computed separately.

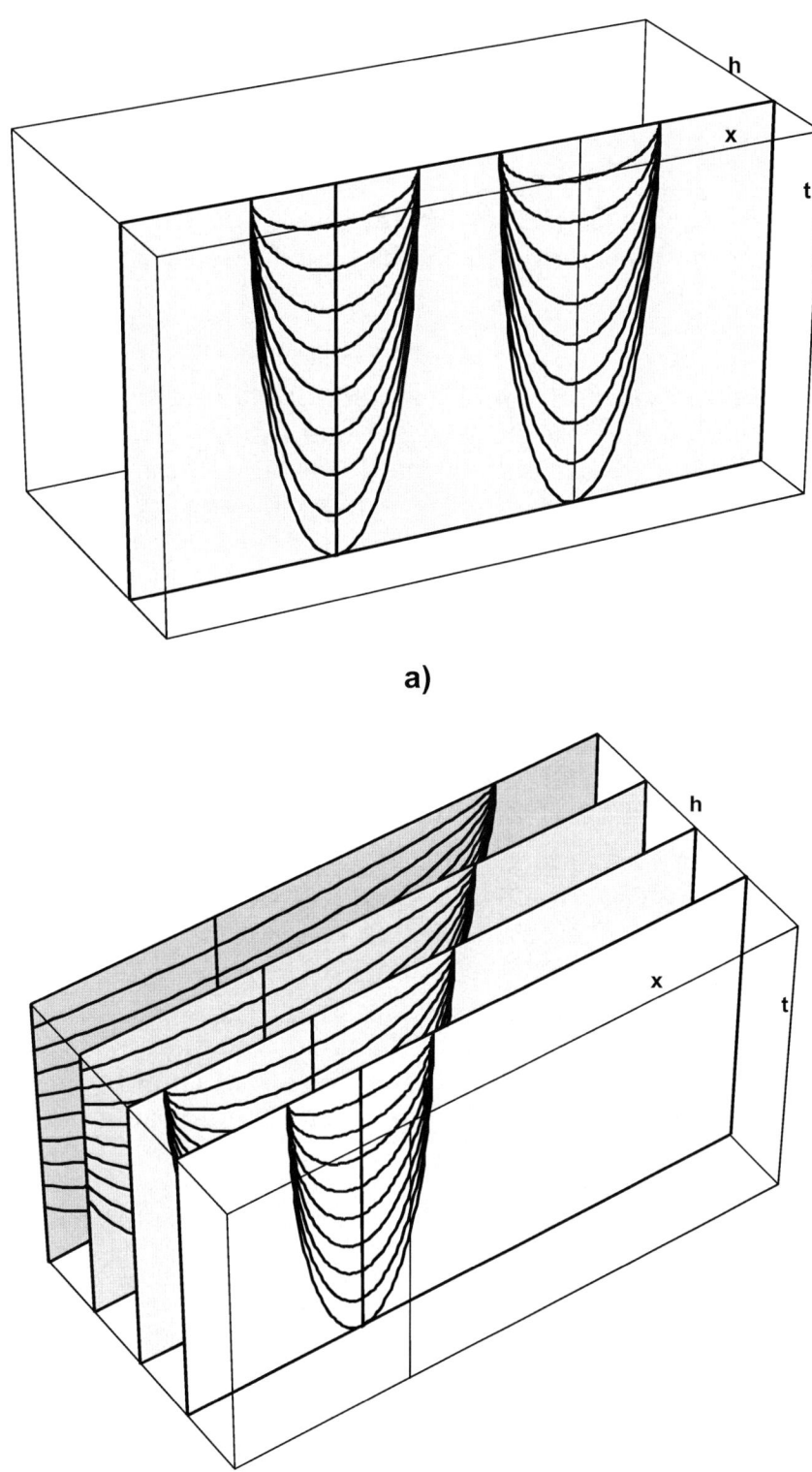

Figure 8.14 Constant offset section and the DMO operators, a) for one offset, and b) for different offsets.

8.3.2 Forming constant offset sections from limited offset stacks

The number of constant offset sections should be defined by the number of offsets in the recording configuration. For example, a split spread of 120 receivers should have 60 offsets.

When sources are located at (for example) every fourth receiver, there are very few traces in each of these constant offset sections, i. e., one live trace every eight traces.

The number of offset sections may be reduced by taking a range of offsets, while at the same time increasing the number of live traces. Reducing the number of constant offset sections requires a "limited offset section" to contain a range of offsets. A large range of offsets may defocus energy when the section is processed at one offset. Recall Figure 7.27 and the stacking smear.

Applying NMO and DMO to each input trace before stacking into an offset section will reduce the defocusing that is due to a large offset range. Other processes such as inverse NMO may then proceed at the constant offset with little error.

Another approach to solving the sparse trace problem is to interpolate traces in the constant offset sections or in MO'd CDP gathers.

Yet another alternative is to start with the maximum number of offset sections, apply MO or a partial MO to each constant offset section, and then reduce the number of offset sections by stacking with a range of offsets.

> DMO is a linear process and will work just as well with sparse traces in a constant offset section before forming a limited offset section.

Data in constant offset sections may also be <u>resorted into CMP gathers</u> for additional processing. This process requires <u>changes to the geometry</u> as DMO creates additional traces. These additional traces are created when seismic energy is moved into areas that contained dead traces on the constant offset section.

After DMO processing the constant offset sections should be <u>considered to have zero-offset</u> and may be stacked to create a final section. Processes such as velocity analysis, however, associate the offset section with the original source-receiver offset, enabling inverse MO to be applied.

What is the optimum offset for velocity analysis?

- **the original half source-receiver offset, or**
- **an offset based on the distance of the source and receiver from a scatter point.**

See Chapter 11 for more information.

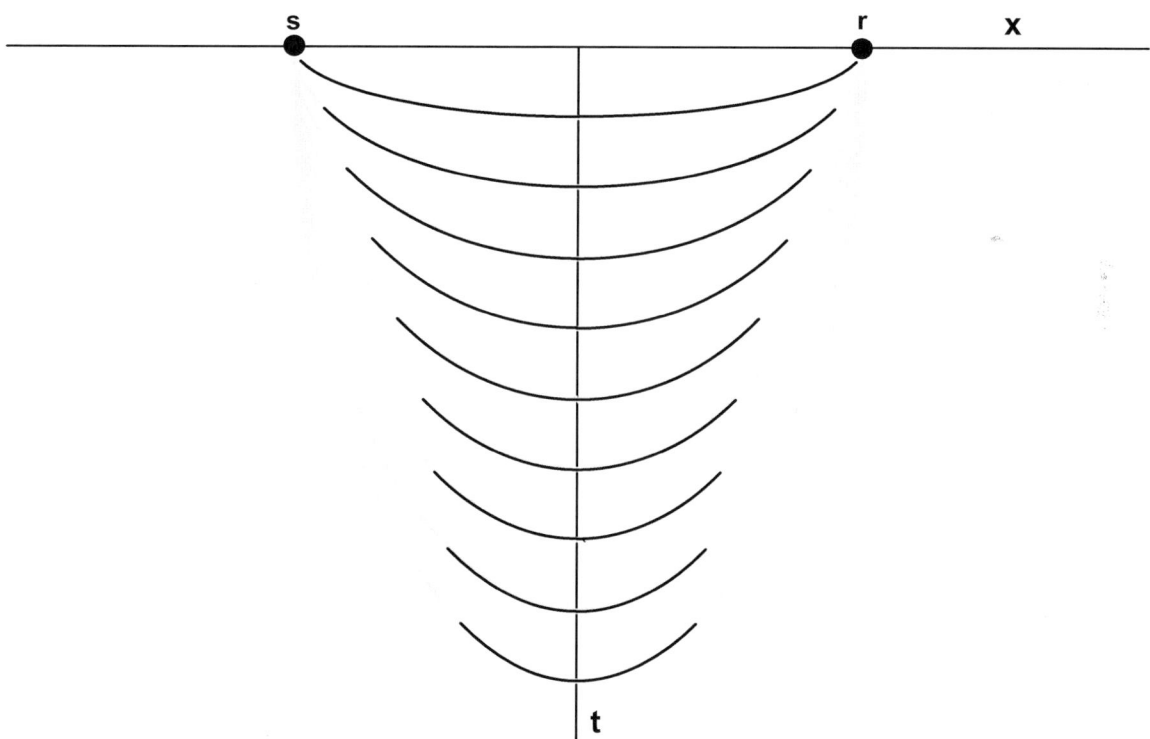

Figure 8.15 DMO curves for construction of Figure 8.16 on following page. (See removable page at the end of this chapter.)

8.3.3 DMO construction from a constant offset section

The main function of conventional DMO is to convert offset data to be <u>equivalent</u> to zero-offset data, i.e. $P(x_{DMO}, h, t_{DMO}) \equiv P(x, h=0, t)$.

The action of <u>DMO on constant offset sections</u> is illustrated in Figure 8.16 with (a) a scatter point diffraction and (b) a dipping reflection.

Responses from the scatter point in (a) and dipping event in (b) are:

- *solid curve*: the offset diffraction or reflection
- *dashed line*: after NMO
- *shaded line*: the zero-offset diffraction or reflection

DMO will convert the N<u>MO'd</u> data to the zero-offset shape.

The kinematics of DMO may be illustrated with the aid of a transparent copy of Figure 8.16 (or by copying Figure 8.15 and Figure 8.16 and using a window). The offset sections have the same offset (indicated by *2h*) as the DMO curves in Figure 8.15.

- Slide the offset sections across the DMO curves of Figure 8.15.
- <u>Maintain time zero</u> on the DMO curves with time zero on the offset section.
- Align the bottom of a DMO curve with a point on the NMO'd diffraction.
- Sketch the energy spread by DMO (the dark portion of the ellipse), and verify that part of the DMO curve will be tangent to the zero-offset diffraction.
- Repeat the process for each DMO curve that intersects the NMO'd diffraction.
- Energy from all points on the NMO'd diffraction will reconstruct on the zero-offset diffraction and cancel elsewhere.
- Repeat the exercise for the scatter point and the dipping reflector.

Recall that DMO uses NMO correction with a RMS type velocity.

If a dip-dependent velocity was used in Figure 8.16b, where would the MO reflection be?

The hyperbolic nature of the offset reflection in Figure 8.16b is illustrated by the gray extension. The peak of the hyperbola is below the intersection of the dipping event with the surface.

Also note the hyperbolic nature of the entire NMO corrected reflection in (b).

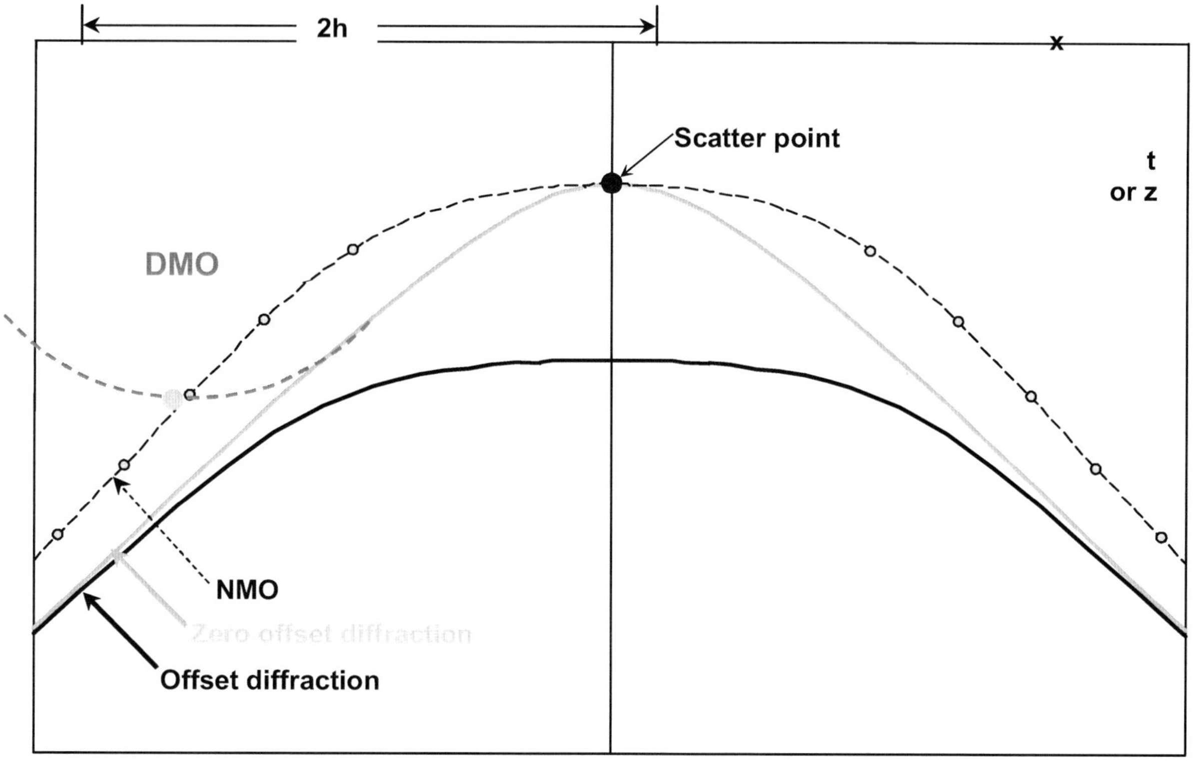

Figure 8.16 DMO construction on a constant offset section for a) a scatterpoint, and b) a dipping reflector using the DMO curves from Figure 8.15.

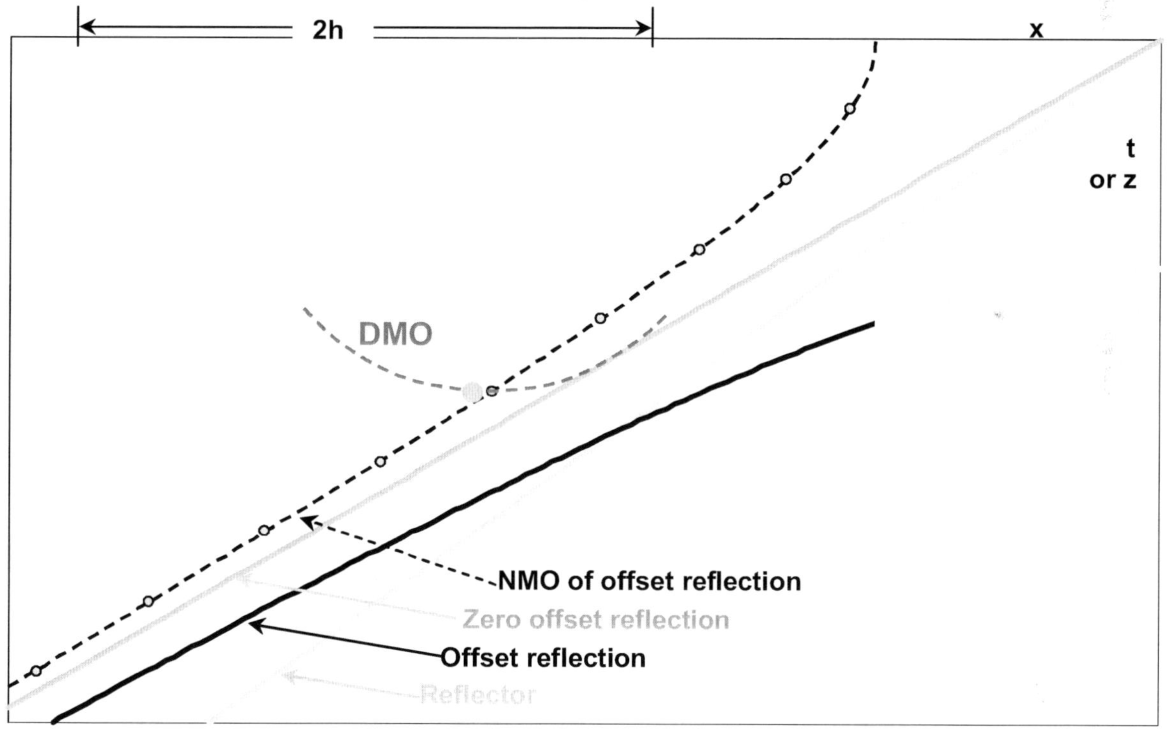

(See removable page at the end of the chapter)

8.3.4 Scatter point NMO and DMO in the prestack volume

The previous figure illustrated DMO on one offset section. The following examples in Figure 8.17 show the same results, <u>with a continuous offset</u> three-dimensional surface in the prestack volume (x, h, t). Two views (front and back) of each surface are included to help define the shape. The surfaces are (a) Cheops pyramid before MO, (b) after NMO, and (c) after NMO and DMO.

Note in (c) that with MO and DMO, all the offsets have the <u>same shape</u> as zero-offset and will produce an <u>optimum stack</u> of a hyperbola at zero offset.

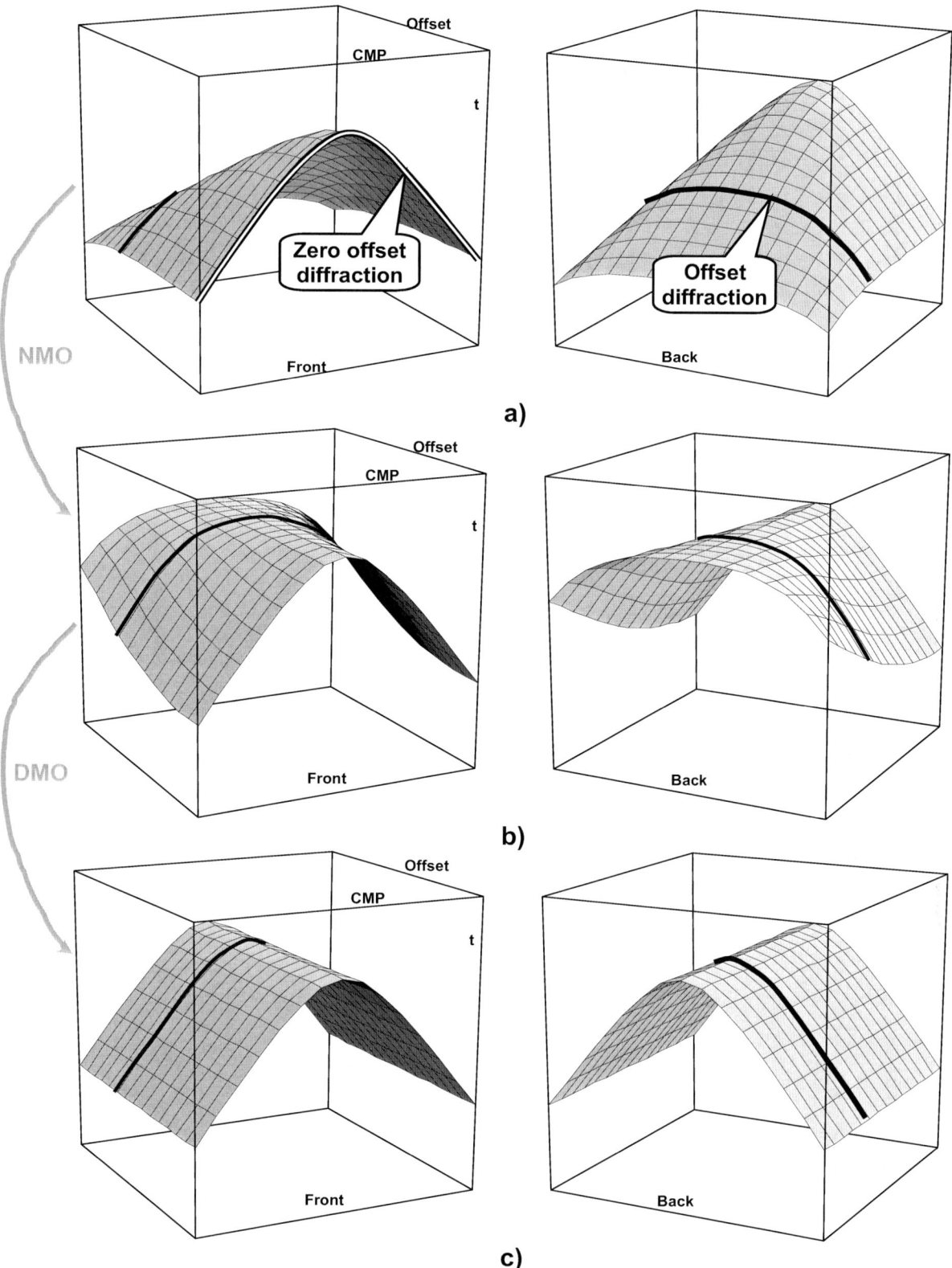

Figure 8.17 Diffraction surfaces from a scatter point at various stages of processing, a) Cheops pyramid, b) after NMO, and c) after NMO and DMO.

8.4 Dip Moveout (DMO) algorithms

8.4.1 Introduction

Dip moveout (DMO) is a term used for processes that converts <u>offset data</u> to be <u>equivalent to zero offset data</u>.

The more correct term of "<u>prestack partial migration</u>" (PSPM) [24], [58] was used in the 1970's and implied a wave theoretical solution. However, Hale's theoretically complete paper in 1984 [57] used the term DMO, which then became the conventional term used by the seismic processing industry.

The term migration to zero offset (MZO) may also be used.

A CMP gather will contain reflections from <u>different reflector positions on a dipping event</u> as illustrated in the traditional Figure 8.18. I emphasize again that the reflection point will be assumed to come from the zero-offset reflection point, especially with DD- MO correction T_{dn}.

The purpose of DMO is to reduce, or remove, this reflector point smear.

In order to remain competitive, some processing companies used the term "DMO" for any process that improve the stacking of dipping events, including DD- MO correction. The following is a list of processes that eventually become true DMO processes.

- **Dip-dependent stack (δt).**   (Normal processing)
- **French method (δt, and δx).**   (Model based)
- **Multiple velocity stacks.**   (Close)
- **Multiple velocity stacks with dip filters.**   (Starting wave theoretical.)
- **Yilmaz method.**   (We are nearly there.)
- **Hale's method.**   (We are there.)
- **Fowler's DMO.**
- **Notfors & Godfrey, and Liner's FK log stretch method.**
- **Time domain methods.**
- **Gardner method (DMO before MO).**
- **NMO-DMO-INMO (NDI).**
- **Radon transforms.**
- **Biondi's log-log source record method.**

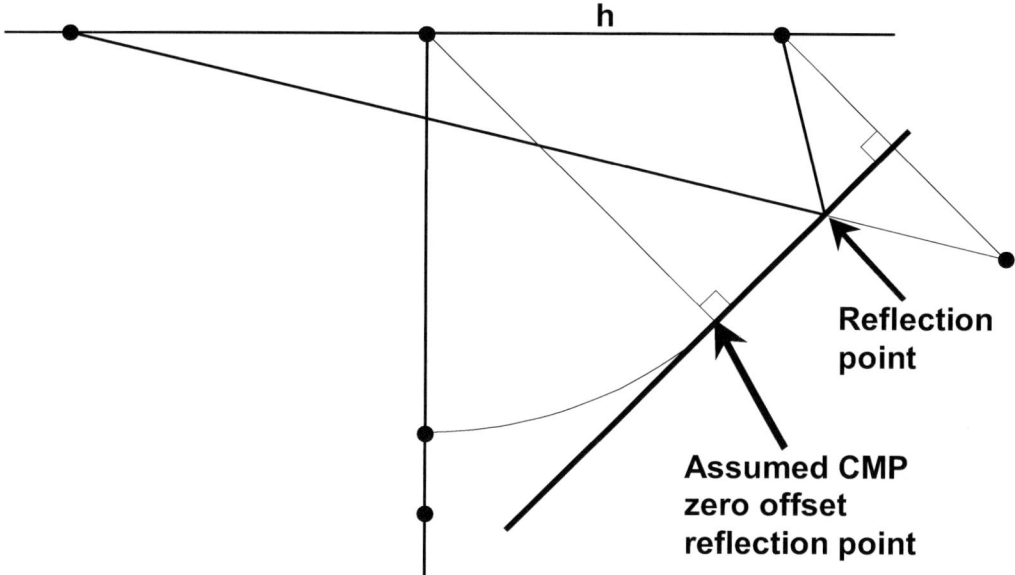

Figure 8.18 Reflection point dispersion from dipping events.

We will take a closer look at some of these DMO methods.

8.4.2 Dip dependent stack

This method, described by Levin [115], is used in normal processing where stacking velocities are increased to stack dipping events (geological dip β) by,

$$V_{stk} = \frac{V_{rms}}{\cos \beta} \tag{8.5}$$

as described in section 7.0.8.

In the 1980's some processing centers used this result to claim they were dip processing, and <u>sold</u> their products as DMO.

This method:

- cannot resolve conflicting dips and
- is model based. (See section 8.4.5 for a variation of this method.)

8.4.3 French's Method

The French method [87] applies a time correction <u>and a lateral or trace correction</u> that moves the energy to the "true" zero-offset location of the reflection point as illustrated in Figure 8.19.

This method <u>required a reasonable knowledge of the structure</u>, and was accomplished by identifying the actual reflection point on the reflector using:

$$d = \frac{h^2 \cos \beta \sin \beta}{D}, \tag{8.6}$$

where the distance between reflection points is d, the half offset is h, and the distance from the CMP that is normal to the reflector is D.

The location of the actual reflection point (RP), and the time T_d may then be estimated.

> Show the maximum displacement in d occurs at a dip of 45 degrees.

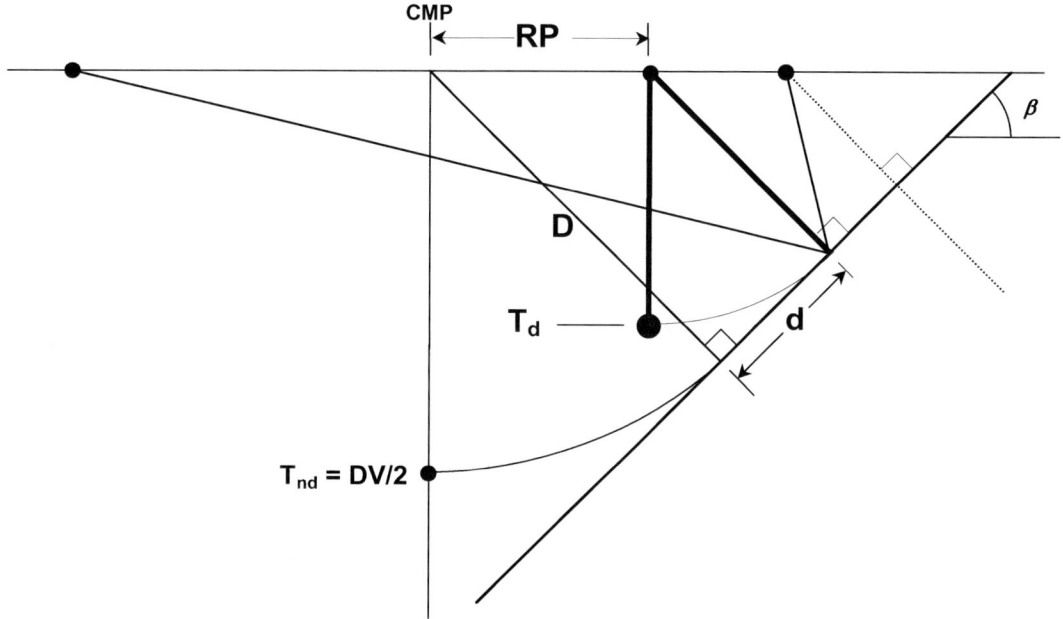

Figure 8.19 French's method of DMO.

Note:

- time correction T_d is different from the previous method,
- the new position <u>removes</u> dip smear,
- requires a complex calculation for each point,
- confusion may still exist with <u>overlapping</u> dips,
- is <u>model based</u>,
- could also be extended to do a prestack <u>migration</u>.

This method:

- does not require wave-front reconstruction to image the reflector,
- requires an accurate knowledge of the geological structure (similar to depth migration),
- was a popular method for 2-D and 3-D seismic in the 1980's.

8.4.4 Multiple constant velocity stack

Multiple constant velocity stacks choose a number of different velocities based on the range of dips in the section as illustrated in Figure 8.20.

- DD-MO correction and stack is applied at each of these velocities,
- the sums of all these sections to complete the process.

The sum of the incoherent noise in each constant velocity section will increase by the square root of the number of sections. The coherent signal will have a value of one. The signal to noise ratio will decrease with an increase in the number of velocities.

The method is:

- fast,
- inexpensive,
- not model based (unless specific dips are defined).

Paper [96] uses this method on two dipping events. Will dip smear still occur?

8.4.5 Multiple constant velocity stacks with dip filter

This is similar to the previous method, but has added the very important step of dip filtering (Jakubowicz 1984). A dip filter is applied to each constant velocity stack at an angle that relates to the dip by the equation,

$$\tan \alpha = \sin \beta . \qquad (8.7)$$

- A dip filter attenuates the dipping noise in each section.
- Only the portion that matches the DD-MO velocity and dip filter will be present.

Figure 8.21 repeats the diffractions of Figure 8.20, but now dip filtering has removed the noise at unwanted dips before summation.

The method is:

- not very fast,
- very good (see the following section).

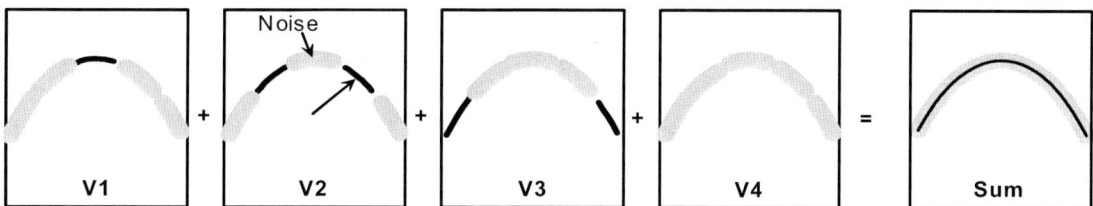

Figure 8.20 Multiple dip stacking of a diffraction showing DD-MO for various velocities and the sum. Noise is shown in gray.

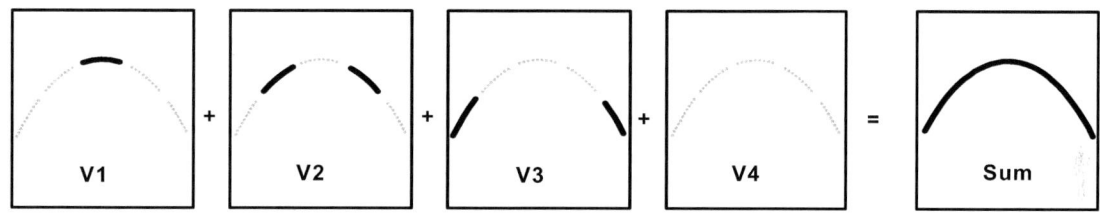

Figure 8.21 Multiple dip stacking of a diffraction with a dip filter added to remove the noise.

8.4.6 Constant velocity stacks with dip filter and the DMO ellipse

Let us pursue this method of multiple DD-MO and dip filtering on a constant offset section a little further.

Figure 8.22 shows an offset time response at a time *T*. Dip-dependent MO correction was applied for various seismic dips α of 0°, 10°, 20°, 30°, and 45°. Note the DD-MO correction for a seismic dip of 45° is actually a geological dip β of 90°, where no DD-MO correction is applied.

Figure 8.23b now shows the effect of including a narrow dip filter at the appropriate dip for each section. The narrow dip filters convert each DD-MO point into a line with a dip of the dip filter.

Each of the dips become tangent to the desired DMO response as illustrated in Figure 8.23a.

> Many DD-MO and dip filtered sections (say 25 to 100) would reconstruct the DMO ellipse.

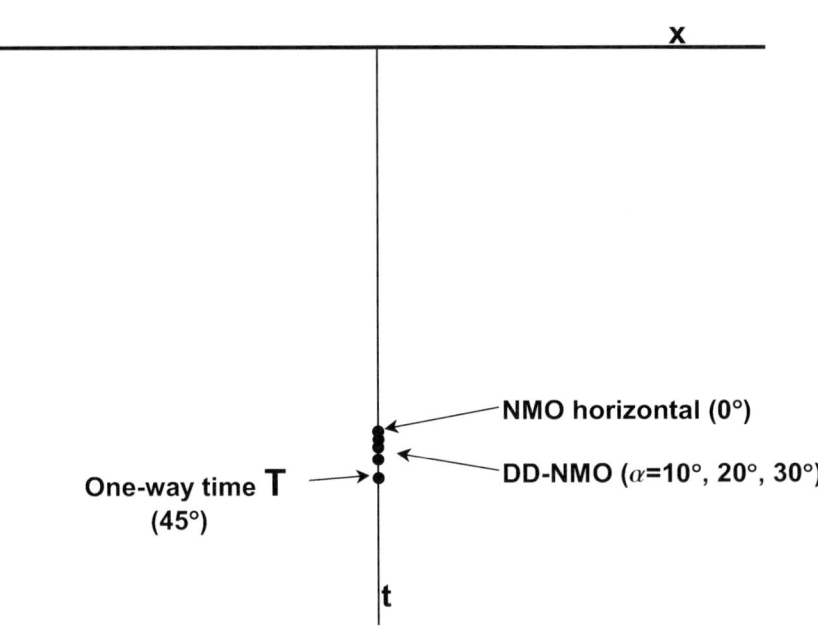

Figure 8.22 Location of a time sample when DD-MO is applied independently for seismic dips α of 0, 20, 30, 40, and 45 degrees.

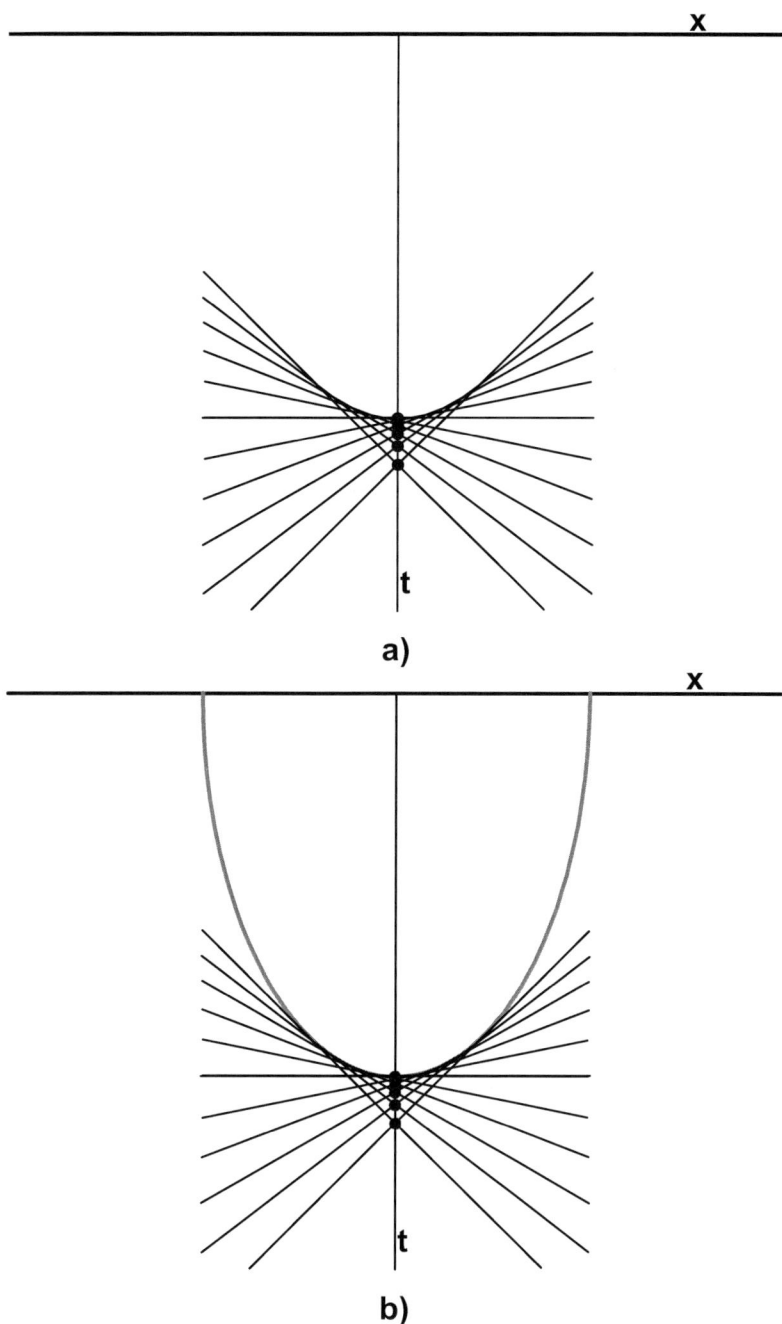

Figure 8.23 Combined effect of applying DD-MO and a narrow dip filter for numerous dips with a) showing the dipping energy, and b) the reinforcement of energy along the DMO ellipse.

8.4.7 Exercise for constant-velocity stacks with dip filters

The previous figures created the DMO ellipse from one input scatterpoint. This exercise shows the same principles may be applied to a constant offset diffraction, and that numerous DD-MO corrections with dip filters will create the zero-offset hyperbola.

Figure 8.24 shows a constant offset section with <u>one scatter point</u>.

The zero-offset diffraction is shown in gray.

The offset diffraction is the lowest and thickest black curve.

The other black curves show the diffractions after repeated application of constant velocity MO corrections.

Each curve represents DD-MO correction that was applied at 5-degree increments of α. (i.e. 0, 5, 10, … 40, 45) where the stacking velocity is defined by equation (7.7) in section 7.5, i.e.

$$V_{stk} = \frac{V_{rms}}{\cos\beta} = \frac{V_{rms}}{(1-\sin^2\beta)^{1/2}} = \frac{V_{rms}}{(1-\tan^2\alpha)^{1/2}}.$$

- Label the constant velocity DD-MO corrected diffractions from 0 to 45 degrees of seismic dip (α).

- Where possible, draw lines that are tangent to both the DD-MO corrected diffractions and the zero-offset hyperbola.

The tangent represents the dip filtered response of a narrow angle dip filter that is applied to the DD-MO corrected diffraction.

The tangent lines will reconstruct the zero-offset hyperbola.

- Use the protractor of Figure 8.25 to verify that the angle of these tangent lines range from 0 to 45 degrees.

- Verify that DD-MO correction, with the corresponding dip filter, can move the energy from a constant-offset diffraction to the zero offset hyperbola.

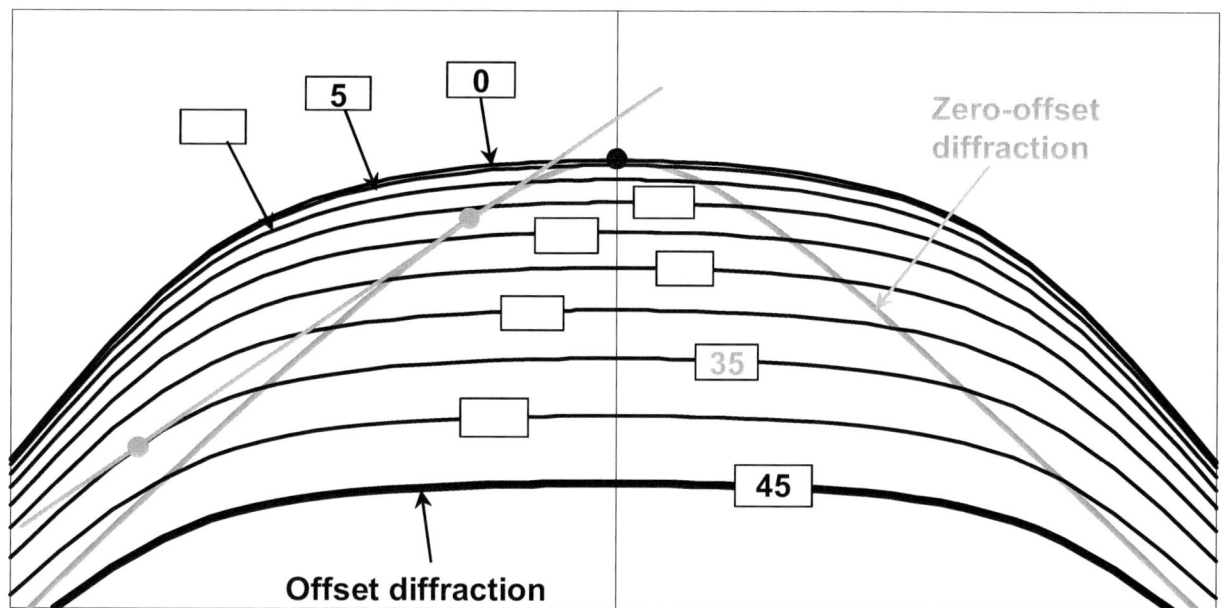

Figure 8.24 Construction of the zero-offset hyperbola from numerous dip-dependent MO corrections applied at five-degree increments ($0.0 \leq \alpha \leq 45.0$).

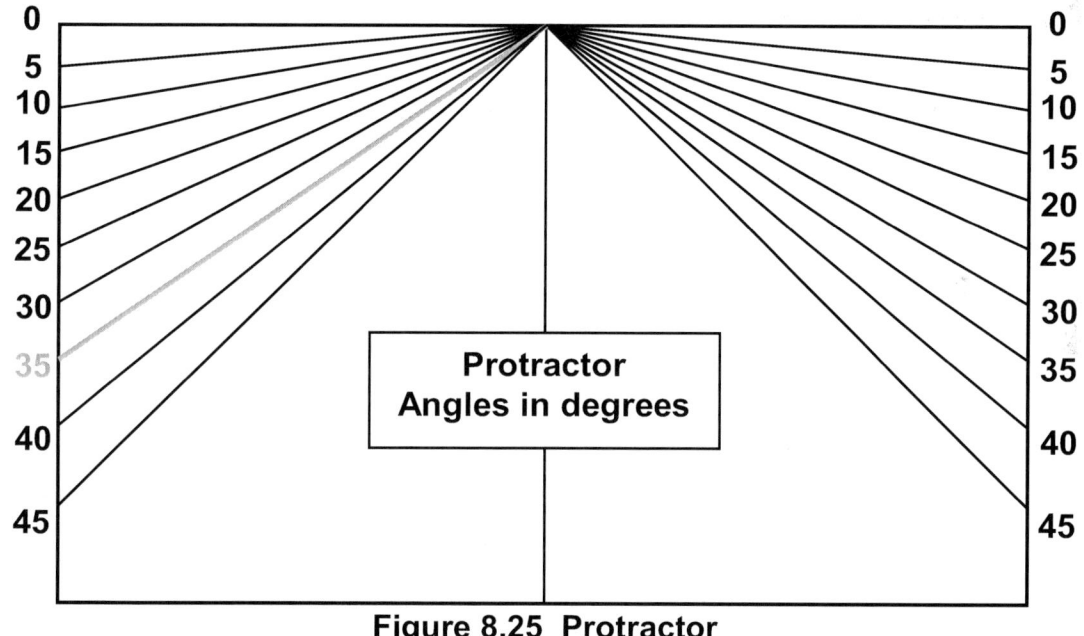

Figure 8.25 Protractor

8.4.8 Wave theoretical DMO methods

- Yilmaz and Claerbout [58] using the finite difference approximations introduced wave theoretical DMO in 1980. They used a more correct term of Prestack Partial Migration (PSPM). See the reference for more details.

- Deregowski [307] introduced the term DMO in 1981 in a time domain approach. See the reference for more details.

- Dave Hale [57], in 1984, introduced a theoretically correct concept using a Fourier transform technique. The algorithm was extremely slow and was not practical for commercial processing, however, it did introduce a new standard.

- Fowler in 1984 [680] presented a method based on many constant velocity stacks. Each stack was 2-D Fourier transformed and events assembled by dip. The inverse 2-D transforms produced a volume (x, t, V) from which a DMO'd section could be interpreted.

- Notfors and Godfrey in 1987 [84] applied a procedure mentioned by Bolondi [82], that stretches the time scale to a log time scale. This method was refined in 1988 by Liner [86, 89] with results that match Hale's Fourier transform method, but are computationally much faster, and commercially practical.

- Time domain methods may also be used to approximate wave equation methods. These methods spread energy to neighboring traces, ensuring that the elliptical paths are followed, the amplitudes are correct, and the frequency and phase match the result of the ideal methods.

- Wang in 1995 showed that a hyperbolic Radon transform may be used to first convert MO'd data to (τ, p) space, which is then followed by an inverse linear Radon transform back to DMO'd (x, t) space.

- Biondi [361, 369] showed in 1986 that DMO might also be performed directly on source records with log stretches in both time and horizontal distance.

DMO can also be applied before MO.
- Gardner developed a method that performs DMO first, assigns new offsets to each DMO'd traces. NMO correction is then applied at the new offsets [111] (see section 8.4.11).

- Constant offset DMO may also be applied before NMO as a combined process of NMO-DMO-INMO or NDI.

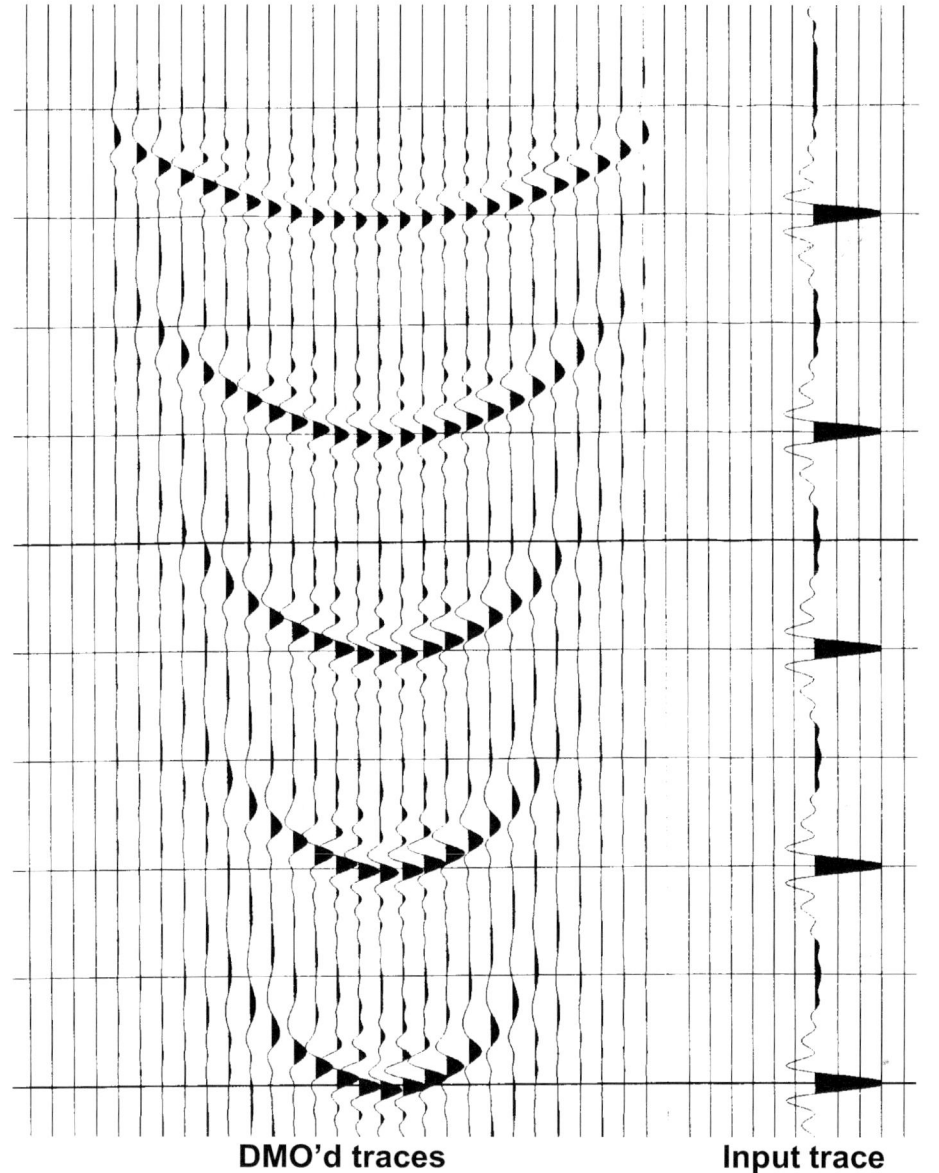

DMO'd traces   Input trace

Figure 8.26 The input trace is on the right, the DMO'd traces are in the center.

Note:
- elliptical shape,
- different amplitudes as the offset and time varies,
- changes in the frequency and phase.

What is happening to the wavelet at the steeper dips?

8.4.9 Hale's DMO

Hale's Geophysics paper in 1984 [57], introduced a theoretically correct method of performing DMO on a constant offset section using a <u>Fourier transform technique</u>.

- The method is illustrated in Figure 8.27, starting with an MO'd constant offset section in part (a) that shows three events with the same geological dip.
- Part (b) shows the 1-<u>D Fourier transform in the spatial direction</u> to produce a section in (k_x, t) where Kx is the wave number in the x direction. The gray area shows the energy of spatial frequencies. The origin is the upper right corner.
- The band of energy in (c) <u>weighted at each frequency (ω) and integrated over t</u>.

$$P(k_x,\omega) = \int p(k_x,t) A^{-1} e^{i\omega tA} dt ,\qquad (8.8)$$

- The sum is placed at (k_x, ω) as shown in (d).
- The process is repeated for all k_x and all ω.
- The inverse 2-D Fourier transform of (d) produces the time section in (e). The data from the constant offset section in (a) is now DMO'd.

The term A in equation (8.8) is given by

$$A^2 = 1 + \frac{k_x^2 h^2}{\omega^2 t^2} . \qquad (8.9)$$

The weighting within the integration of equation (8.8) does not allow a Fast Fourier Transform (FFT) to be used for ω and is a slow process. Energy will be concentrated at the appropriate frequency of the dipping event but will integrate to zero at other frequencies.

The algorithm is very slow and was not practical for commercial processing

It did, however, introduce a theoretically correct standard by which all other DMO's are compared.

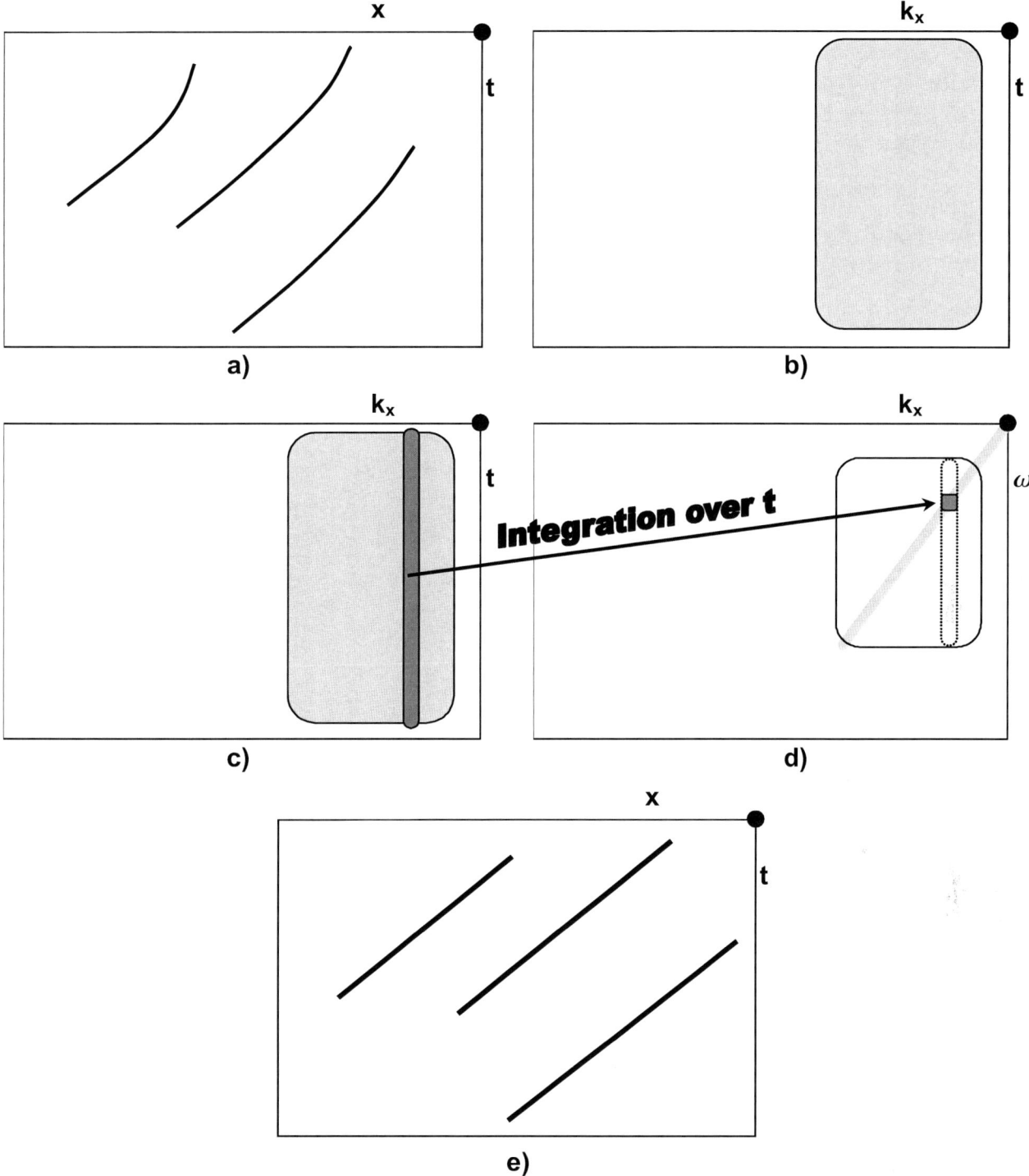

Figure 8.27 Development of Hale's DMO, a) the input constant offset section, b) the Fourier transform in x, c) shows the weighted integration over t to get one point in the (k_x, ω) domain in d). The inverse 2-D Fourier transforms results in e) is the DMO'd section.

8.4.10 Fowler's DMO

Fowler presented an interesting method of DMO in 1984 [680]. Figure 8.28 contains a <u>horizontal and a dipping event</u> (geological dip β) in a constant velocity (*V*) environment. Creating many constant velocity stacked sections over a range of velocities will create a 3-D volume (*x, t, V_{stk}*).

- The horizontal event will <u>stack best</u> at *V* as in (a).
- The dipping event will <u>stack best</u> with velocity $V_{stk\beta}$ = *V/cos(β)* at the zero-offset dip α as illustrated in (b).

Energy from all sections could be stacked to form a multiple velocity stack as described in section 8.4.3. However, Fowler used the Fourier transform domain (*kx, ω*), to "cut and paste" the dipping energy from the appropriate constant velocity sections. The selective "cutting and pasting" in (c) provides accurate dip filtering of the data to become equivalent to the multiple constant velocity stacks with dip filtering discussed in the previous sections 8.4.4 and 8.4.5.

The 2-D Fourier transform of each constant velocity section will align dipping events to their common location in each (*kx, ω*), with the optimum stacks still at their appropriate stacking velocity.

The dip α in the transform domain is defined by

$$\tan\alpha = \frac{Vk_x}{2\omega} = \sin\beta \tag{8.10}$$

and when combined with
$$V_{stk} = \frac{V}{\cos(\beta)} \tag{8.11}$$

produces
$$V_{stk} = \frac{V}{\left(1-\sin^2(\beta)\right)^{1/2}} = \frac{V}{\left(1-\frac{V^2 k_x^2}{4\omega^2}\right)^{1/2}} \tag{8.12}$$

For a given velocity *V*, we can locate the optimum dipping energy at V_{stk} and move it back to *V*, as illustrated in part (d), for an <u>ideal "cut and paste"</u>.

<u>Inverse transforming</u> back to (*x, t*), combines the horizontal and dipping event optimally stacked and filtered at the original velocity *V* as illustrated in part (e).

In the RMS velocity world, the volume of data may be interpreted to select a <u>velocity surface</u> that provides the best stack as shown in (f).

Each constant velocity section could be poststack migrated (using the FK method), to remove the problem of conflicting dips, and an <u>optimum poststack migration</u> may be interpreted from the volume.

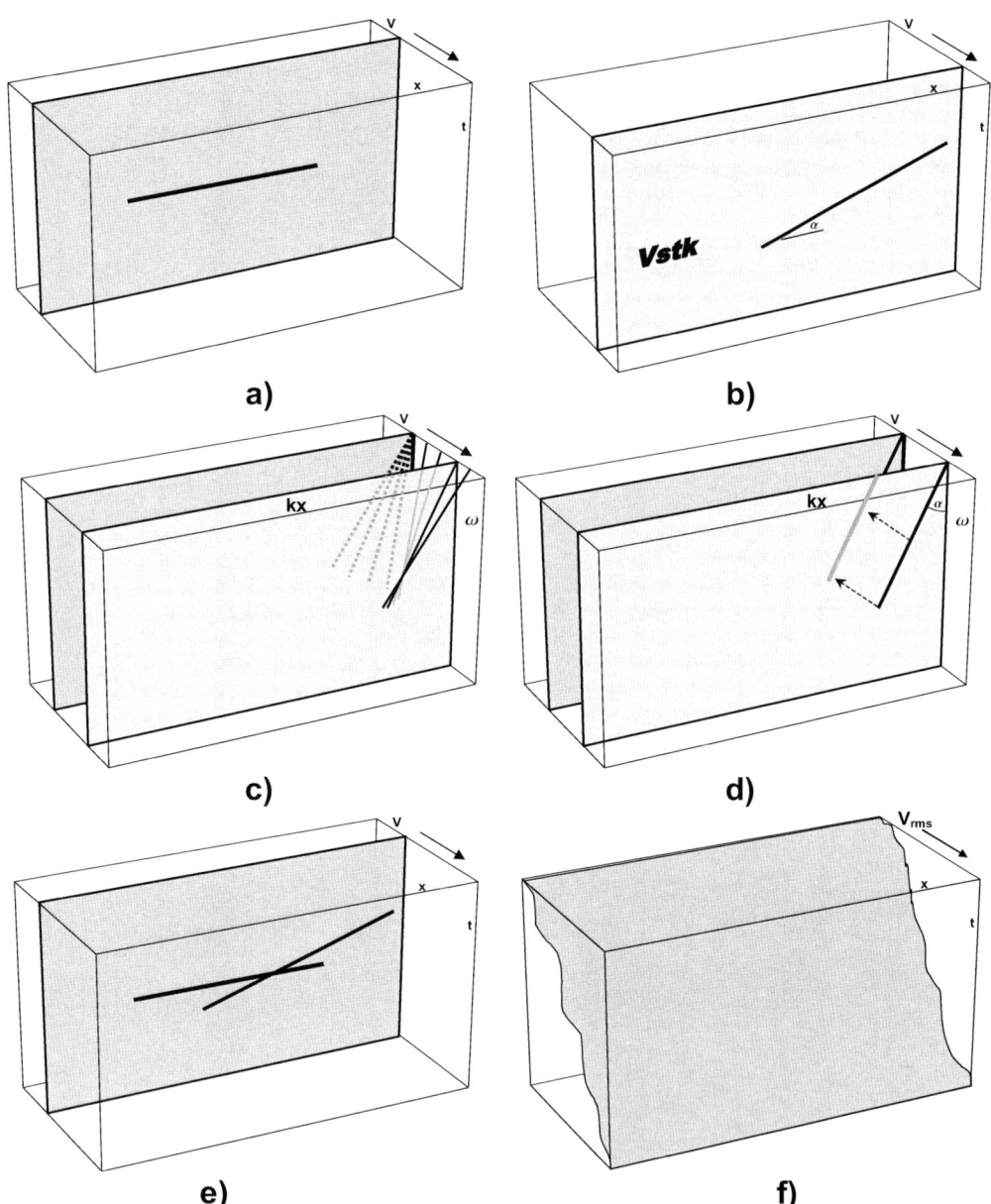

Figure 8.28 Fowler's method of DMO showing a) a horizontal event stacked at velocity V, b) a dipping event stacked at velocity V_{stk}. Part c) shows the 2-D Fourier transform of a number of dipping slopes that stack at different velocities relative to the velocity of a horizontal event. Part d) shows the 2-D Fourier transform of the volume showing the movement of dipping energy in (b) from V_{stk} to V and e) the resulting stack at velocity V. Part f) shows a velocity surface interpreted from the volume that contains the best poststack migration.

8.4.11 Log stretch method

The log stretch method is most efficient for constant offset sections. The kinematics of the process may be followed by studying the shape of ellipses, and the effect of converting the time scale to a log value. A complete description is found in Liner [89].

- An ellipse may be formed from a circle by linearly stretching the *t* scale as illustrated in Figure 8.29.
- The ellipses have been formed by multiplying the *t* value on the semi-circle by the constants 2 and 3,

$$t_c = circle = \frac{1}{v}(r^2 - x^2)^{1/2} \qquad t_2 = 2 \times t_c \qquad t_3 = 3 \times t_c \qquad (8.13)$$

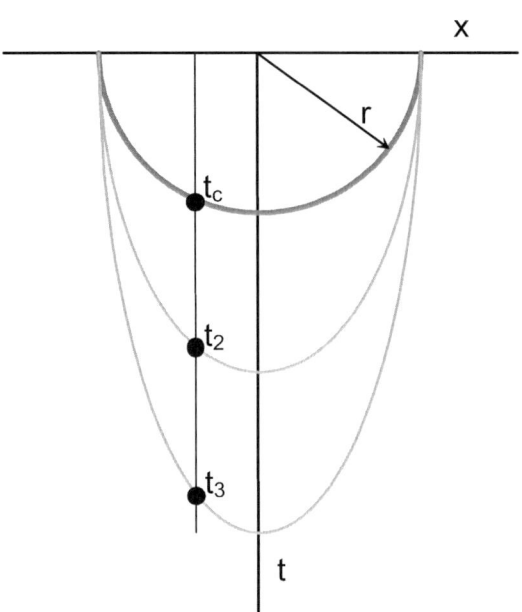

Figure 8.29 Ellipses formed by scaling a circle.

Taking logs of the *t* scale, the multiplication becomes an addition:

$$\log t_2 = \log 2 + \log t_c \qquad \log t_3 = \log 3 + \log t_c. \qquad (8.14)$$

The only difference between log *t₂* and log *t₃* is the addition of a constant, log2 or log3, while the shape log t_c will remain the same as illustrated in Figure 8.30.

The first logarithm sample must start at the first sample or an appropriate scale value of 1.0. Why???

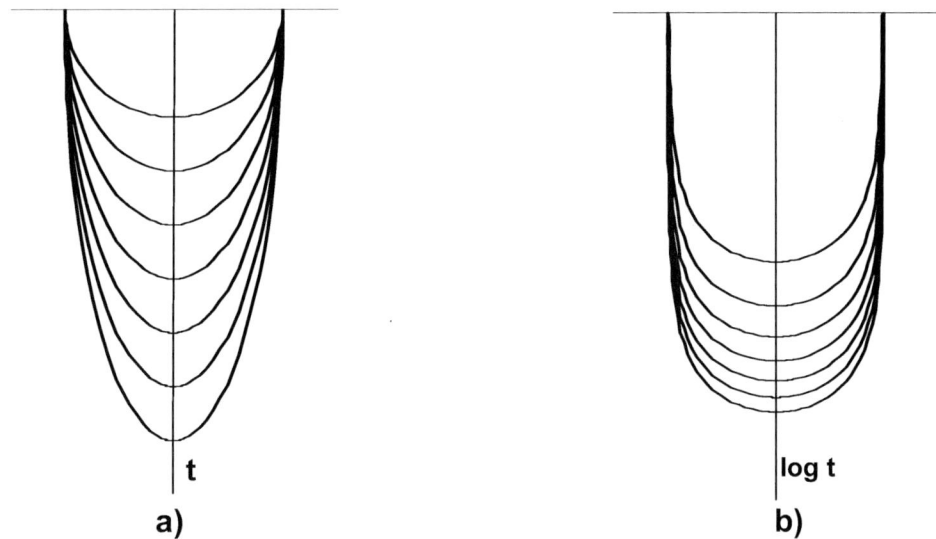

Figure 8.30 Ellipses in a) time and b) log-stretched time.

Note that the log stretched data has the same shape for each curve.

Figure 8.31 illustrates a processing sequence using a few kinematic samples.

- An input time trace (*Time*) is log stretched to time (*Log time*).
- The trace (or <u>constant offset section</u>) is then convolved or filtered using the FK domain to create the log-stretched-ellipses as illustrated by (FK filter).
- The filter is a complex multiplication for each point (k_x, ω) (Liner [89]).
- An inverse log stretch (*Inv. Log*) yields the DMO ellipses.

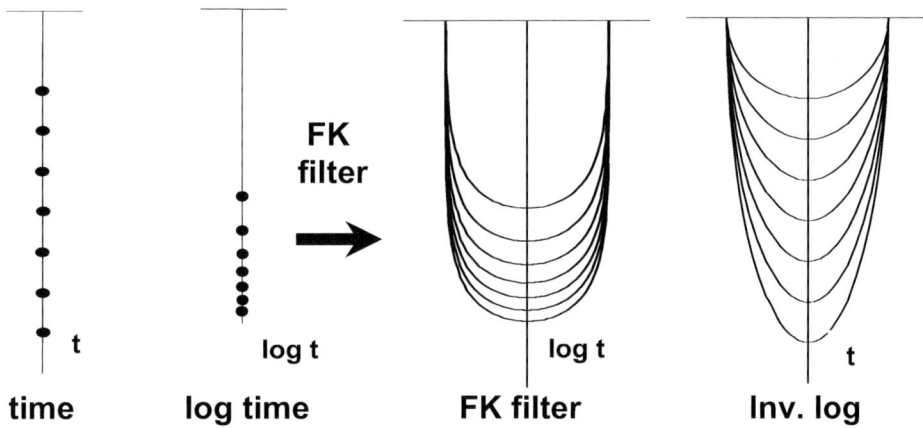

Figure 8.31 Log stretch DMO method.

8.4.12 Hagedoorn time domain method

The time domain application of wave equation DMO is similar to Hagedoorn migration in which energy from the input trace is spread along curves to the neighboring CMP traces. This method is possible because <u>DMO is "independent" of velocities.</u>

This Hagedoorn type method is also the dual of the Kirchhoff method, and the root $j\omega$) or <u>45 degree phase shift filter</u> must be applied.

Data moved along the ellipse eventually has a steep dip and requires the input traces to be low-pass filtered to <u>prevent aliasing</u>. These dips, however, are limited to 45 degrees. Limiting the energy and frequency of the steeper dips results in a <u>dip filter</u> that reduces noise and also reduces run times.

On Figure 8.32, construct the shape of the summation curve at the time T indicated.

- Note the slope of the "diffraction".
- Identify the 45 degree limit.

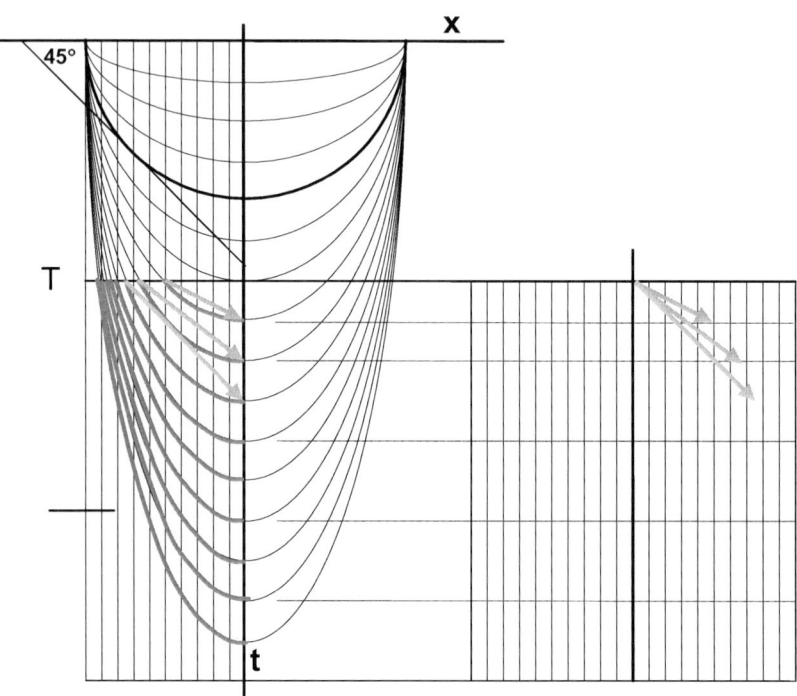

Figure 8.32 Time domain DMO curves for construction summation curve.

8.4.13 Hale's time domain method

There are many time domain methods in use. One that preserves the correct amplitude and phase without a special filter was developed by Hale [101], in which trajectories of equal time shifts ($\delta\tau$) are computed, as illustrated in Figure 8.33.

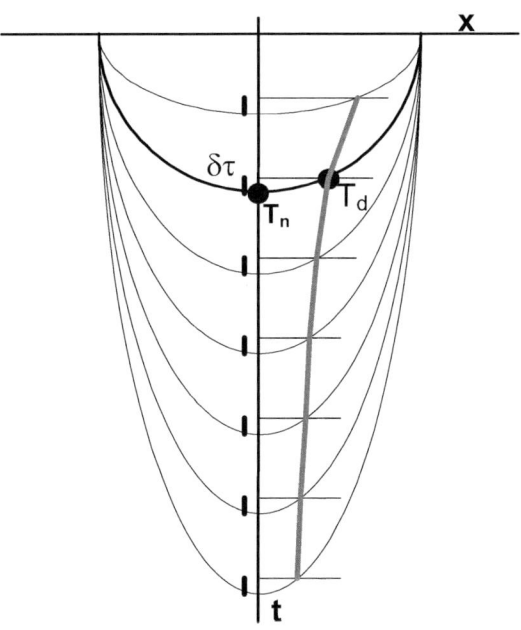

Figure 8.33 Hales DMO trajectories (or stems).

Data from time T_n is added to the appropriate trace on the trajectory at time T_d. All samples from the input traces are moved to various traces on the trajectory. Many trajectories, say 100, are used.

The movement of data to the trajectories will apply the correct amplitude and phase changes to the DMO'd traces.

8.4.14 Bancroft time domain method

A method by Bancroft [91] makes use of the Hale trajectories to simplify the amplitude calculation.

- **Times** along the DMO ellipse are computed from scaled values on the semicircle defined with radius in time $T_r = 2h/V$ as shown in Figure 8.34. The time on the semicircle T_c is given by

$$T_c = \frac{2}{V}\left(h^2 - b^2\right)^{1/2}, \quad (8.15a)$$

and the time on the ellipse T_d by

$$T_d = \frac{T_c T_n}{T_r} = T_n\left(1 - \frac{b^2}{h^2}\right)^{1/2}. \quad (8.15b)$$

- **Amplitudes** along the trajectories are related with the amplitude on the semicircle.

The amplitude is estimated from the slope S_T at T_{dt} and the slope S_c where the trajectory crosses the semicircle at T_{ct}.

$$amp_a(T_{dt}) = amp(T_n)\frac{S_c}{S_t} \quad (8.16a)$$

An additional amplitude factor that compensates for the half source-receiver offset h, the trace spacing δx, and the depth increment δz is found from:

$$amp_b = \frac{0.5\delta_x}{(2h\delta_z)^{1/2}}. \quad (8.16b)$$

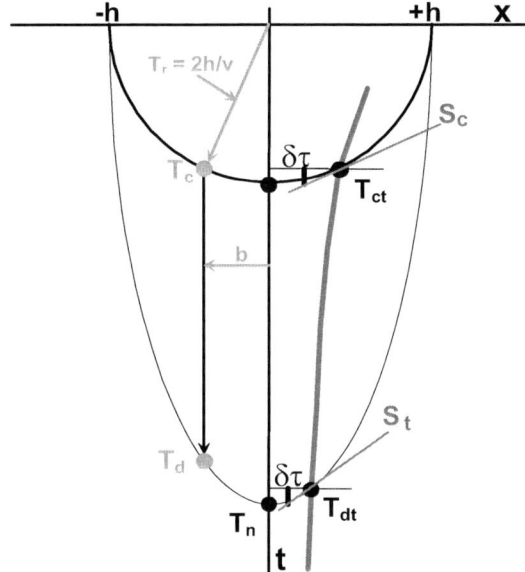

Figure 8.34 Time and amplitude calculations for the DMO ellipse.

The Kirchhoff $rj\omega$ amplitude and phase correction must also be applied with this method.

8.4.15 Radon DMO

The Radon domain may also be used to perform the DMO process (Wang [678], and [679]). Recall the constant velocity stacks with dip filters in section 8.4.5 where the energy portion of the DMO ellipse was constructed from straight lines. These straight lines may be transformed with a linear Radon transform, (also referred to as the τ–p transform), to become points in τ–p space. While in τ–p space, the amplitudes of these points may be manipulated to enhance linear reflections.

The DMO process itself may be accomplished by mapping input samples directly to curves in the τ–p space or by a <u>hyperbolic Radon transform</u>. Once the data is in τ–p space, a linear inverse τ–p transform produces the DMO'd result.

Consider Figure 8.35 in which (a) shows a number of straight lines are tangent to the DMO ellipse. For convenience, the input samples are located at x = 0, with the DMO ellipse symmetrical about the time axis. The slope of the lines define the "p" value, and their intersection with the time axis (x = 0) define the τ value. A mapping of the seven input lines from (a) are shown mapped to seven points in the τ–p space of (b). A continuum of lines, tangential to the DMO ellipse in x-t space, will produce a hyperbola in τ–p space that passes through the points as illustrated in (b).

Expressed directly, the DMO ellipse (centered at x = x_c) is defined in x-t space by

$$\frac{t^2}{T_n^2} + \frac{(x - x_c)^2}{h^2} = 1, \tag{8.17}$$

where h is half the source-receiver offset, and T_n the NMO'd time of the input sample. This <u>ellipse is mapped directly to a hyperbola</u> in τ–p space by

$$(\tau - px_c)^2 = T_n^2 + p^2 h^2, \tag{8.18}$$

A series of DMO curves from one input trace ($x_c \neq 0$) are shown mapped to hyperbolas in τ–p space of Figure 8.36. One possible method of Radon DMO would map an input trace (x = x_c) of (a) to the area covered by the hyperbola in (b).

The tangential lines in Figure 8.35 are restricted to the DMO'd energy which is confined to ± 45 degrees. In a similar manner, the "p" parameter in τ–p space will also be limited to a maximum slope of ± 1.0 as indicated by the range of points. Energy on curves for p up to ±5.0 is shown for kinematic clarity.

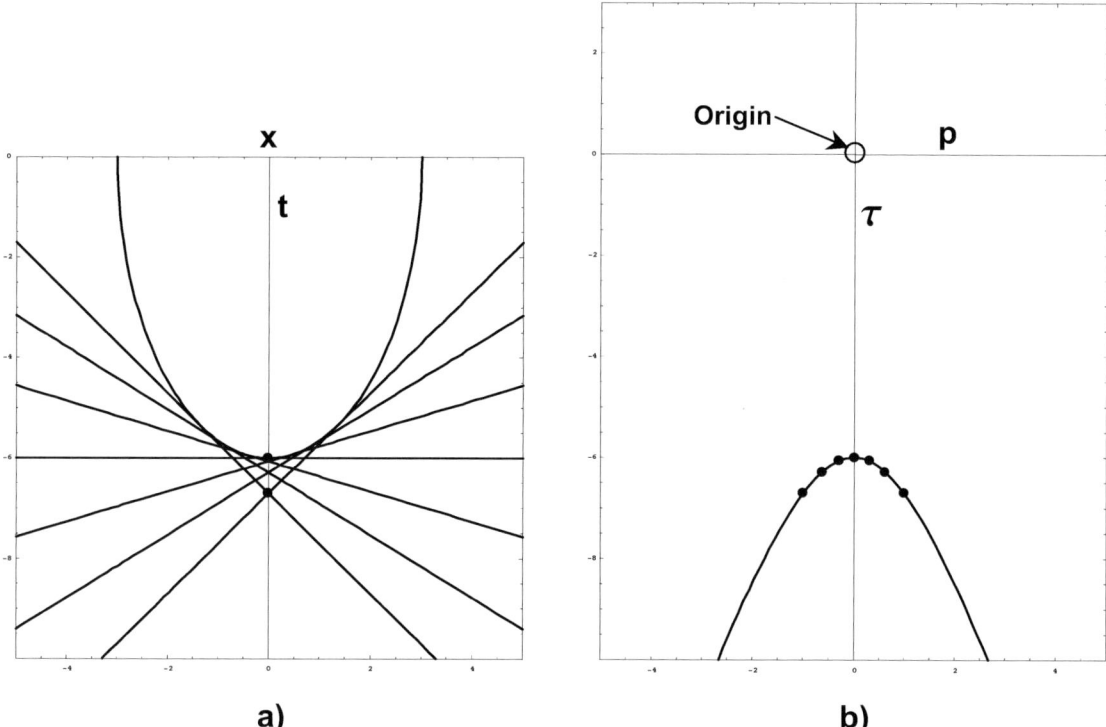

Figure 8.35 Illustration of Radon DMO by mapping in a) where lines tangent to the DMO ellipse are mapped in b) to points on a hyperbola in τ–p space.

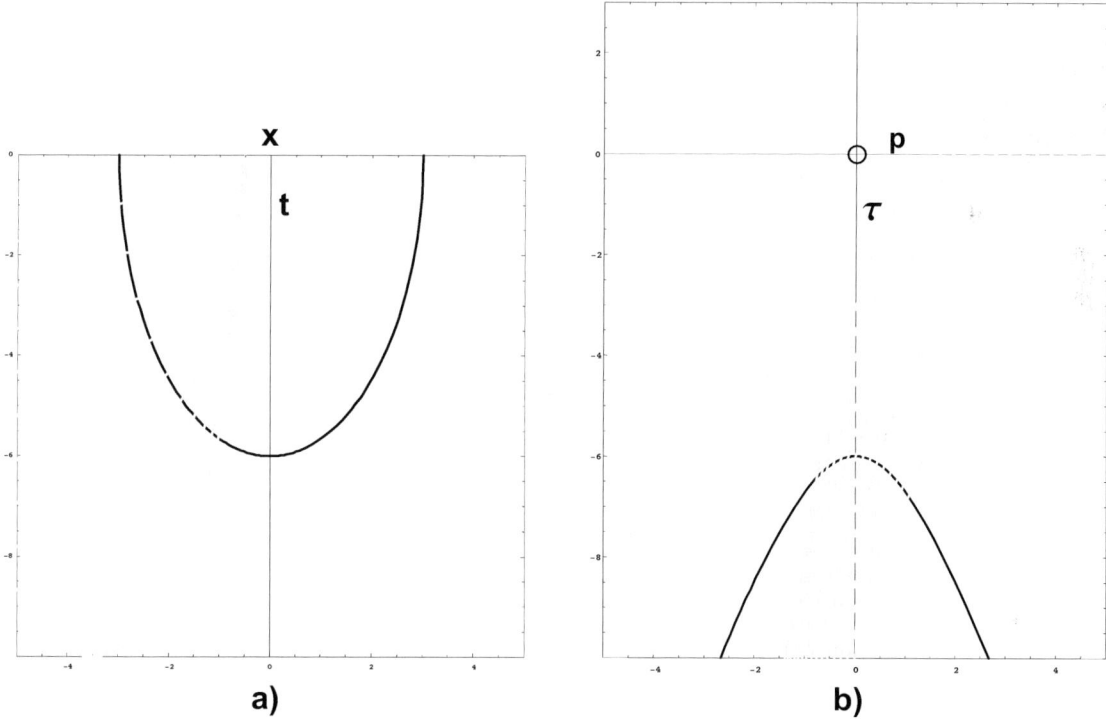

Figure 8.36 A series of a) time domain DMO ellipses mapped in b) to a series of hyperbolae in τ–p space.

An alternate view of Radon DMO is gained by considering reflections from a dipping event with a geological dip β. In a constant velocity environment, the zero-offset reflection on a time section is a linear event with dip α, where $\tan\alpha = \sin\beta$.

DMO of any <u>finite offset</u> reflection point from the dipping reflector will reconstruct energy on the zero-offset dipping event as illustrated in Figure 8.37. Shown in (a) are four finite offset samples, which have DMO ellipses that are tangent to the dipping event. In (b), all hyperbolae intersect at a common point.

Consider the following:

- all DMO ellipses with arbitrary offset will be tangent to the zero-offset event with slope p_1 and t-axis intercept τ_1,
- the dipping event is a single sample (τ_1, p_1) in τ–p space , <u>therefore</u>
- <u>all the DMO hyperbolae in τ–p space must intersect</u> the point (τ_1, p_1) as illustrated in Figure 8.37b

An interesting consequence of this property allows all offsets, with the same 3-D azimuth, to be DMO'd to the same τ–p space.

Direct computation to τ–p space using the general radon transform

Figure 8.38 contains an event with dip β in solid gray, and assuming velocity of 0.5, a zero-offset reflection with dip α in dashed gray. An offset reflection is shown by the black curve <u>below</u> the zero-offset reflection. NMO with dipping velocities, $V_{RMS}/\cos(\beta)$, will move this energy vertically to the zero-offset reflection. However, NMO with the actual velocity, V_{RMS}, will move energy to the black curve <u>above</u> the zero-offset reflection.

Using the image point of a receiver, as in Figure 7.6a, the reflection T may be shown to have a hyperbolic shape defined by

$$T^2 V^2 = 2x_c^2(1 - \cos 2\beta) + 2h^2(1 + \cos 2\beta). \tag{8.19}$$

The offset reflected energy lies only on a portion of this hyperbola as indicated in black: the portion that contains no energy is shown in gray.

After NMO with V_{rms}, the energy moves to T_n on a new hyperbola defined by

$$T_n^2 V^2 = 2x_c^2(1 - \cos 2\beta) - 2h^2(1 - \cos 2\beta). \tag{8.20}$$

This hyperbola is symmetric about the $t = 0$ axis and is used with the <u>generalized</u> Radon transform to <u>map NMO'd data directly to a point in τ–p space</u>. This point is then mapped to x-t space using an inverse <u>linear</u> Radon transform to produce the zero-offset reflection, Wang 1996 [679].

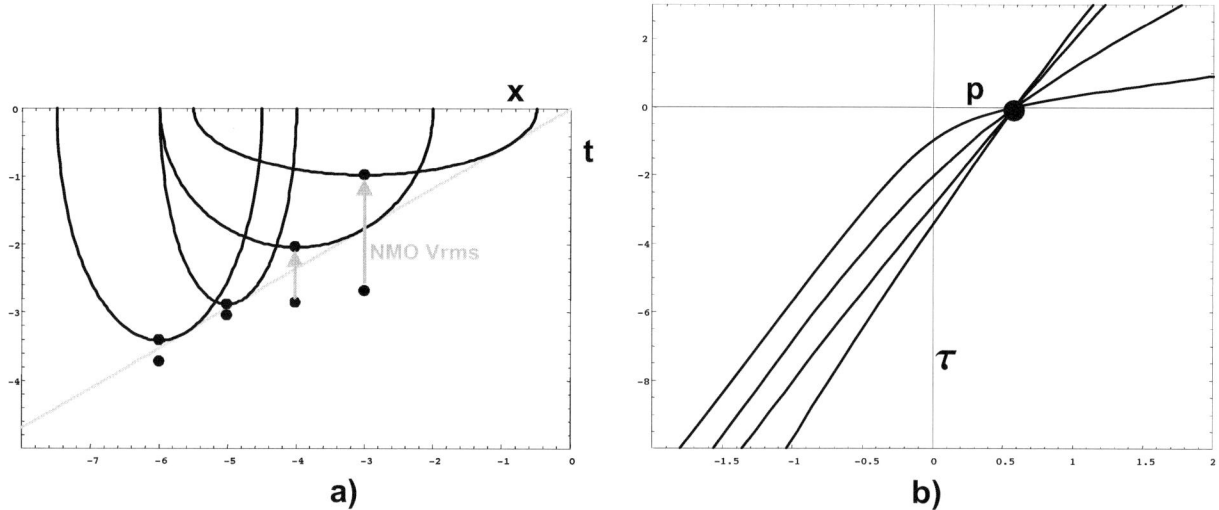

Figure 8.37 DMO energy for a dipping reflection in a) the time domain with the zero-offset reflection in gray, and b) the τ-p domain showing the DMO'd energy intersecting at the point defined by the zero-offset reflection in (a).

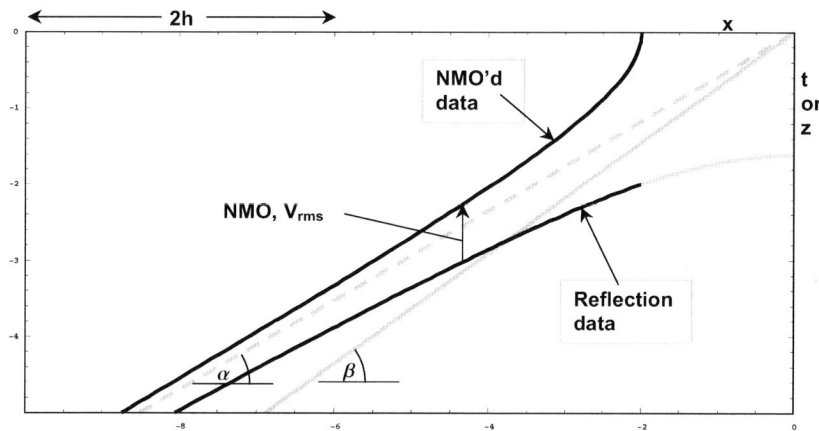

Figure 8.38 A dipping reflector (solid gray), its zero-offset reflection (dashed gray), its offset reflection (black) below the zero-offset reflection, and its NMO'd reflection (black) above the zero-offset reflection.

The <u>NMO'd hyperbola</u> in equation (8.20) is always symmetric about the $t = 0$ axis and is thus in a form suitable for mapping directly to τ–p space using a <u>generalized form of the Radon transform</u>.

The reflection hyperbola in equation (8.19), however, varies in space and time with the location of the dipping event. It is therefore <u>unsuitable</u> for use with the generalized form of the Radon transform.

> How is the generalized Radon mapping of NMO'd data to τ–p space similar to producing a semblance plot for velocity analysis?

8.4.16 DMO of source record by log-log method

Biondi [361, 369], introduced source record DMO by applying <u>log scaling to both time and offset space</u>. Applying log-log stretching makes the shape of the DMO impulse response independent of position as illustrated in Figure 8.39.

When the log-log operator is both time and space invariant, it may be <u>efficiently generated</u> by 2-D convolution in the Fourier transform domain.

Care must be taken when creating the log-log mapping of the input data to account for time and space at <u>zero values</u>, as well as prevent <u>aliasing</u>.

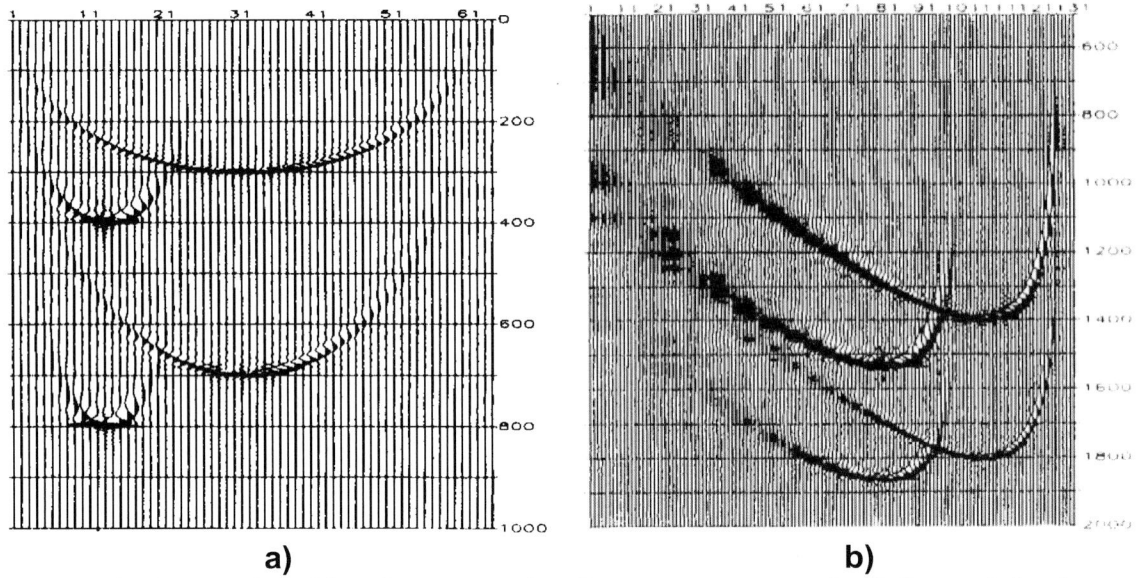

Figure 8.39 Example of log-log stretch of a source record, a) the original source (x, t), and b) the log-log mapping from Ohanian [287].

Additional material may be found in [156] and [278].

8.4.17 Introduction to Gardner's DMO

Gardner's DMO (GDMO) is <u>DMO performed before NMO</u> [111, 322]. Performing DMO first requires the DMO'd traces to be assigned (or moved to) a new offset prior to NMO. Moving the data to new offsets distributes the energy along an ellipse in a radial plane, (see Ottolini [373]).

The 3-D diagrams in Figure 8.40 illustrates:

(a) conventional DMO after NMO

(b) GDMO with new offsets on a circular cylinder, and

(c) GDMO curve lies on <u>radial planes</u> through the ($h=0$, $t=0$) axis.

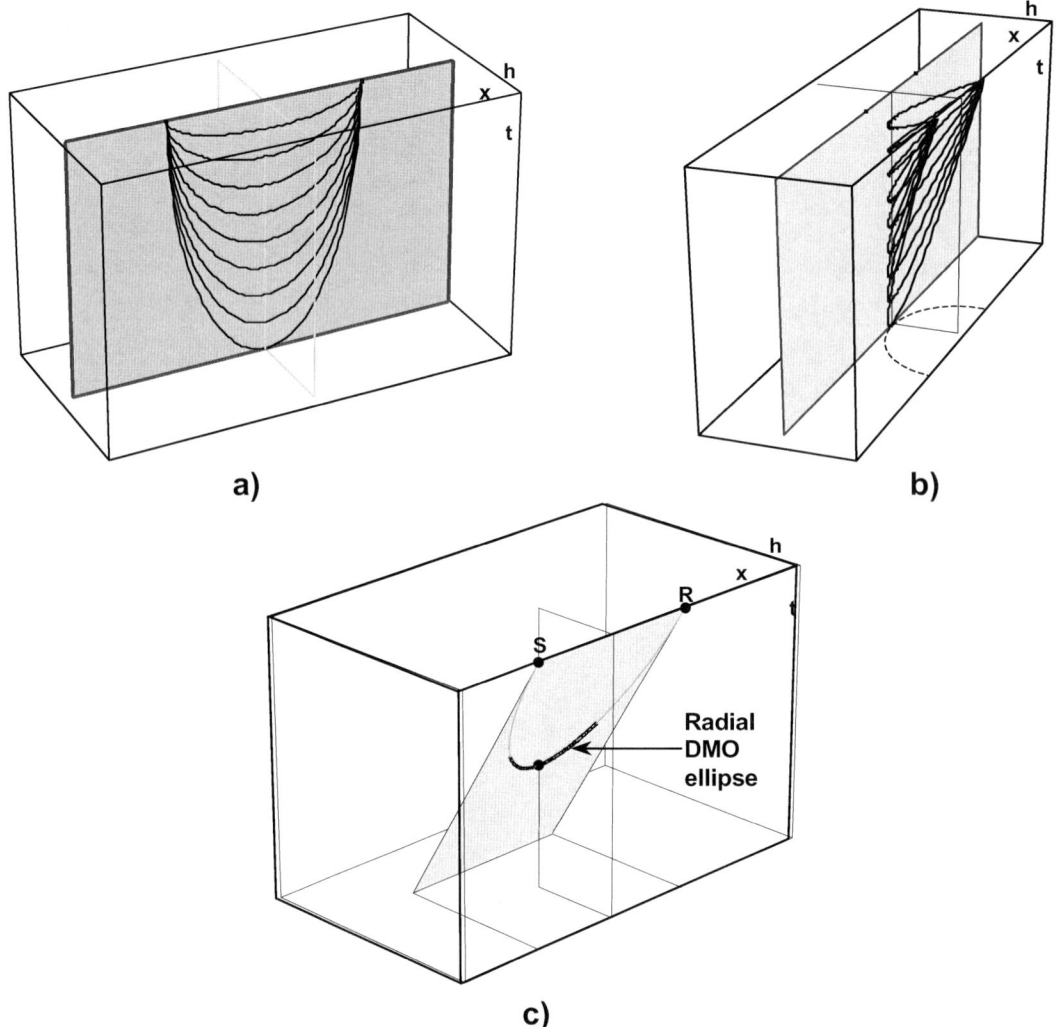

Figure 8.40 Comparison of a) conventional DMO after NMO, and b) using GDMO. Part c) shows one input sample spread along an ellipse in a radial plane with the darker portion showing the energy confined to dips less than 45 degrees.

The benefit of GDMO is visualized in Figure 8.41 where input energy from Cheops pyramid in (a) is reconstructed to the hyperboloid of (b). Compare the shape of the moveout curves on the left side of each figure:

- The moveout on Cheops pyramid is <u>non-hyperbolic</u>, and is approximated using $V_{stk} = V_{RMS}/\cos(dip)$.
- The moveout on the hyperboloid is <u>hyperbolic</u>.
- The GDMO <u>process is independent of velocity</u>.
- GDMO is based on constant velocity assumptions.

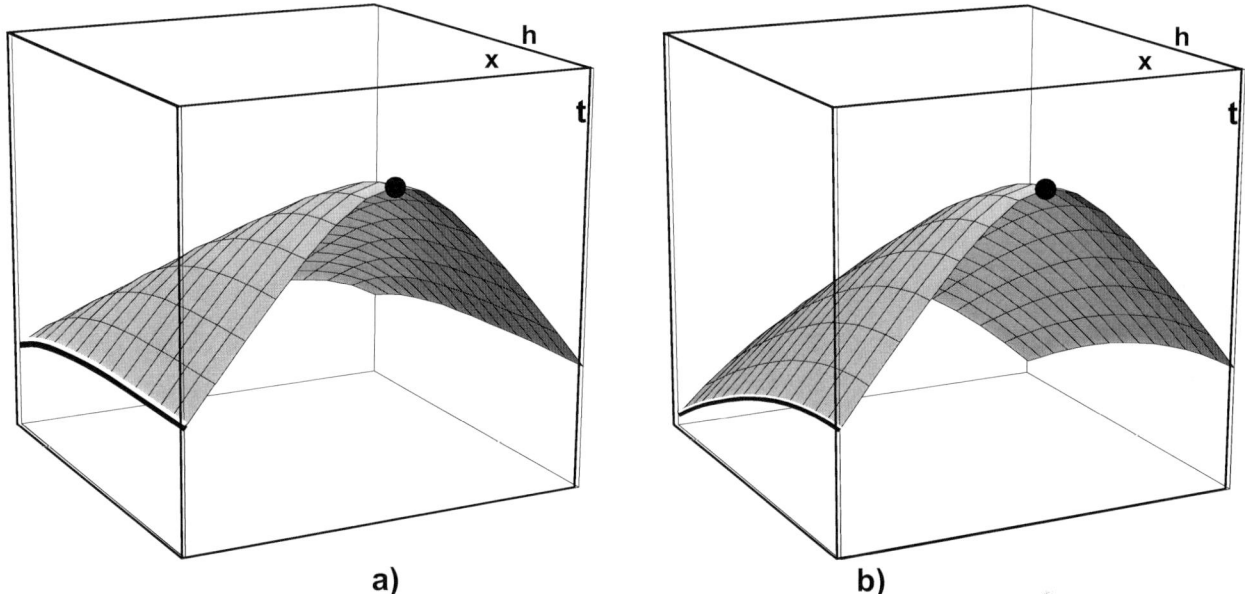

Figure 8.41 Applying GDMO to Cheops pyramid in a) reconstructs the hyperboloid in b).

NMO of the hyperboloid of **Figure 8.41**b would have the same zero-offset shape at all offsets (hyperbolic cylinder), and would stack to the same zero-offset hyperbola. See Figure 8.16c for an example.

Unfortunately, the <u>moveout velocity</u> is defined at the <u>apex of the hyperboloid</u> or scatter point and the problem of <u>conflicting dips remains</u>.

Gardner's DMO followed by PSI [322] eliminates the overlapping dip problem and produces prestack migration gathers that may be NMO'd to produce a prestack migration. See Chapters 9 and 11 for more details.

8.4.18 Derivation of modified offset (k) for Gardner's DMO

Gardner's DMO may be understood by comparing the paths shown in Figure 8.42.
- Normal processing in (a) applies NMO (T to T_n) then DMO (T_n to T_d).
- Gardner's method (b) applies DMO (T to T_g) then NMO$_k$ (T_g to T_d).
- The NMO$_k$ (T_g to T_d) requires a new offset k on the radial plane.

A. <u>The conventional NMO-DMO path</u> first gives the NMO time T_n from,

$$T_n^2 = T^2 - \frac{4h^2}{V^2}. \qquad (8.21)$$

The time T_d (after DMO) may be found using T_c on the unit circle, i.e.,

$$\frac{T_d}{T_c} = \frac{T_n}{h/V}, \quad \text{where} \quad T_c = \frac{\sqrt{h^2 - b^2}}{V}, \quad \text{giving} \quad T_d = T_n \frac{\sqrt{h^2 - b^2}}{h}. \qquad (8.22)$$

Combining NMO then DMO we get;

$$T_d^2 = \left(\frac{h^2 - b^2}{h^2}\right)\left(T^2 - \frac{4h^2}{V^2}\right). \qquad (8.23)$$

B. <u>Alternatively using Gardner's method</u>, apply DMO first,

$$T_g = T\frac{\sqrt{h^2 - b^2}}{h}, \quad \text{and NMO}_k \text{ with new offset } k, \quad T_d^2 = T_g^2 - \frac{4k^2}{V^2}, \qquad (8.24)$$

Combining these equations we get:

$$T_d^2 = \left(\frac{h^2 - b^2}{h^2}\right)\left(T^2 - \frac{4k^2}{V^2}\frac{h^2}{h^2 - b^2}\right). \qquad (8.25)$$

Equating both methods of DMO, i.e. equations (8.19) and (8.21), yields the definition of the new offset k:

$$k^2 = h^2 - b^2. \qquad (8.26)$$

These new offsets lie on a semicircle as shown in Figure 8.43.

See Section 9.4 for more information on DMO-PSI.

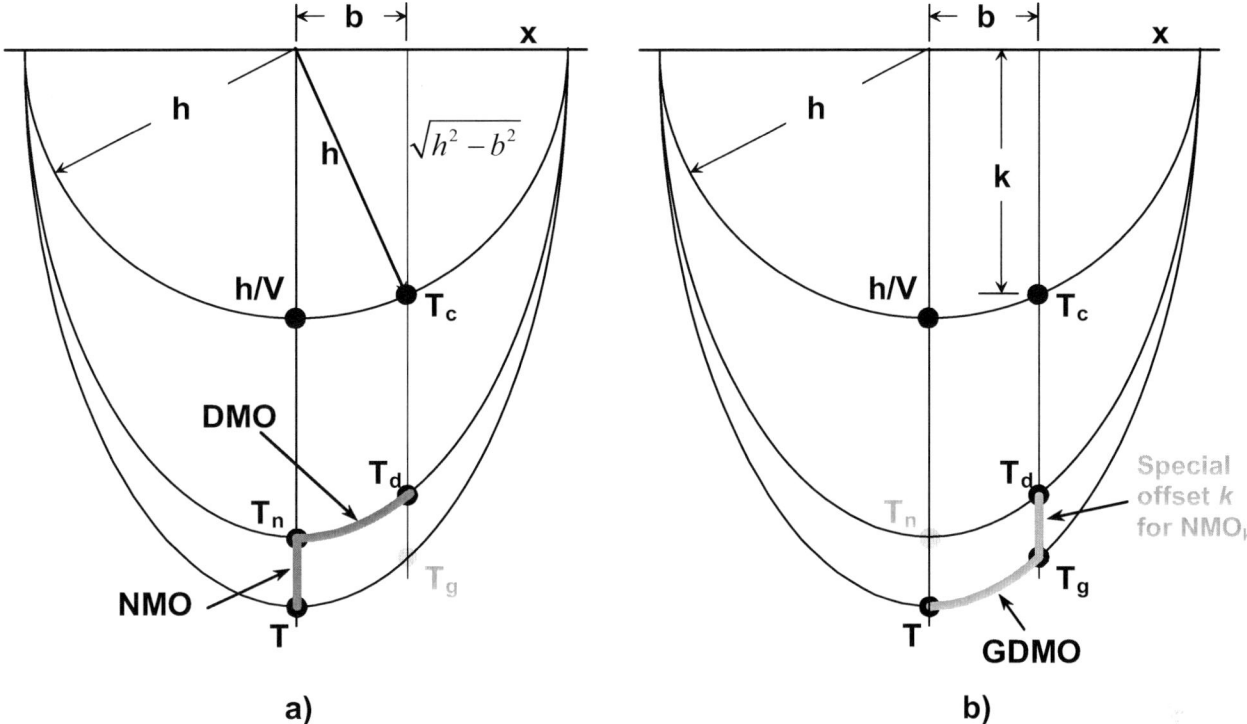

Figure 8.42 DMO ellipses for defining a) conventional NMO and DMO, and b) Gardner's method of DMO and modified NMO_k.

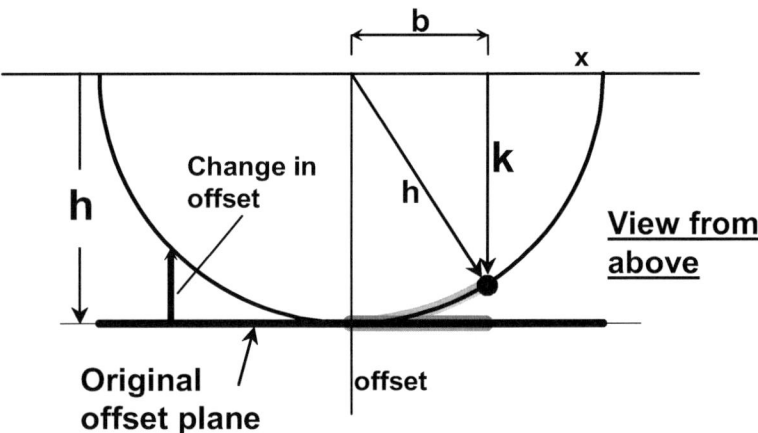

Figure 8.43 Semicircle defining new offset k to be used for NMO_k after Gardner's method of DMO. This view is the top surface of Figure 8.31b.

The kinematics of both methods, after NMO and DMO, or DMO and special NMO_k, will produce identical results when projected to zero-offset.

8.4.19 NMO-DMO-INMO (NDI); constant offset DMO before NMO

When the velocities are unknown, a <u>constant offset process</u> that combines conventional DMO with inverse NMO (INMO) can be used to obtain improved CMP gathers for velocity analysis (Yilmaz [83] p340). The NMO and INMO process uses the same guess velocity V_g. The new CMP's will tend to RMS type velocities that have dipping effects of V_{stk} reduced.

The three-step process of NMO-DMO-INMO may be combined into one process (NDI). Starting with the three equations of NMO, DMO and INMO,

NMO $\quad T^2 = T_n^2 + \dfrac{4h^2}{V_g^2}$, **DMO** $\quad T_d = \dfrac{T_n k}{h}$, and **INMO** $\quad T_I^2 = T_d^2 + \dfrac{4h^2}{V_g^2}$,

where k is defined from $k^2 = h^2 - x^2$, with the two-way time T being the input time, T_n the NMO'd time, T_d the DMO'd time, and T_I the INMO time.

Substituting the DMO time into INMO

$$T_I^2 = \frac{T_n^2 k^2}{h^2} + \frac{4h^2}{V_g^2}, \tag{8.27}$$

and then into NMO we get

$$T_I^2 = \frac{T^2 k^2}{h^2} + \frac{4x^2}{V_g^2} \tag{8.28}$$

This equation may be written in the form of <u>an ellipse</u>, i.e.,

$$\frac{t_I^2}{T^2} + x^2\left(\frac{1}{h^2} - \frac{4}{T^2 V_g^2}\right) = 1 \tag{8.29}$$

where the time variable $t_I = T_I$, the bottom of the ellipse is defined by the input time T, and the sides intersect the x-axis at x_e i.e.,

$$x_e = \left(\frac{1}{h^2} - \frac{4}{T^2 V_g^2}\right)^{-1/2} \tag{8.30}$$

These NDI ellipses are shown by the black curves in Figure 8.44.

- At large T and V, x_e tends to h, and the NDI ellipse is similar to a conventional DMO ellipse.
- At shallower times, x_e increases and tends to infinity when T tends to $2h/V$.
- <u>Energy at times less than 2h/V will be zero</u>: it is included to define the ellipses.

Ellipses from conventional DMO are also shown as gray curves in Figure 8.44a and match well at larger times. The shallower times have a better match when the DMO ellipses are shifted in time by $T_{shift} = 2h/V$, as shown in (b). A better overall fit between NDI and DMO ellipses are observed in (c) where the time shift of the DMO curves is half that of (b).

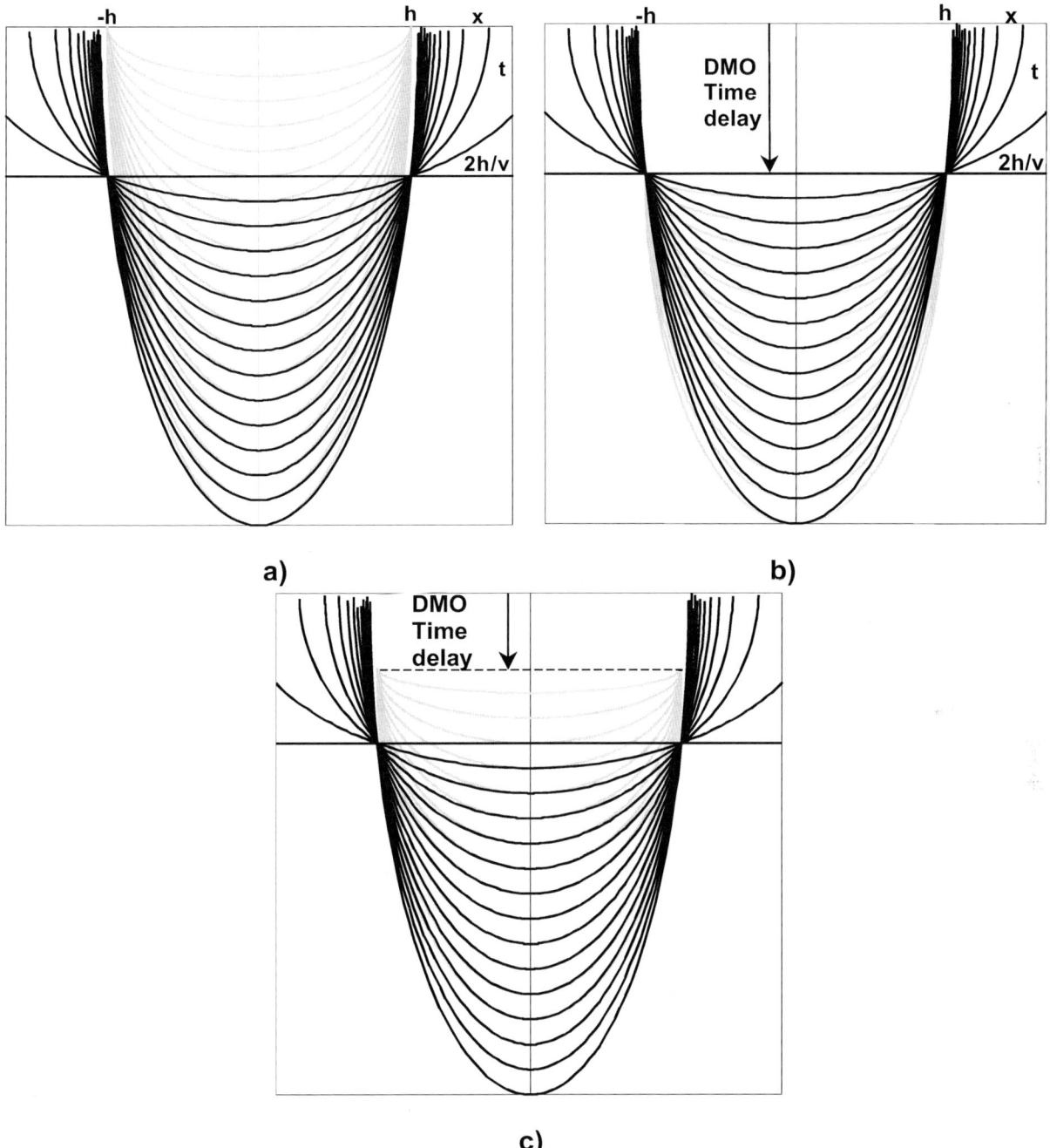

Figure 8.44 Comparison of NDI elliptical curves with a) conventional DMO, b) with a time delayed DMO of 2h/V, and c) with a DMO delay of h/V.

> The similarity of the NDI ellipses with the time shifted DMO ellipses in Figure 8.44d demonstrates the possibility of using conventional DMO to approximate NDI.

The previous examples in Figure 8.44 used the <u>exact</u> constant velocities.

NDI, however, was developed for <u>unknown velocities</u>.

It is anticipated that velocity analysis after NDI will produce improved velocities over those that were initially used.

Figure 8.45 compares the shape of NDI ellipses when the NMO and INMO velocities were (a) 90% and (c) 110% of the ideal. They are compared to the <u>DMO ellipses with a 66% time shift</u> (b), which provides a better match at shallower times. A combination of these three figures is shown in (d) where the relative time errors may be observed.

<u>Note: energy after DMO is confined to dips less than 45 degrees</u>.

The overall good fit of the NDI ellipses with the time shifted DMO ellipses in Figure 8.45d indicates NDI is fairly insensitive to velocities (<10% error), and that NDI could possibly be replaced with a more computationally efficient time shifted DMO.

> Conclusion:
>
>   Constant offset <u>DMO before NMO</u> could be <u>approximated</u> with a time shifted conventional DMO.

> Rather than vary the start time, the width of the family of curves could be varied.

> Constant offset DMO algorithms use the Fourier transform after log-stretching, and it may appear that a <u>phase-shift</u> could be used to apply the time shift. That is <u>not the case</u> as the time shifting must be performed before the log-stretching.
>
> The log-stretching could incorporate the time stretching by choosing the first sample at the time shift.

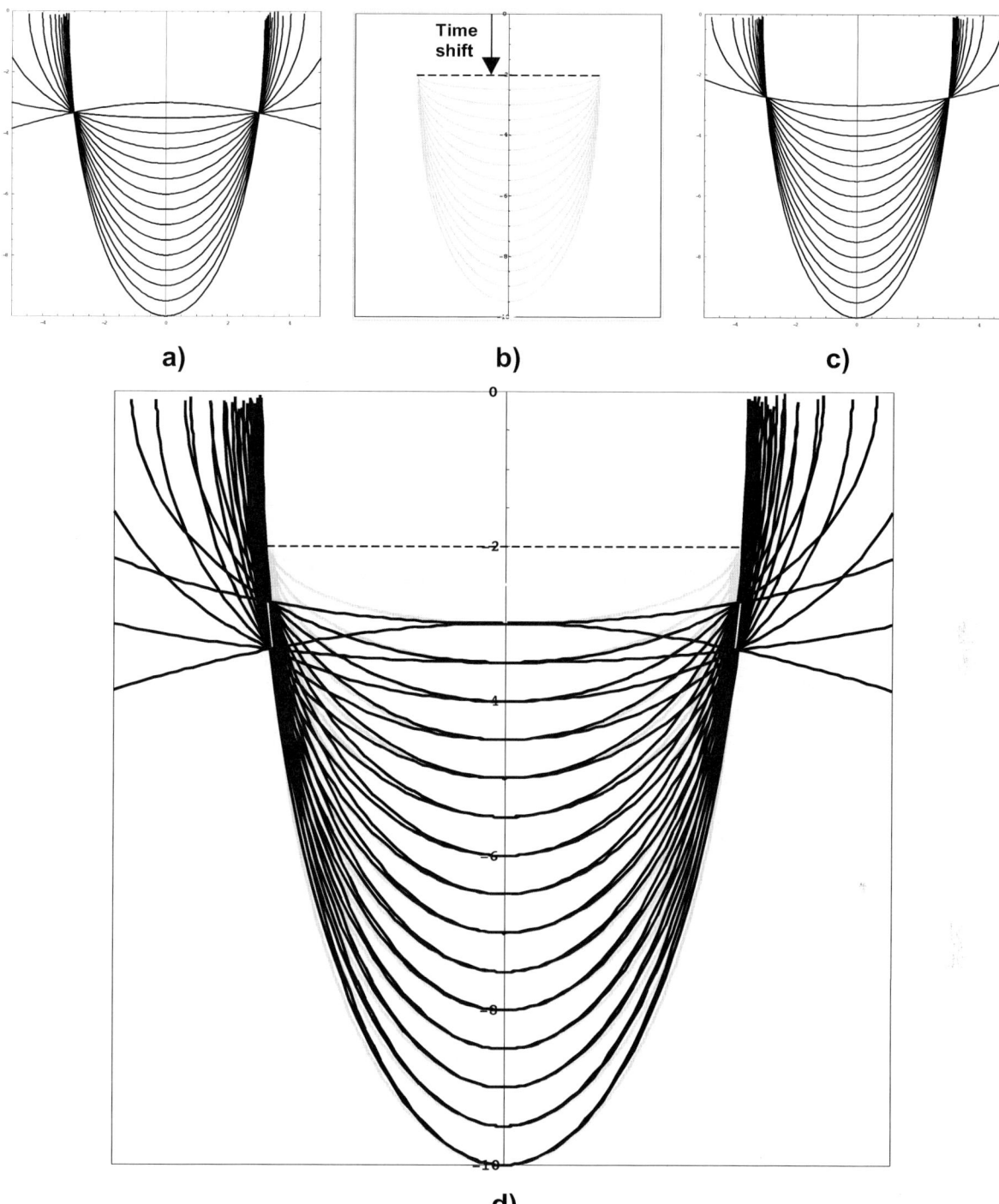

Figure 8.45 Velocity sensitivity of NDI ellipses (*x*, *t*) by comparing a) 90% velocities for NDI, b) DMO ellipses with a 66% time shift, and c) with 110% NDI velocities. All three figures are enlarged and superimposed in d).

When large dips are present, and the velocity errors are also large, NDI will produce positioning errors.

The separate kinematic steps of the NMO, DMO, and INMO are illustrated in Figure 8.46 for a velocity that is <u>30 percent too low</u>. The result is compared with ideal NMO-DMO using the correct velocity.

Figure 8.46 shows:

    (a) Ideal NMO-DMO using the correct velocity.

    (b) NMO with a velocity 30% too low, then DMO.

    (c) NDI or INMO of the DMO curve in (b), using the initial NMO velocity.

    (d) Applying NMO with the correct velocity to (c).

NDI with NMO in (d) allows comparison with the correct NMO-DMO. Errors in horizontal events will be insignificant, and new velocity estimates will tend to be more accurate. However, the deviations at larger offsets are significant, and will lead to positioning errors and incorrect velocity estimations. A number of iterations may be required to obtain an optimum velocity model.

In structured areas, errors in the initial velocities should be less than 30% to allow convergence to the RMS velocities.

Conclusions:
- NDI is suitable for <u>horizontally layered</u> media.
- NDI is velocity <u>dependent</u> as indicated by equation (8.24).
- NDI will mis-position structured data when incorrect velocities are used.
- NDI will reconstruct the energy from Cheops pyramid to the corresponding hyperboloid (using the velocity of the scatter point).
- Estimated velocities tend to a RMS type by removing dip compensation.
- A number of iterations may be required to converge to a more accurate velocity model.
- The problem of different velocities for overlapping reflections remains.
- DMO must be recomputed once an accurate velocity is established.
- When applicable, the NDI ellipses may be approximated by a time shifted conventional DMO.
- NDI may not converge in highly structured data.

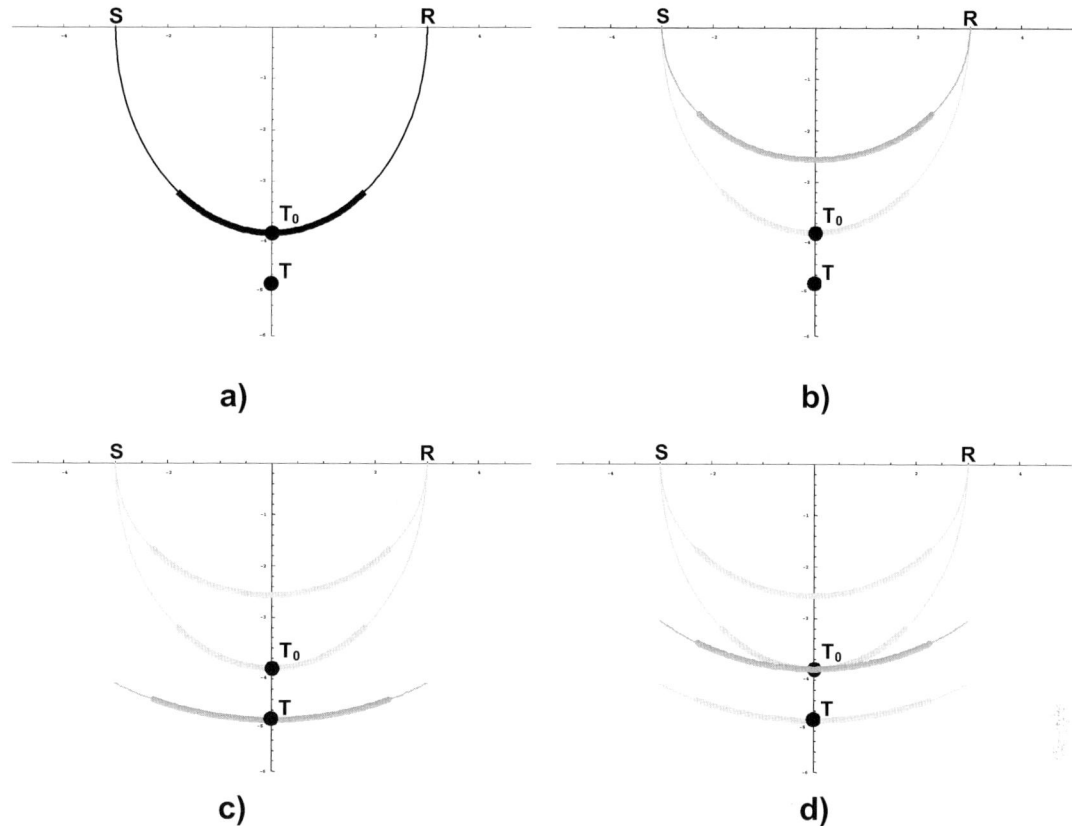

Figure 8.46 Sequential processes showing a) the correct DMO, b) NMO and DMO with velocities 30 percent too low, c) the subsequent INMO, and d) the resulting NMO with the correct velocities.

8.4.20 Summary of "Wave theoretical" methods.

- Log stretched method is an <u>efficient</u> method for a <u>number</u> of traces in a constant offset section.
- A different shaping operator is required for each offset.
- No model of dips is required.
- Independent of <u>velocity</u>.
- Inherent <u>noise filter</u> (45 degrees dip limit).
- Handles all dips.
- No dip smear, will <u>recreate</u> anomalies on a dipping event.
- Time domain methods, that match the quality of transform methods, must take special care of the <u>amplitude</u> distribution, the <u>phase</u> changes, and <u>aliasing</u> of steeper dips.
- <u>Single trace DMO</u> is very flexible and may be used for all applications such as source gathers, constant offset sections, or 3D data sets.
- Conventional DMO requires NMO first.
- <u>Gardner's DMO</u> is applied before NMO, but requires new offsets for the DMO'd traces.
- Constant offset NDI is <u>velocity dependent</u>.
- Where applicable, NDI may be approximated by a <u>time shifted conventional DMO</u>.
- Gardner's DMO and NDI reconstruct energy from <u>Cheops pyramid to an equivalent hyperboloid</u>.
- NDI DMO may be approximated by conventional DMO.
- The radon transform may be used for DMO.

8.5 Pre-Processing for DMO (and prestack migration)

8.5.1 Trace amplitudes

The amplitude of stacked sections is partially controlled by the time varying fold in CMP gathers. DMO spreads energy into neighboring traces with differing times and amplitude weightings and invalidates a simple fold computation.

DMO also requires a full and <u>even coverage</u> of all traces at all offsets for the appropriate reinforcement or cancellation. This is rarely the case in seismic data and special care must be taken to minimize problems with geometry and variations in trace amplitudes.

One method of reducing this problem is to scale the traces in each CMP gather by the time varying fold as illustrated in Figure 8.47.

- (a) illustrates a conventional CMP gather with balanced traces, a summed trace, and stacked trace.
- (b) illustrates a fold compensated CMP gather in which the amplitudes have been divided by the <u>time varying fold</u>.
- The sum of input traces in (b) is equivalent to the stacked trace in (a).

Once the fold compensation has been applied to all input traces, DMO or prestack migration may be applied with the resulting summed section equivalent in amplitude to the original stack.

Basic definitions of <u>sum</u>ming and <u>stack</u>ing relative to a gather of traces:

Sum: Adding the amplitudes of all traces at each time sample.

Stack: Adding the amplitudes of all traces at each time sample, then dividing by the time varying fold.

   or

   Average the traces in a gather.

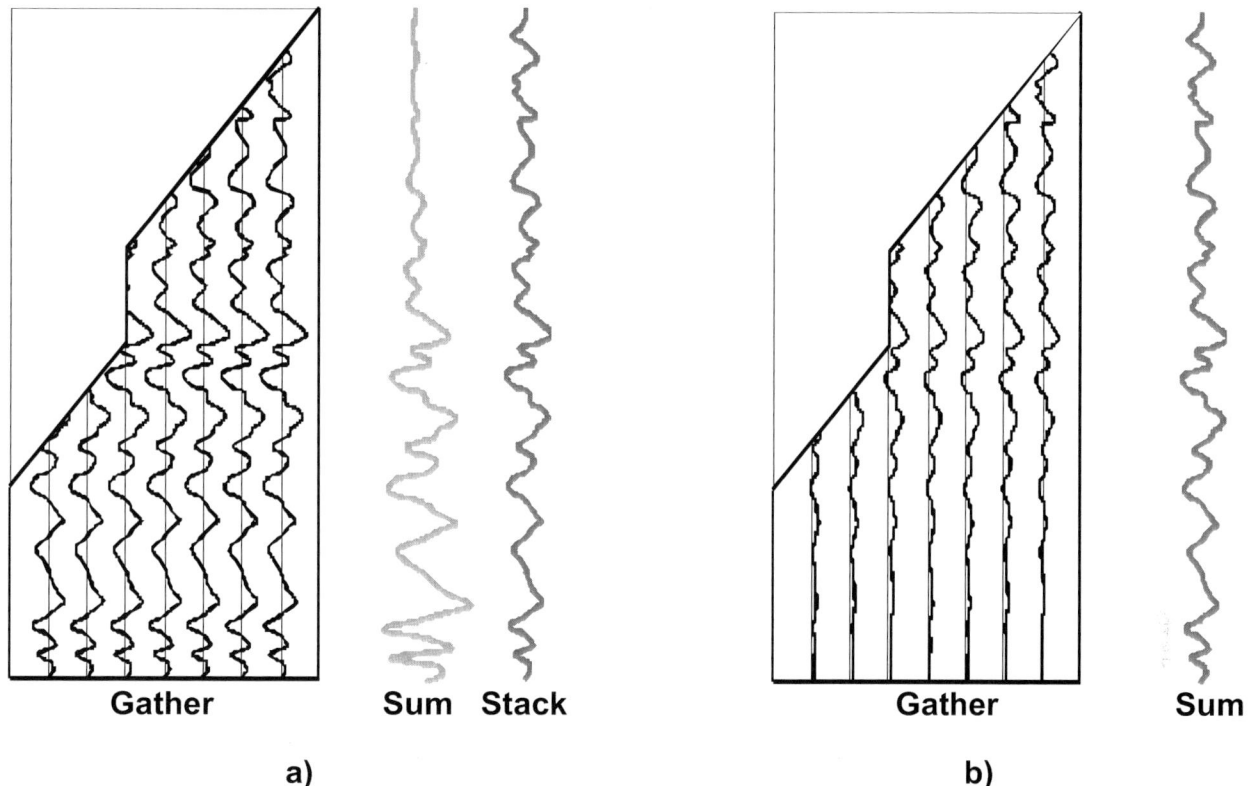

Figure 8.47 Trace stacking and summing, a) shows the conventional method, and b) scaling the traces in the CMP gather before summation for use with DMO.

8.5.2 Statics and Datums

As with most migrations, DMO and NMO should be performed from a horizontal datum that is close to the surface [see Section 6.11]. The hyperbolic assumptions of RMS velocities assume the asymptotes of the hyperbolas intersect at the surface [Section 2.1.1].

A floating datum is formed by filtering (or smoothing) the elevations with a spatial low pass filter. The width of the filter typically spans three to five acquisition spread lengths. Elevation statics shift the traces to the new floating datum. The floating datum is nearly constant over the offset range of the CMP gathers to approximate a horizontal datum.

The DMO aperture is the same as the CMP gathers, therefore DMO may be applied with the same floating datum.

The offset range of the migration aperture is much greater than that of CMP gathers, and the elevation changes of the floating datum may be too severe for the migration algorithms. Therefore, after NMO and DMO, the data should be converted to a horizontal datum, or, algorithms used that compensate for a floating datum. This may be accomplished by:

- simple time shift,
- wave equation datuming,
- zero velocity layer migration, or
- migration from surface.

After migration, the data may be time shifted to the client's datum for comparison with other seismic data in the same area.

Surface consistent statics are evaluated at each source and receiver location and have greatly improved the quality of seismic processing. These statics compensate for errors in the subsurface model and errors in the replacement velocities. They are estimated and applied before any migration step.

Trim statics are applied to CMP gathers after conventional NMO. The trim time is estimated by correlating each trace in the CMP gather with a model trace. It may be possible to apply these statics to traces before DMO.

Caution must be used in estimating trim statics before DMO or prestack migration. If using the RMS velocities (not stacking) for the NMO, the trim statics may trim on dipping events.

After DMO, or prestack migration, these reconstructed reflections in constant offset sections may not be fully formed due to sparse traces, and introduce trim static errors.

> In general, trim statics should not be applied after DMO or any migration algorithm.

> Some processors "feel" trim statics may be applied with caution after source gather DMO. These statics are not the same as those without DMO.

Statics and equivalent offset migration (EOM)

> Residual statics are estimated as part of the equivalent offset method of prestack migration (EOM). These statics are estimated with no NMO applied to the data. The CSP gathers, (see Chapter 11), provide model traces for correlation.

8.6 DMO Processing Loops

8.6.1 Method 1; velocities are known

If the RMS Velocities are known (or stacking velocities for flat events) then a direct method of processing may be used.

- Floating datum, statics, gain recovery and trace equalization, deconvolution, NMO, etc.
- DMO
- Final Stack
- Convert to horizontal datum (or migration aperture floating datum)
- Poststack migration
- Client datum correction

Floating datums

A floating datum for CMP processing requires elevations that are smoothed with a spatial operator that is 3 to 5 spread lengths. (i.e. a box car filter)

Elevation on the floating datum will be approximately linear over a CMP gather.

DMO is bound by the acquisition aperture and may be applied to data at the floating datum used for standard processing.

The migration aperture is usually much larger than the spread length.

A floating datum for migration would require a spatial filter that is 3 to 5 migration apertures.

The elevation changes on the migration aperture will be approximately linear and the migration will be as close to the surface as possible.

Wave-equation datuming may be required to convert a processing floating datum to a suitable migration datum.

(Some floating datums are computed by spatially filtering statics.)

8.6.2 Method 2; approximate velocities

DMO was first introduced as a processing step before migration. It soon became apparent that it had other benefits, especially for <u>velocity analysis</u>.

When DMO is used with <u>velocity analysis</u>,
- the problem of <u>dip-dependent velocities</u> was removed, and
- the estimated <u>velocities are of an RMS type</u> which are <u>suited for migration and inversion</u>.

When the <u>stacking velocities for flat events (RMS type) are approximately known</u> (velocity error < 10%), then it is easier to estimate more accurate velocities after some form of DMO i.e.,

- Floating datum, statics, gain recovery and trace equalization, deconvolution, NMO, etc.
- DMO (in constant-offset sections)
- <u>Inverse NMO</u> (INMO) with previous NMO velocities, (in CMP gathers)
- Pick <u>new velocities</u>
- NMO with new velocities on same data
- Final Stack
- Convert to horizontal datum
- Poststack migration
- Client datum correction

8.6.3 Method 3; Using a DMO loop to estimate velocities

This method is similar to that proposed by Deregowski in his excellent paper "What is DMO?" [76]. It is often referred to as the <u>Deregowski loop</u>. This method is used when there are large errors in the initial velocity model, however, if they are too large, then the estimated velocities may not converge.

Starting with the best available velocities and statics:
- Floating datum, statics, gain recovery and trace equalization, deconvolution, etc.
- Loop start
    - NMO (on INMO data or original data)
    - DMO (on source gathers, or constant offset sections)
    - INMO CMP gathers
    - Pick new velocities (and statics where appropriate)
- End loop
- NMO with new velocities on original input data
- DMO (direct to stack)
- Convert to horizontal datum
- Poststack migration
- Client datum correction

Yilmaz text "Seismic Data Processing" [83], pages 336 and 338, have excellent <u>examples</u> on the power of DMO to estimate velocities using the INMO step.

When using DMO, the initial NMO velocities must be reasonably close to the desired RMS type velocities to allow <u>velocity convergence</u>.

Some form of time or depth migration may be included after DMO, and before INMO, to aid in estimating velocities which are defined in time or depth.

These methods are often referred to as prestack migration.

8.6.4 Method 4; Processing with Gardner's DMO

The main benefit of using Gardner's DMO (GDMO) is the elimination of the velocity estimation loop. Once the data has been GDMO'd, and sorted into the new offsets (defined by k) one velocity analysis session provides accurate RMS velocities.

Starting with the best available velocities and statics:

- Floating datum, statics, gain recovery, trace equalization, deconvolution, etc.
- (Convert to horizontal datum)
- <u>GDMO</u> and move traces to new offset location
- Velocity analysis (only once)
- NMO with new velocities on DMO'd data
- (Convert to horizontal datum)
- Poststack Migration
- Client datum correction

This method, along with the previous methods, resolve the dip-dependent velocity problem, however, the <u>conflicting dip problem remains</u>. The difference in conflicting dip velocities is reduced (much was due to different dips), however, the difference in the RMS velocities at the geological locations still remains.

The conflicting RMS velocities are resolved with prestack migration.

Gardner's DMO combined with another process, PSI (prestack imaging), provides the basis for a prestack migration [322], and is discussed in Chapter 9.

8.7 Real world DMO

The DMO process provides a <u>great foundation</u> for improving the quality of a seismic section and improving velocity estimation. This improvement has been shown many times in the <u>literature</u>, <u>but may not be seen</u> in sections about an office. The problem has often been referred to as the big promise.

Consider three theoretical sections, A, B, and C.

A stack: This stack represents the <u>best possible</u> conventional processing (no DMO) where dips were stacked with dipping velocities.

B stack: This stack may be produced as part of the DMO process after NMO has been applied with <u>RMS velocities</u> (not dip compensated). A stack of this data prior to DMO will be inferior, especially for the dipping data.

C stack: Full DMO processing.

- Initial enthusiasm "honestly" compared B stack with C stack (i.e. only include DMO) and the apparent improvement was usually spectacular
- A <u>more realistic comparison</u> should have been A stack with C stack.

When problems with DMO still remain, consider the following:
- inaccurate velocities,
- use of stacking velocities which include dip compensation,
- datum not close to surface,
- offset range too large,
- presence of sideswipe or oblique reflections,
- need for a depth DMO,
- data may be too complex and a full prestack migration must be used,
- 3-D data is required, or
- Large velocity contrasts.

Sufficient time and effort must be extended to accurately define the velocities for DMO. (I have seen too many examples where processors have used stacking velocities in a processing flow described in Method 1).

The DMO described so far has been based on constant velocities, and is applied to areas where the velocity function is generally quite smooth, (also successfully in areas with moderate velocity variations).

In areas where the velocity is too complex, a depth DMO may be required. Numerous papers address this issue, such as [336] and [333]. Paper [333] shows substantial improvement in turning wave migration when using a depth variable DMO.

In some areas with complex geology, DMO may actually harm the data and produce a section inferior to a conventional NMO and stack. Prestack migration is required in these areas.

8.8 DMO Examples on Modelled Data

The examples on the following pages show the improvement that DMO has on constant offset sections.

Figure 8.48 and Figure 8.49 compare NMO'd sections with NMO-DMO'd sections at short and far offsets. DMO has little effect on the short offsets, but significant effects on the large offsets in Figure 8.49. Note the change in diffraction shape, and how the linear reflection becomes linear with DMO.

The two stacked sections in Figure 8.50 show the effect of NMO with velocities chosen for (a) horizontal and (b) dipping event. Note the sections only stack energy at the dips at which the NMO was designed. Some CMP gathers for these sections are shown Figure 8.50 at a spatial location similar to the data in the sections. Energy in the CMP gathers are only suitable for stacking flat events in Figure 8.51a and only for the dipping events in Figure 8.51b.

Figure 8.52 and Figure 8.53 show a similar section and CMP gathers after the application of DMO. Note the improved stacked image, and that the CMP gathers have energy that tends toward the horizontal at all locations.

Figure 8.54 contains a poststack migration of the DMO'd data in Figure 8.52 and may be compared with a prestack migration in Chapter 9.

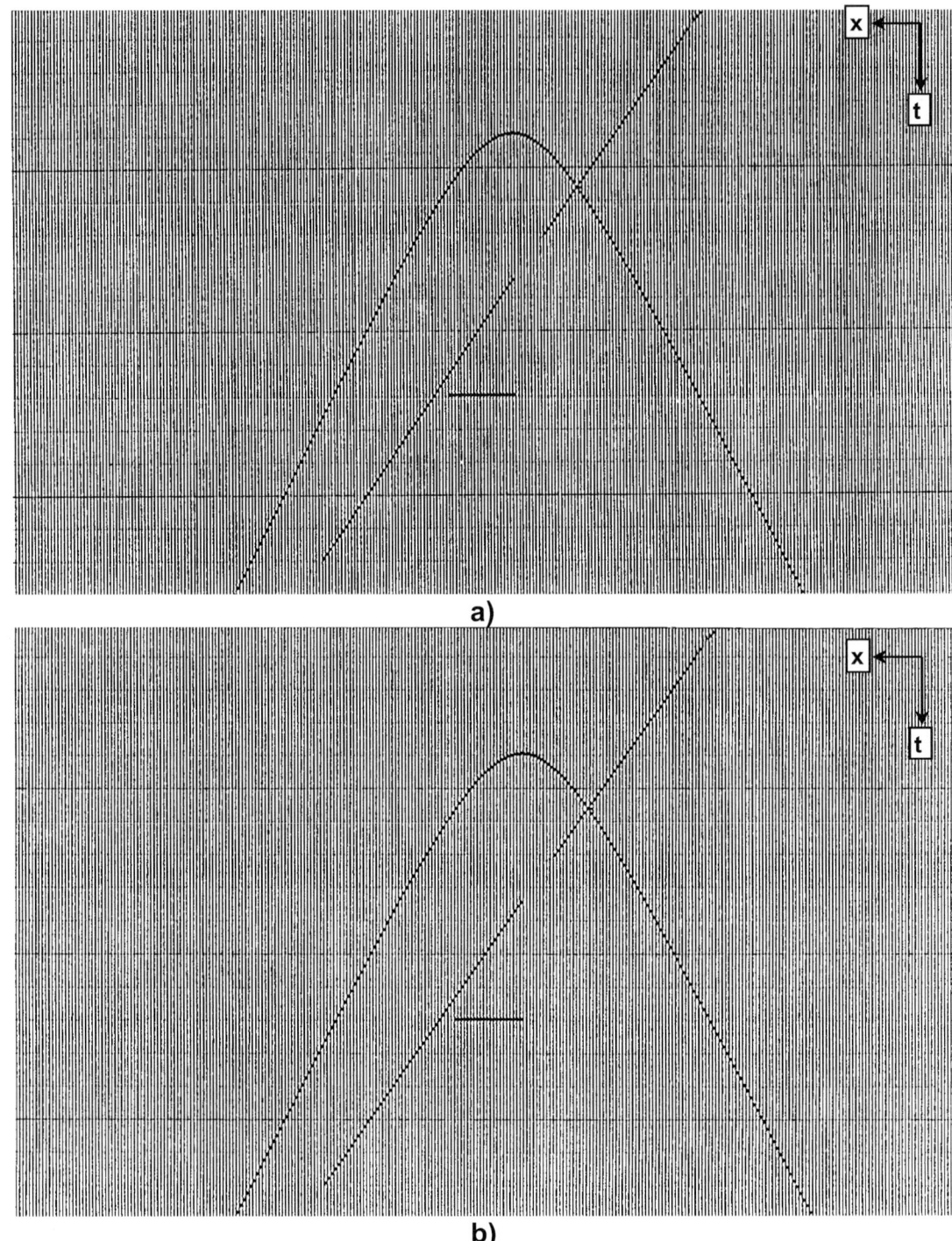

Figure 8.48 Limited offset stacks (x, $h = h_{no}$, t) for <u>near offset</u>, a) after NMO only has been applied, b) after NMO and DMO. Negligible changes are observed.

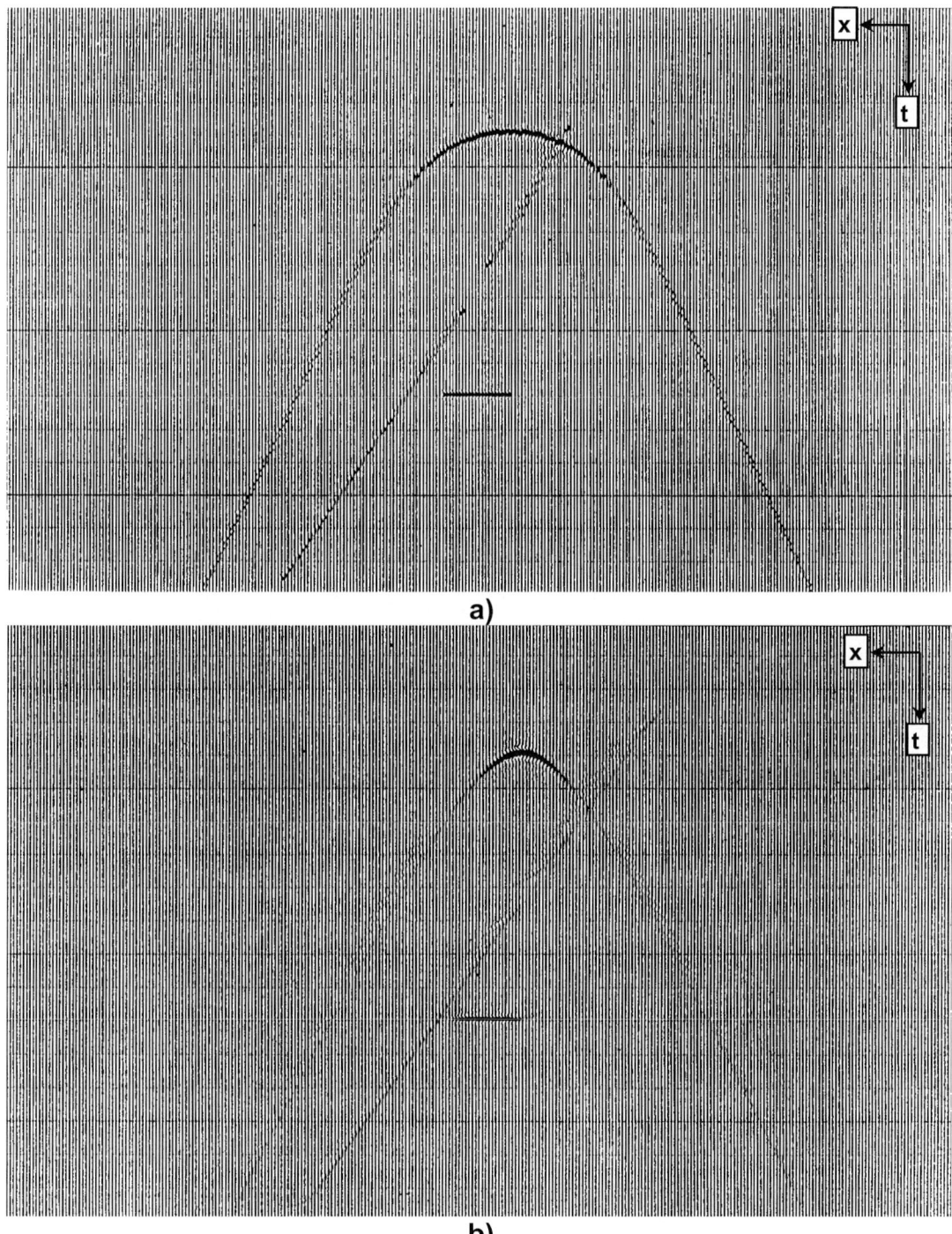

Figure 8.49 Limited offset stack(x, $h = h_{fo}$, t) for <u>far offset</u>. a) after NMO only has been applied, b) after NMO and DMO have been applied. Note change in diffraction shape, and movement of the gap along the dip.

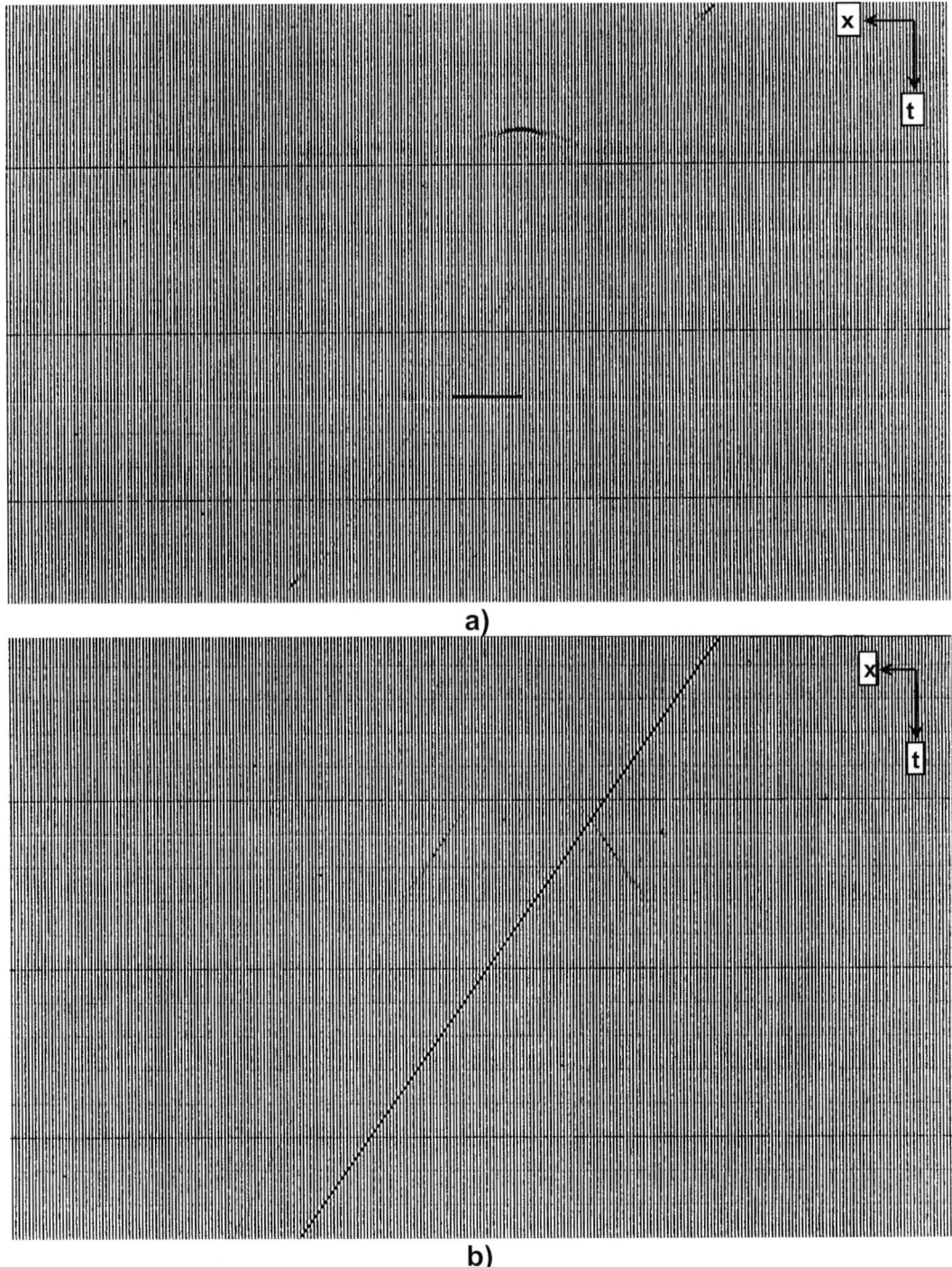

Figure 8.50 Sections (*x, t*) of model which a) shows stack with NMO velocity for flat event, b) shows stack with NMO velocity for the dipping event.

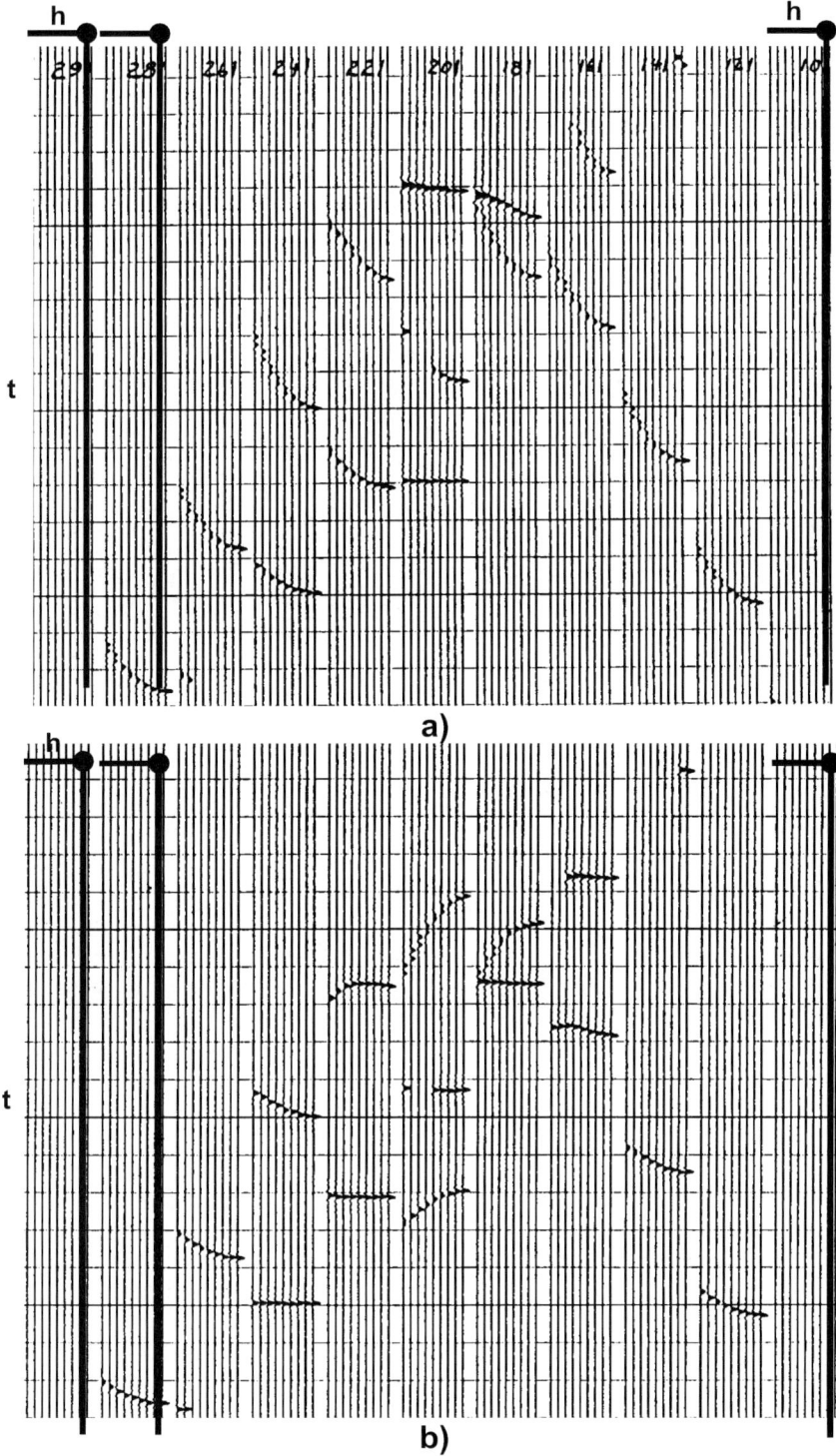

Figure 8.51 Eleven CMP gathers at incremental locations across the sections in Figure 8.48, showing a) NMO correction for a horizontal event, and b) NMO correction for a dipping event.

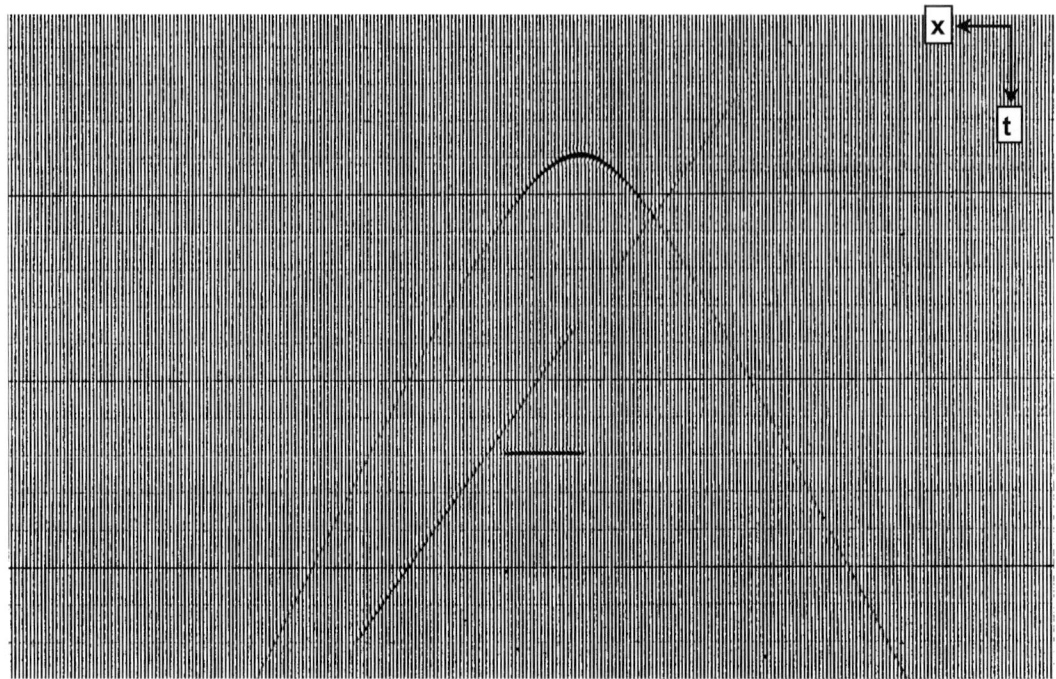

Figure 8.52 A stacked section (x, t) after DMO.

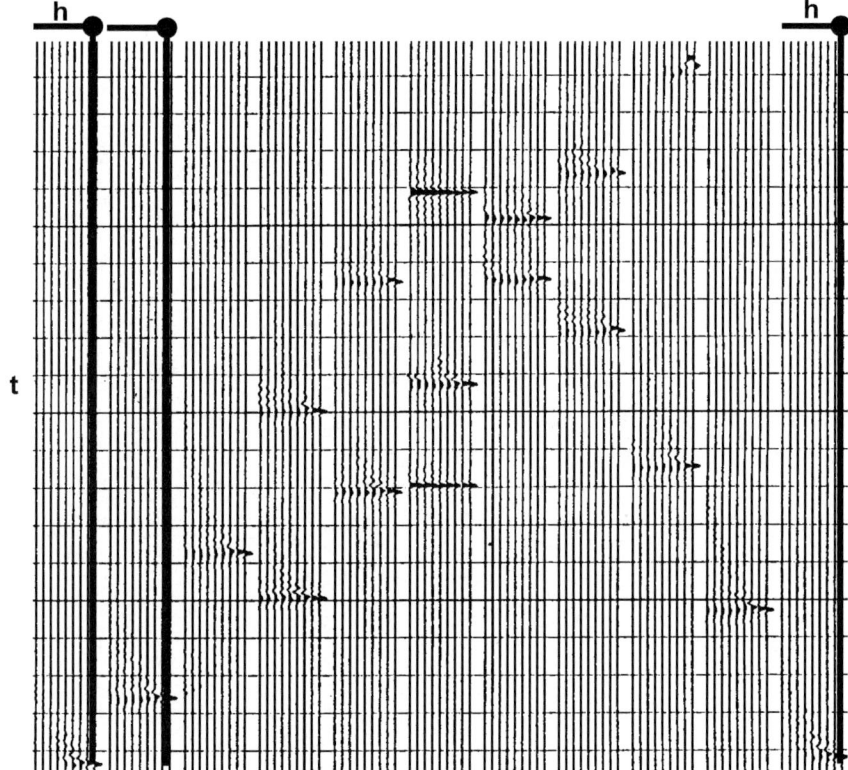

Figure 8.53 DMO'd CMP gathers at incremental locations that correspond to Figure 8.51.

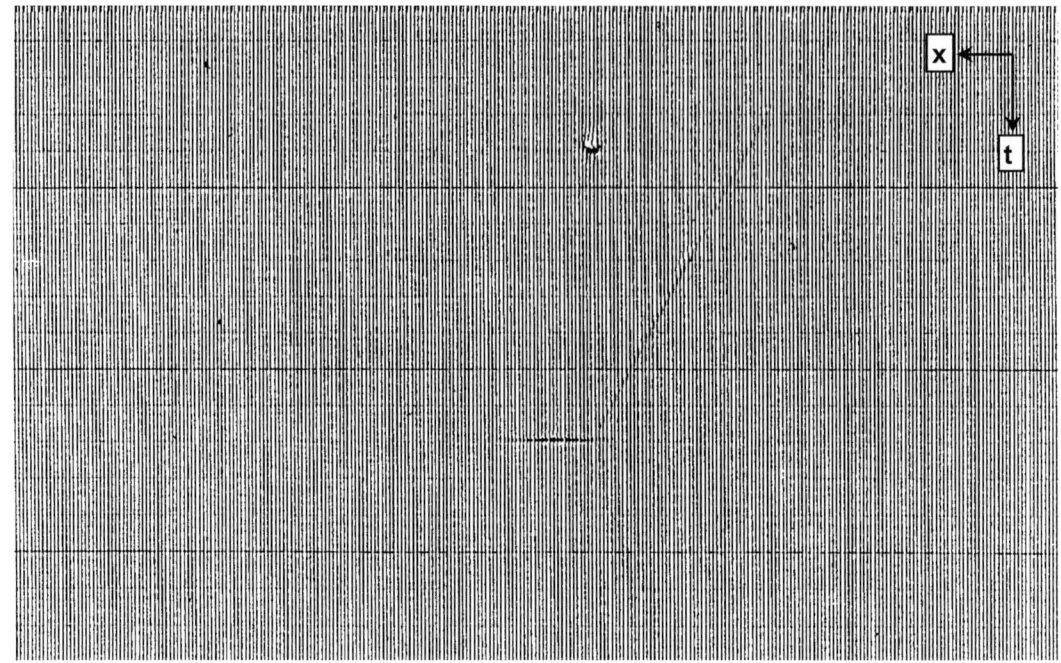

Figure 8.54 A poststack migration (*x, t*) of the DMO'd section in Figure 8.51.

The seismic model did not include diffraction of the ends of each reflector, or in the gap. Migration effects can be observed at these ends, and the gap has appeared to fill in the missing data.

Figure 8.53 may be compared with other migrated sections in Section 9.2.3.

8.9 Kinematics of DMO on Cheop's pyramids

This section compares the effect of DMO on three scatter points in a constant and variable velocity environment. The figures below show the location of the three scatter points in the prestack volume. The locations of the scatter points were chosen to have the <u>same zero-offset traveltime</u> at a CMP located at the left of the prestack volume. One scatter point is also located at this position. Figure 8.55 shows raypaths and traveltimes for (a) zero-offset, and (b) finite offset.

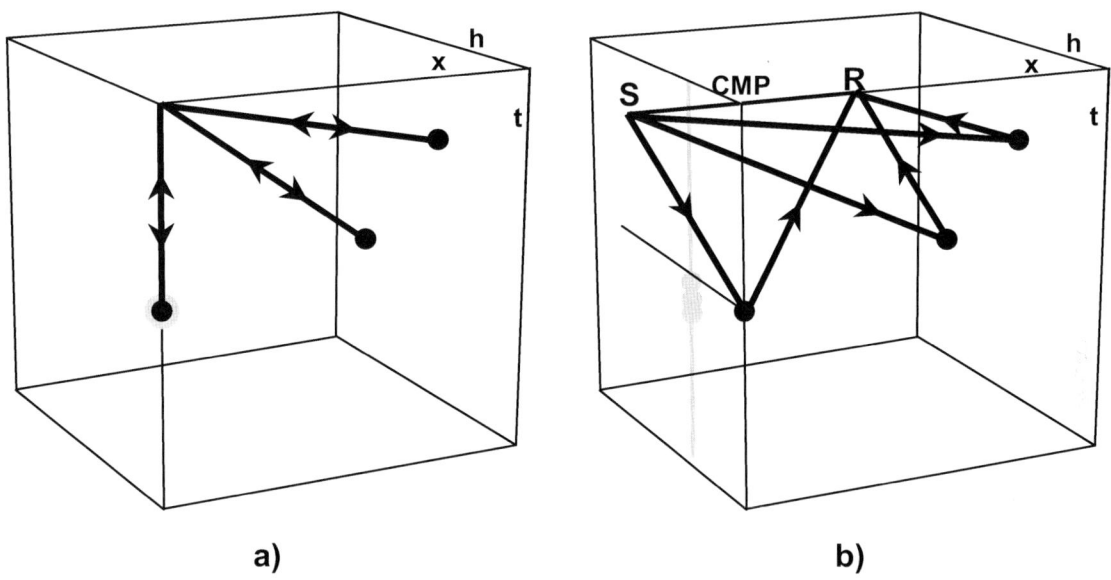

Figure 8.55 Prestack volumes showing the location of the scatter points and raypaths for a) zero-offset and b) finite offset.

The following Figure 8.56 and Figure 8.57 compare the kinematic differences between the constant and variable velocity effects on the above three scatter points.

We imply, in these notes, that NMO-DMO is exactly the same as GDMO-NMO. The kinematics to final stack are identical. However, the distribution of energy in the prestack volume before stack is slightly different. Amplitude scaling applied to the data in the prestack volume may produce slightly different results after stack.

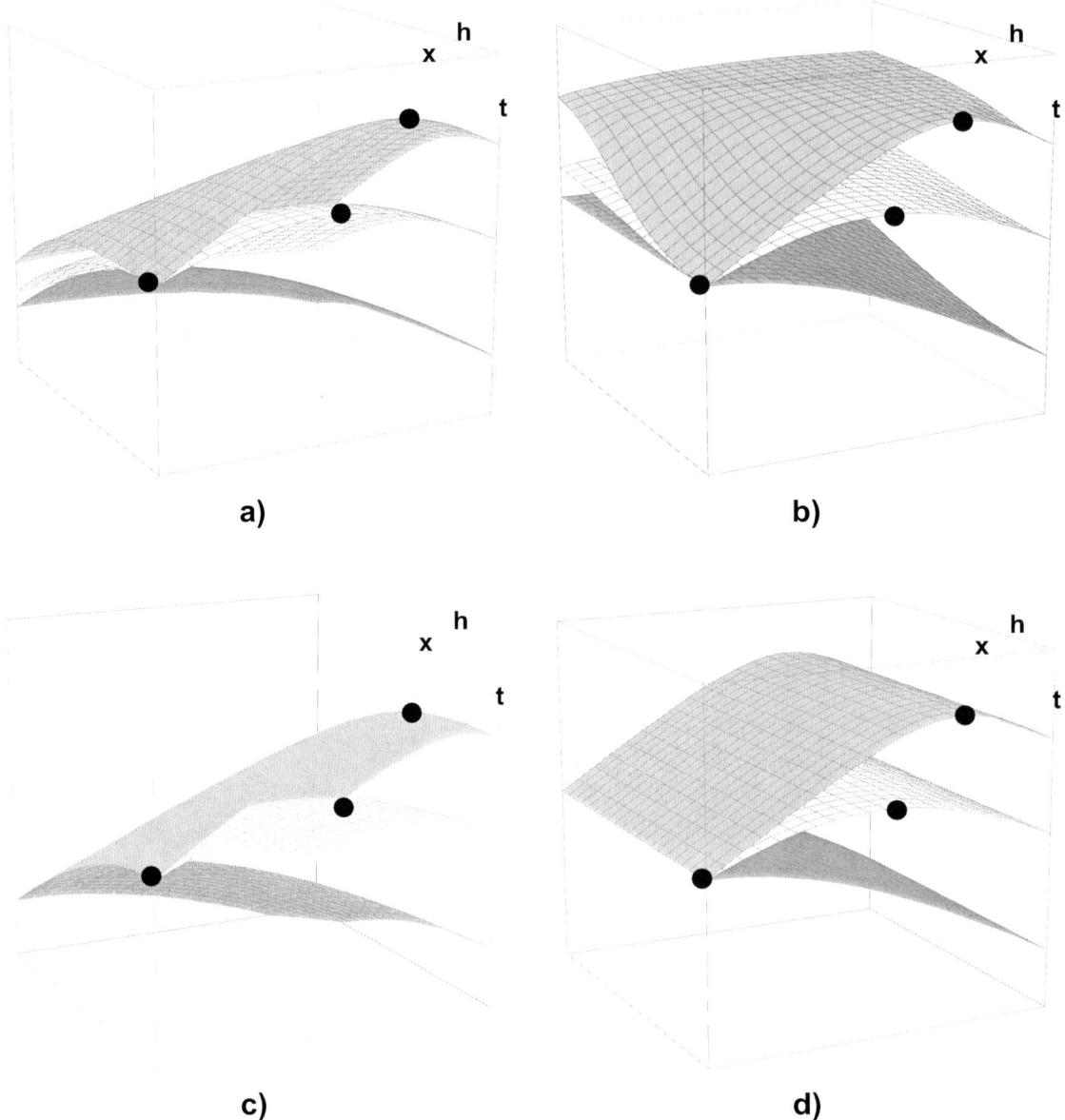

Figure 8.56 Three scatterpoints in a <u>constant velocity model</u>, with a) the Cheops pyramids, b) with NMO, c) with GDMO, and d) with either NMO-DMO, or GDMO-NMO$_k$.

Note:

- **NMO-DMO = GDMO-NMO$_k$.**
- **NMO-DMO results in three hyperbolic cylinders.**
- **A perfect stack should be obtained.**

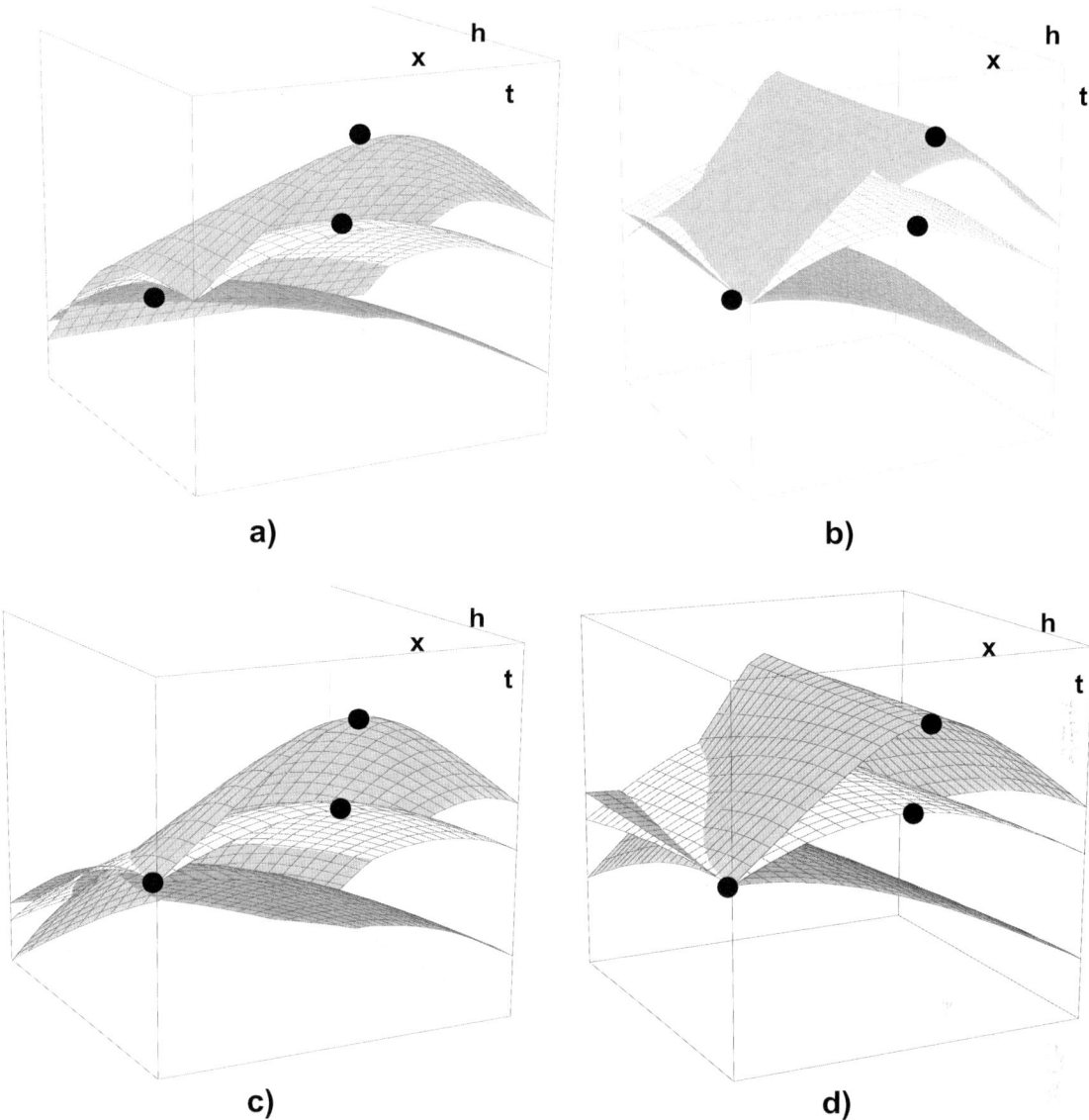

Figure 8.57 Three scatterpoints in a <u>variable velocity</u> model $V=V_0(1+0.5t)$, with a) the Cheop's pyramids, b) with NMO, c) with GDMO, and d) with either NMO-DMO, or GDMO-NMO$_k$.

Note:

- NMO is correct at each scatter point.
- GDMO still forms <u>hyperboloids</u> for each scatter point.
- NMO-DMO or GDMO-NMO$_k$ does not produce the desired surfaces.
- DMO may possibly produce <u>worse results</u> than NMO alone.
- Will any process fix the variable velocity problem? (Yes, prestack migration).

8.10 Summary of Points to Note in Chapter 8

- Stacking is only valid for <u>flat data</u>.
- <u>Diffractions</u> don't stack.
- Constant velocity NMO, DMO, and post stack migration, is the <u>same</u> as prestack migration.
- There are <u>many processes</u> that use the name DMO.
- <u>Prestack partial migration</u> (PSPM) may be used to refer to wave equation DMO methods.
- When the DMO process is applied to a constant offset section, the resulting DMO section will be equivalent to a <u>zero-offset section</u>.
- <u>Time domain</u> DMO is fast and efficient (for a single input trace).
- <u>Log-stretch methods</u> will be faster for gathers of traces (i.e. source or constant offset).
- DMO algorithms that handle variable velocities are available.
- DMO <u>does</u> resolve the problem of <u>dip compensated velocities</u>.
- DMO does <u>not</u> resolve <u>conflicting dips</u>.
- DMO has a <u>smaller aperture</u> than migration and may be applied with a <u>floating datum</u>.
- <u>Two velocity functions</u> are required: one for NMO and another for poststack migration.
- Gardner's DMO is done before NMO correction, independent of velocities, each DMO'ed trace has a new offset.
- NDI is a constant offset DMO before NMO correction, slight velocity correction

---

The term migration to zero-offset (MZO) is often used for the combination of

NMO, DMO, and stack,

GDMO, NMO_k, and stack,

NDI, NMO and stack.

Chapter 8 2-D Dip Moveout (DMO)

Removable page

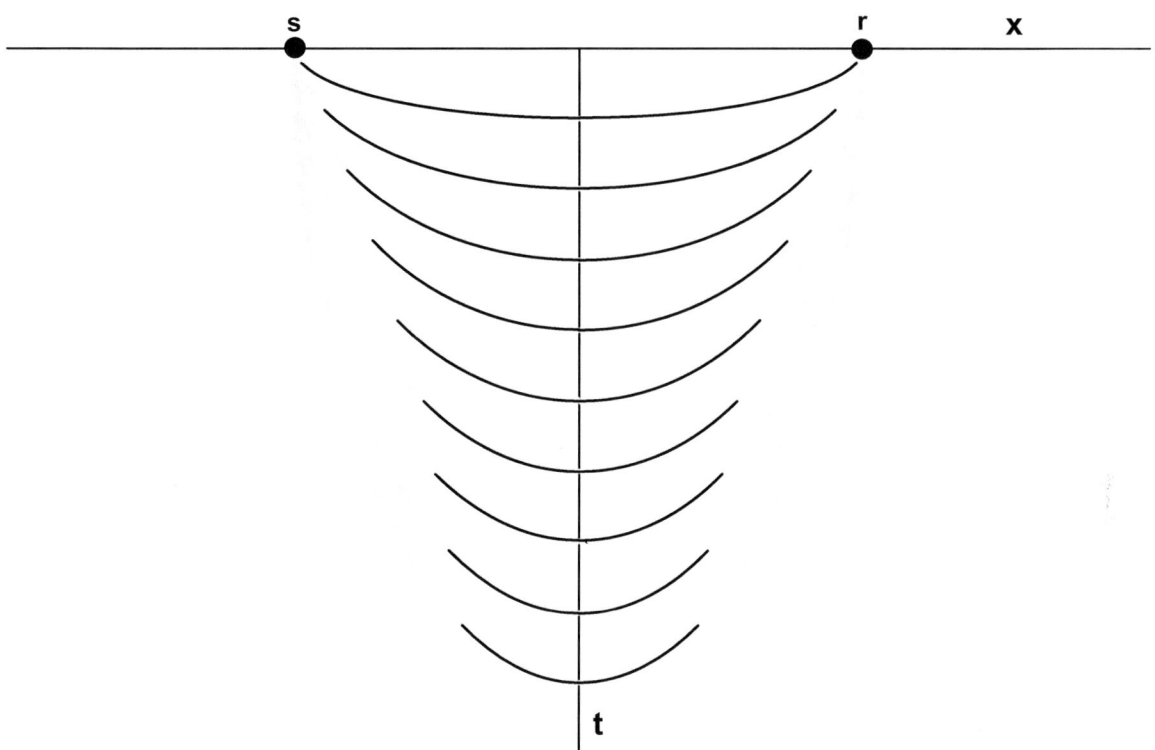

Figure 8.58 DMO curves for construction of Figure 8.16 on following page.

Left blank for removable page (tracing curves in Figure 8.16).

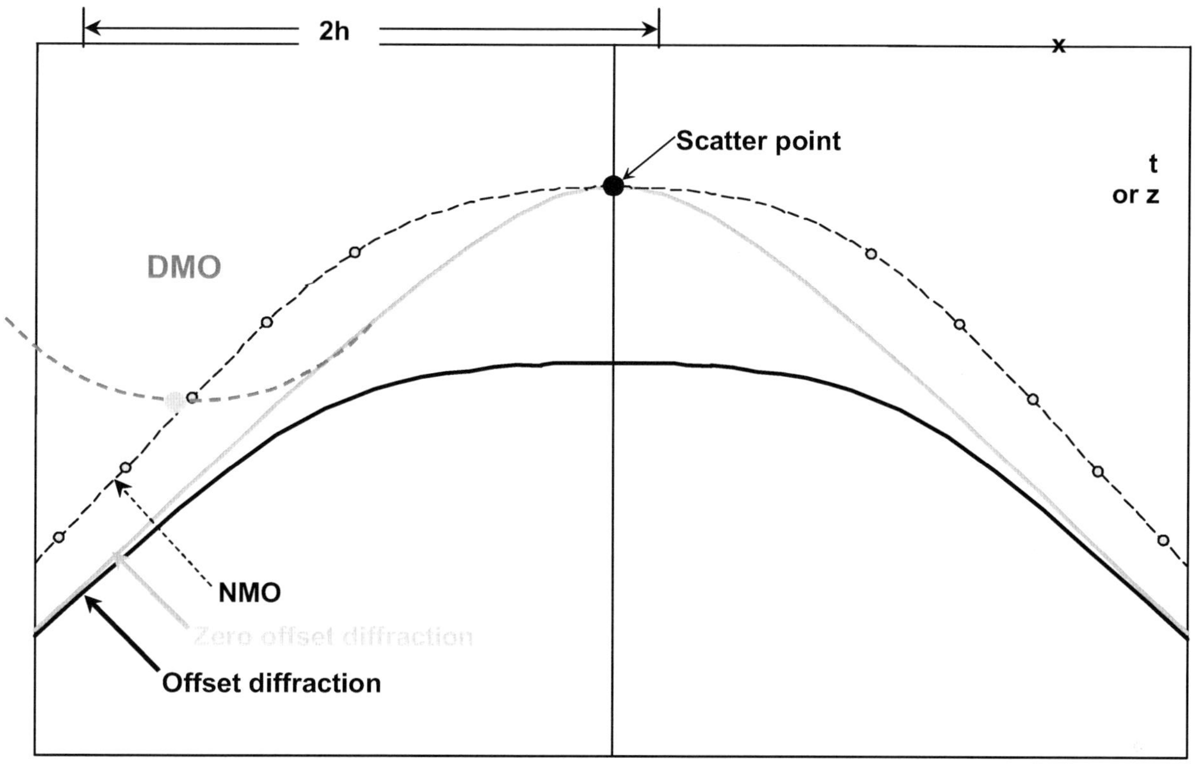

Figure 8.59 DMO construction on a constant offset section for a) a scatterpoint, and b) a dipping reflector using the DMO curves from Figure 8.15.

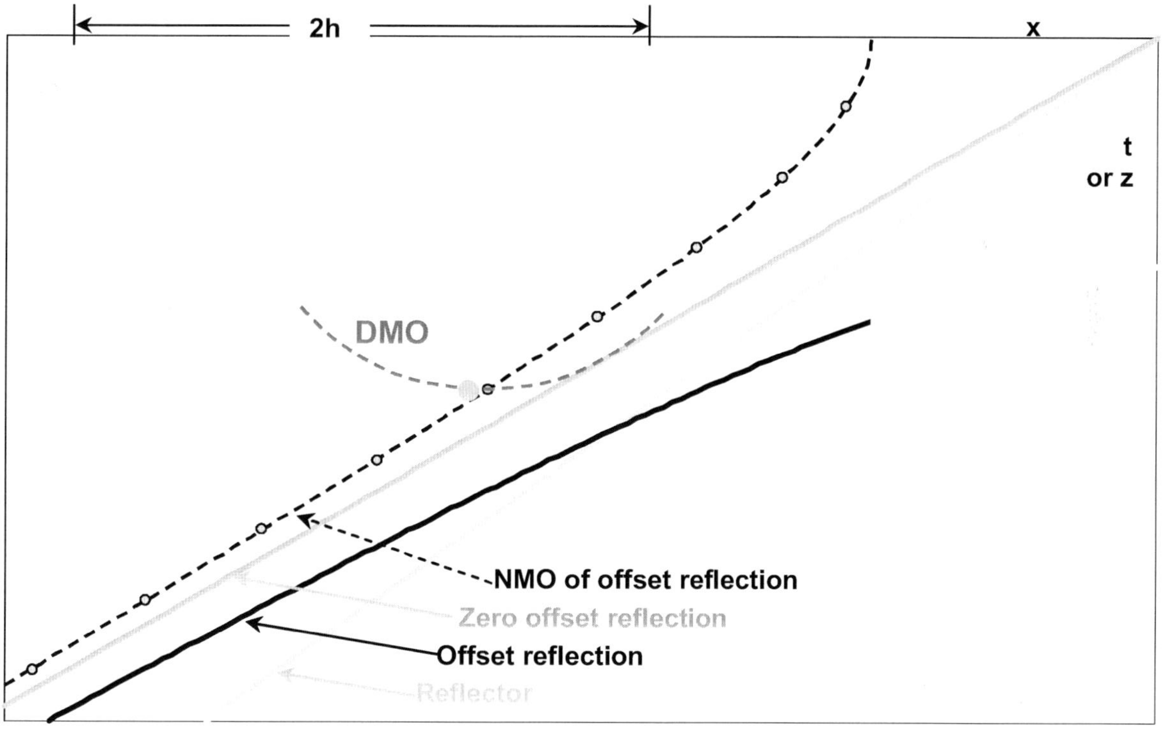

Chapter Nine

Prestack Migration (and 3-D DMO)

Introduction

Objectives

- Know that prestack migration can eliminate errors of conflicting dips.
- Know that NMO, DMO, and poststack migration is considered by some to be prestack migration.
- Identify the main methods of prestack migration.
- Understand the iterative nature of prestack migrations that are based on CMP concepts.
- Understand the process of GDMO-PSI.
- Introduction to EOM (covered in detail in Chapter 11).
- Know that constant velocity 3-D prestack migration may be accomplished with 2-D DMO and 3-D poststack migration.

9.1 Introduction to Prestack Migration

9.1.1 Various views of the objectives of prestack migration

1. Put reflection back at the reflector position.
   - Directly (model-based method)
   - Using DMO and poststack migration
2. Collapse Cheops pyramid to scatterpoint
3. Distribute energy from one input sample to the prestack migration ellipse or ellipsoid.
4. Energy in a trace could come from any scatterpoint in the surrounding volume.

5. Formation of prestack migration gathers for velocity analysis:
   - Using inverse NMO (INMO)
   - Directly (Gardner's or Bancroft's method)

Depth
- Most desirable when the velocity field can be estimated.
- More accurate positioning of energy, especially if anisotropic affect are considered.
- Expensive.
- Requires an interpreter to maintain a reasonably accurate geological model for the velocities.
- Will produce the best focusing if the velocity model is known (exactly).

Time
- Better focusing than <u>poststack</u> migration.
- Possibly better focusing than prestack depth migration with estimated velocities.
- Inexpensive.
- Focusing is independent of the velocity model.
- Reasonable results from a processor.

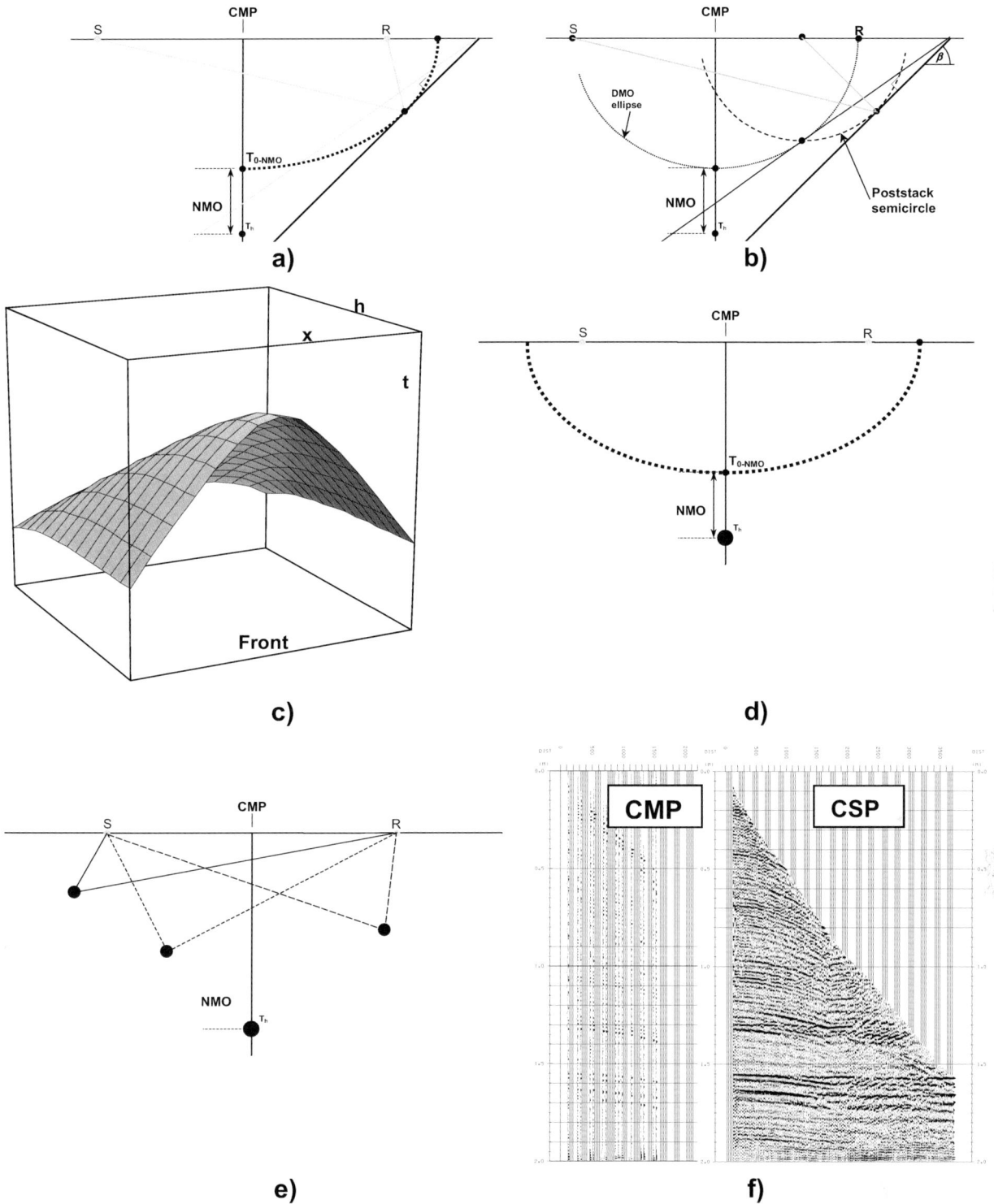

Figure 9.1 Approaches to prestack migration, a) input direct, b) using DMO, c) summing Cheops pyramid, d) forming an ellipse, e) reflections come from anywhere, and f) CMP and CSP gathers.

9.1.2 Prestack migration techniques

DMO and poststack migration (two velocity functions)
- NMO-DMO-Constant offset poststack migration
- NMO-DMO-Source record migration
- NMO-DMO-MIG-INMO-velocity analysis-NMO
- NMO-DMO-TimeMIG, VA-Stack -> InvTMIG->Post-DepthMIG

Prestack TimeMIG, -> Inv-Po-TimeMIG, -> Post-DepthMIG

Direct Kirchhoff (Cheops summation)

Source record migration
- Direct Kirchhoff
- Downward continuation of receivers
    - Kirchhoff imaging condition
        - RMS velocities (time mig.)
        - traveltime on grid or raypaths (depth mig.)
    - Forward modelling from source location
        - cross correlation
        - inversion

SG (shot/geophone) method: Downward continuation of source and receiver records.

Constant offset migration

Stolt 3-D transform of 2-D data

GDMO-PSI

EOM

Tau-p methods

I usually use the term "full" or "true" prestack migration" for those methods that use one velocity function.

DMO methods use two; one for MO correction and another for migration. I consider these methods to be a "pseudo" prestack migration.

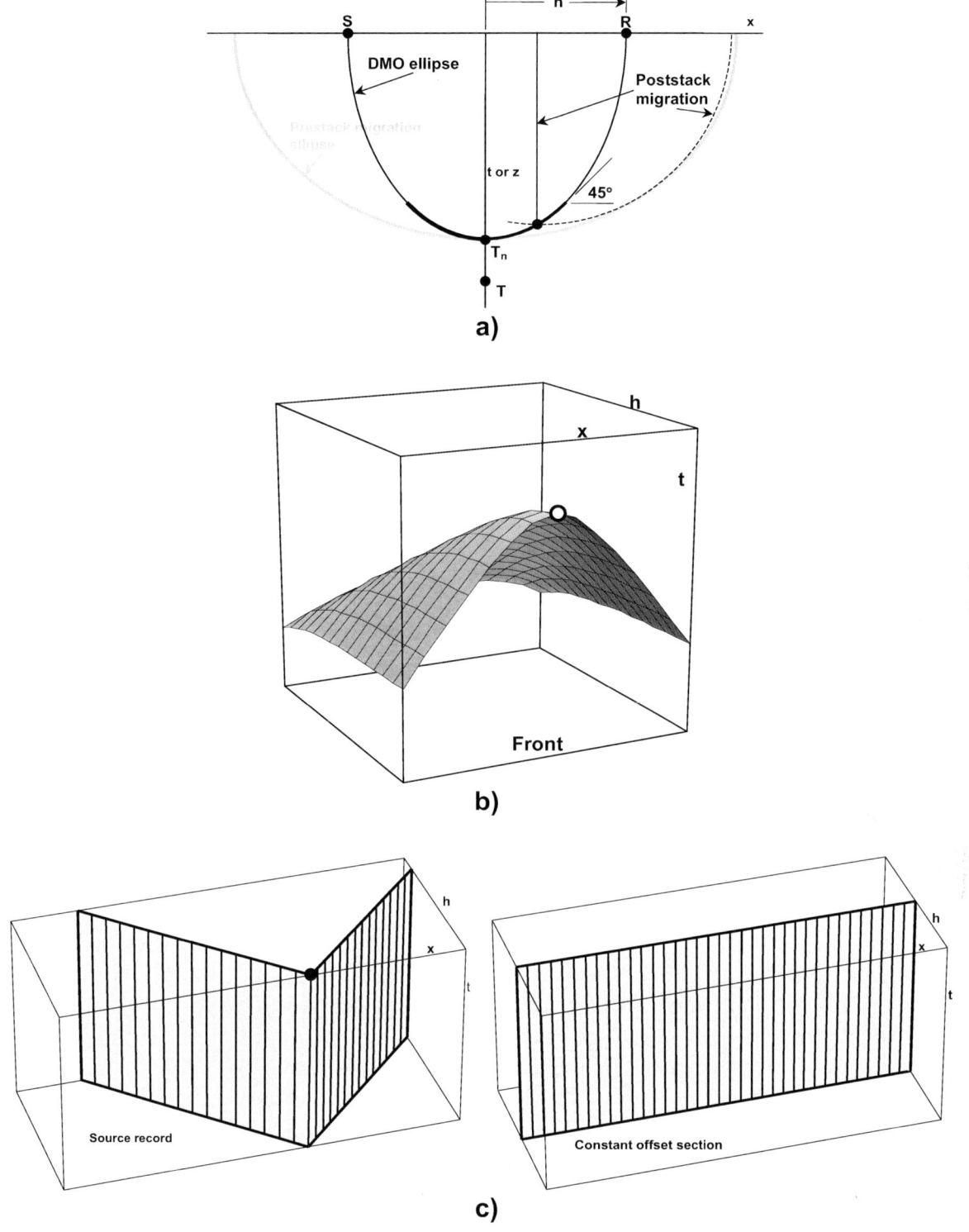

Figure 9.2 Various methods of prestack migration, a) using DMO, b) full Kirchhoff migration, and c) using source records or constant offset sections.

9.2 DMO and Prestack Migration

In a constant velocity environment, MO, DMO, and poststack migration are kinematically equivalent to prestack migration (Hale [101]). However, in a variable velocity environment, possible conflicts in velocity still occur when reflections overlap.

- DMO removes the dip effect on stacking velocities, i.e. no inverse cosine effect.
- DMO does not correct the problem of time varying velocities with conflicting dips.

After DMO a single trace, source record, or constant offset section is essentially equivalent to zero-offset, i.e. $P(x, h, t_{MO-DMO}) = P(x, h = 0, t)$.

- After DMO, traces in the same CMP gather can be summed to produce a stacked section.
- The DMO'd source records or constant offset sections may be poststack migrated to create the prestack migration ellipse as illustrated in Figure 9.3.
- After poststack migration, they may be called "prestack migrated source records" or "prestack migrated constant offset sections".
- DMO and poststack migrated data may be sorted into CMP gathers, which may be referred to as common reflection point (CRP) gathers.
- In CRP gathers, energy from reflectors should be flat or have constant time with offset. Velocity analysis is required to ensure the flatness on events.
- Stacking the CRP gathers completes the "prestack migration".

The result would be identical if the DMO'd data was first stacked then one poststack migration applied to the stacked section. Consequently there is some resistance to calling these DMO processes a prestack migration. Because of the DMO limitations to constant or smoothly varying velocities, this method of processing may be inferior to prestack migration methods that don't use DMO.

The DMO method of analysis uses one application of velocities for MO and one for the poststack migrations. A number of iterations will be required to converge to one stable velocity.

> DMO and poststack migration computes much faster than full prestack migration and more time can be spent estimating NMO and migration velocities.

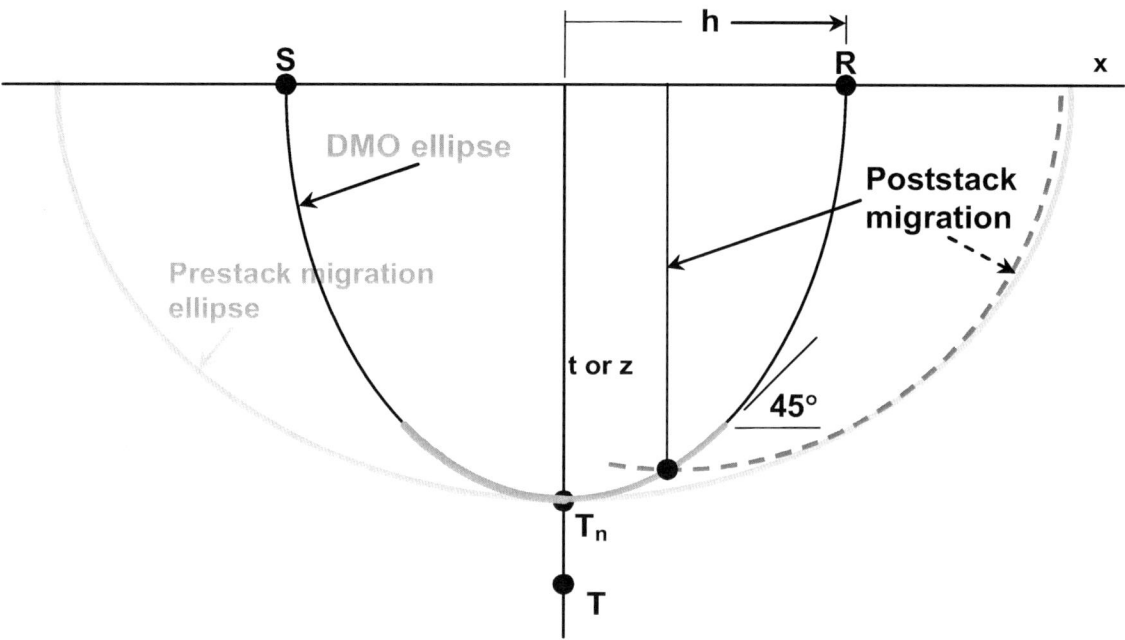

Figure 9.3 DMO and poststack migration recreate the prestack migration ellipse.

> When migration is applied to <u>constant offset sections</u>, there are no overlapping reflections as they are now in their approximate geological location. The problem of overlapping reflections is therefore resolved.

The main purpose in forming migrated constant offset sections is <u>velocity analysis</u>.

Inverse moveout (IMO) is applied before stacking to allow a refined velocity to be estimated. Although <u>each DMO'd trace is essentially zero-offset</u>, an offset from the original geometry is assumed. The offset of the DMO'd traces may be different, and depend on the <u>processing geometry</u>, either a source record, or a constant offset section. <u>Any offset may be used for IMO</u>, but it should be chosen to enhance the velocity analysis.

After velocity analysis, the data may be stacked, or the entire process repeated for optimal imaging.

9.3 DMO Prestack time, IMO to stack, poststack depth migration

DMO and poststack time migration may be used to prepare an optimum stacked section for poststack depth migration.

- MO-DMO constant offset sections.
- Time migrate each constant offset section.
- IMO, estimate an improved velocity, and then MO.
- Stack the offset sections.
- Inverse time migrate (or model) the stacked section with the same velocities and algorithm as in the second step.
- Poststack depth migration.

These processing schemes enable a more accurate positioning in space and time of the input data prior to a final velocity analysis. The stacked section is formed from data that is focused at the scatterpoint. The inverse time migration <u>forms a zero-offset section</u> that has prestack focusing, however, the approximations of time migration have been effectively removed (or reduced). Poststack depth migration completes the process and tends to an image produced by prestack depth migration.

The extensive use of time migrations permit simplified velocity analysis and faster algorithms to be used, reducing the cost of processing.

The DMO prestack migration steps are illustrated in Figure 9.4. With a constant velocity, the scatterpoint energies in Figure 9.4d are ideally focused to a constant time and position (x). When the velocities are in error, the energy will form on a surface that is spread laterally, and will curve away from constant time, as shown in gray in (d). IMO, velocity analysis, and NMO, (or residual MO correction) will realign the energy to a constant time for improved imaging. The lateral spread of energy will be reduced when the improved velocities are applied to the original input data and the procedure iterated.

An improved version of the above would be the use of <u>prestack time migration</u>, followed by inverse poststack time migration to form an optimal stacked section. A poststack depth migration would complete the process.

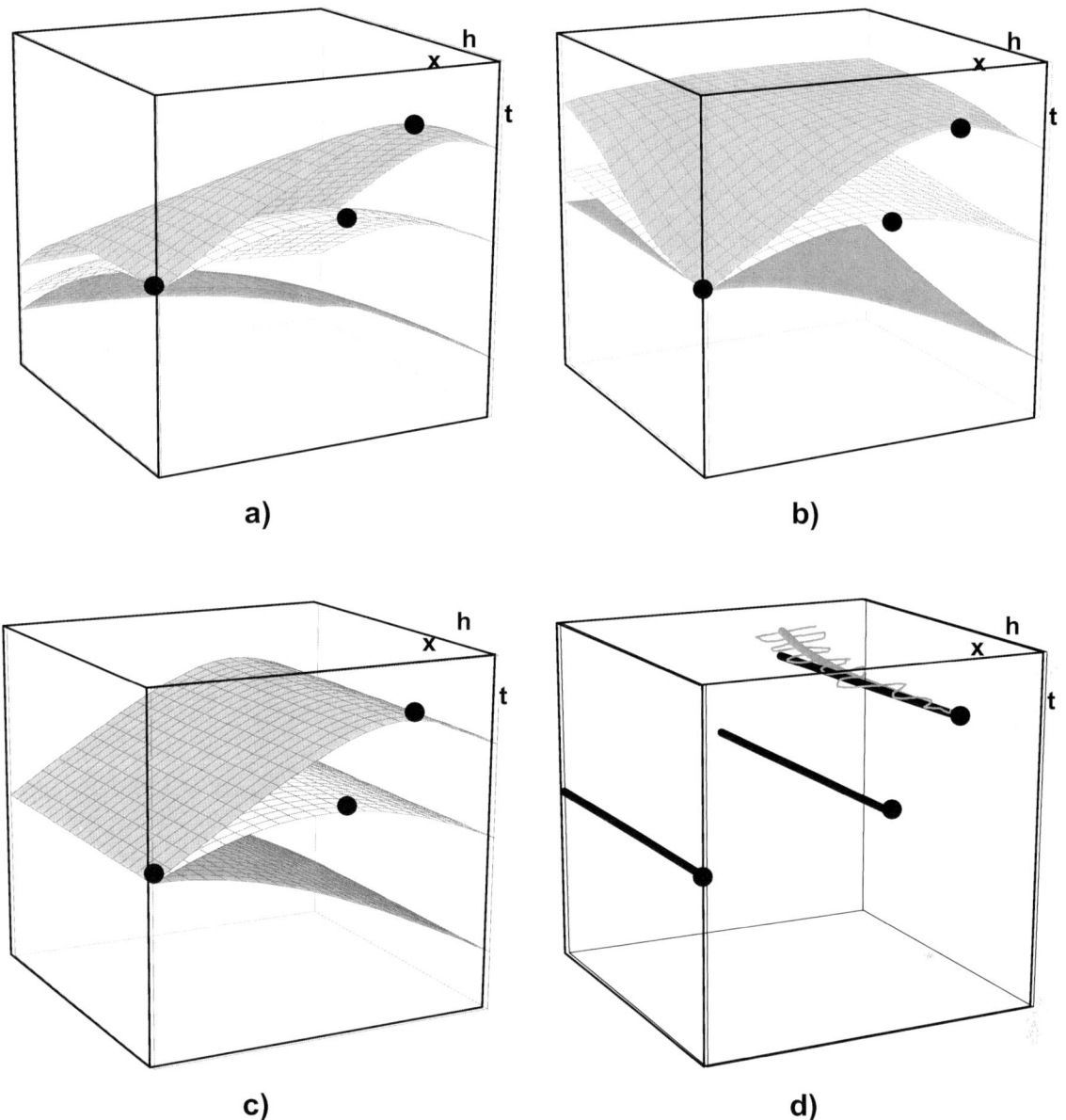

Figure 9.4 Constant velocity DMO prestack migration with a) three Cheops pyramids, b) after NMO, c) after DMO, and d) after poststack migration of each constant offset section.

9.4 Direct Kirchhoff migration from the prestack volume

We can prestack migrate data directly from the prestack volume to a 2D line using Kirchhoff migration. Recall that a scatterpoint below a 2D line produces a Cheops pyramid shape in the prestack volume. Summing the energy on that surface produces the migrated sample at the location of the scatterpoint.

This procedure could be repeated for all the 2D scatterpoints.

However, using the more efficient method where we loop over the input and output traces we use similar pseudo code as the post-stack migration, but we now use the Double Square Root (DSR) equation to define the "diffraction" shape, i.e.,

$$T = \left(\frac{T_0^2}{4} + \frac{(x_s - x_m)^2}{V^2} \right)^{1/2} + \left(\frac{T_0^2}{4} + \frac{(x_r - x_m)^2}{V^2} \right)^{1/2} = \left(\frac{T_0^2}{4} + \frac{h_s^2}{V^2} \right)^{1/2} + \left(\frac{T_0^2}{4} + \frac{h_r^2}{V^2} \right)^{1/2} \quad (9.1)$$

where we now use the location of the source x_s, the receiver x_r, and the migrated trace x_m.

The input traces can be in any order, i.e., source records, or CMP gathers, as eventually all input traces will be summed into all migrated traces. We only need the input geometry of the input trace, i.e. the source and receiver location.

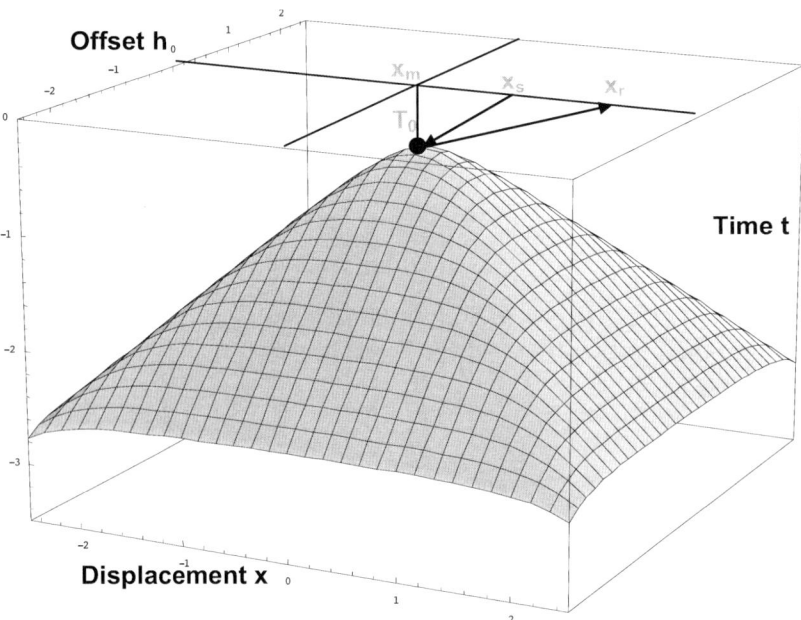

Figure 9.5 Cheops pyramid, the summation surface in the prestack volume for a 2D Kirchhoff prestack migration.

Pseudo code for direct Kirchhoff prestack migration

Loop <u>*migrated trace*</u>	`for i = 1 : I`
Migrated trace geometry	`x`$_m$`, V  (y`$_m$` for 3D)`
Loop <u>*input traces*</u>	`for k = 1 : K`
Input trace geometry	`x`$_s$`, x`$_r$`, (y`$_s$`,y`$_r$` for 3D)`
Migration geometry	`h`$_s$`, h`$_r$
Loop <u>*migrated samples*</u>	`for j = 1 : J`
Define time on input trace	`T = DSR-eqn(T`$_0$`, h`$_s$`, h`$_r$`, V)`
Define time sample number	`m = T/dt`
Check bounds of the input data	`if j < J`
Cosine weighting, plus others	`W = To/T`
Sum data to migrated trace	`Migdata(j,i)+=W*Indata(m,k)`
End	`end`
End	`end`
End	`end`

The same code can be used for a direct 3D <u>**prestack Kirchhoff migration**</u> by including the *y* coordinate in computing the h_s and h_r terms in the DSR equation, i.e.,.

$$h_s^2 = (x_s - x_m)^2 + (y_s - y_m)^2, \qquad h_r^2 = (x_r - x_m)^2 + (y_r - y_m)^2 \tag{9.2}$$

In this case the prestack volume is three dimensional and can't be visualized as in the 2D case.

Recall that Kirchhoff migration can use independent geometries for input and migrated data. For 3D data, it can directly combine overlapping input data set, say with imperial or metric geometries or different orientations into a migrated data set with a another orientation.

9.5 Prestack Migration of Source Records

9.5.1 Direct Kirchhoff migration of the source record

In a source record, each <u>scatterpoint</u> has:

- **<u>One</u>** *source-scatterpoint* **raypath with traveltime** t_s.
- **<u>Many</u>** *scatterpoint-receiver* **rays with traveltimes** t_r.
- **The diffraction <u>shape</u> is defined by the** *scatterpoint-receiver* **raypath.**
- **The <u>time at the "apex"</u> of the diffraction is defined by the** *source-scatterpoint* **raypath and the** *depth* **of the scatterpoint.**

Figure 9.5 shows modelling with raypaths to diffraction patterns. The traveltimes on the diffraction is defined by $t_s + t_r$.

Kirchhoff prestack migration is accomplished by:
- assuming a scatterpoint location,
- defining the diffraction shape,
- summing the energy on the diffraction shape, and
- placing the amplitude of the summed energy at the <u>scatterpoint</u>.

The diffraction shape may be defined using RMS velocities for a time migration, or from raytracing or wave front modelling for a depth migration.

> For an in-depth analysis of Kirchhoff migration see Lumley [54].
> Additional references on source record migration are Schultz and Sherwood [27], van der Schoot [197], and Lee [194].

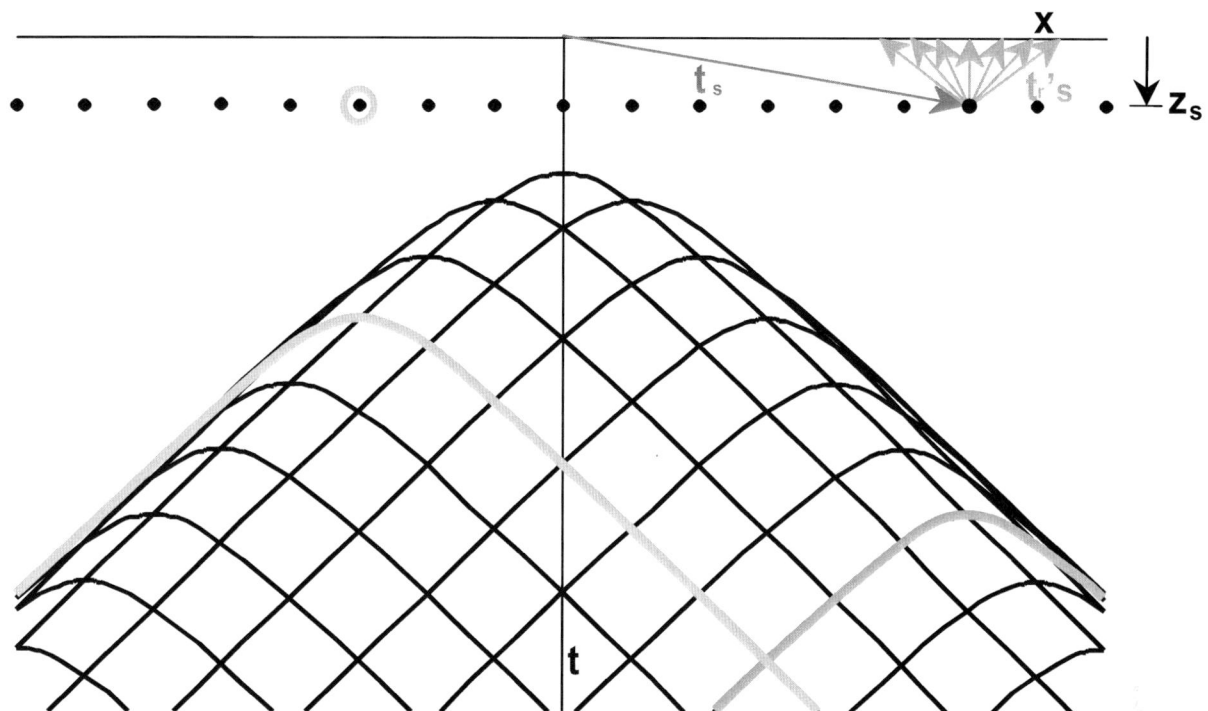

Figure 9.5 Example of ray modelling on a source gather used to define the diffraction shapes of prestack migration.

As is common with many diagrams in these notes, the above figure contains two images that are super imposed. The first a depth cross-section (x, z) with scatterpoints at a depth of z_s and raypaths to and from one of the scatterpoints. The second image is the recorded time section (x, t) that contains diffractions from each of the scatterpoints.

Note the <u>input traces are located at the surface position</u> of the receivers, and not at the midpoint location. Migrated traces can be located at CMP locations and care must be taken to prevent aliasing.

This concept of locating the source record traces at the receiver location is consistent with nearly all prestack source record migrations. Some algorithms require the traces to be interpolated before migration to achieve a trace spacing that is compatible with the conventional CMP trace interval.

9.5.1.1 Pseudo code for source record Kirchhoff migration

The pseudo code for a source record migration is similar to the poststack code and identical to the direct Kirchhoff prestack migration. We only use part of the input data that corresponds to a source record.

Pseudo code for source record migration.

Loop *migrated trace*	`for i = 1 : I`
Migrated trace geometry	x_m, V (y_m for 3D)
Loop *input traces*	`for k = 1 : K`
Input trace geometry	x_s, x_r, (y_s, y_r for 3D)
Migration geometry	h_s, h_r, x
Loop *migrated samples*	`for j = 1 : J`
Define time on input trace	`T = DSReqn(`T_0`, `h_s`, `h_r`, V)`
Define time sample number	`m = T/dt`
Check bounds of the input data	`if j < J`
Cosine weighting, plus others	`W = To/T`
Sum data to migrated trace	`Migdata(j,i)+=W*Indata(m,k)`
End	`end`
End	`end`
End	`end`

The only difference between source record migration and th full prestack Kirchhoff is the input data in now a source record.

9.5.2 Spatial extent of source record migration.

The migrated space will be considerably larger than the input source record as the diffraction (or migrated smiles) extend well beyond the recorded offsets as illustrated below. In Figure 9.6a the one input sample, in midpoint and oneway time, produce the red ellipse to migrate to the blue point. In (b) the Kirchhoff approach assumes the trace is located at the receiver location with two-way time.

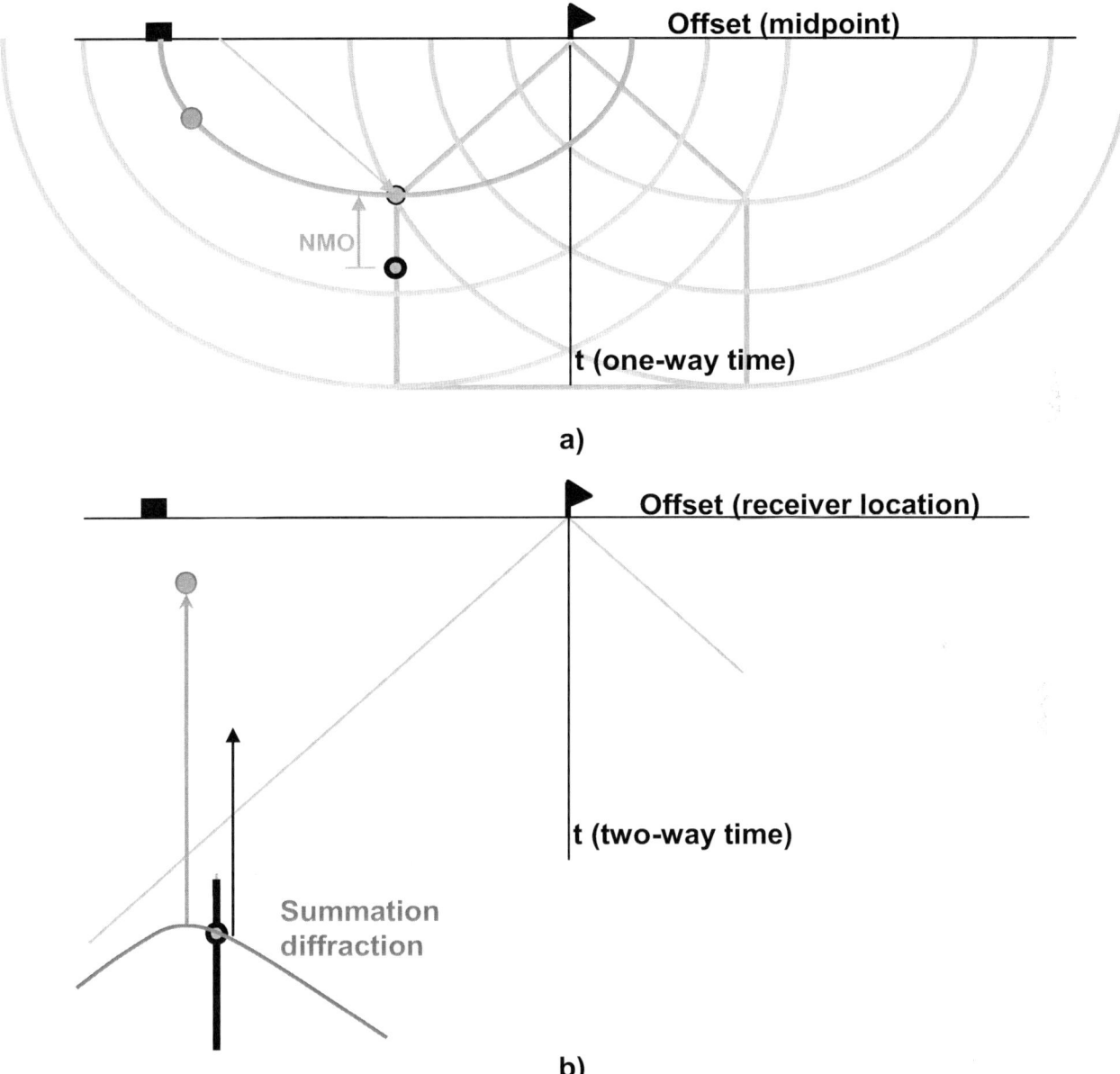

a)

b)

Figure 9.6 Illustrations of a) source record (*x, t*) with summation diffractions, and b) prestack migrated source record.

9.5.3 Fast zero-offset migration of source records using a post-stack migration (Junk)

The construction of the source record in Figure 9.7 shows <u>linear reflectors</u>, horizontal or dipping, produced <u>hyperbolic reflections</u> with asymptotes that intersected at the surface, (similar to zero-offset migration).

A <u>poststack type of migration</u> on the source record will collapse these <u>hyperbolic</u> reflections to a point. (Velocities and or offsets may require adjustment as the <u>trace geometry may give the receiver location</u>).

The migrated point from a linear reflector is located at the mirror image of the source relative to the reflector surface.

Moving these migrated points to one-way time, and then to the CMP position, will place the energy at the <u>zero-offset reflection point</u> of each linear reflector.

Locate these points on Figure 9.7.

This <u>appeared</u> to be a promising <u>prestack</u> technique, however:

- The final image will be composed of <u>energy points</u> spread along a reflector. If the sources are located at multiple receiver intervals, the energy point will also be at multiple CMP intervals; the <u>linear event will not be reconstructed</u> evenly.

- Another problem is <u>noise</u>. The enhancement and cancellation of energy is inferior and the noise was often greater than the signal.

- This method of migration is a linear process with respect to CMP position, and the exact <u>same final result will occur if the sources are first stacked (without MO etc.) and then migrated as one stack</u>, as illustrated in Figure 9.8.

This method of processing which stacks data without MO is also referred to as <u>infinite velocity stacking</u>. Recall the intersection of Cheops pyramid with CMP gathers.

- As the CMP gather moves away from the scatterpoint the zero-offset dip increases

- As the CMP gather moves away from the scatterpoint, the moveout curve tends to be horizontal.

- Stacking without MO will tend to stack the steeper dipping events (but not the shallow dips).

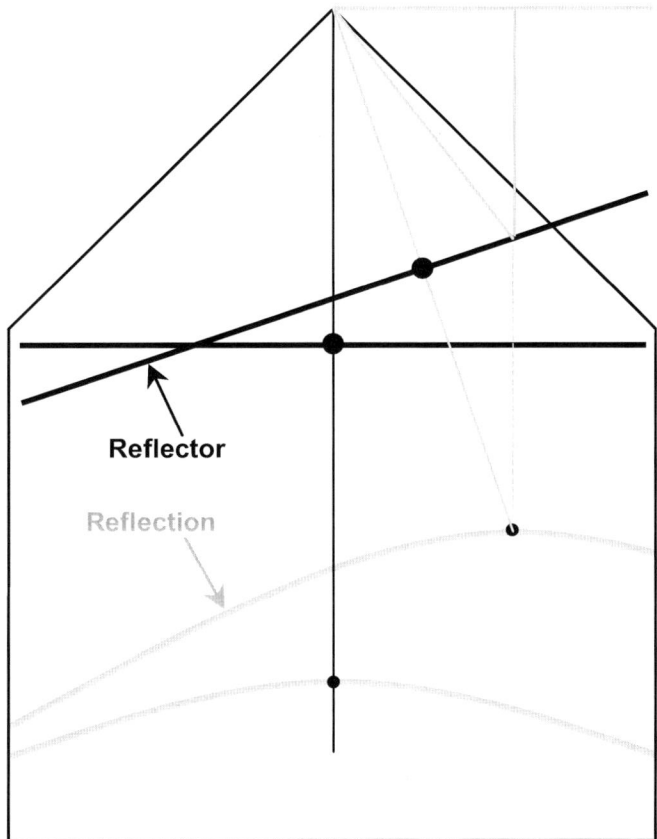

Figure 9.7 Source record with linear reflectors. The hyperbolic reflections could be migrated to a point at the zero offset location on the reflector.

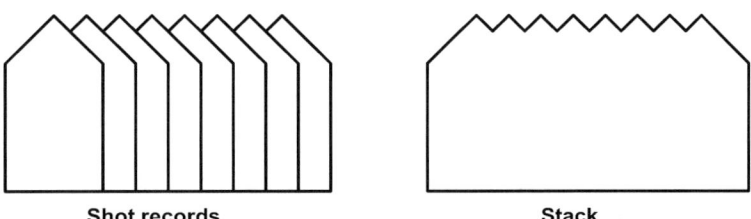

Figure 9.8 Illustration of linearity of source records when offset is ignored. Migrating each source record with a conventional post-stack migration and then stacking is identical to migrating a poststack migration when the stack was formed with no moveout correction.

This method has been used to quickly produce sections of highly structured data without any velocity information. The <u>prestack migration part is not required</u> but was a great revenue generator for some companies.

9.5.4 Downward continuation migration of source records

9.5.4.1 The focusing of energy when downward continuing a source gather

Consider the following figure that contains a number of scatterpoint aligned along four reflectors. The source record (seismic response) for a centered shot is shown on the right. This source record will be downward continued (DC) using a time migration algorithm. At the appropriate depth, the diffraction energy will focus at its apex.

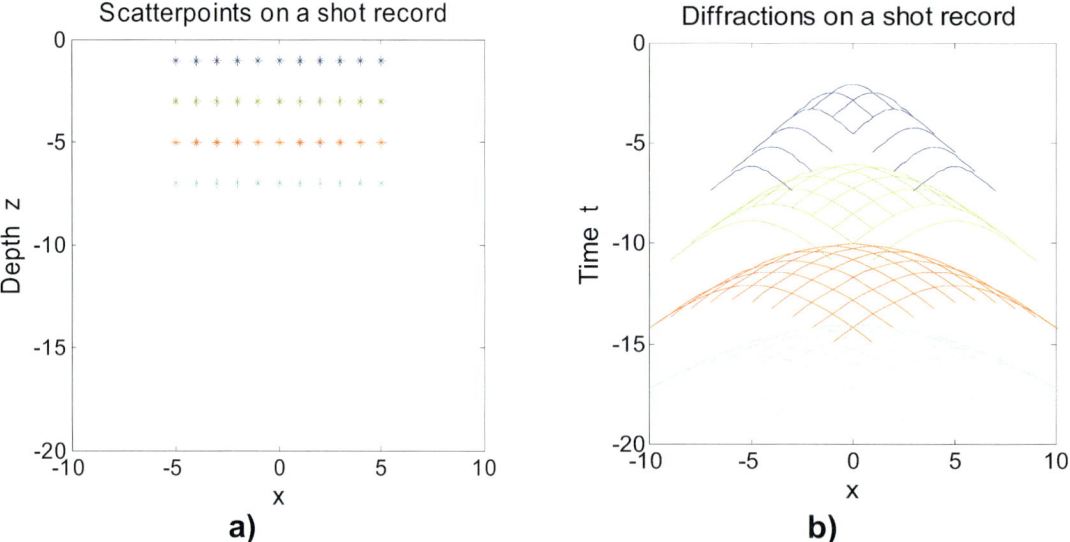

Figure 9.9 A constant velocity model a) with scatterpoint on four reflectors, and b) the corresponding source record. (2004\ShotDif.m)

- Downward continuation of the source record to a **specific depth** z_A will **focus** some of the diffraction energy to a point.
- This focused energy is associated with **scatter points** at that depth z_A.
- The next step is to **locate this focused** energy and **placing its amplitude** at the location of the corresponding scatterpoint.
- This is loosely referred to at this time as the "**imaging condition**".

(In poststack migrations, this focusing of the diffraction energy was much simpler as it occurred at a fixed time in a time migration, or at the surface or $t = 0$ for a depth migration.)

Figure 9.10 illustrates the focusing of the energy when the depth of the DC reaches the depths of the corresponding reflector levels in Figure 9.9a.

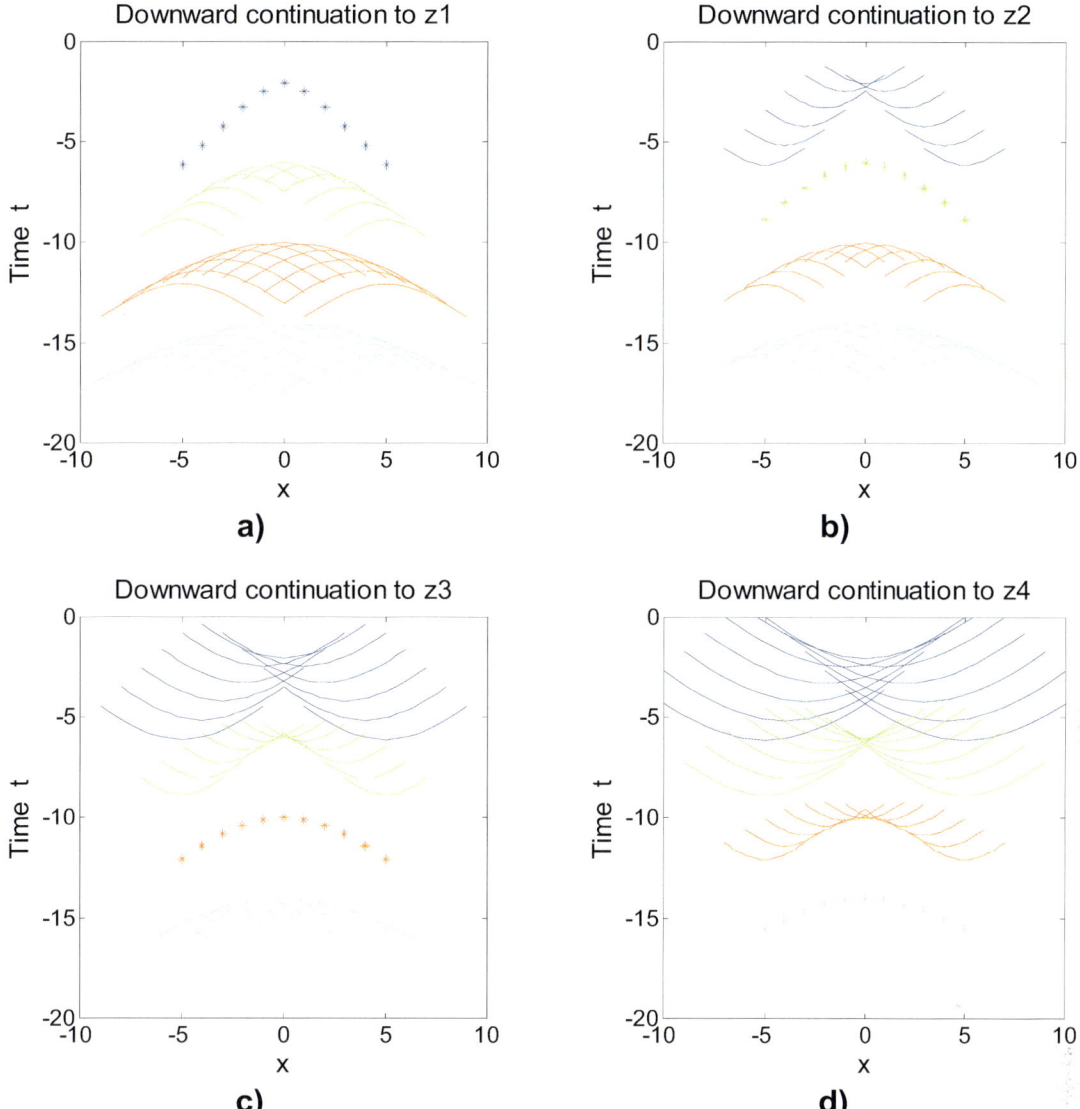

Figure 9.10 Four time section at different downward continuation levels that match the depth of the corresponding reflecting layers using a time migration.

The diffractions below the imaging condition are partially collapsed or under migrated, and they will continue to collapse until the DC depth reaches their depth. Energy above the imaging condition will be over migrated.

Conventional downward continuation method are used such as finite difference, phase shift, PSPI, ωX, etc.

The above figure represents a time migration of the source records in which the diffraction energy focuses at the apex of the diffraction.

In a depth migration, the apex of the diffractions will move toward $t = 0$, proportional to the depth of the scatterpoints.

The images from the previous figure are now plotted in a 3D volume with the additional parameter of depth.

Note the focussing of the diffractions at the corresponding depths.

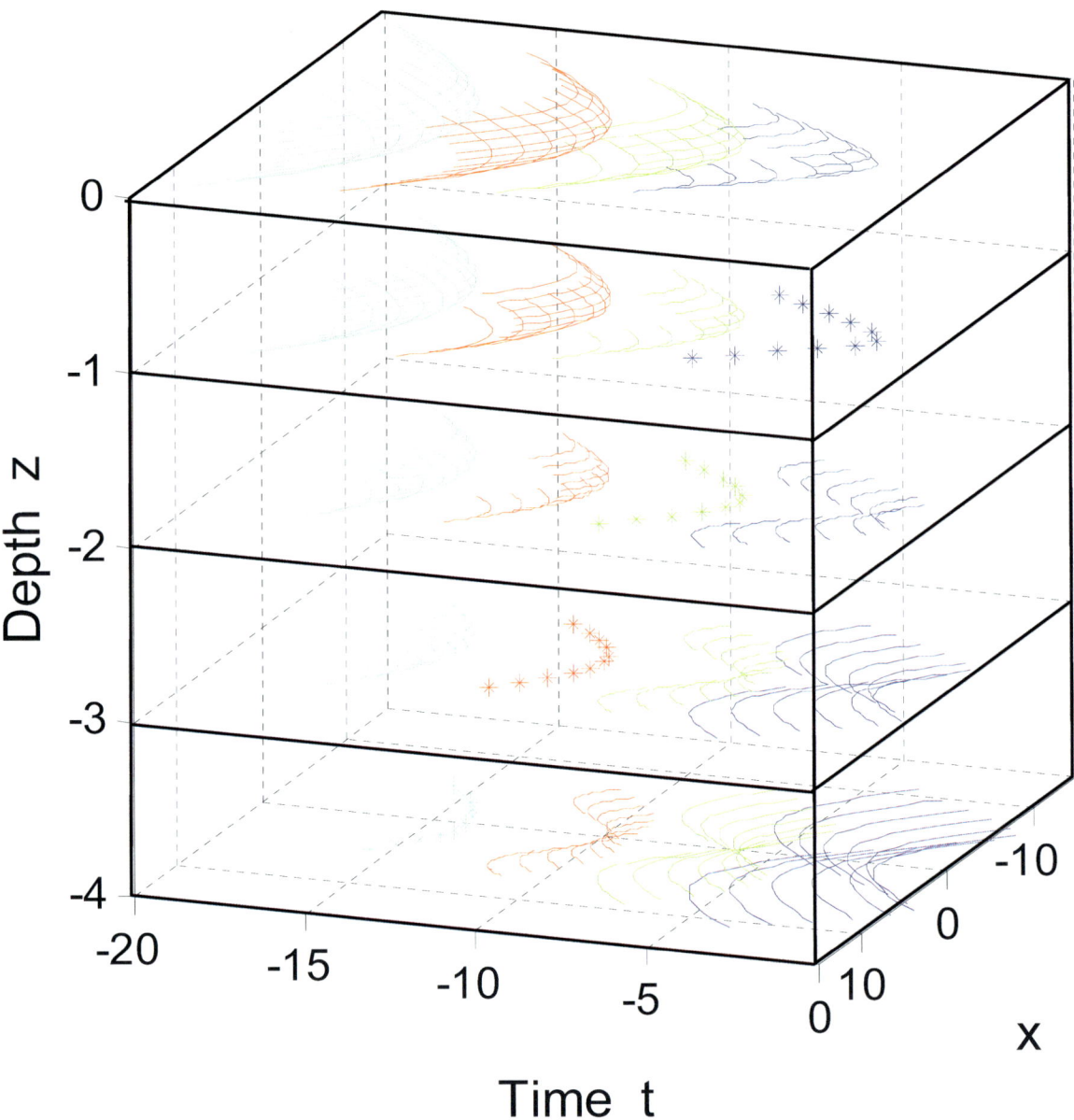

Figure 9.11 Downward continuation of the diffractions in the 3D volume.

We now know that when the downward continuation reaches a specific depth z_{DC}, the energy of some diffractions are focus to a point. This <u>focussed energy</u> corresponds to <u>scatterpoints</u> that are on a reflector at that depth z_{DC}.

This focussed energy, on the time section, is located at the correct spatial location, however they are located at a time that is considerably different from the corresponding depth. In a constant velocity medium, the times of this focussed energy lie on a hyperbolic path.

The following section is very important because it shows how to map this focussed energy to its corresponding scatterpoint.

i.e. the imaging condition

Moving this energy to the scatterpoint location <u>completes the prestack migration at this depth</u>.

This process in <u>analogous to moveout correction</u> on a source record.

There are a number of ways to define this mapping, or to define the imaging condition.

Kirchhoff (traveltimes)
    Time migration V_{rms}
    Depth migration
        Gridded traveltimes
        Raypath traveltimes,
            First arrivals
            Maximum energy
            Multiple arrivals

Wavefront modelling
    First arrivals
    Maximum energy
    Full waveform (Inversion)

9.5.4.2 Kirchhoff "imaging condition"

One method of source record prestack migration combines a downward continuation algorithm with a Kirchhoff algorithm (Reshef [109], and Ng [243]).

- Downward continuation (DC) is used to lower the receivers into the subsurface which will focus the diffraction energy to scatterpoints at that depth.

- The location of the "imaging condition" on the DC time section is found using Kirchhoff migration principles. The apex of the diffraction is in the correct spatial position. The vertical time is estimated by using rays from the source (and receiver) to the scatterpoint. The amplitude of the focused energy is then mapped to the corresponding scatterpoint location.

- All points at the imaging condition are mapped, one to one, to the corresponding scatterpoint locations.

- The process is repeated for each depth level.

Depth migration: Assume the DC level is at depth z_A. Raytracing or traveltime mapping is used to define the time t_s from the source to the scatterpoint. The amplitude at the imaging condition (x, t_s) in Figure 9.12a is then mapped to the migrated location (i.e. the location of the scatterpoint), (x, z_A).

Time migration: The pseudo "depth" z_A of the DC may be defined by the two-way DC time $t_{DC} = V*z_A/2$. The two-way time t_{DC} is equal to twice the vertical travel time t_{r0} from the scatterpoint to the surface. The source ray time t_s may also be defined to give the imaging condition time t_I is than found using $t_I = t_s + t_{r0}$. Energy along this imaging condition at $(x, t_s + t_{r0})$ in Figure 9.12b is mapped to the time migrated source record at $(x, 2t_{r0})$.

Exercise: Figure 9.12 shows diffractions on a source record that represent the reflections from a horizontal reflector. Use (a) for a depth migration, and (b) for a time migration. On these figures:

- Are the <u>times</u> of the diffractions one way or two-way times? _____
- Where are the <u>traces positioned</u>? CMP or surface station (STN)? _____
- Sketch the result of downward continuation <u>migration</u> to the depth of the scatterpoints.
- Define the <u>shape</u> of the imaging condition. _____
- What is the maximum <u>offset</u> of the migrated traces? _____

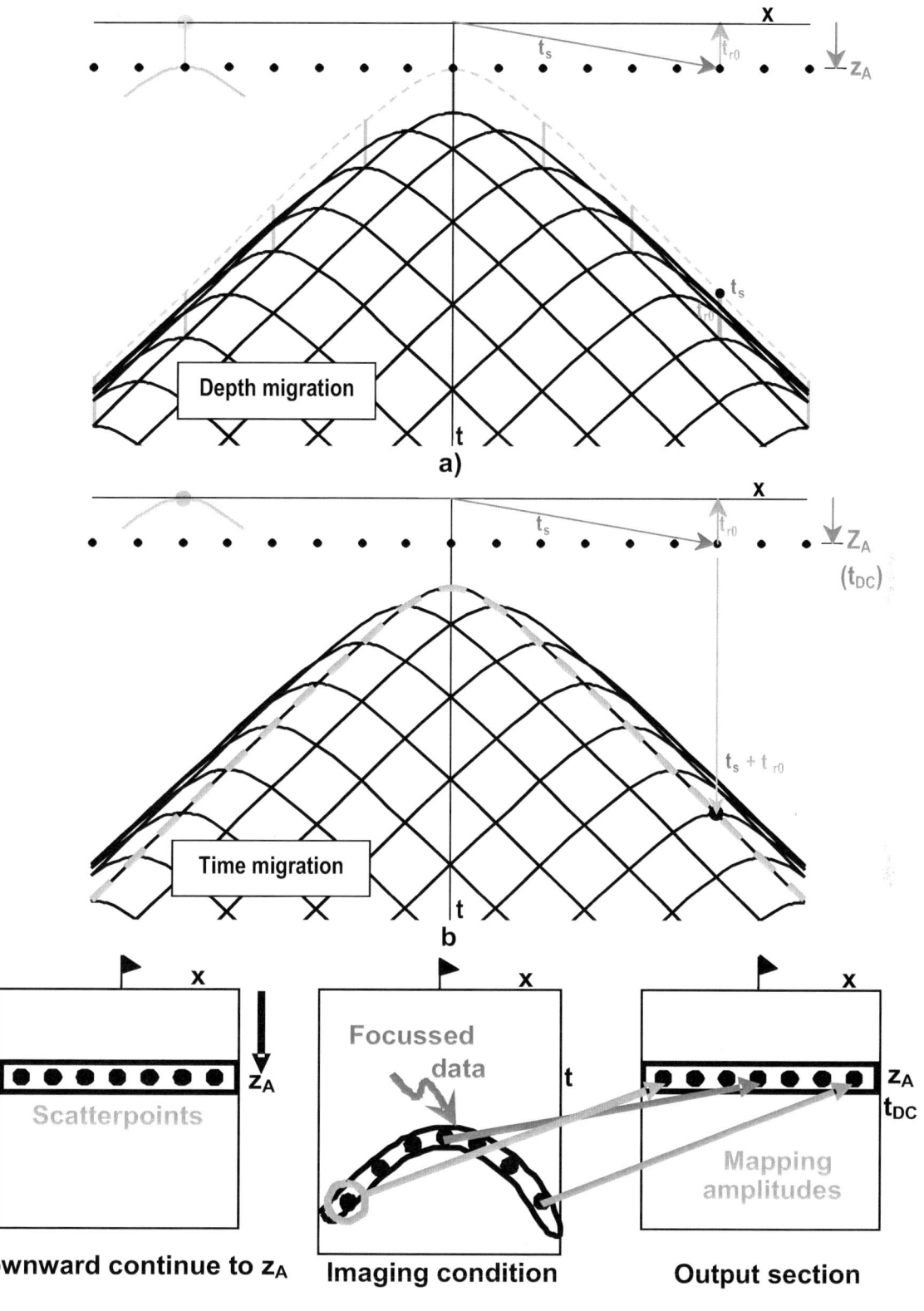

Figure 9.12 Source records used to illustrate combined algorithm migration. Use a) for a depth migration, (b) a time migration, and c) the imaging condition.

9.5.4.3 Defining the Kirchhoff time of the "imaging condition" for a <u>time</u> migration

We assume that a downward continuation time migration algorithm was used, where the diffraction energy is focused at the apex of the diffraction. We use the RMS velocity defined at the two-way DC time t_{DC} and spatial location *x* to get the traveltimes of the source ray t_s.

$$t_s^2 = \frac{t_{DC}^2}{4} + \frac{x^2}{V_{RMS}^2} \qquad (9.3)$$

The time of the imaging condition t_I for a time migration also requires the addition of the time from the scatterpoint vertically to a receiver at the surface $t_{r0} = t_{DC}/2$, i.e.,

$$t_I = t_s + t_{r0} \qquad (9.4)$$

Energy along this imaging condition at (x, $t_s + t_{r0}$) in Figure 9.12b is mapped to the time migrated source record at (x, t_{DC}).

Note that the shape of the "imaging condition" is the shape of a <u>reflector</u> at a specific depth.

It may appear to be a diffraction, but that is not the case.

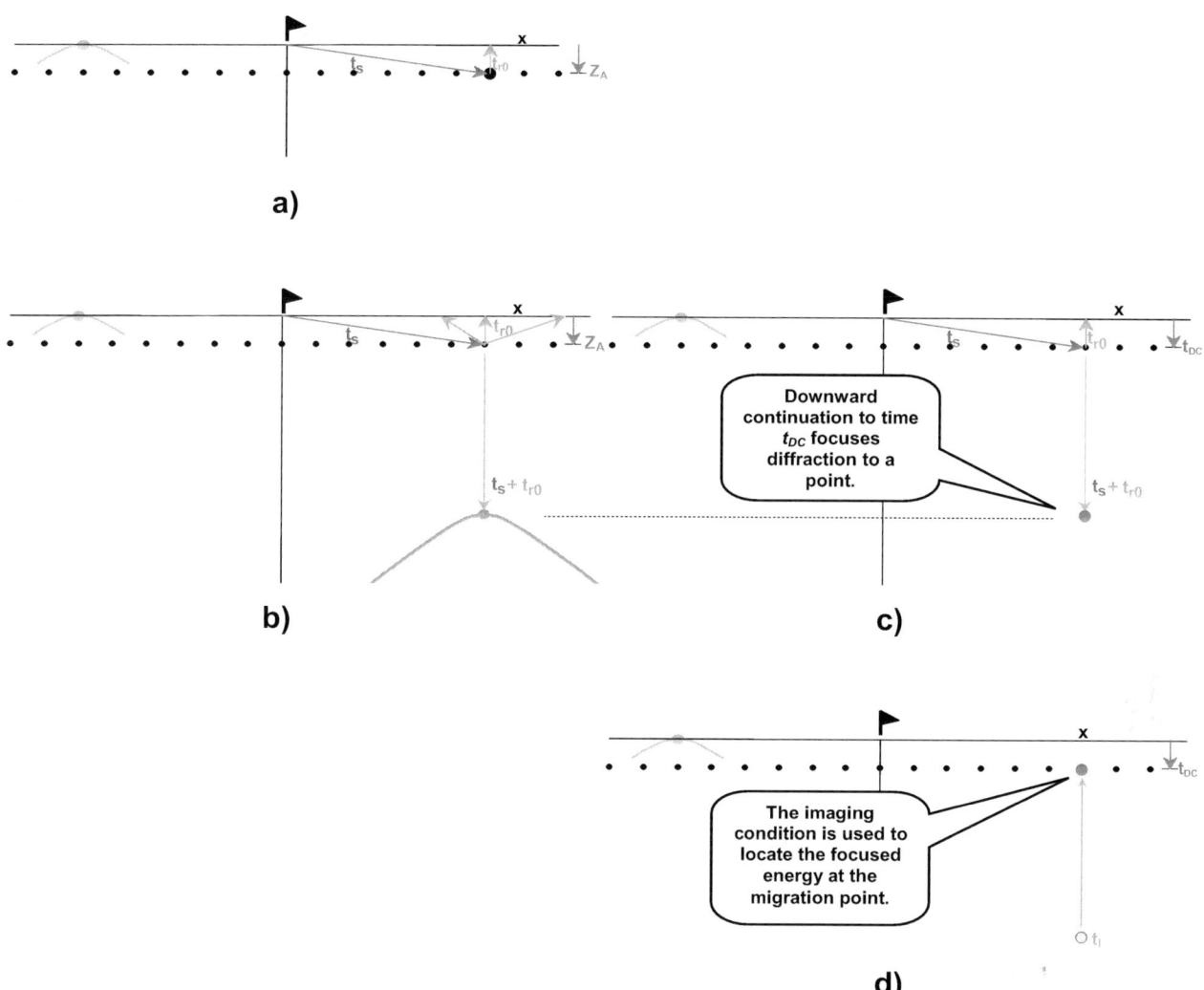

Figure 9.13 Kirchhoff <u>time migration</u> sequence of a source record. Part a) shows a particular scatterpoint with the source and vertical receiver raypaths, b) adds the time section and the diffraction from the scatterpoint in (a). Time migration in c) focuses the energy at the apex of the diffraction in (b). The imaging principle locates the migrated point in (c), and in d), moves it to the migrated location, (i.e. the original scatterpoint location).

9.5.4.4 Defining the Kirchhoff time of the "imaging condition" for a <u>depth</u> migration

For a <u>depth migration</u>, the location of the focused energy in Figure 9.13c will move vertically by the traveltime from the scatterpoint to the surface t_{r0}. Now the "imaging condition" will be defined by the traveltime from the source to the scatterpoint, i.e.,

$$= t_s \qquad (9.5)$$

This simplifies the imaging condition for depth migrations to be the traveltimes **emanating from a source point**.

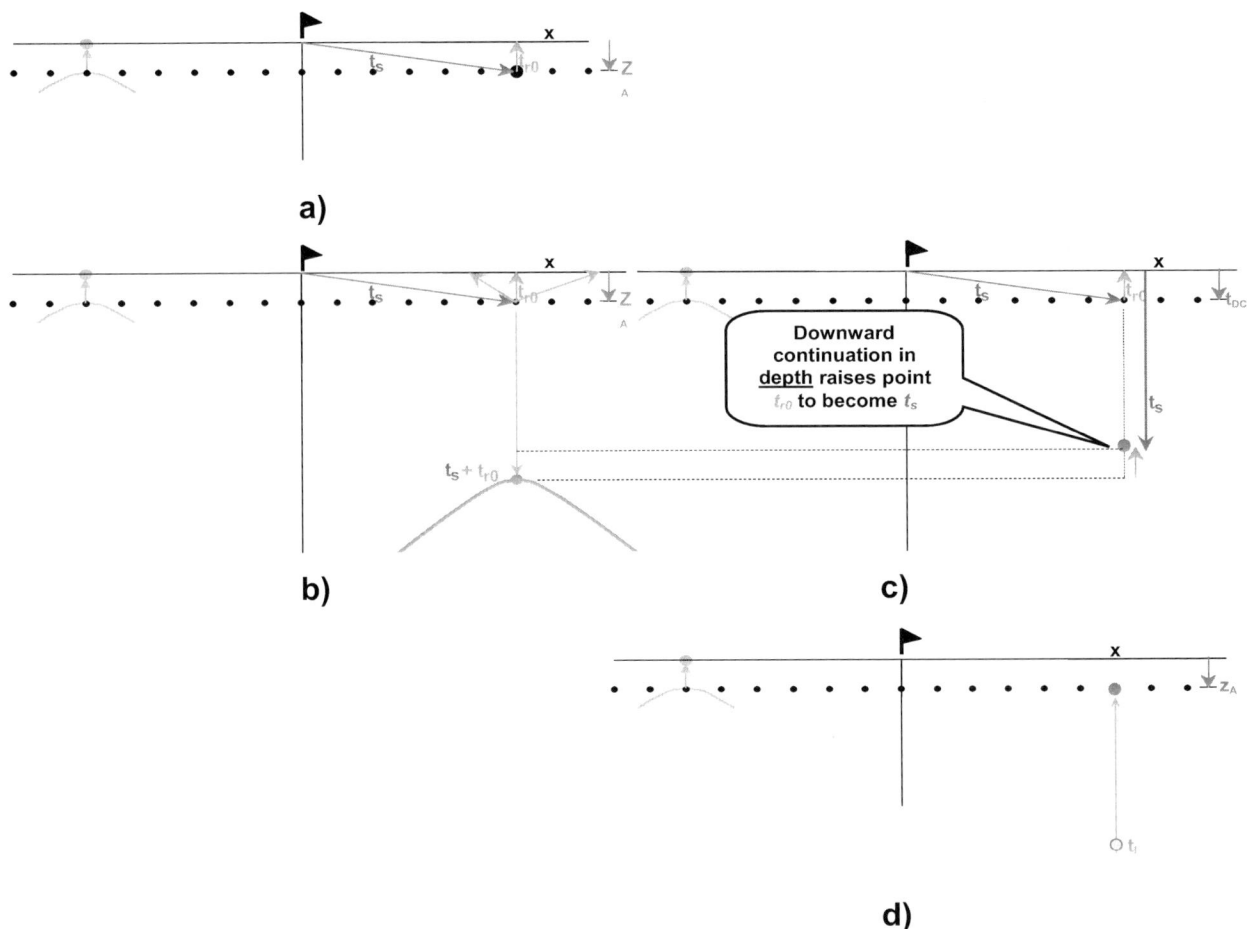

Figure 9.14 Kirchhoff <u>depth migration</u> sequence of a source record. Part a) shows a particular scatterpoint with the source and vertical receiver raypaths, b) adds the time section and the diffraction from the scatterpoint in (a). Depth migration in c) focuses the energy above the apex of the diffraction in (b). The imaging principle locates the migrated point in (c), and in d), moves it to the migrated location, (i.e. the original scatterpoint location).

9.5.4.5 Defining the Kirchhoff time of the imaging condition for a depth migration

Determining the time of the imaging condition for a depth migration is considerably much more complex that the previous time migration case.

These traveltimes are computed buy a number of techniques using:
- Gridded traveltime mapping,
- Ray tracing, or
- Wave propagation.

Gridded traveltimes are computed as expanding wavefronts on a subsurface grid, where the grid is defined for each migrated sample. These grid points are computed using the techniques described in Section 2.8.2 using the Eikonal equation or similar technique. A traveltime grid (or traveltime map) is computed for each source location. The traveltime t_s from a given source to a scatterpoint is then found from the corresponding traveltime grid.

However, this method usually computes the first arrival times, which may not be the best for imaging. An example would be the earlier arrival of head-waves before reflected energy.

Raytracing methods shoot a bundles of ray from each source into the subsurface. Some form of interpolation is then required to map the traveltimes along the rays to all gridded migration points.

Each source and a scatterpoint may have a single raypath or may have multi raypaths, leading to numerous algorithms. Some algorithms allow the choice of the first arrival, or the raypath that arrives with the maximum energy.

Wave propagation methods use forward modelling to compute an expanding wave field from the source location. It is based on a "full wave equation" that can include all modes of propagation, multiples, attenuation effects, various wavelets, etc.

This method allows the selection of various imaging times such as:
- first arrivals,
- maximum energy arrival of the wavefront,
- multiple arrivals of the wavefront energy,
- and other methods that …

This method may be quite expensive relative to the previous methods but does provide the Kirchhoff methods with significant and unique capabilities. This method is also very similar to inversion techniques that will be discussed in the following section.

9.5.4.5.1 Gridded traveltimes

Gridded traveltimes are computed from a source location using methods similar to those described in Section 2.8.2. These methods typically compute the first arrival and are referred to as a traveltime map. Contours of the traveltimes are often displayed as illustrated below in Figure 9.15a. The same information from the first layer is plotted in a 3D view of part (b) where the traveltimes become the third dimension.

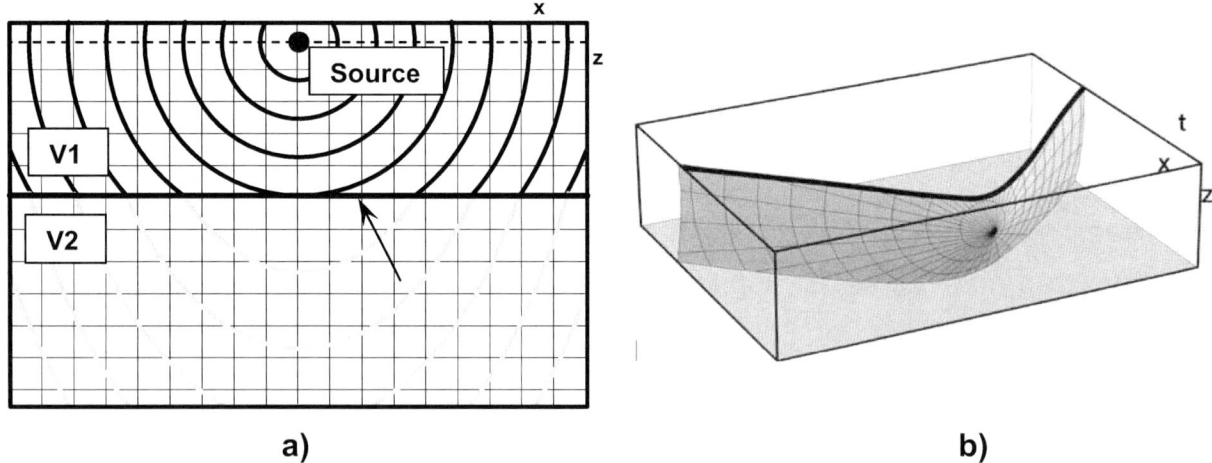

a)   b)

Figure 9.15 Traveltimes from a source plotted in a) as contours and in b) as a three dimensional image with traveltime as the third dimension.

The reason for showing the perspective view of the traveltimes in the upper layer is to identify the <u>arrival times on a time section</u> at a given depth, corresponding to the times and depth of a downward continued source record, defining the "<u>imaging condition</u>".

The gridded traveltimes are straight forward to compute and usually only compute the first arrival times.
In contrast, a Kirchhoff time migration computes the traveltimes from the pseudo depth layer using RMS velocities.

The following example uses a smoothed Marmousi velocity model to compute a traveltime grid using the Eikonal finite difference method described in section 2.8.2. After the gridded times were computed, contours of the traveltimes were estimated and displayed on the Marmousi model, (complements of Chad Hogan).

The traveltimes at a depth of 1km are mapped to a time section to become the imaging condition for one depth of a downward continued source record.

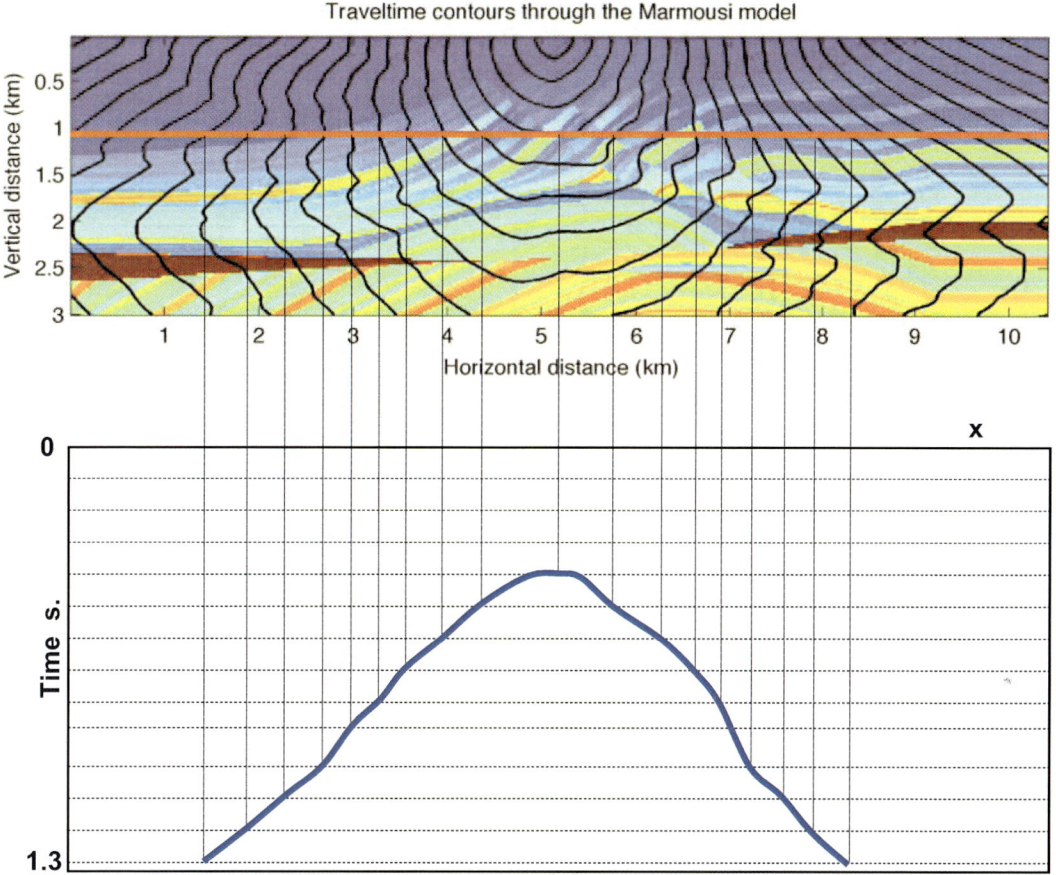

Figure 9.16 Contours of gridded traveltimes on the Marmousi model using the Eikonal finite difference method. The traveltimes at one depth (red line) are then mapped to a time section to defined the "imaging condition".

Note the simple shape of traveltimes that are estimated using the first arrival times.

Also note the arrival of refraction times on the upper left side of the contoured traveltimes.

9.5.4.5.2 Ray tracing

Traveltimes may be computed using ray tracing using the Eikonal equation as defined in section 2.5.2. A bundle of rays are "shot" from the source location and the traveltimes estimated along the raypaths. Traveltimes on the raypaths are then mapped to a grid for use in migration algorithm. The density of the rays in the bundle depends on the complexity of the geology and ability to interpolate to a grid.

Amplitude information may also be computed along the raypaths using the "transport" equation.

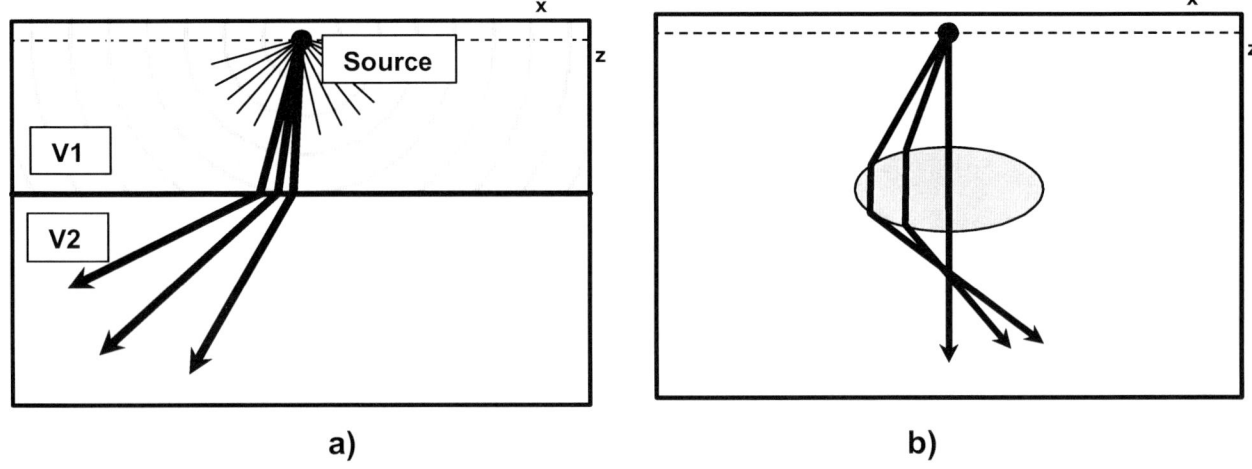

Figure 9.17 Raypath diagrams with a) showing a bundle of rays leaving the source point and three highlighted rays in a simple medium, while b) show three rays in a more geologically complex medium in which the rays cross each other.

The raypaths in a simple medium are illustrated in Figure 9.17a. In a more complex medium in (b), a seismic lens refracts rays so that they cross over, producing multiple arrival times (and amplitudes) at certain locations. From the multi arrivals, we could choose the first arrival, the ray with the maximum amplitude, or a combination.

It should be noted that the <u>first arrival is a logical choice in a simple medium</u> where there is no multipathing in the areas of interest. However, multipathing will exist in horizontally layered media where refraction energy may arrive before reflection energy, and limit the use of shallow reflection energy at larger offsets.

In complex media, the first arrivals may follow a higher velocity layer and produce a head wave in a lower velocity layers that arrives before the desired incident energy. This effect is also evident in Eikonal gridded time solver.

9.5.4.5.3 Full wavefront modelling

Modelling with the full wavefield also allows the choice of detecting the time of the first arrival or the time of the maximum amplitude when there are multiple arrivals. Other features of the wavefront may also be extracted for defining the imaging condition.

Note that in these cases, one should be viewing the data using the 3D view of Figure 9.15b, where the wavefront with an associated wavelet will intersect the horizontal reflector at a continuum of times. The times of the maximum arrival will be different from that of the first arrival and may cause slight changes in the migrated result as illustrated below.

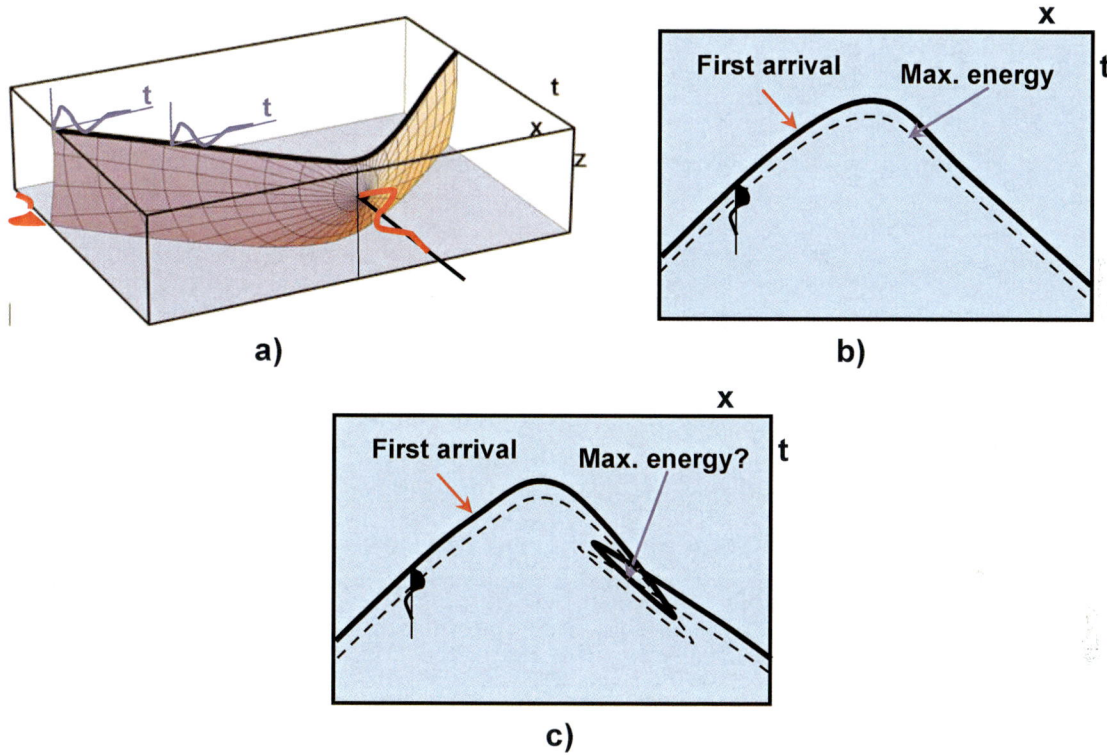

Figure 9.18 The wavefront in a) intersecting the horizontal reflector with a simplified wavelet, b) the time section at depth with a wavelet illustrating different times for the first arrival and the arrival of maximum energy, and c) the time section from a complex structure showing a triplication in the waveform as it passes the horizontal reflector.

At a given depth, the downward continued time section should be "focussed" at a location similar to the location defined by the forward model in Figure 9.18.

Can we do more with this information that simply defining a mapping "imaging condition" ???

9.5.4.5.4 Examples using the full waveform modelling of the Marmousi model

The following images were estimated on the full waveform to simulate the first arrival, maximum energy arrival, and multiple energy arrival to illustrate the complexity in estimating the Kirchhoff "imaging condition".

These figures represent the wavefield at one fixed time. Defining the "imaging condition" requires mapping one depth level from these figures to a time section, and repeating the process for all the constant time wavefields, i.e. Figure 9.18c.

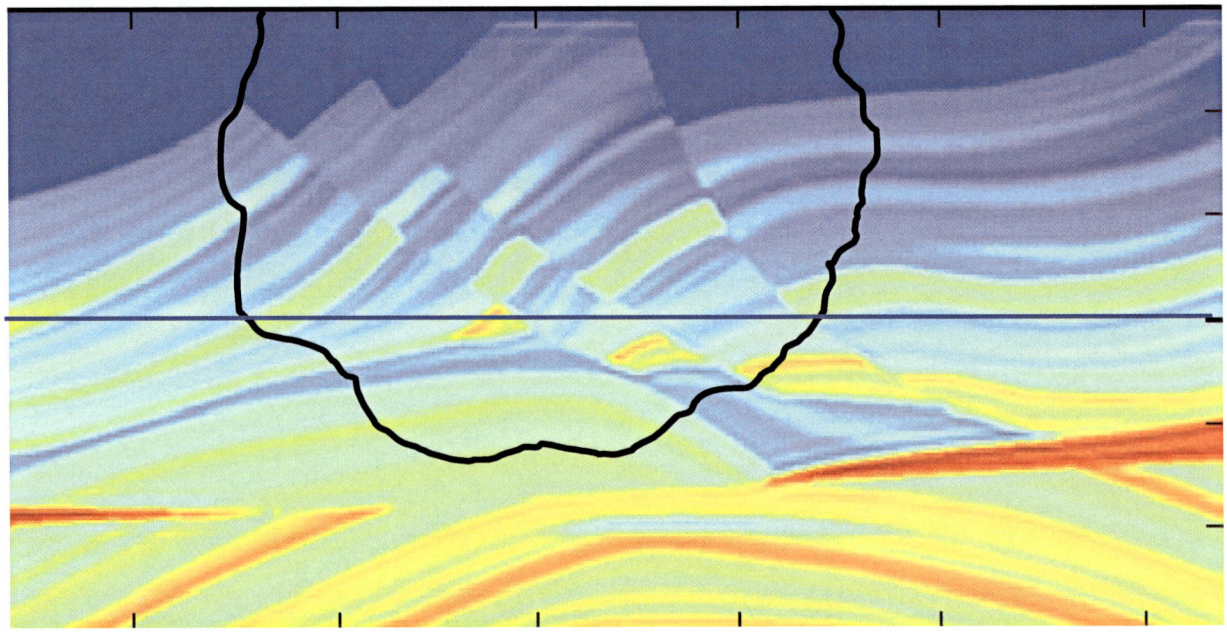

a) First arrival at time t_A on the depth section.

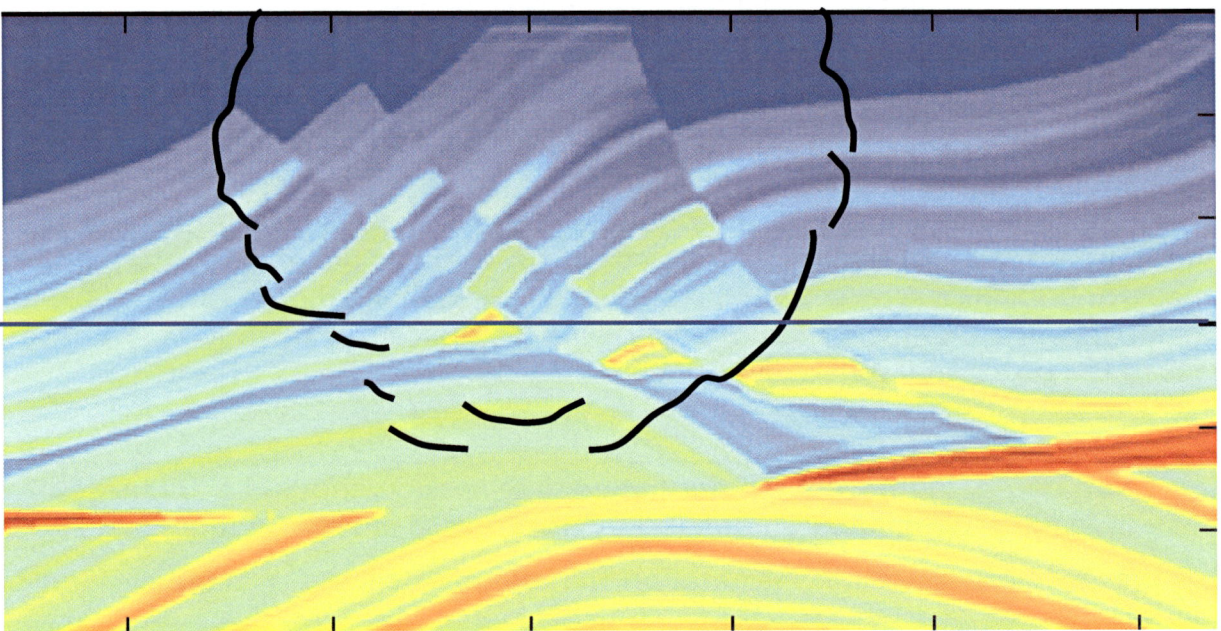

b) Maximum energy arrival at time t_A on the depth section.

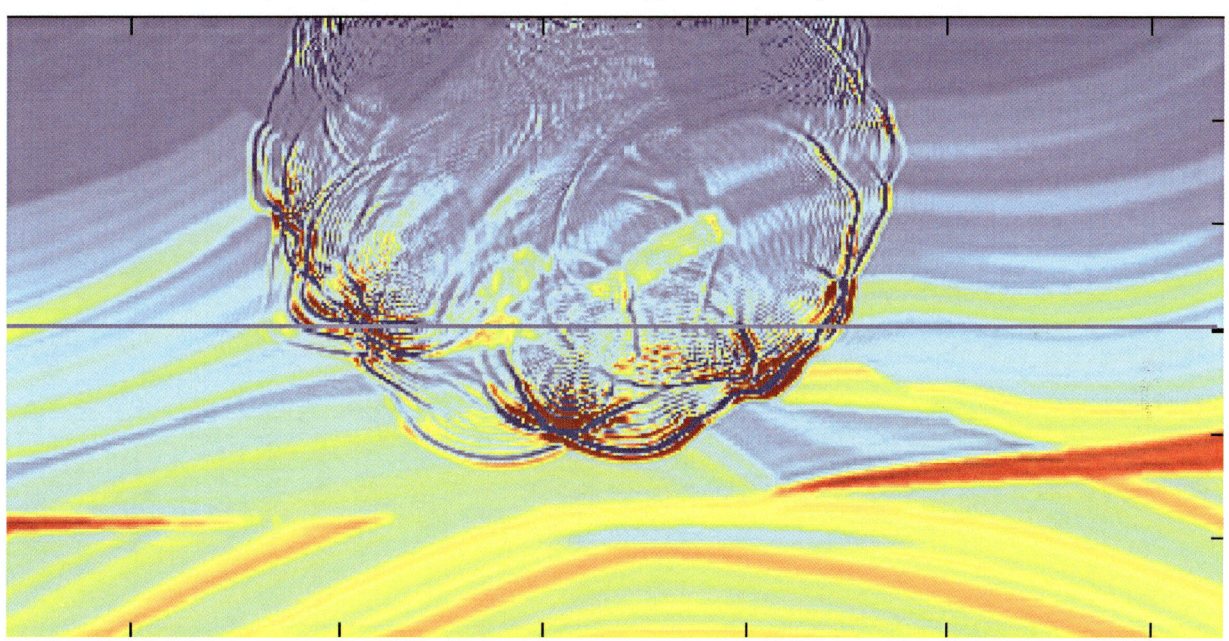

c) Multiple arrival at time t_A on the depth section

d) Full waveform at time t_A on the depth section (from Margrave movie Ch. 7)

Figure 9.19 Picks of the estimated wavefields at time t_A on depth sections, illustrating the difficulty in estimating the "imaging condition".

These wavefields when created at all times create a volume $W(x, z, t)$. A constant depth slice through this volume will define the time section $D(x, z = z_A, t)$.

9.5.4.6 Exercise:

The previous figures show picks for the first arrivals, maximum energy, and multiple arrivals, along with the full waveform, all for one given time t_A of the wavefield.

Use the information on these wavefields at the middle depth (1500 m, the blue line) to sketch the $D(x, z = z_A, t)$ image on the following time sections (similar to that in Figure 9.16. These sketches represent the "imaging condition".

You will need to estimate the wave fields at different times to complete these images (see Figure 7.47).

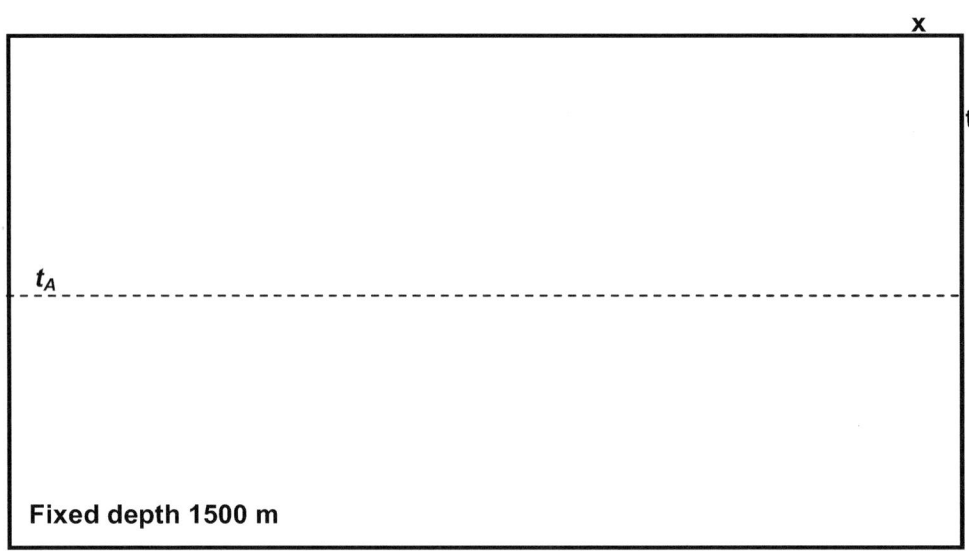

a) First arrivals

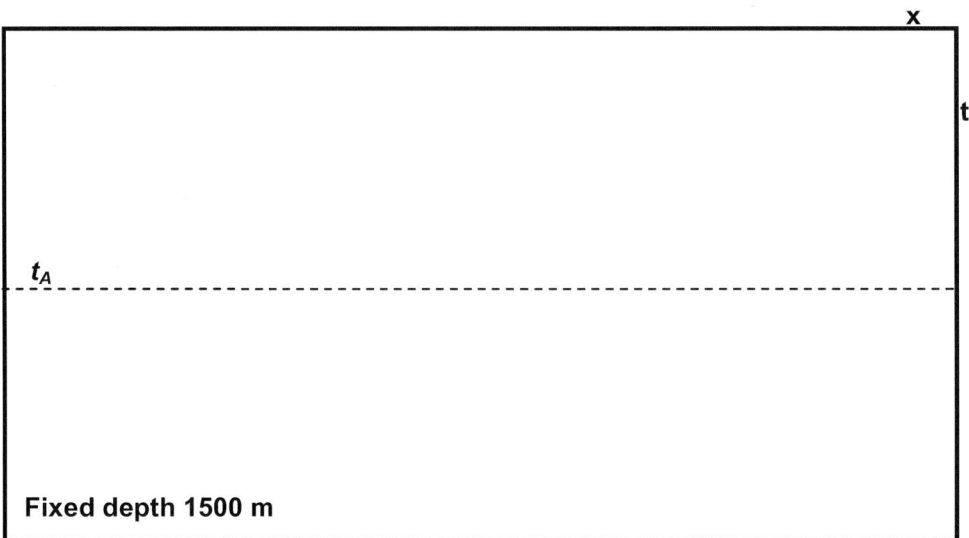

b) maximum energy

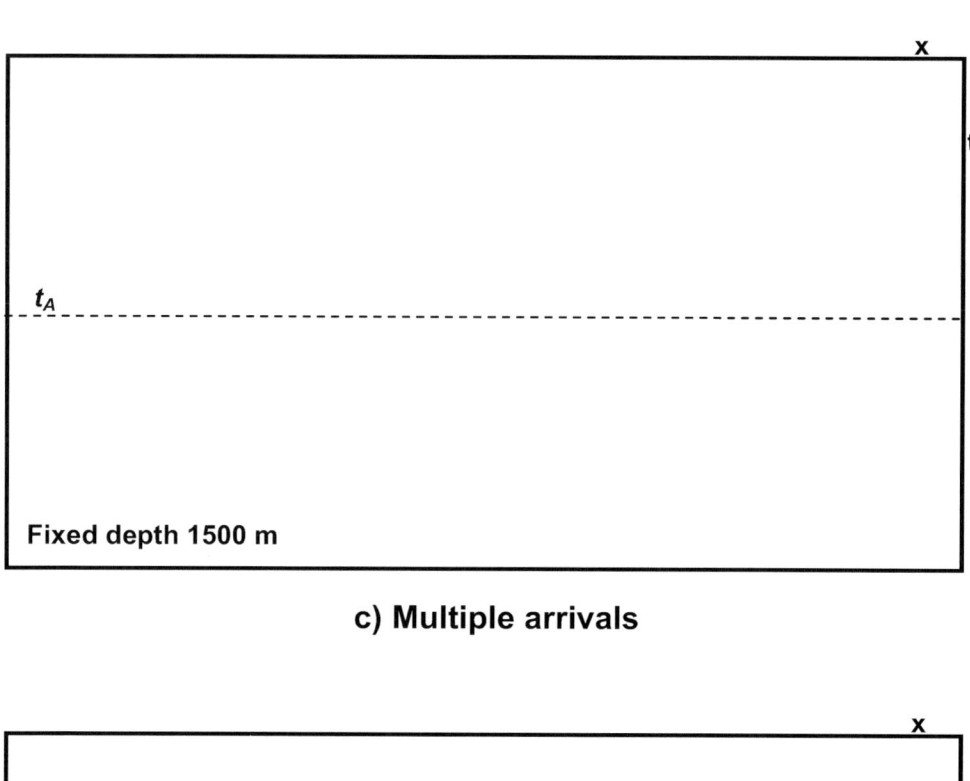

c) Multiple arrivals

d) Full wavefield (including the wavelet)

Figure 9.20 Representations of the "imaging condition" for various assumptions of the D wavefield.

The intent of this exercise is to emphasise the difficulty in choosing an "imaging condition" using the Kirchhoff approach, and to start thinking about an alternate approach that leads to *inversion*.

9.5.5 Model examples of prestack migrated source records

The following three examples in Figure 9.1 illustrate three different methods of migration on the hockey stick model.

The first figure (a) is a poststack migration of a stack in which the NMO was applied for horizontal events.

The second method in (b) was produced using the fast source record method described in section 9.3.1. The input data had a source at each station.

The third figure (c) used prestack source record Kirchhoff migration.

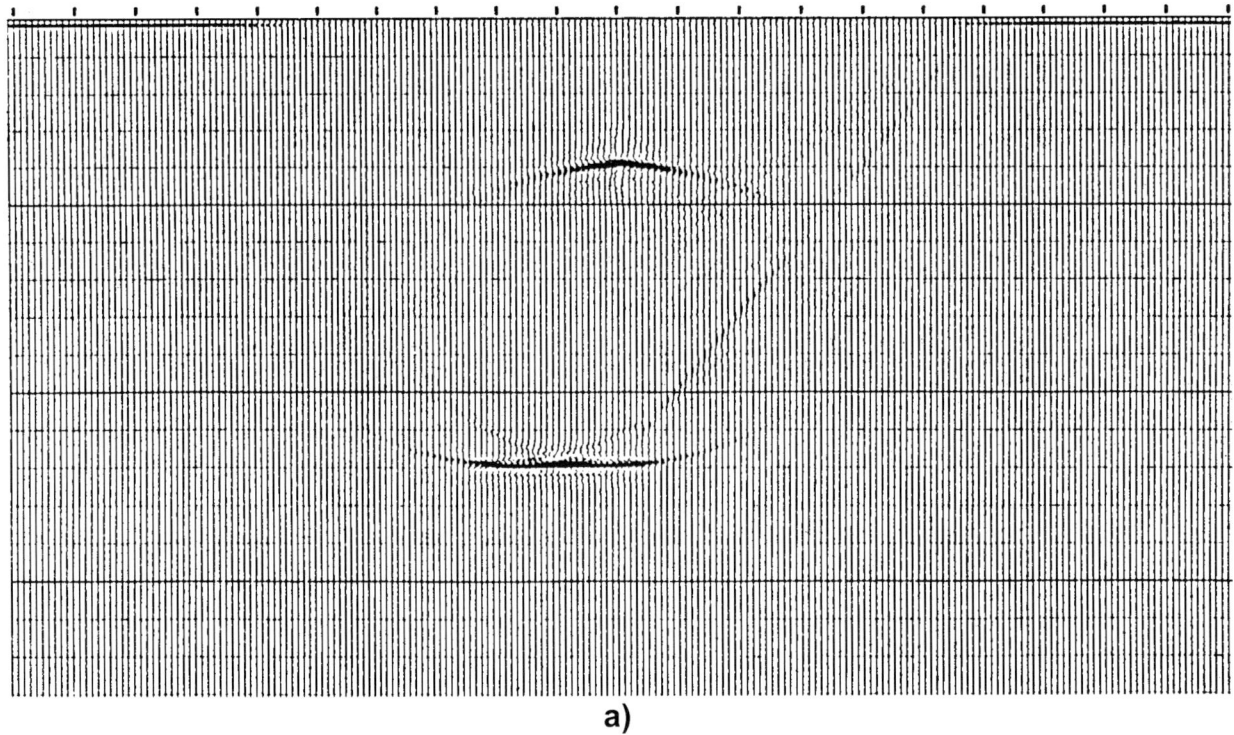

a)

Figure 9.21 Various migrations of hockey stick model, a) poststack migration, b) fast shot migration with no MO, and c) full Kirchhoff prestack migration of source records.

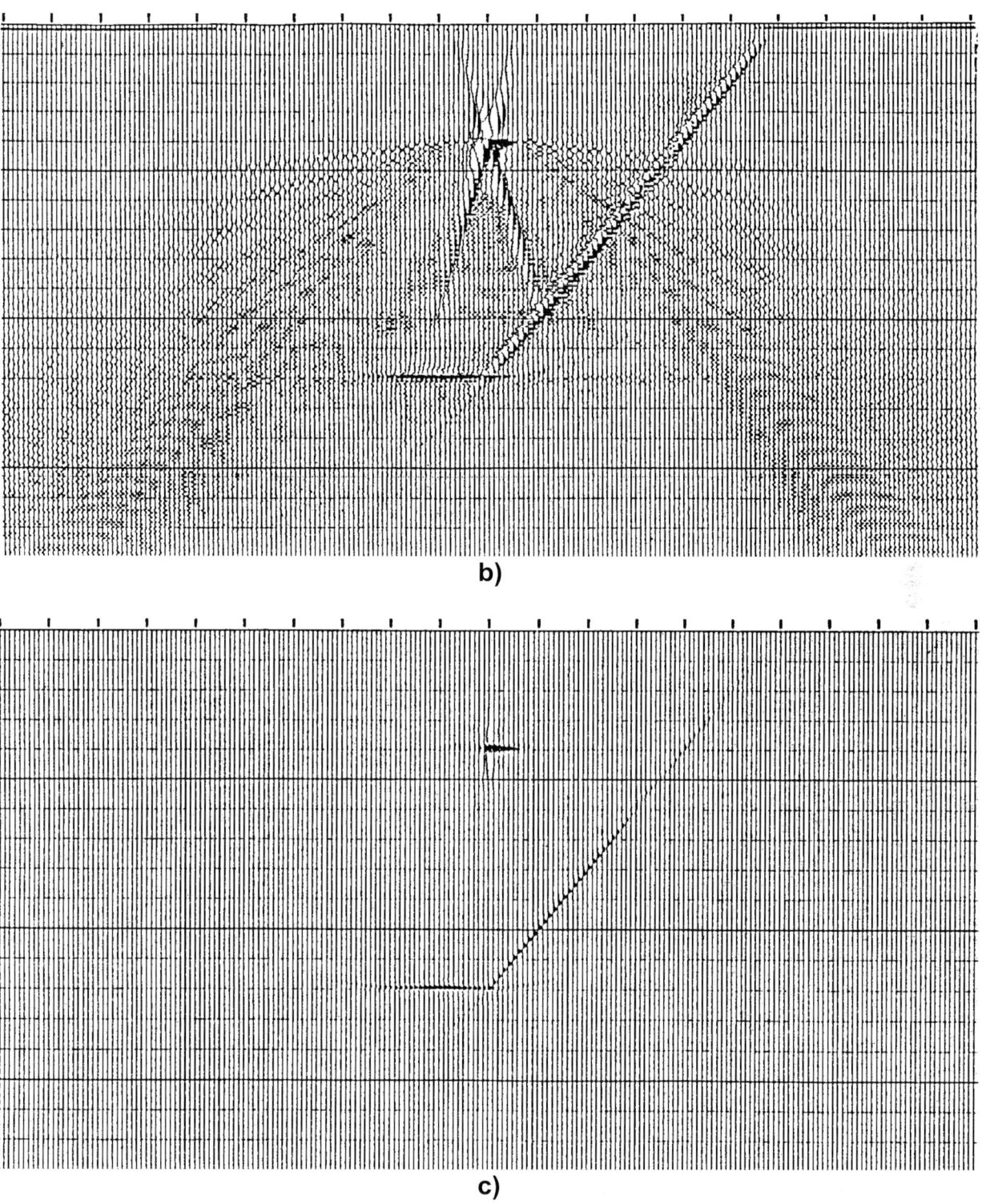

Figure 9.21 continued.

9.6 Inversion Method of Imaging a Source Record

9.6.1 Preamble

The previous Kirchhoff method that used forward modelling has lead us to a completely new imaging method that was introduced by Claerbout 1971 [7]. It may be more correctly defined as inversion. (Yes, the date is correct.)

As in the previous case, we still use the downward continuation of the source record along with the forward model of the source point, but we now define the "imaging condition" at a reflector to get the <u>reflectivity</u>. This is obtained by dividing the <u>reflected wavefield</u> just above the reflector by the <u>incident wavefield</u> just before the reflector. In Claerbout's words.

… to calculate a depth section: reflectors exist at points in the ground where the first arrival of the downgoing wave is time coincident with an upgoing wave.

This definition is appropriate for the previous methods, but Clarebout went on to define the "reflectivity" as outlined in this section.

<u>The incident wave field</u> is defined by the forward modelled energy from a source location to the reflector.

<u>The reflected wavefield</u> is defined using the downward continuation the source record.

At this point we abandon the concept of diffractions, and go straight to the <u>concept of reflectivity</u>.

To emphasize the significance of this concept, note that amplitude of the reflectivity will be independent of the energy in the modelled or DC energy.

In the following figures, we change from examining diffractions from scatterpoints to reflection energy from a reflector at a given depth z_A. A horizontal reflector at that depth will have a hyperbolic shape, which is the envelope of a horizontal array of scatterpoints, as illustrated in Figure 9.12.

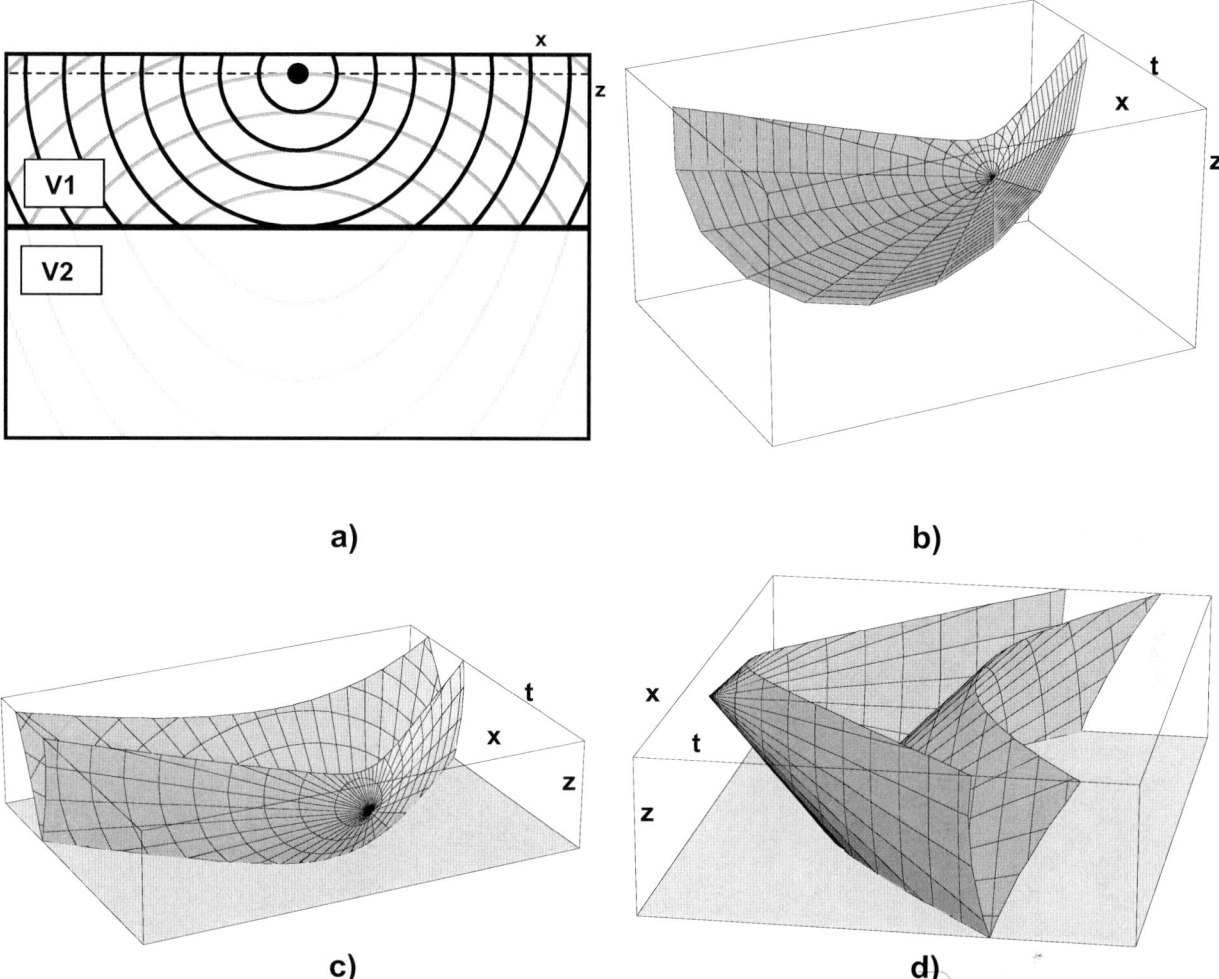

Figure 9.22 Wavefront model for a horizontal reflection with a) showing a multiple exposure of the wavefront emanating from the source and the returning reflection, b) the source wavefront in 3D view with time as a coordinate, c) the combined source and reflected energy in perspective view, and d) an alternate perspective view of (c).

9.6.2 Defining inversion at the imaging condition

Consider the downgoing wave (*D*) from modelling a shot, and the up-going wave (*U*) from the reflector as illustrated below. Both waves occupy the "same space" at the reflector.

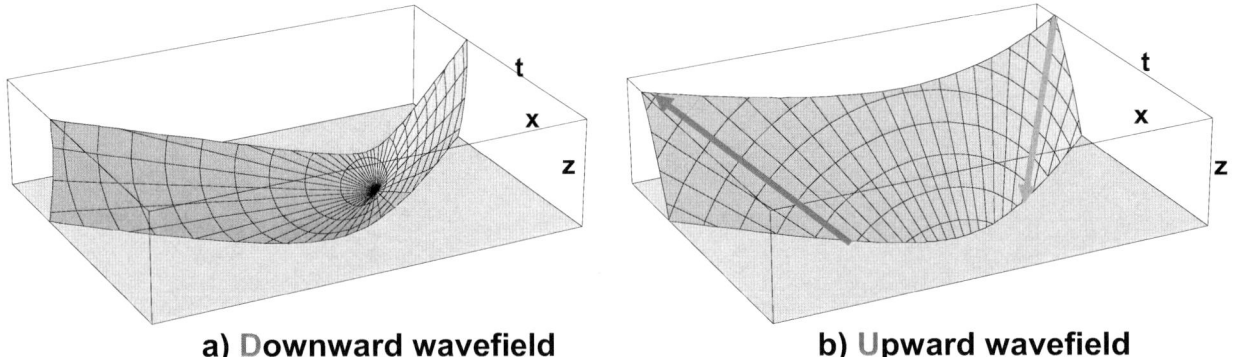

a) **D**ownward wavefield   b) **U**pward wavefield

Figure 9.23 Down and up-going wavefields from a single reflector, with the blue arrow representing the upward reflection energy, and the red arrow the downward continuation of the surface reflecting energy down to the reflector.

Now consider the downgoing wavefield D just above the reflector, and the up-going wavefield U just above the reflector as illustrates below.

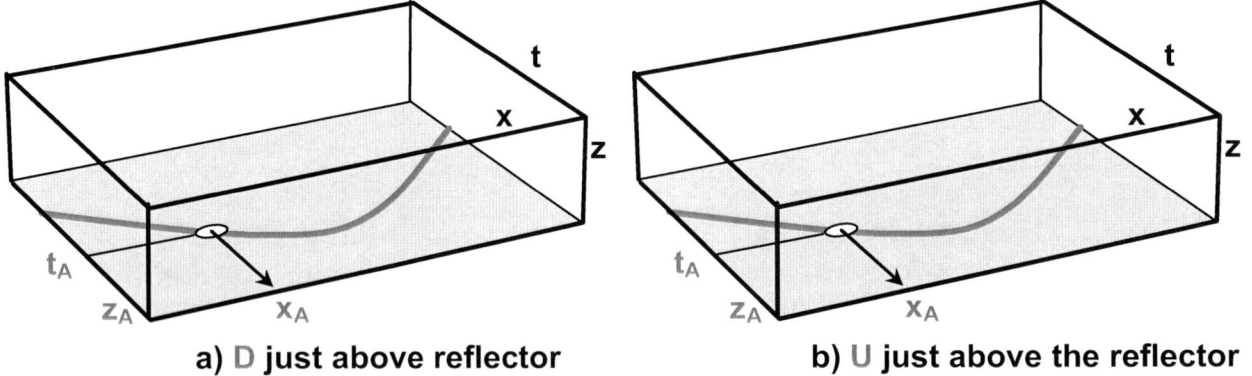

a) **D** just above reflector   b) **U** just above the reflector

Figure 9.24Fig Down and up-going wavefields just above the reflector.

Suppose we could divide U by D, to get the reflectivity R at each spatial location x_A. That would be inversion.

$$R(x_A, z_A) = ? \frac{U(x_A, z_A, t_A)}{D(x_A, z_A, t_A)} \tag{9.6}$$

Do you realize the significance of this concept? It is huge. It is the foundation of imaging inversion to get "rock properties", in addition to using inversion as a mathematical process.

Consider a more realistic case where the wavefields contain multiples or mode converted energy as illustrated below.

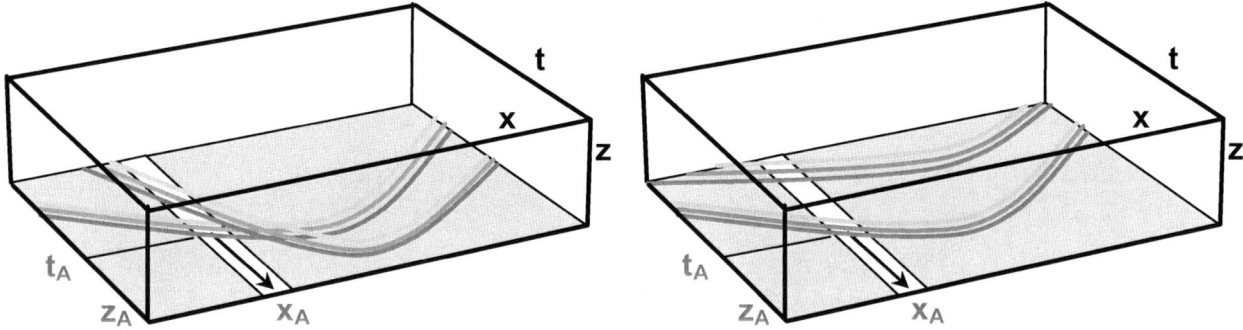

Figure 9.25 Cartoon representation of the D and U wavefields at depth z_A

Note that a yellow band at x_A has been added to indicate that all the energy in this band contains the energy of the D and U wavefield of the reflecting element at (x_A, z_A). However it is <u>only the energy that correlates</u> between D and U that is indicative of the reflectivity.

How can we do the division as desired? Claerbout 1971 told us to use the Fourier transform to accomplish the desired division..

$$R(x_A, z_A, f) = \frac{U(x_A, z_A, f)}{D(x_A, z_A, f)} \tag{9.7}$$

All is OK unless there potential zeros in the denominator of D. We can improve the stability by multiplying the numerator and denominator by the conjugate of D, D^*, i.e.,

$$R(x_A, z_A, f) = \frac{U(x_A, z_A, f) D^*(x_A, z_A, f)}{D(x_A, z_A, f) D^*(x_A, z_A, f)} \tag{9.8}$$

Note that $D \times D^*$ is a real number and just a scale factor in the equation while $U \times D^*$ is the guts (phase or timing) of the whole process. Now we have

$$R(x_A, z_A, f) \approx k\, U(x_A, z_A, f) D^*(x_A, z_A, f), \tag{9.9}$$

which is the <u>product of two frequency domain arrays</u>. In the time domain that <u>becomes a cross-correlation</u>.

$$R(x_A, z_A, \tau = 0) \approx k\, U(x_A, z_A, t) \otimes D(x_A, z_A, {}_a t), \tag{9.10}$$

where I use $\otimes$ to indicate a cross correlation. The <u>zero lag</u> of the cross correlation gives the amplitude of the reflectivity, which is essentially sum the dot product of each corresponding traces in U and D^*.

Even thought the "time" of the imaging condition is curved with spatial position, the peak of the cross correlation will be placed at x_A and z_A, as an estimate of the reflectivity $R(x_A, z_A)$.

The cross correlations are repeated for all x locations, and for all depths.

9.6.3 What if the velocity field is not accurate?

Figure 9.26 shows the two time sections $D(x, t)$ and $U(x, t)$ for a given depth z_A, but with an <u>error in the velocity</u> producing different times of the "imaging condition".

These errors may result form an incorrect geological structure, or in errors of the velocity in the layers.

A partial focusing of energy may occur at a slightly different time and depth, and the peak of the cross correlation will <u>not</u> occur at zero offset.

Mapping the lag and depth of the cross-correlation provides an indicator of the error in the velocities, leading to numerous velocity analysis techniques. (See Al-Saleh 2006, PhD Dissertation)

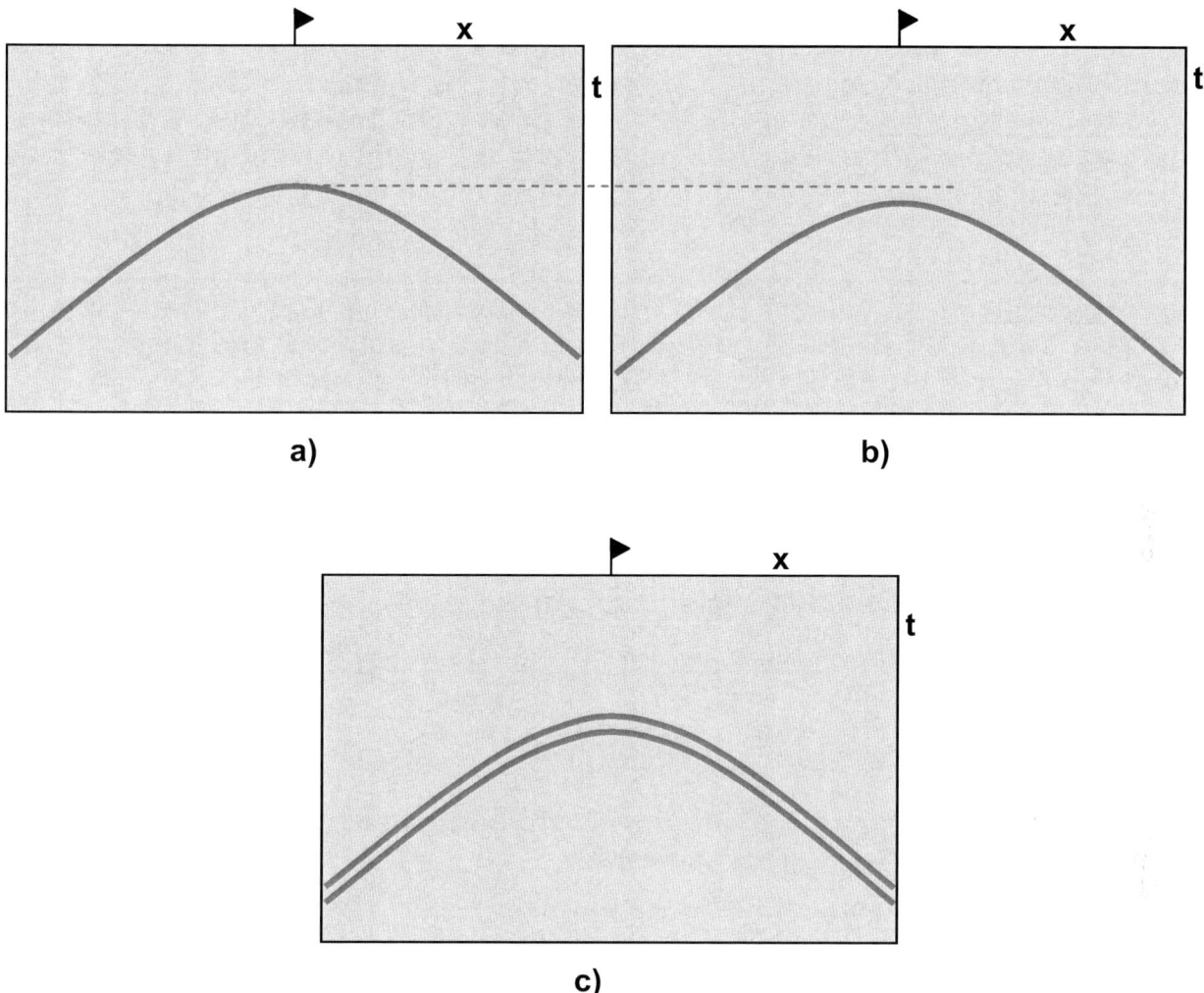

Figure 9.26 Up and downgoing wavefields at z_A with a slight velocity error. The figure in a) shows the up-going wavefield, b) the downgoing wavefield with a slight time shift, and c) with the times of the up and down going wavefields combines on the same time section.

9.7 Source-Receiver (Shot-Geophone) Method

9.7.1 Basics

From the previous sections, we are aware that the <u>shape of a diffraction</u> (impulse response) on a source record is determined by the location of the scatterpoint <u>relative to the receivers</u>. We also know that downward continuation will collapse the energy in those diffractions to a point when the downward depth reaches the depth of the scatterpoints.

<u>Reciprocity principle</u>: A seismic trace from a source at *A* to a geophone at *B* is the same as a source at *B* and a geophone at *A* if the sources and receivers are similarly coupled to the earth Sheriff [543]. (No mode conversion)

From this principle:

- Source records could be considered <u>common receiver gathers</u>.
- Diffractions are defined by <u>traveltimes from the sources to scatterpoints</u>.
- Downward continuation <u>collapses the diffracted energy from the sources</u>.

<u>The s-g method</u>:

- <u>Source records</u> in Figure 9.27 are downward continued one depth increment to focus the energy diffracted to receivers.
- Data is <u>sorted to receiver gathers</u> of Figure 9.27.
- Receiver gathers are downward continued <u>one depth increment</u> to focus the energy "diffracted" from sources.
- Energy from scatterpoints located at this depth will be <u>focused</u> at $t = 0$ (depth migration) and copied to the migrated section.
- The process continues by <u>alternating downward continuation steps</u> between source of (a) and receiver gathers of (b) until the migration is complete.

A tremendous amount of data <u>sorting is required between source and receiver gathers</u>. This sorting runs more efficiently on computers with <u>large CPU memory</u>.

The sorting of traces may be reduced by using "C" type programming languages that allow traces to be located by address.

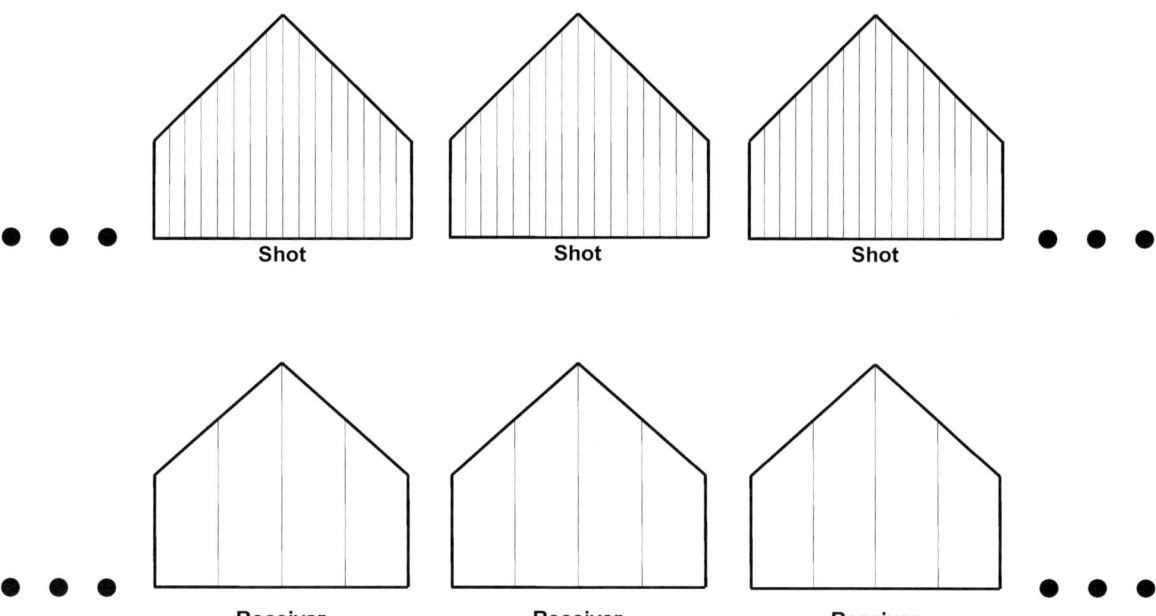

Figure 9.27 Source and receiver gathers for s-g migration.

When the velocities are known, the action of alternate migration of the source and receiver gathers <u>moves data toward the zero-offset position</u> of each gather.

- For <u>depth migration</u>, the output point occurs at t = 0.
- For <u>time migration</u>, the output is at the time of the tau step.

When velocities are in error, the data will <u>not</u> focus at zero-offset or at $t = 0$ for depth migration. Use of this focusing condition may help improve the estimate of velocities.

A major problem with this method is the large trace spacing in the receiver gathers that result from shooting at multiple station intervals. This large trace spacing will contribute to <u>aliasing</u> of the receiver gathers.

What could be done to eliminate the problem of aliasing in the receiver gathers?

9.7.2 Construction example for s-g migration.

The objective of this construction is to understand the kinematics of s-g prestack migration. Energy from Cheops pyramid will be focused to the scatterpoint at zero-offset on the source record or receiver gather.

Figure 9.28a shows a geological cross section with one scatterpoint and seven source locations.

- Figure 9.28b shows the source records corresponding to sources in (a).
- There are nine receivers activated symmetrically about each source.
- The width of the <u>source records has been compressed</u> to accommodate the far offset of Figure 9.28a.
- Figure 9.11c is a plan view of Cheops pyramid showing the time contours and locations of two-sided source records.

Exercise

- Identify the result of downward continuation <u>time migration</u> to the depth of the scatterpoint in the source records in Figure 9.28b and c.
- With the aid of Figure 9.28c, resort the data to receiver gathers then sketch the results on Figure 9.28b.
- Apply the same downward continuation to the receiver gathers.
- What is the offset and time of the energy after s-g downward continuation time migration to the depth of the scatterpoint?

Note:

- After s-g migration, all energy will have moved to zero offset.
- Each source gather and each receiver gather will only have one live trace.
- In a real migration, the ability to focus energy to zero offset will depend on the accuracy of the velocity model.
- The miss-positioning of energy may indicate how the velocity model should be modified [787].

> S-g migration requires sources at each station location.

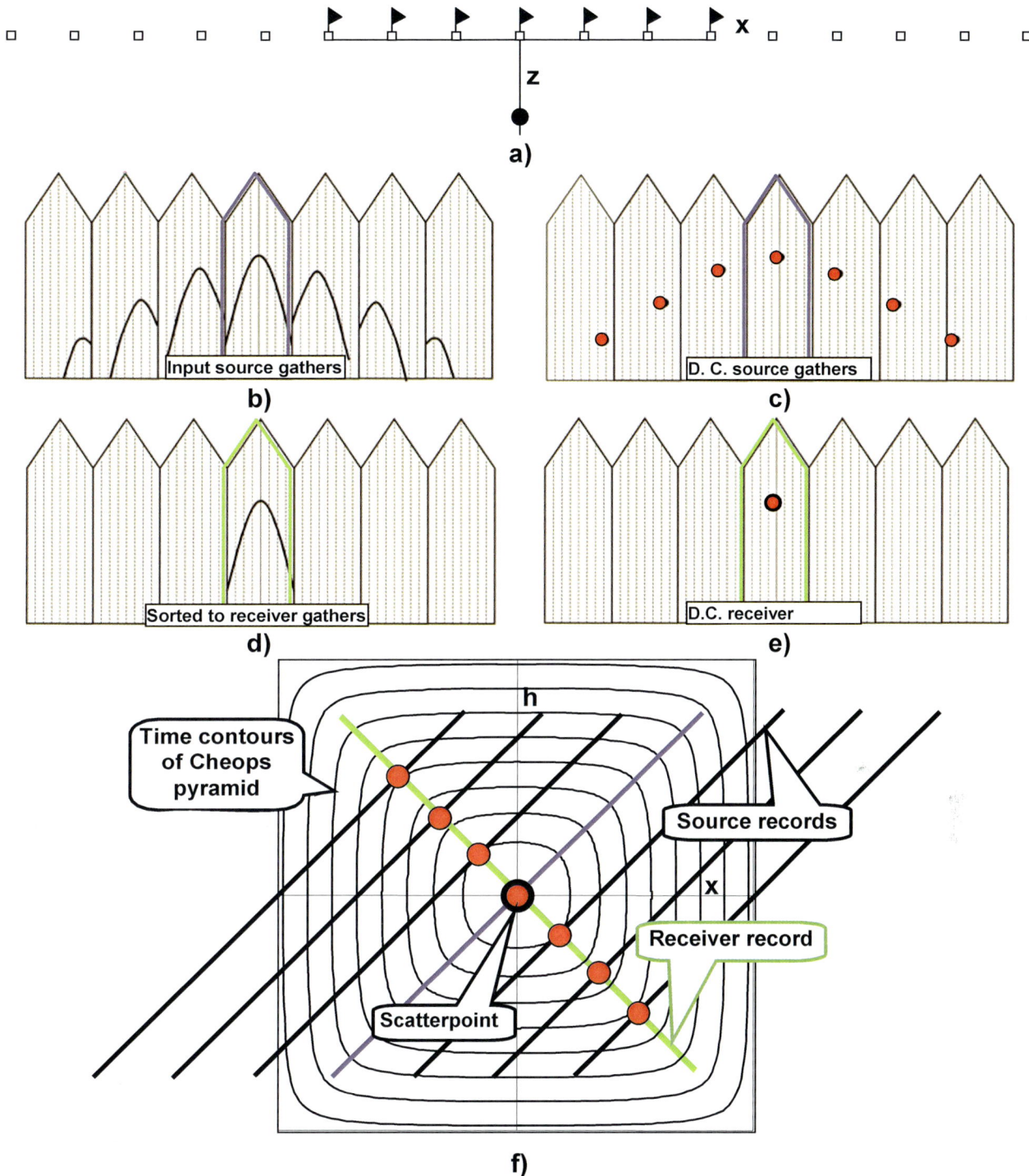

Figure 9.28 Views of s-g migration, a) shows a scatterpoint in a depth cross-section, b) source records, c) receiver records (after source records are migrated), and d) plan view of Cheops pyramid with the scatterpoint and source records.

9.8 Prestack Migration of a Constant-Offset Section

9.8.1 Kirchhoff Method

Another common method of prestack migration is performed on constant offset sections. A number of algorithms are available, however Kirchhoff appears to be more popular.

Figure 9.29 shows the diffraction shapes from <u>scatterpoints</u> on a constant offset section. Shown in gray are the zero-offset diffractions. Kirchhoff migration computes the shape of these diffractions from RMS velocities, or from traveltime estimates. The energy is placed at the scatterpoint location on the migrated output section, (<u>not at the apex</u> of the offset diffraction).

For a scatterpoint at $x = 0$ and vertical two-way travel time T_0, the shape of the diffraction (T, x) for time migration is given by the double-square-root (DSR) equation

$$T = \left[\left(\frac{T_0}{2}\right)^2 + \frac{(x+h)^2}{V^2}\right] + \left[\left(\frac{T_0}{2}\right)^2 + \frac{(x-h)^2}{V^2}\right], \qquad (9.11)$$

where the offset is h, and the RMS velocity V is defined at T_0.

Figure 9.30a illustrates the <u>dual process</u> of migrating a <u>single input sample at time T</u> to all possible scatterpoints at times T_0. For <u>constant</u> velocity, the curve is an ellipse with the source and receiver locations at the foci, the bottom of the ellipse is defined by the NMO time T_0, and the lateral extent at $t = 0$ defined by $TV/2$, i.e.,

$$\frac{t^2}{T_n^2} + \frac{4x^2}{T^2 V^2} = 1 \qquad (9.12)$$

When the <u>velocity varies in space and time</u>, the curve is no longer an ellipse, but is defined by the RMS value at each scatterpoint (similar to Hagedoorn poststack migration). This approach is therefore limited to constant velocity sections.

In contrast, the Kirchhoff summation approach is based on a single RMS velocity that is defined at the scatterpoint, from which all diffraction times are computed. Since a different RMS velocity may be defined at each scatterpoint, the Kirchhoff approach is applicable to areas where the RMS assumptions hold.

<u>Constant offset depth migrations</u> may use a Kirchhoff method that computes wavefront traveltimes for the source and receiver raypaths. Downward continuation of a constant offset section is difficult.

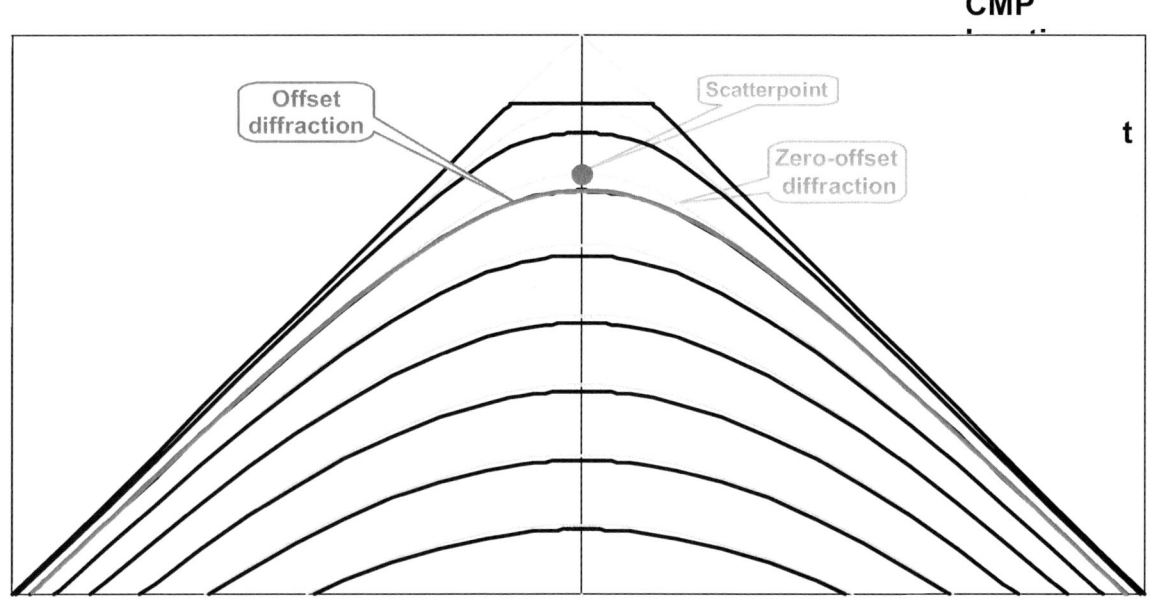

Figure 9.29 Diffractions from scatterpoints on a constant offset section in black, and the corresponding zero-offset diffractions in gray.

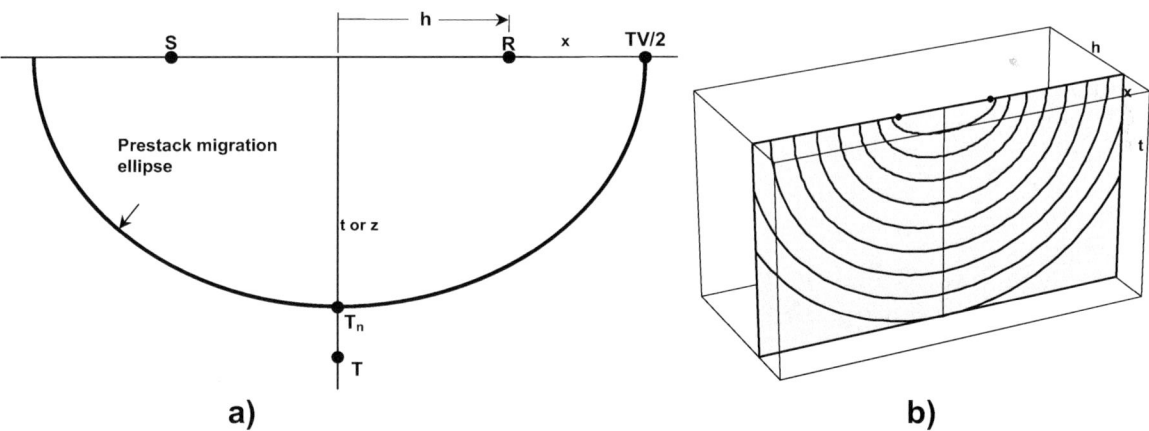

a)   b)

Figure 9.30 Constant offset prestack time migration with a) showing the prestack migration ellipse, and b) the prestack migration of one trace in the prestack volume (x, h, t). These curves are elliptical when the velocity is constant.

References for prestack migration of constant offset sections are Notfors[141], Sattlegger [210], Deregowski [211], Schleicher [266], Kim [276], Trorey [420], Sullivan [444], Dubrulle [454], Jannaud [514], Popovici [522] and section 33.3 in Claerbout's book [294].

9.8.2 Pseudo code for constant offset migration

Pseudo code for constant offset section migration

Loop *migrated trace*	`for i = 1 : I`
Migrated trace geometry	`x_s, V`
Loop *input traces*	`for k = 1 : K`
Input trace geometry	`x_s, x_r, x_mp,`
Migration geometry	`h_s, h_r, x`
Loop *migrated samples*	`for j = 1 : J`
Define time on input trace	`T = DSReqn(T_0, h_s, h_r, V)`
Define time sample number	`m = T/dt`
Check bounds of the input data	`if j < J`
Cosine weighting, plus others	`W = To/T`
Sum data to migrated trace	`Migdata(j,i)+=W*Indata(m,k)`
End	`end`
End	`end`
End	`end`

Compare the pseudo code for a constant offset migration with the pseudo code for a source record migration.

Explain the differences.

9.8.3 Downward continuation approach to constant offset (???)

In the constant offset modelling section, we saw that it is <u>not possible</u> to <u>propagate energy</u> from a scatterpoint directly to an offset diffraction.

I will indicate the same problem exists in downward continuing constant offset data. It will turn out that we can downward continuation one increment, but in doing so, the energy is moved from the original single offset to all smaller offsets.

In the zero offset case, downward propagation moves the energy down the raypaths toward the scatterpoint as illustrated below.

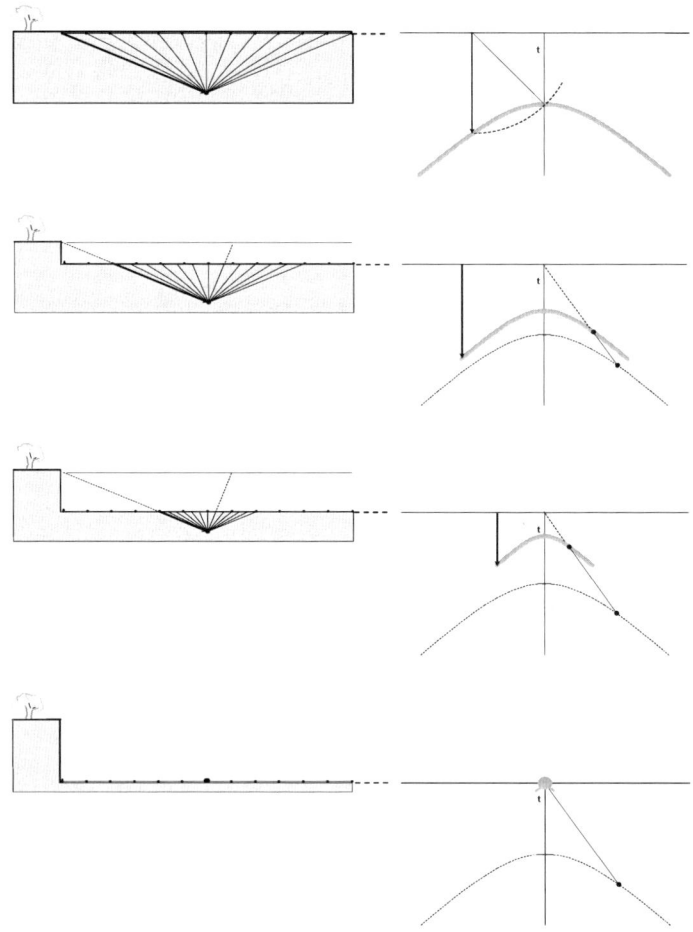

Figure 9.31 Zero offset downward continuation, illustrating the diffraction energy on the right propagates to a point when the downward propagation, on the left, reaches the depth of the scatterpoint.

In the offset case, I use Figure 9.32 to illustrate the movement of energy back along the raypaths toward the scatterpoints. We see energy propagating towards a number of scatterpoints. The relative location of the source and receiver to each scatterpoint is changed with one downward increment, i.e. the data is no longer at a constant offset.

When the downward continuation is at the depth of the shallowest scatterpoint, the new offset for the source and receiver (S_0, R_0) is zero. The deeper scatterpoints have offsets that now range from short, then to tending toward the original offset. Consequently, the data is then spread over a range of offsets.

(In the zero-offset case, the data remains at zero offset.)

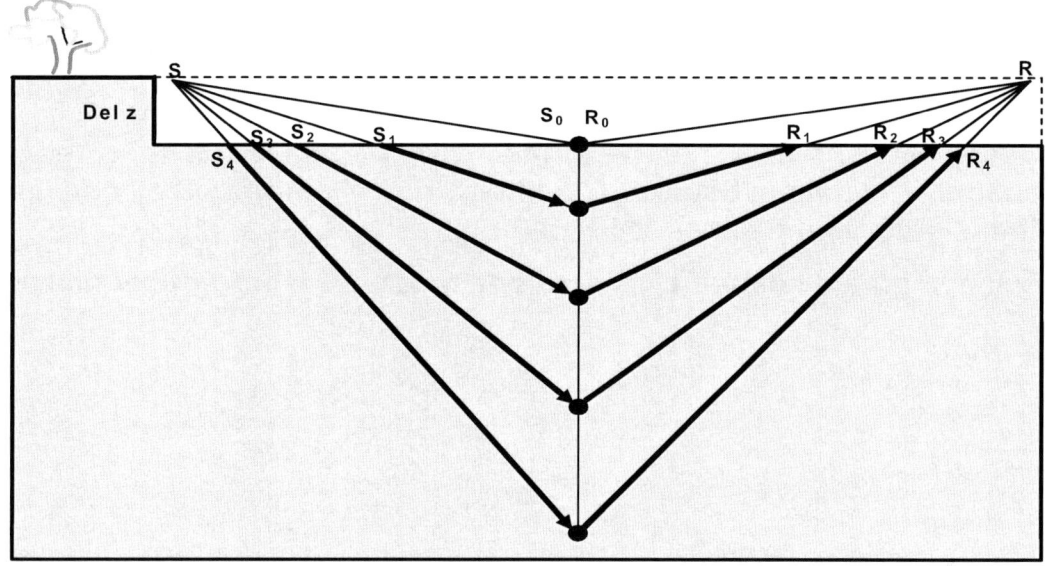

Figure 9.32 Raypaths illustrating the change in the source-receiver offset after one downward increment on constant offset data.

Prestack downward continuation or wave propagation migrations are only performed on source records.

Recall that a DMO process converts an offset section to zero offset, enabling a zero-offset downward continuation.

The Kirchhoff approach remains the only practical algorithm for migrating a constant offset section.

9.9 Constant Angle and τ-p Migration

An interesting method of prestack migration was presented by Ottolini [373] [452] in which the prestack volume (CMP, offset, and time) was sliced at constant angles to form <u>radial sections</u> as shown in Figure 9.33. These radial sections intersect Cheops pyramid to form hyperbolic curves as indicated in Figure 9.34.

> **Prestack diffractions on radial sections are hyperbolic.**

These properties are only valid for <u>constant velocities</u>. Ottolini and Claerbout [452] described a similar section that uses a constant ray parameter for variable velocities.

On a CMP gather, a radial line intersects MO hyperbolas at a slope defined as "p" as illustrated in Figure 9.35. These "p" values may be used to create τ-p transforms (or slant stacks) of the CMP gathers.

Constant "p" sections may then be formed from the slant stacks, created at each CMP location. The data in these sections may now be migrated by conventional 2-D algorithms such as the phase-shift method.

Stacking the migrated constant "p" sections produces a zero-offset time section.

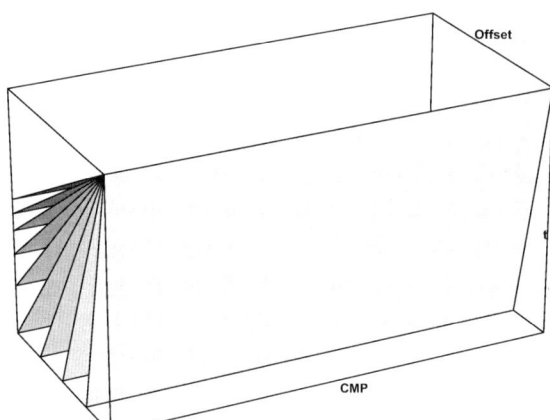

Figure 9.33 Radial trace sections.

> A conventional 2-D Kirchhoff migration will collapse the radial diffractions on the surface of Cheops pyramid to the NMO hyperbola on a single CMP gather. This gather is located at the center of Cheops pyramid or scatterpoint. Velocity analysis would allow a more accurate estimate of the velocity for NMO.

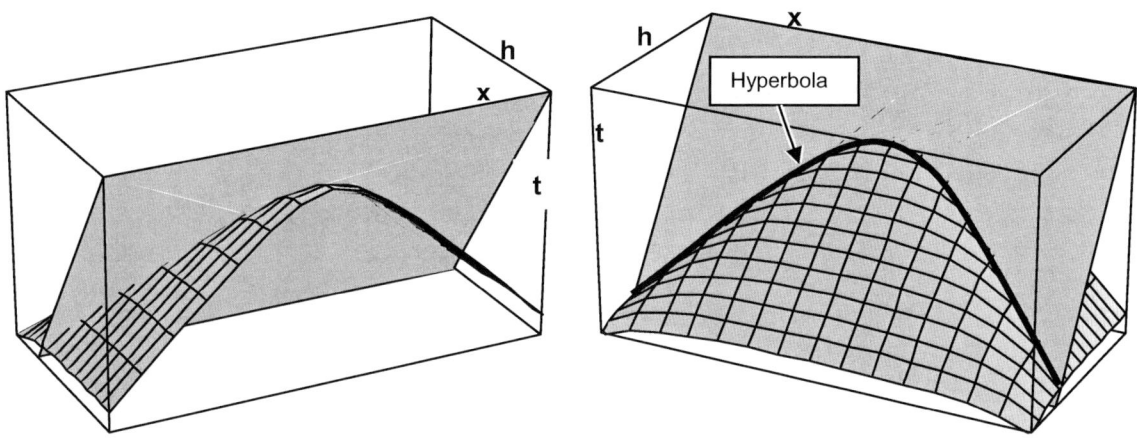

Figure 9.34 Front and rear views of the Cheops pyramid intersected by a radial plane, illustrating the intersection to be hyperbolic.

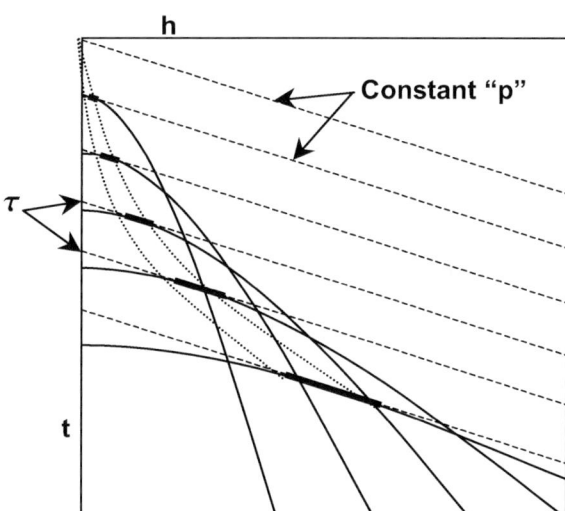

Figure 9.35 A CMP gather (*x, h*) showing various MO hyperbola with an area of common slope highlighted. The areas of common slope "*p*", and the intersection of the common slopes with *h = 0*, defines the various times of τ for the τ-*p* transform.

9.10 Stolt Prestack Migration of 2-D Data

A three dimensional volume (x, h, t) has been used to display the prestack energy of 2-D data and is repeated in Figure 9.36a.

This 3-D volume may be <u>3-D Fourier transformed</u> to a new volume with co-ordinates *Kx*, *Kh*, and ω, where the K's represent wave number, and ω the frequency, as shown in Figure 9.36b.

Stolt [21] described a method for converting the 3-D volume *(Kx, Kh, ω)* into an output section *(x, t)* at a <u>constant velocity</u>. This process may be repeated many times to build a new 3-D volume *(x, t, V)* illustrated in Figure 9.17c where *V* is the velocity.

Data in this 3-D volume *(x, t, V)* will <u>focus the energy on a surface</u> that matches an <u>optimum velocity profile for the line</u>. This surface may be interpreted and picked on a workstation to extract a velocity profile and to produce a prestack migrated section.

The shape of Cheops pyramid in Figure 9.36a is dependent on velocity.

Prestack Kirchhoff migration sums the energy in Cheops pyramid to the scatterpoint located at the apex of the pyramid.

The above Stolt process is similar to a time domain <u>prestack Kirchhoff migration</u> over the range of velocities.

The optimum velocity would match the summation surface with the input surface, producing a maximum at that point in the *(x, t, V)* volume.

Data from all scatterpoints will thus form the desired optimum surface.

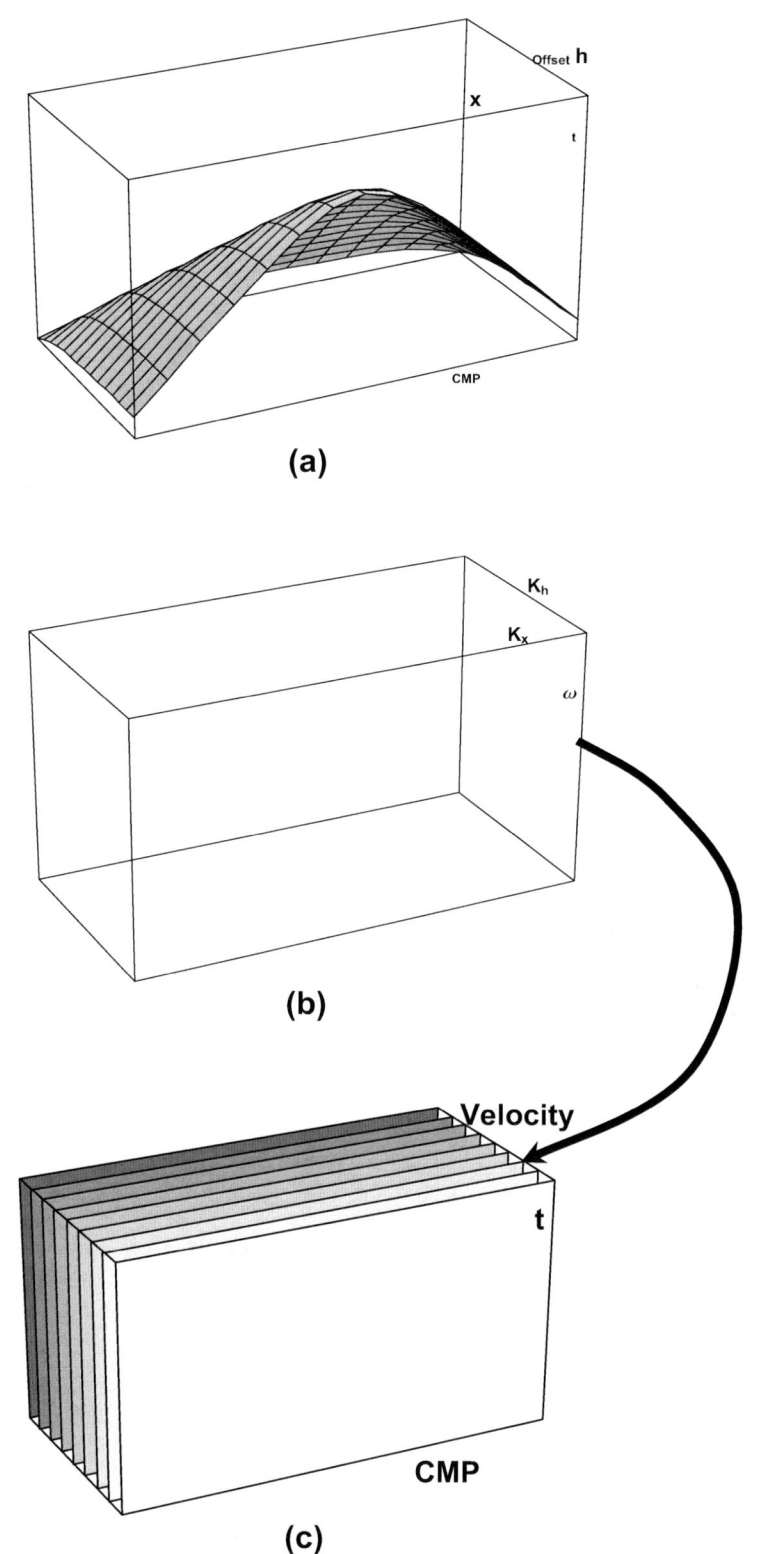

Figure 9.36 Stolt migration of 2-D data showing a) the input volume (x, h, t), b) the Fourier transform volume (Kx, Kh, ω), and c) a number of constant velocity prestack migrated sections.

9.11 Gardner's GDMO-PSI

9.11.1 Introduction

This method of prestack migration is one of the premier methods of prestack migration. An intermediate step in the processing forms prestack migration or CSP gathers, independent of any velocity. After the gathers have been formed, velocity analysis may be performed, and the prestack migration process completed with Kirchhoff MO and stacking.

The previous Chapter on DMO introduced Gardner's DMO (GDMO) as a method of DMO that is applied before MO. Although GDMO is velocity independent, it is (by itself) unable to resolve the problem of different velocities for overlapping events.

Application of GDMO, followed by prestack imaging (PSI), does resolve the problem of overlapping reflections to create an optimum prestack migration.

The process of GDMO, followed by PSI is referred to as GDMO-PSI.

GDMO-PSI is totally independent of velocities.

The prestack migration gathers formed by GDMO-PSI are referred to as common scatterpoint (CSP) gathers as the energy is optimally positioned along hyperbolic paths for velocity analysis.

Figure 9.37 illustrates the GDMO-PSI method:

(a) Energy from a scatterpoint lies on the surface of Cheops pyramid.

(b) After GDMO, the energy is reconstructed on the surface of a hyperboloid.

(c) PSI rotates the energy from the hyperboloid to a hyperbola in a CSP gather that is located at the scatterpoint.

After the CSP gathers have been formed, Kirchhoff MO and stacking complete the prestack migration process.

The energy from the scatterpoint will also be summed into other CSP gathers that are formed away from the scatterpoint. Energy from the hyperboloid cancels by destructive interference. The process thus focuses the scattered energy at the scatterpoint in both time and space.

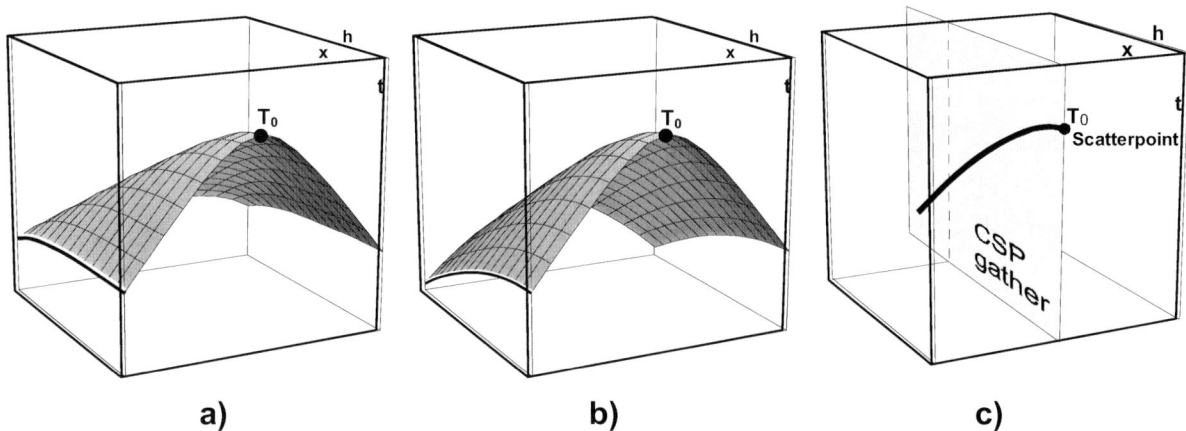

Figure 9.37 The GDMO-PSI process showing a) a Cheops pyramid from a scatterpoint, b) the hyperboloid formed from applying GDMO to (a), and c) the result of PSI, where the energy from the hyperboloid is summed to a hyperbola on a CSP gather.

The kinematic equations presented in the following section 9.7 are derived in Appendix 4.

9.11.2 GDMO and energy from a scatterpoint

- One scatterpoint distributes energy on Cheops pyramid in the prestack volume.
- The unique sample on the radial DMO that is tangent to the hyperboloid is defined as Tg as illustrated in Figure 9.38a and (b).
- GDMO reconstructs the energy from Cheops pyramid to a hyperboloid.

In (b), a line, normal to the prestack migration ellipse at T_0, intersects $t = 0$ at b. This spatial position b is the center for the poststack migration semicircle and also defines the intersection on the GDMO ellipse at T_g.

For all input points (x, h, T), this point of tangency (b, k, T_g) is defined by

$$b = x - \frac{4xh^2}{T^2 V^2} \tag{9.13}$$

$$k = h\left(1 - \frac{16x^2 h^2}{T^4 V^4}\right)^{1/2} \tag{9.14}$$

and

$$T_g = T\left(1 - \frac{16x^2 h^2}{T^4 V^4}\right)^{1/2} \tag{9.15}$$

It is convenient to consider only the points of tangency on the hyperboloid when discussing the PSI process.

When GDMO is applied to all energy on Cheops pyramid, the energy will, through reinforcement, reconstruct on the surface of a hyperboloid. Energy above the hyperboloid (shorter times) will contribute energy, which forms a wavelet that must be corrected with some form of shaping filter.

The equation of the time T_g on the hyperboloid is given by

$$T_g^2 = T_0^2 + \frac{4(b^2 + k^2)}{V_0^2}, \tag{9.16}$$

where T_0 is the time of the scatterpoint, b the new spatial location, and k the new offset.

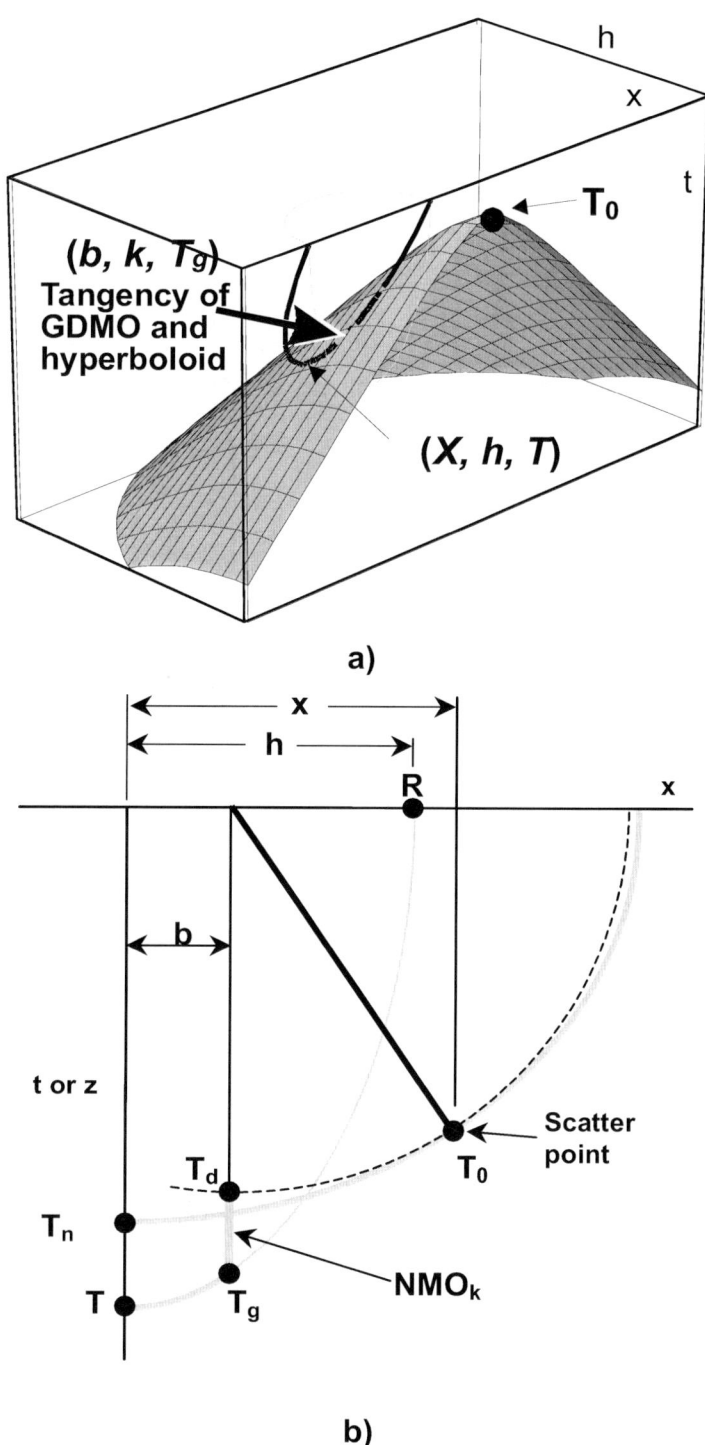

Figure 9.38 Showing a) the point of tangency between the GDMO ellipse and the hyperboloid and b) geometrical relationships giving b.

9.11.3 GDMO and poststack migration (one input time sample)

Gardner's DMO moves energy from an input sample to a radial DMO as illustrated in Figure 9.39a. Points on the GDMO curve move to smaller offsets and times. Each of these points may be MO corrrected to zero-offset where the energy is identical to that from conventional NMO and DMO.

Both DMO processes (for a constant velocity) may be poststack migrated to construct the prestack migration ellipse as illustrated in Figure 9.39b.

- **NMO-DMO (black) then poststack migration.**
- **GDMO- MO_k (dark gray) then poststack migration.**

A third method PSI (section 9.7.3) provides an alternate path for the energy at T_g to move to the scatterpoint.

- The sample at T_g is moved <u>at the same time</u> to a CSP gather located at the scatterpoint location as shown by the large horizontal dashed line. MO with a new offset moves the energy to the scatterpoint.

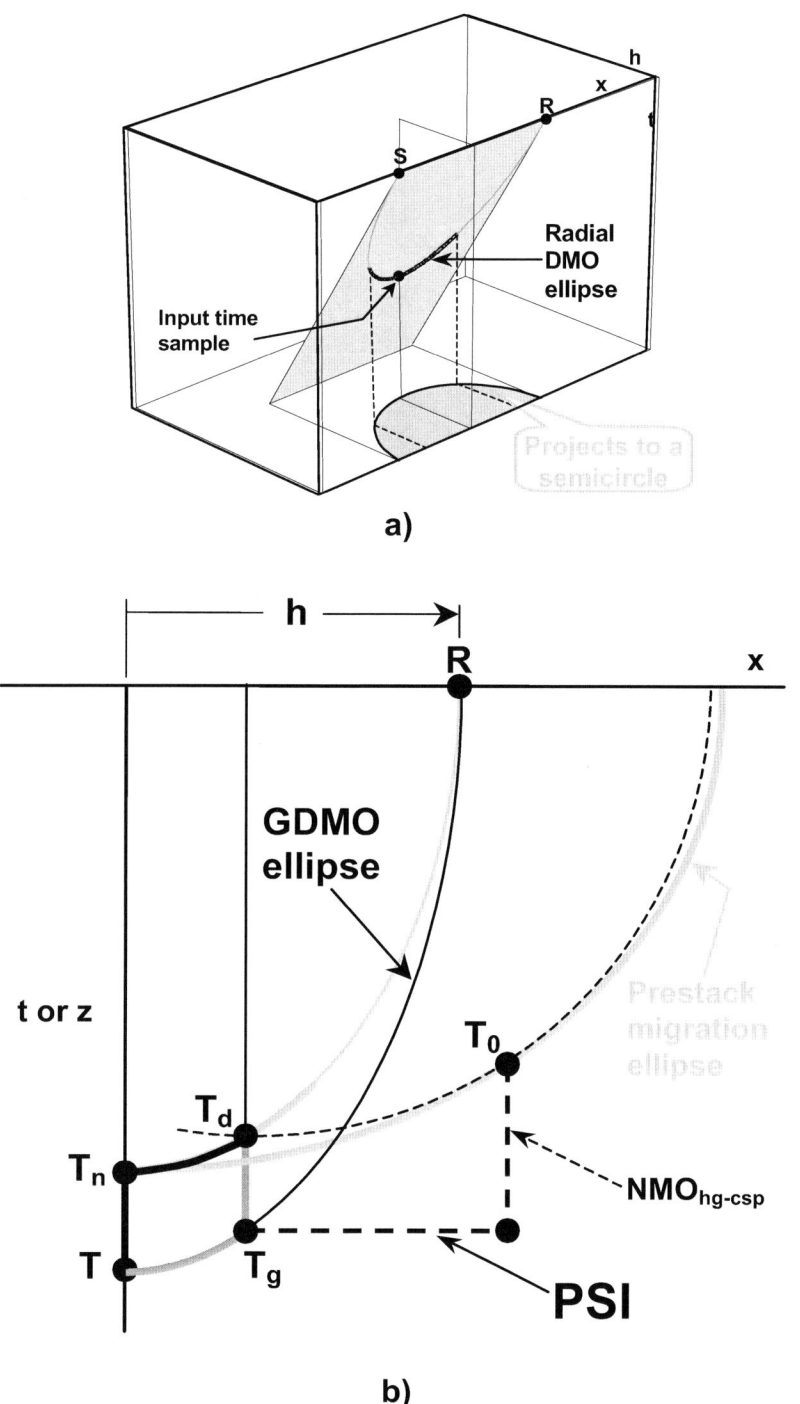

Figure 9.39 GDMO is illustrated in a) by spreading the energy from an input sample along an ellipse in the radial plane. Part b) illustrates prestack migration using GDMO-MO and poststack migration, or using GDMO-PSI-MO$_{hg\text{-}csp}$.

9.11.4 Prestack Imaging (PSI)

Prestack imaging (PSI) is a process that gathers energy from the hyperboloid and moves it to a hyperbola on a CSP gather that is located at the scatterpoint as illustrated in Figure 9.40. The hyperbola has the same shape as the hyperboloid at *x = 0*.

The energy gathering process is (conceptually) performed in constant time planes (time slices) as illustrated in Figure 9.41a.

- The slice through each hyperboloid will produce a semicircle with the center at the CSP location as illustrated in Figure 9.41b.

- A modelling program (inverse of migration) would collapse the energy of these semicircles to a point on the CSP gather as illustrated in Figure 10b.

- Modelling of all time layers would reconstruct the energy from the hyperboloid to the hyperbola on the CSP gather.

Modelling is performed on a slice with dimensions *(x, h)* and is independent of time.

Fourier transforming the prestack volume *(x, h, t)* to the *(K_x, K_h, t)* domain allows for fast PSI (similar to FK modelling). Inverse transforms back to the *(x, h_{CSP}, t)* domain yield the CSP gathers.

The point of tangency (for a given velocity) will be mapped through PSI to the CSP gather at the same time T_g and to an offset $h_{g\text{-}CSP}$ given by

$$h_{g-CSP}^2 = x^2 + h^2 - \frac{8x^2 h^2}{T^2 V^2} \tag{9.17}$$

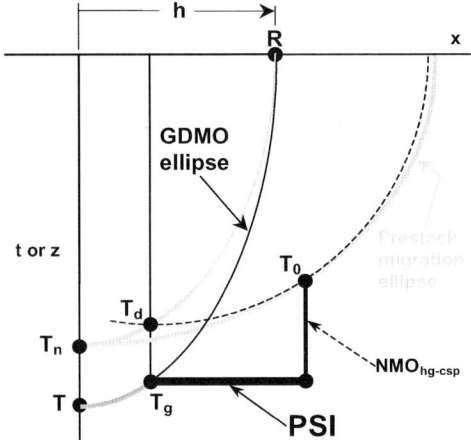

Repeat of Figure 9.20b

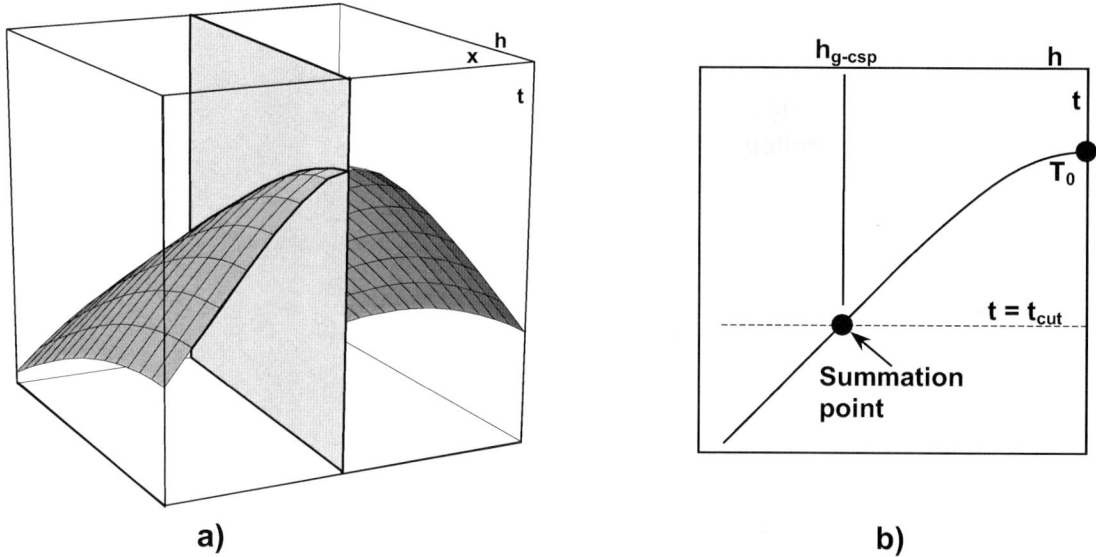

Figure 9.40 The CSP gather located at the scatterpoint in a) the hyperboloid, and b) with the energy from the hyperboloid rotated to the CSP hyperbola.

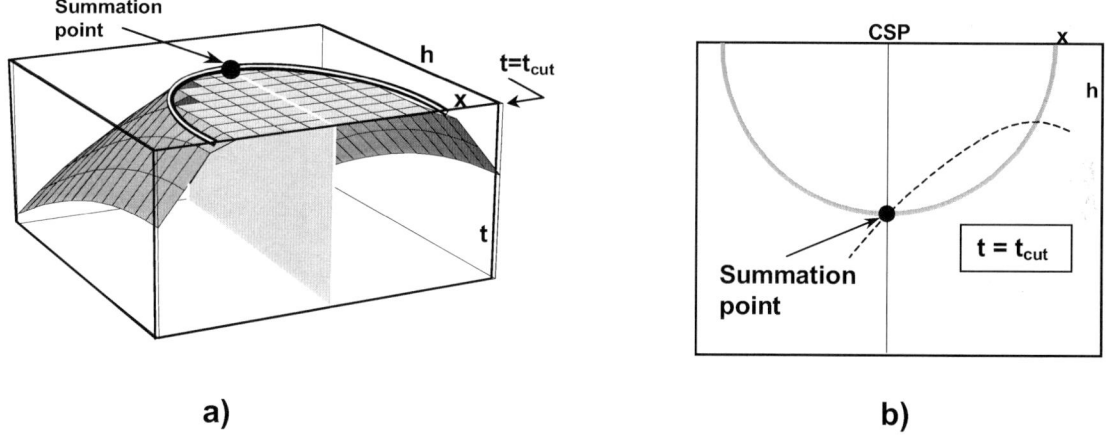

Figure 9.41 A time-slice of a hyperboloid in a) the prestack volume and b) the corresponding time slice. Energy on the semicircle will rotate (reconstruct) at the summation point through a modelling process.

9.11.5 Combining GDMO and PSI

Consider one input sample and mapping it to the zero-offset plane.

The cascade of GDMO-PSI maps a single input point:

- first to the radial GDMO ellipse in Figure 9.42a and then
- maps the <u>radial GDMO</u> to the <u>GDMO-PSI curve</u> in any neighboring CSP gather as illustrated in Figure 9.42b.

The GDMO-PSI curve in Figure 9.42b and Figure 9.43 are formed from one input sample at T, and are <u>independent of velocity</u>.

If the <u>input sample</u> came from a scatterpoint located below the defined CSP location, then the CSP hyperbola (path of MO) will be tangential to the GDMO-PSI curve as illustrated by the solid black CSP hyperbola in Figure 9.43.

If the velocity of the scatterpoint was different from above, then the T_0, and CSP hyperbola would adjust to maintain a point of tangency as illustrated by the gray curves.

Figure 9.44a shows a family of GDMO-PSI curves from one input time sample. The point of tangency on each curve, is also shown (for an assumed velocity).

Figure 9.44b shows the hyperbolic moveout of each point of tangency to the <u>prestack migration ellipse</u>.

Energy on the actual prestack migration ellipse is a reconstruction of NMO'd energy from Figure 9.44a and will form a wavelet that extends above (shorter time) the prestack migration ellipse.

The point of tangency in the CSP gather between the GDMO-PSI curve and the CSP hyperbola (T_g, $h_{g\text{-}CSP}$) is given by

$$T_g = T\left(1 - \frac{16x^2 h^2}{T^4 V^4}\right)^{1/2} \qquad (9.18)$$

and

$$h^2_{g\text{-}CSP} = x^2 + h^2 - \frac{8x^2 h^2}{T^2 V^2}. \qquad (9.19)$$

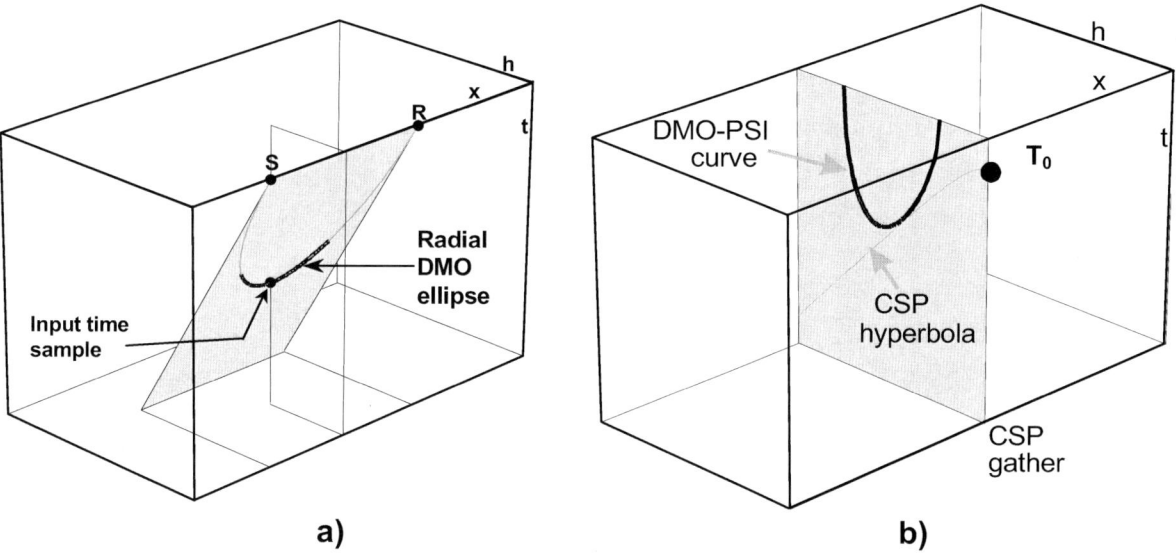

Figure 9.42 Formation of the GDMO-PSI curve onto any CSP gather.

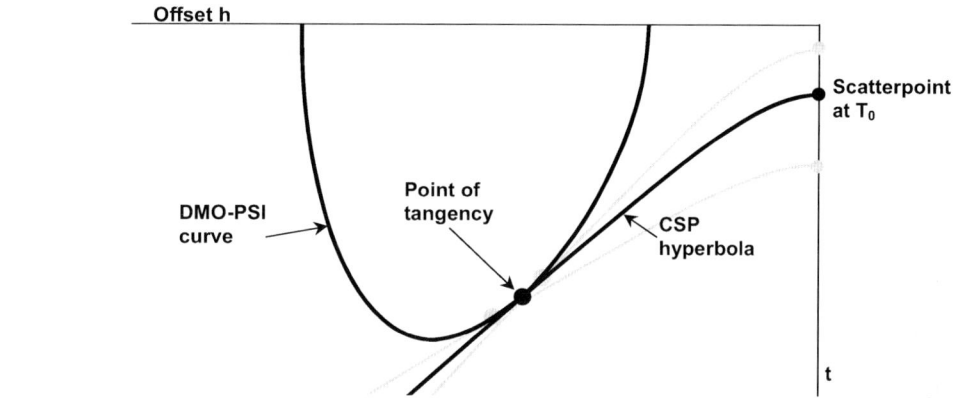

Figure 9.43 GDMO-PSI curve on a CSP gather showing three possible scatterpoints and their corresponding hyperbolas.

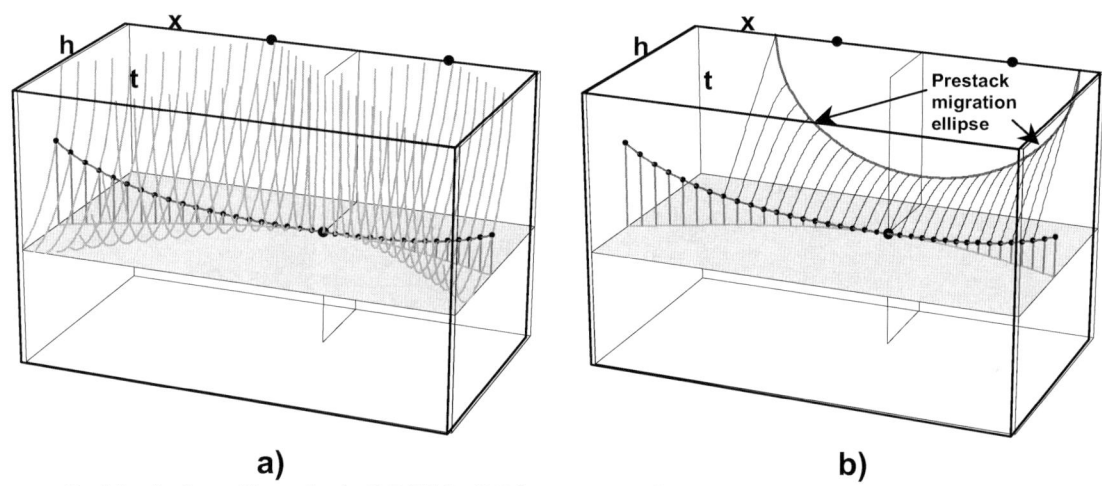

Figure 9.44 A family of a) GDMO-PSI curves from one input sample, and b) the corresponding NMO hyperbola and prestack migration ellipse.

9.12 Equivalent Offset Migration (EOM)

Equivalent offset migration (EOM) is a <u>modified form of prestack Kirchhoff migration</u>. It can be a time or depth migration.

EOM contains an intermediate step that forms common scatterpoint (CSP) gathers prior to time-shifting the input data.

9.12.1 Definition of the equivalent offset

Equivalent offset migration is the topic of Chapter 11; only a brief introduction is presented in this section.

Equivalent offset migration (EOM) produces similar results to GDMO-PSI as both produce CSP gathers (Bancroft and Geiger, 1994). However, the method of producing those gathers is significantly different. The EOM method moves an input sample with one-to-one mapping directly to the CSP gather.

There is

- <u>no time shifting</u> and
- <u>no DMO</u>.

The method computes the location of a <u>collocated source and receiver</u> that <u>maintains the same traveltime</u> to a scatterpoint as the original source and receiver, as illustrated in Figure 9.45.

The offset from the CSP to the colocated source and receiver is defined as the equivalent offset h_e. In essence, the equivalent offset allows the double square root equation to be expressed as a hyperbola, i.e.,

$$T = \left(\frac{T_0^2}{4} + \frac{(x+h)^2}{V^2}\right)^{1/2} + \left(\frac{T_0^2}{4} + \frac{(x-h)^2}{V^2}\right)^{1/2} = 2\left(\frac{T_0^2}{4} + \frac{h_e^2}{V^2}\right)^{1/2} \quad (9.20)$$

where T_0 is the vertical two-way time from the scatterpoint to the surface, V the velocity, x the distance from the CSP to the CMP, and h the half offset. The equivalent offset term h_e may be solved exactly (see Appendix A4.7) to give

$$h_e^2 = x^2 + h^2 - \frac{4x^2 h^2}{T^2 V^2} \quad (9.21)$$

The equivalent offset is both time and velocity dependent as defined by the cross term ($4x^2h^2/T^2V^2$) in equation (9.11). An input trace (which has a constant x and h) will span a range of equivalent offsets, or cover a range of offset bins in the CSP gather.

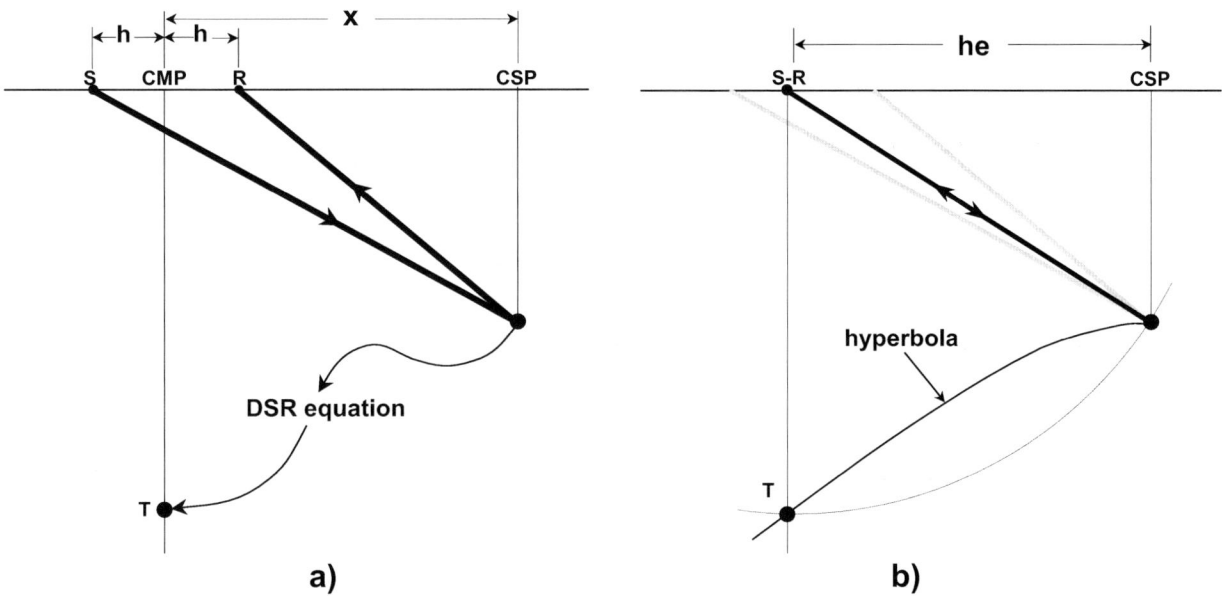

Figure 9.45 Raypaths for a) the actual source and receiver, and b) the equivalent offset with collocated source and receiver.

9.12.2 EOM and Cheops pyramid

A constant time slice through Cheops pyramid in (x, h) space is shown in Figure 9.46a. All intersection points on the slice have the same equivalent offset for a CSP gather located at the scatterpoint. The shape of this intersection tends from a semicircle just below the scatterpoint to a square at large times.

EOM sums energy on the intersection to all CSP gathers.

However, the particular CSP gather that passes through the scatterpoint will coherently sum all the energy to one point as illustrated in the plan view of Figure 9.46b. Energy at all other CSP gathers will destructively cancel.

At zero displacement (i.e. $x = 0$) Cheops pyramid is a hyperbola in (h, t) space and aligns with the hyperboloid from GDMO-PSI and the scatterpoint hyperbola.

Therefore, EOM maps all point on Cheops pyramid directly to the CSP hyperbola as shown in Figure 9.47. All points on the 3-D surface of Cheops pyramid (x, h, T) will therefore map to the hyperbola on the CSP gather (h_e, T).

Kirchhoff MO and stacking of the CSP gathers complete the prestack migration.

In practice, the actual movement of data to the CSP gathers is much simpler than indicated above. Each input trace is simply summed (with no time shifting) into appropriate bins of the CSP gather. The only required computations are the first offset bin and sample, and the transition times where that data is summed into the next offset bin.

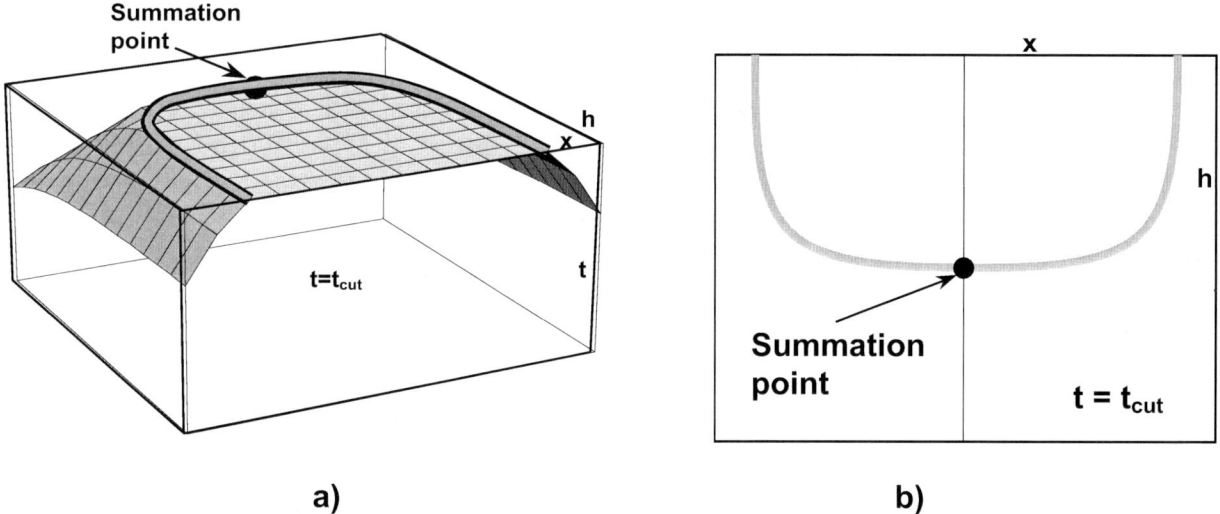

Figure 9.46 EOM at a constant time T showing a) a time section through Cheops pyramid, and b) the resulting summation of energy to the CSP gather.

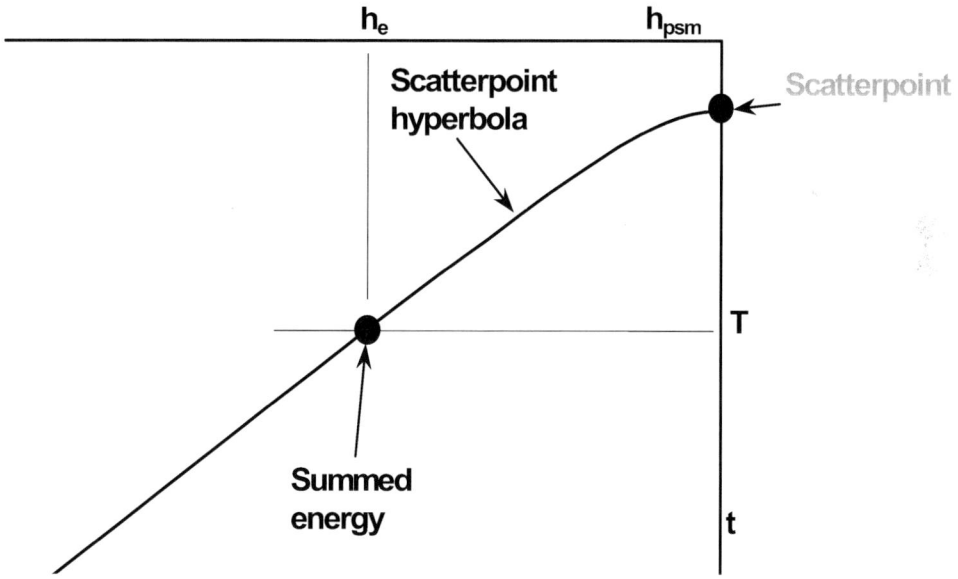

Figure 9.47 The CSP gather located at the scatterpoint showing the summed input energy from the time slice. All energy from Cheops pyramid will be summed to the scatterpoint hyperbola.

9.12.3 Kinematics of EOM for *one* input sample

The previous discussion viewed EOM from a scatterpoint perspective with many input samples spread over Cheops pyramid.

We now consider the distribution approach where we evaluate the energy movement from <u>one input sample</u>. To accomplish this, we consider equation (9.11) at a constant time *T*, offset *h*, and velocity *V*. The *x* variable now represents the displacement from the input CMP location (also constant at *x = 0*) to any CSP gather at *x*. Equation (9.11) may now be written as

$$h_e^2 = x^2\left(1 - \frac{4h^2}{T^2V^2}\right) + h^2 \qquad (9.22)$$

The term in brackets can be replaced by a constant *c*, giving

$$h_e^2 - x^2 c = h^2 \qquad (9.23)$$

Equation (9.13) is a hyperbola on a plane (x, h_e) at constant *T*, as illustrated in Figure 9.48a. Kirchhoff MO applied to this equivalent offset hyperbola will now form the <u>prestack migration ellipse</u> as illustrated in Figure 9.48b.

Note:

- The prestack migration ellipse <u>limits the extent of the energy</u> in the equivalent offset hyperbola.

- There is a <u>one-to-one mapping</u> of the input sample to each point on the prestack migration ellipse.

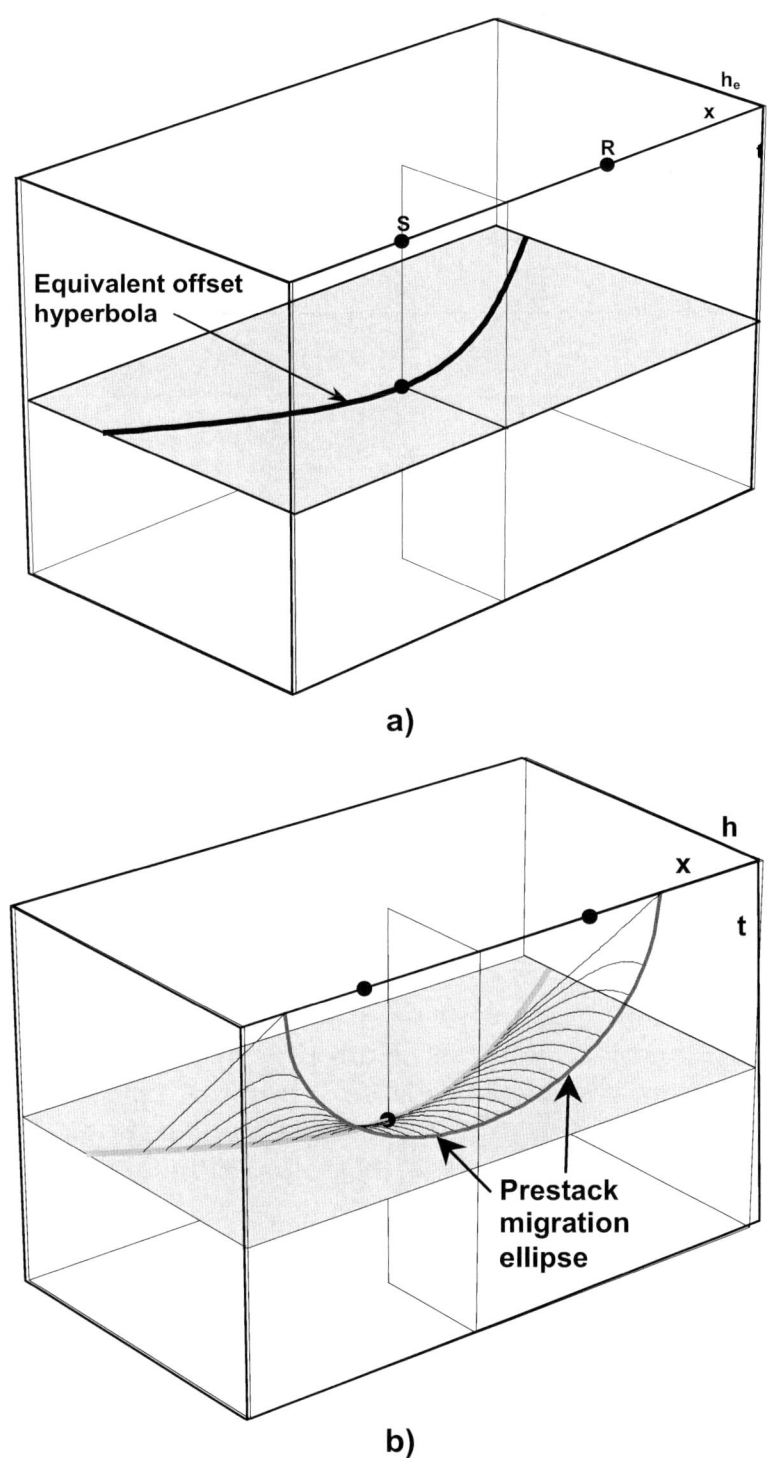

Figure 9.48 Prestack view of a) the equivalent offset hyperbola at constant time and b), forming the prestack migration ellipse with Kirchhoff MO.

9.13 Kinematic Comparison Between GDMO-PSI and EOM

The kinematics of GDMO-PSI and EOM can be compared using the equations that define the mapping of energy to the CSP gather. <u>For a given velocity V</u>, the GDMO-PSI point of tangency at time T_g and offset $h_{hg\text{-}csp}$ is compared with the EOM point at time T, and at the equivalent offset h_e. The relevant equations are given in Table 1.

Table 1. GDMO-PSI point of tangency and EOM point on a CSP gather.

	Time	Offset
GDMO-PSI point of tangency	$T_g^2 = T^2 - \dfrac{16x^2h^2}{T^2V^4}$	$h_{hg-csp}^2 = x^2 + h^2 - \dfrac{8x^2h^2}{T^2V_{sp}^2}$
EOM	T (no time shift)	$h_e^2 = x^2 + h^2 - \dfrac{4x^2h^2}{T^2V_{sp}^2}$

The GDMO-PSI curve and point of tangency is compared with the EOM point in Figure 9.49. Note the EOM point is even with the bottom of the GDMO-PSI curve. Both the point of tangency and EOM point lie on the scatterpoint hyperbola.

Figure 9.50a shows a comparison between the GDMO-PSI point of tangency and the EOM point in the prestack volume. Note the EOM point remains on the plane defined by the input time while the GDMO-PSI point of tangency is shifted in both offset and time. Part (b) of this figure includes the Kirchhoff MO hyperbola that maps both sets of points to the prestack migration ellipse.

Note that Figure 9.50 only shows the point of tangency of GDMO-PSI. A more accurate comparison is made between the full GDMO-PSI curves in Figure 9.44a, with the simple EOM hyperbola in Figure 9.48a.

The above comparison required a velocity V.
- GDMO-PSI requires no velocity information.
- EOM is slightly velocity dependent.

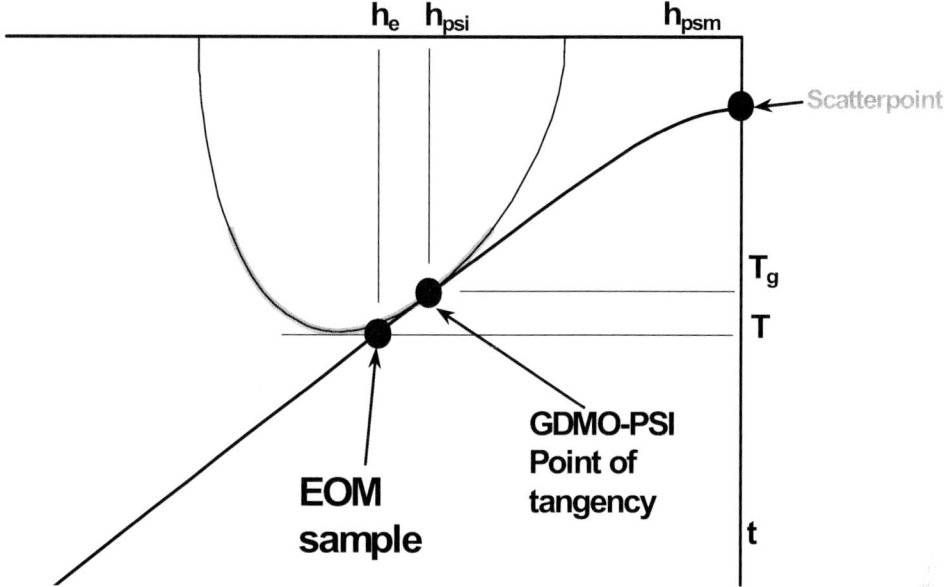

Figure 9.49 A CSP gather comparing the kinematics of GDMO-PSI and EOM.

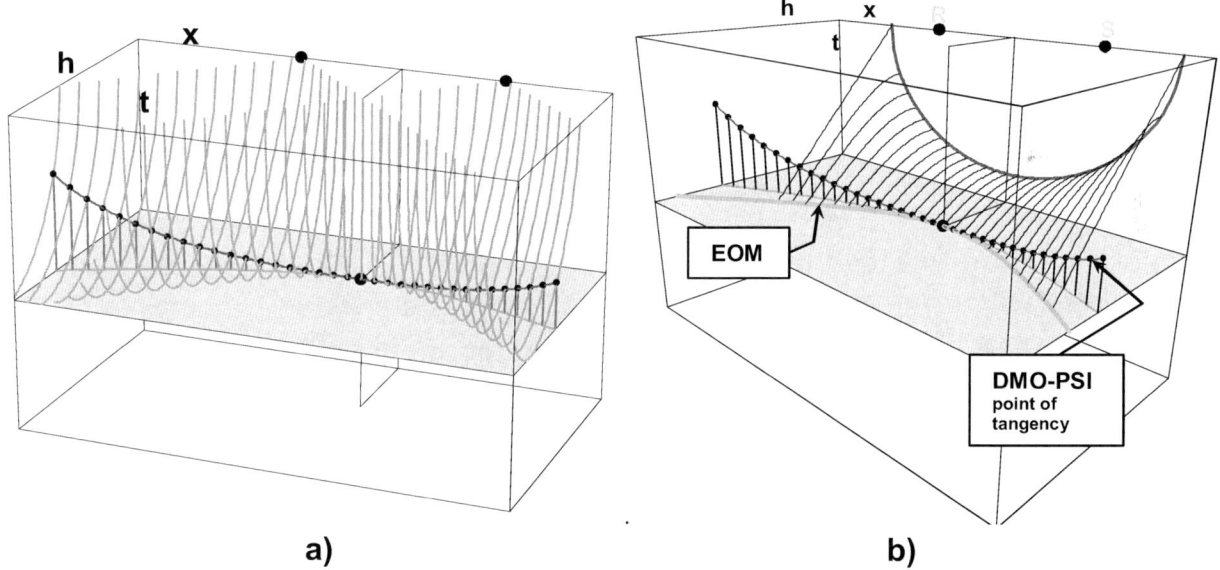

Figure 9.50 Comparison of a) GDMO-PSI (showing only the point of tangency) with EOM at various SCP locations, and b) with Kirchhoff MO added to illustrate both methods move energy to the prestack migration ellipse.

9.14 Prestack Migration Aliasing and Irregular Geometry

Many of the previously mentioned methods of prestack migration have problems with spatial aliasing. This usually results when the source locations are at multiple station intervals. Some methods suffer these effects more than others.

Example:

- Typical source record, δx = 30, V=3000, and maximum geological dip β of 40 degrees, spatial aliasing will start to occur at frequencies around 40 Hz.
- When the sources are located at every fourth station, then the trace spacing in a receiver gather is four times greater than the source record.
- The receiver gather will start to spatial alias (the dipping data) at frequencies above 10 Hz.

Examples of aliasing problems are illustrated in Figure 9.51:

(a) S-g migration, the source record may be OK but the trace spacing in a receiver gather is too large.

(b) Source record migration may have no aliasing, but there must be enough source records when stacked to reconstruct the structure of the data.

(c) Constant offset record may have only one line trace in every eight traces (when shot with a four-station interval).

A typical solution to these problems is to interpolate additional traces to fill gaps of say the receiver gathers or constant offset sections.

Another solution is to combine a number of offset sections to increase the ratio of live traces. There must be a sufficient number of remaining constant offset sections to constructively and destructively interfere when creating the data.

A major concern of each prestack migration user should be...

How is aliasing handled?

Uneven geometry is not an issue for poststack migration. The energy in each stack is first balanced by fold, and then by "trace conditioning" schemes.

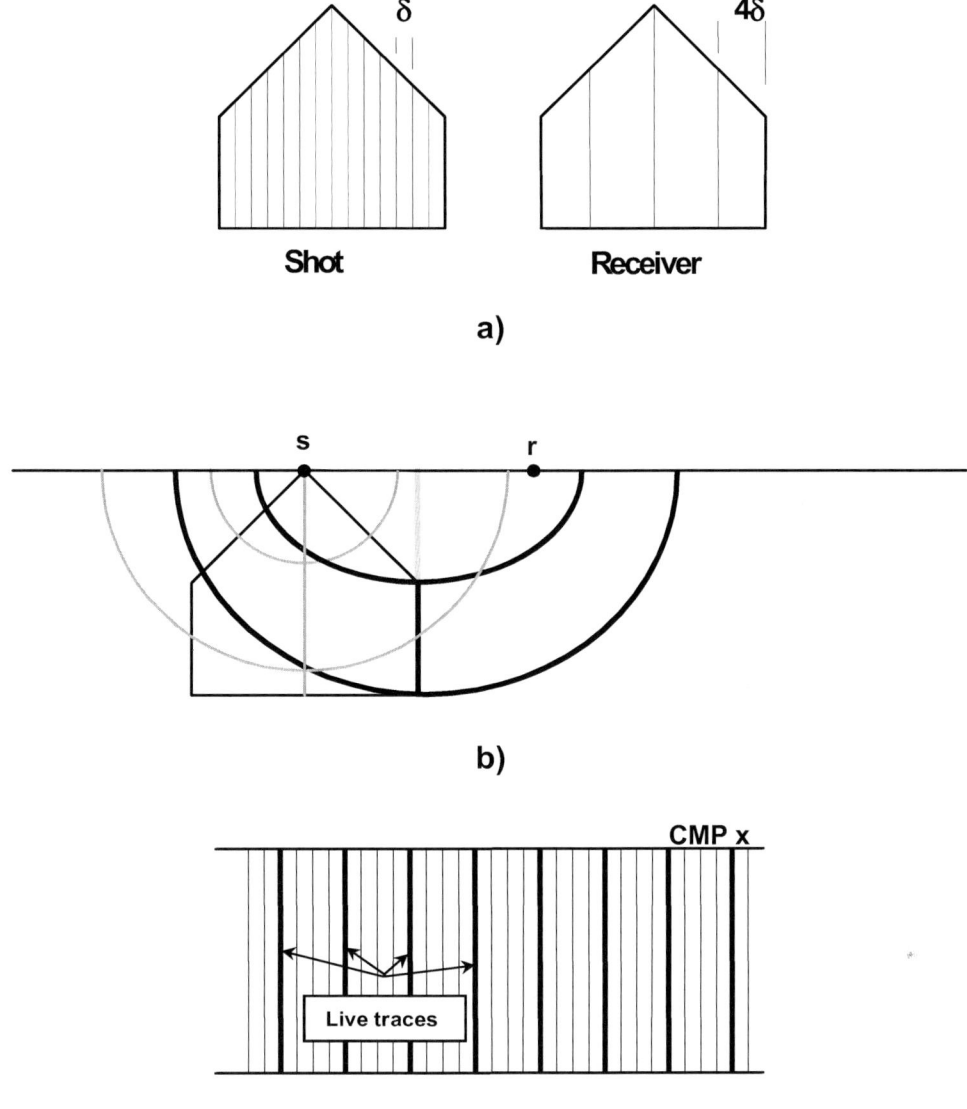

Figure 9.51 Aliasing in prestack migration, a) for s-g gathers, b) for full source gathers, and c) for constant offset sections.

9.15 Prestack Migration of 3-D Data Volumes

The examples in Figure 9.52 show the difference between (a) 2-D prestack migration <u>ellipse</u>, and (b) 3-D prestack migration <u>ellipsoid</u> (prolate spheroid).

- A single input trace in 2-D prestack migration distributes energy to traces in the <u>plane</u> of the section (say 200 traces).
- A single input trace in 3-D prestack migration distributes energy to traces in <u>all the neighboring traces in the 3-D region</u> (say 40,000 traces). The equation of this surface is given by:

$$\frac{x^2}{a^2} + \frac{y^2}{b^2} + \frac{z^2}{c^2} = 1, \text{ where } b = c. \tag{9.24}$$

Not only is the energy spread to more traces, there are also more input traces.

We will assume a <u>small 3-D</u> has dimensions similar to the size of the migration aperture, say ± 100 traces.

When the dimension of a small 3-D is increased by 2, the number of acquired traces increases by 4, and the number of prestack migrated traces is also increased by a factor of 4, for a total of <u>16 times the number of traces</u>. The increase in traces to be handled is the ratio of the sizes raised to the <u>fourth power</u> (i.e., 2^4 = 16).

This rate of increase is reduced to the second power when the size of the 3-D is much greater than the prestack migration aperture.

When the processing of 3-D's became commercial in the early 1980's, the algorithms for 3-D DMO and prestack migration followed rapidly. However, they were <u>not used on a regular basis</u> due to the costs of long run times.

In the early <u>1990's 3-D DMO</u> became a viable part of 3-D processing and 3-D prestack migration was limited to special applications.

Current trends now find <u>3-D prestack migration</u> in greater use.

These trends have been aided by the continued <u>improvement</u> in computer hardware and in algorithms that allow the storage of large data sets and faster run times.

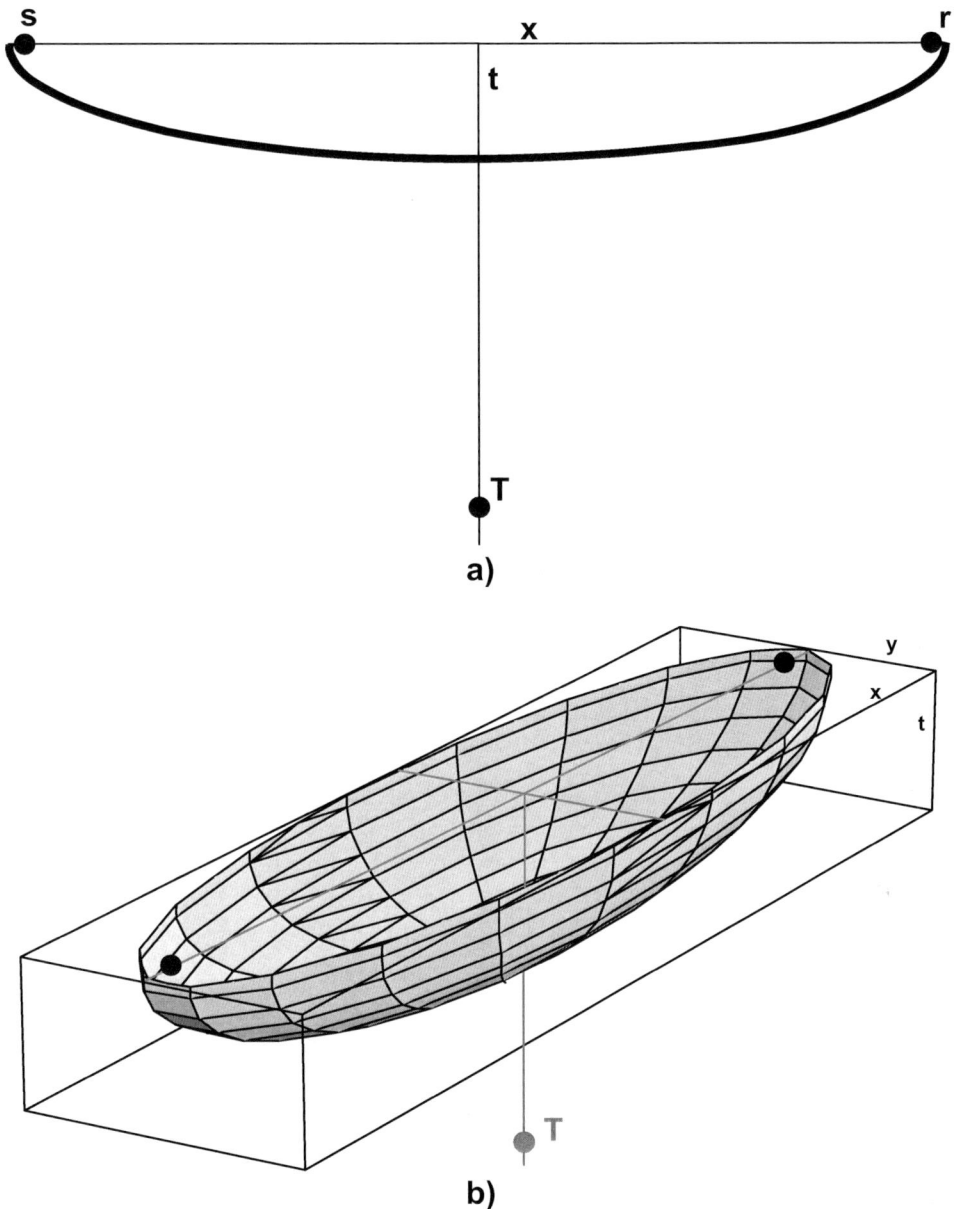

Figure 9.52 Prestack migration showing a) 2-D ellipse, and b) 3-D ellipsoid (or more specifically, a prolate spheroid).

The above figures are an <u>exaggeration</u> as shallow reflectors are used with a large offset. It does, however, illustrate the <u>elliptical</u> features. As the reflection time increases, the elliptical shape will tend more to a <u>hemi-sphere</u>.

9.16 3-D prestack time migration

There are a number of techniques used for prestack 3-D migration. Some use reverse time migration techniques, while others use Kirchhoff methods.

The figures of the ellipse on the preceding page are related to Hagedoorn migration, and are only valid for constant velocity. The dual operation is to sum energy in the diffraction shape that uses RMS velocities.

Typically, <u>one source record</u> is migrated at a time, into all the surrounding bins that are within the 3-D migration aperture.

The following will concentrate on the Kirchhoff method as it is more intuitive, and provides excellent results. It is extremely computer intensive, and requires a <u>$j\omega$ filter</u> (not $rj\omega$) for phase and scale correction.

The procedure is very similar to the 2-D case:

- An output bin will search all the neighboring traces for energy in the shape of a hyperboloid as is illustrated in Figure 9.53 (which assumes a continuum of receivers).
- The <u>time at the hyperboloid's apex</u> is defined by the raypaths from the source to scattering point T_s and the vertical time to a receiver T_{0r}.
- The <u>shape of the hyperboloid</u> is determined from the traveltimes between the scatterpoint and receivers.

Figure 9.53b shows the <u>same scattering point</u> in (a) but with a different source location. In (a) the source is directly above, and in (b), the source may come from various locations as indicated by the gray curve. As the distance from the scatterpoint to the surface are the same, the shape of the hyperboloids are the same, while the depth to the apex depends on the source distance.

When the 3-D hyperbolic diffraction shape (hyperboloid) is computed from RMS velocities, the result is a <u>time</u> migration.

When the non-hyperbolic diffraction shape is computed from ray tracing, or wave front analysis, the result is a <u>depth</u> migration.

Figure 9.34 assumes a continuum of receivers on the surface. A land 3-D will have a few lines of receivers for one shot. Consequently the shot record will only contain <u>intersections of the hyperboloid</u> below the receiver lines.

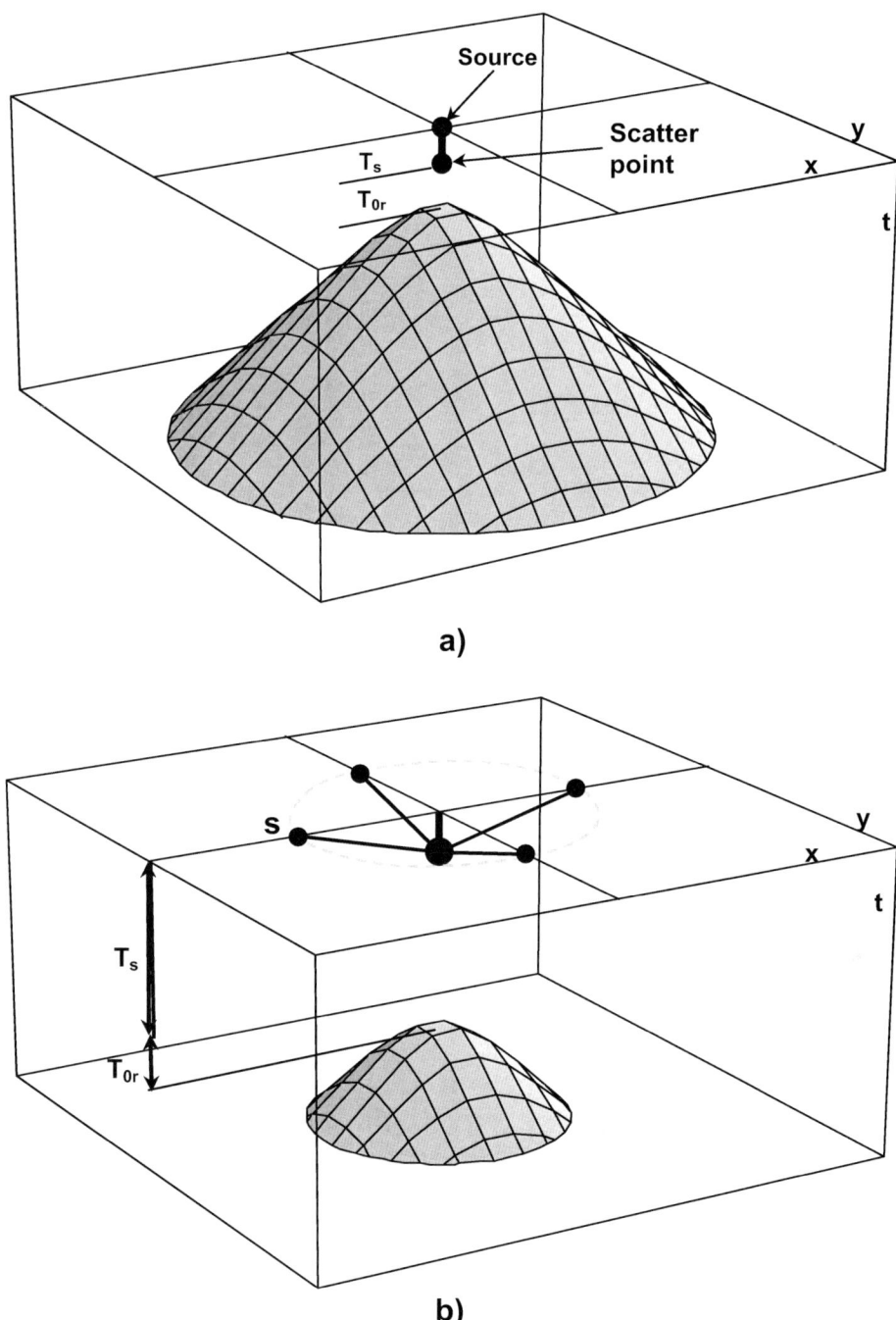

Figure 9.53 3-D prestack hyperboloids with a) the source is directly over the scatterpoint, and b) the source locations as indicated.

9.16.1 3-D prestack depth migration

3-D prestack depth migration may be derived from ray tracing or <u>wave front traveltime analysis</u>. Reliable solutions are available, but once again the most difficult part of the process is the construction and visualization of the 3-D depth model [247].

The following is a brief description of a wave front method as reported by Reshef [142], and illustrated in Figure 9.54.

- A full 3-D volume of <u>wave front traveltimes</u> is computed for each source and each receiver.

- However, the full 3-D traveltimes are not required as a complete volume; <u>only one depth layer at a time</u> is required.

- The <u>total traveltimes</u> for each input trace to one <u>output point</u> on that depth layer may be estimated from the traveltimes of the appropriate sources and receivers.

- The <u>sample on each input trace</u> is then located, and <u>summed</u> to the output location.

- This process is <u>repeated</u> for each output sample at the given depth.

- The wave front <u>traveltimes are propagated down</u> to the next depth layer, and the process repeated until the desired output depth has been reached.

The example in Figure 9.54 shows one source and three receivers with traveltimes computed to a depth layer.

The cut away diagram in (a) shows the location for one sample, where the traveltime from the source and the traveltimes to the receivers may be defined. Energy on the appropriate trace in (b) is copied to the output sample in (c).

> Note the entire volumes of the traveltime maps are not required to be saved. Only enough layers are required to compute the traveltimes on the next depth layer. Possibly only <u>two depth layers</u> may be required for each source and each receiver. In typical land 3-D's, the source locations do not correspond to the receiver locations.

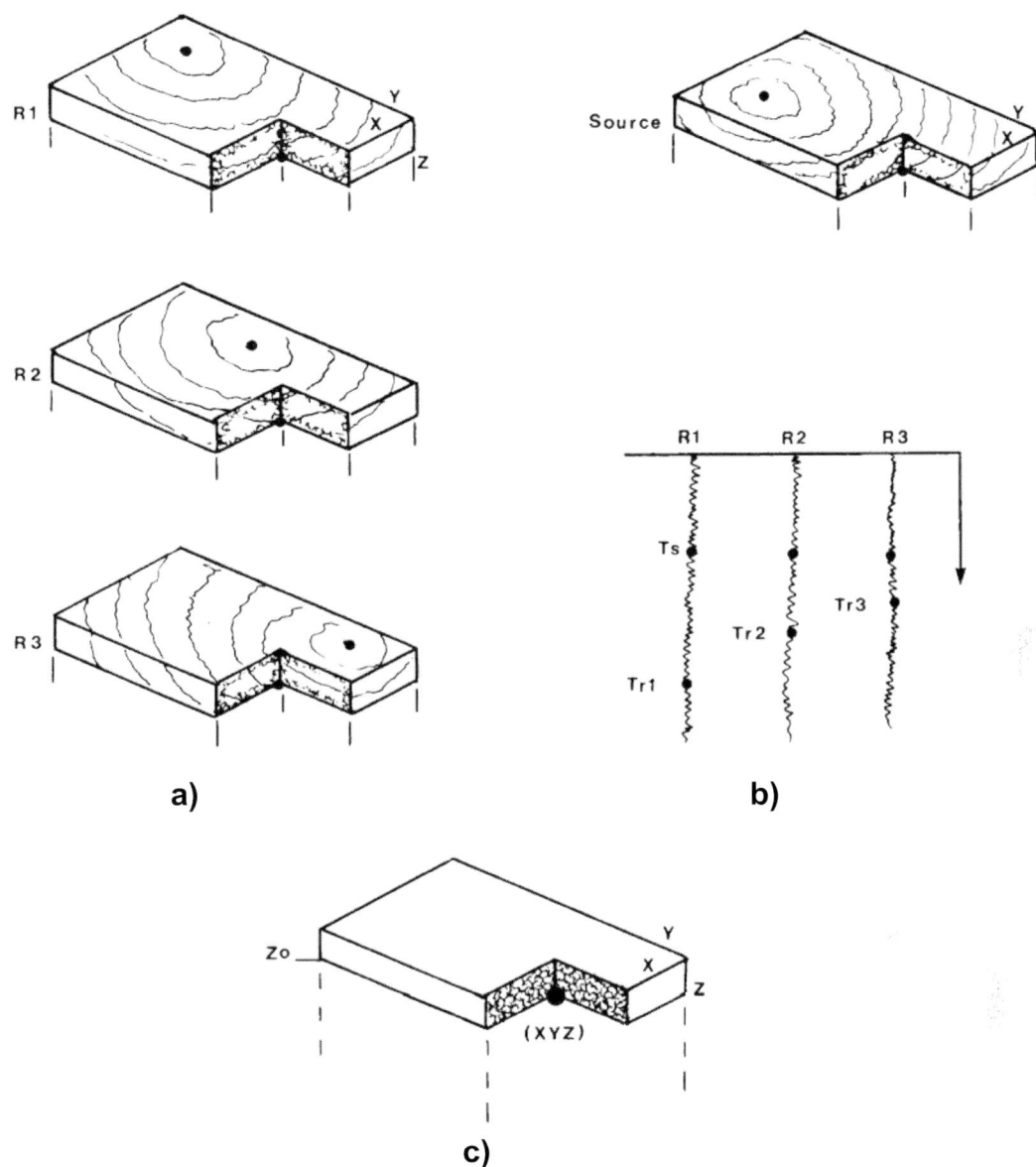

Figure 9.54 Illustration of 3D prestack depth migration. Part a) shows the traveltime volumes for one source and three receivers, b) the corresponding traces, and c) the output volume.

9.17 3-D DMO

9.17.1 Early methods of 3D DMO

The objective of 3-D DMO is to create a stacked volume that is zero-offset and suited for poststack migration.

As with the 2-D case there were many methods of processing that were called 3-D DMO. Various approximation methods evolved that were replaced in time with the wave theoretical type solutions. These earlier methods paralleled the 2-D algorithms and should be recognized as:

- dip-dependent stacks,
- French method,
- multiple constant velocity stacks,
- multiple constant velocity stacks with dip filter,
- and wave equation DMO.

<u>Dip-dependent Stacks</u>

When the velocity of a 3-D volume is known, the dipping information may be applied to the moveout correction in a manner that is similar to the 2-D case, but an additional parameter of azimuth must be used for each trace [115].

$$V_{NMO}^2 = \frac{V_{RMS}^2}{1-\sin^2\beta\cos\phi} \qquad (9.25)$$

where β is the dip, and ϕ the azimuth.

This method would be considered normal 3-D processing:

- requires a model of the data
- and will smear dipping data.

French Method

It is similar to the 2D case in which the equivalent zero-offset reflection point is found by computing δx, δy and δt.

- It will not smear dipping data,
- does not require wave form reconstruction, but
- it is model based.

Multiple Velocity Stacks

As with the 2D case some methods stacked the 3D volume with a number of velocities and after combining them, claimed DMO processing.

- Signal to noise (SNR) is reduced
- Not model based
- Will smear data

Multiple velocity stacks with dip filter

This method is also similar to the 2D case.

- It can achieve results if a sufficient number of constant velocity stacks are used.
- Not model based,
- Will not smear data
- Does have trouble with dip filters.

> All these methods have now been replaced with wave equation solutions.

9.17.2 Wave theoretical 3-D DMO

Wave equation <u>3-D DMO</u> combined with <u>poststack</u> 3-D migration should recreate the prestack migration ellipse. Figure 9.55 illustrates 3-D DMO, and how it recreates the prestack migration ellipsoid:

(a) shows a 2-D DMO ellipse between a source and receiver.

(b) poststack migration creates hemi-spheres from each point on the DMO ellipse; a <u>few</u> of these hemispheres are shown for a few points on the DMO curve.

(c) the prestack migration ellipse is reconstructed from DMO and poststack migration.

The difference in the number of traces generated by DMO and prestack migration becomes a factor close to three orders of magnitude.

i.e. 3-D DMO may be <u>two orders of magnitude faster</u> than a similar algorithm for prestack migration.

Marine 3-D

Marine 3-D's are usually acquired as a multitude of closely spaced 2-D lines. If there is no feathering (bending due to tides, etc.) of the towed cable, then each line can be <u>processed as a 2-D line</u>, including the DMO; i.e. all source/receiver azimuths are parallel. The processing becomes more complex when a number of receiver cables are towed, forming 2-D swath patterns.

Land 3-D

Land 3-D's are usually acquired with a cross spread formation where receiver lines are normal (at right angles) to source lines. Consequently, there are very few traces that have the same azimuth.

As a result, 3-D DMO requires <u>each input trace to be DMO'd separately</u>. The advantage of the 2-D common offset section is not available (where all traces had the same azimuth). A 3-D constant offset volume could be created, but the azimuths are still random.

In land 3-D, the sources and receivers are at <u>various azimuths</u> to the CMP (or binned) location. Each input trace is DMO'd separately, and the created traces are inserted into the 3-D volume between the source and receiver locations.

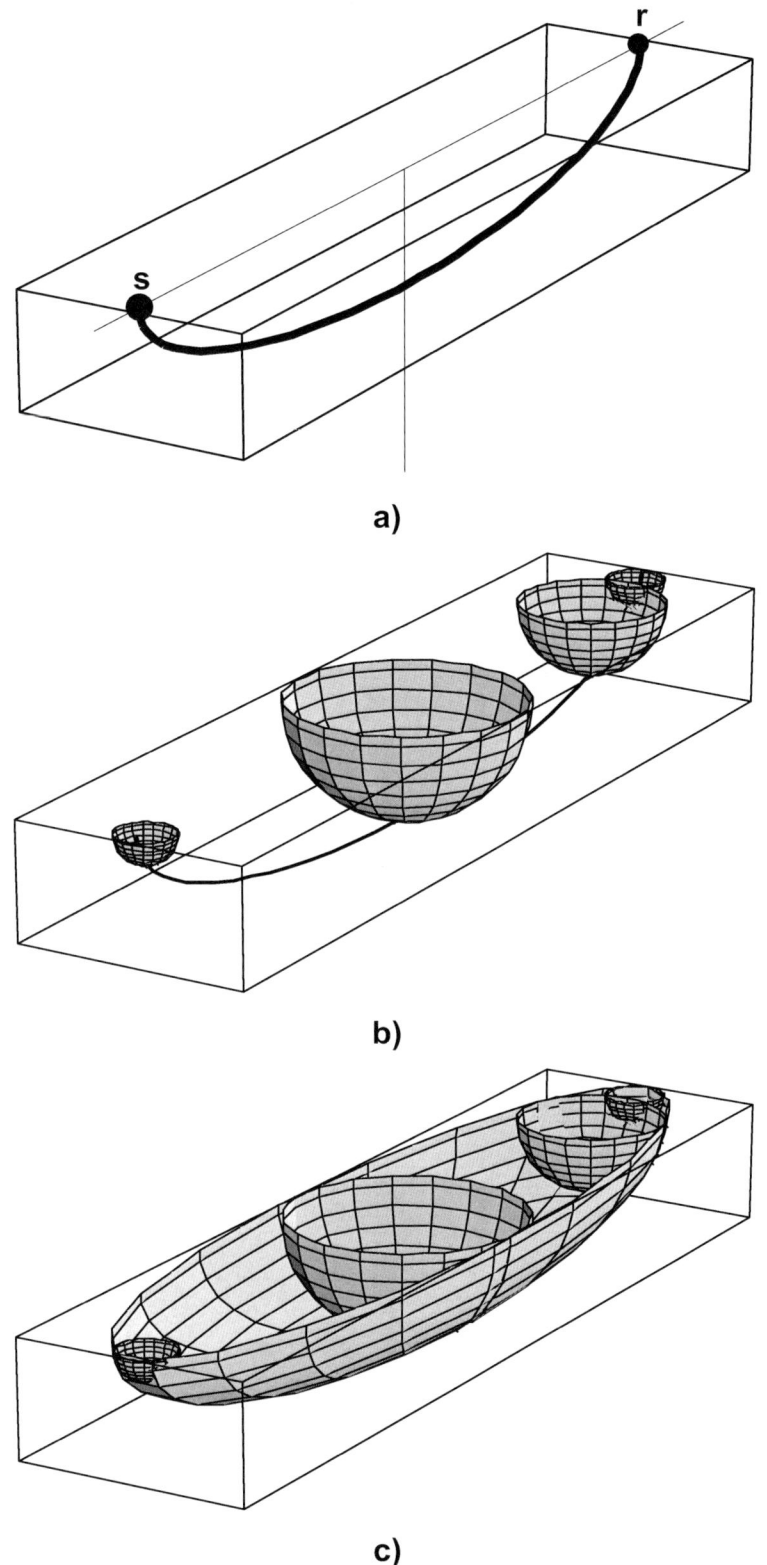

Figure 9.55 3-D DMO in (*x, y, t*) volume showing a) the DMO curve, b) poststack migration of DMO, and c) the formation of the prestack migration envelop.

9.17.3 Super bins for 3-D land DMO

A <u>super bin</u> is a group of bins that are added together for velocity analysis. All the traces with CMP's that lie within the super bin are used to increase the fold.

DMO allows many more traces to be included in the super bin.

These DMO'd traces come from any input trace in which a <u>line between the source and receiver</u> passes through the super bin as illustrated in Figure 9.56.

These DMO'd traces greatly increase the fold for velocity analysis, and reconstruct to give <u>RMS type velocities</u> that are independent of dip.

Use of 3-D DMO also simplifies the <u>velocity</u> being estimated to a RMS type that is <u>independent of azimuth and dip</u>.

The process does not need to compute all the traces of the DMO response. Only those traces that lie <u>within the bin</u> need to be calculated.

The time domain method is ideally suited for this purpose.

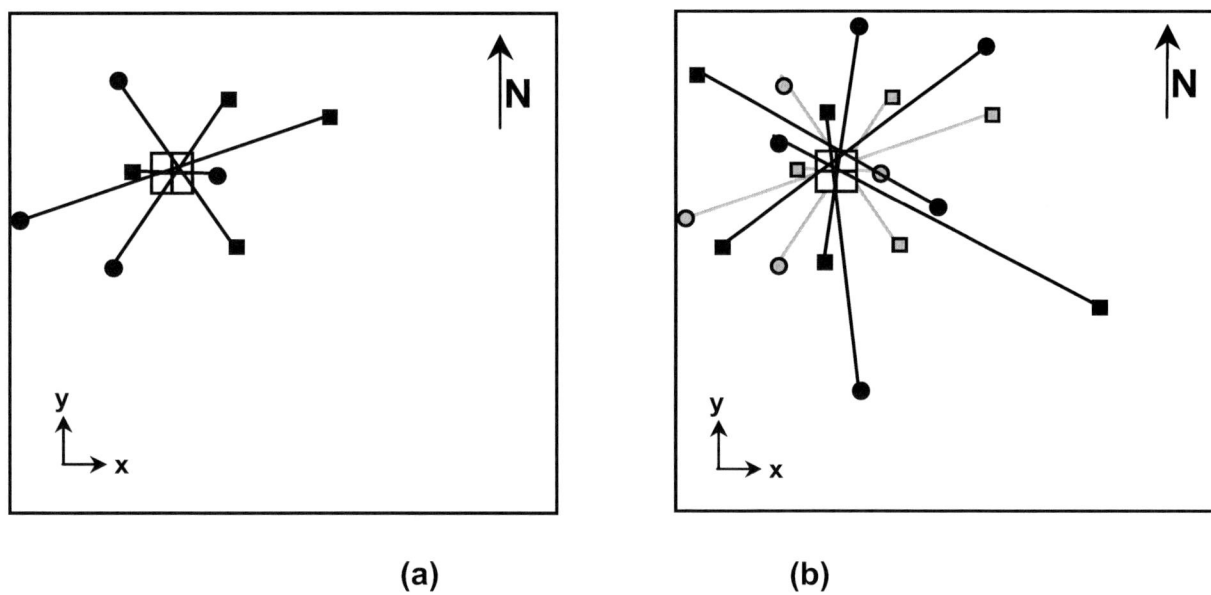

Figure 9.56 Traces in a super bin for a) CMP traces, and b) DMO'd traces.

9.17.4 3-D DMO with linear increasing velocities

Most of the discussions in the notes have concentrated on constant velocities, with extensions that apply for smoothly varying RMS velocities.

A 1990 development by Perkins and French [153] extended 3-D DMO to a case in which the velocities increase linearly with depth. The results are interesting because the simple 2-D DMO curve is replaced by a 3-D <u>saddle shape</u> surface. It is displayed in Figure 9.57, which was taken from [153]. See also [248] and [249].

This process is also referred to as migration to zero-offset or MZO.

The traditional DMO curve is part of the new saddle shape.

Many 3-D DMO implementations use this concept of MZO to partially spread energy on an elliptical band (out of the source receiver plane) to provide a more even subsurface coverage.

DMO does not work in local areas where the velocity varies rapidly, such as shale-carbonate interfaces. In these areas the DMO'd sections are often worse than processing without DMO. DMO will improve these sections in areas well away from the sharp velocity transitions.

Hale's course notes [637] describe a DMO algorithm for depth-variable velocities. In addition, the notes provide a number of "rules of thumb" that give limits of cable feathering, or when constant velocity DMO is worse than no DMO.

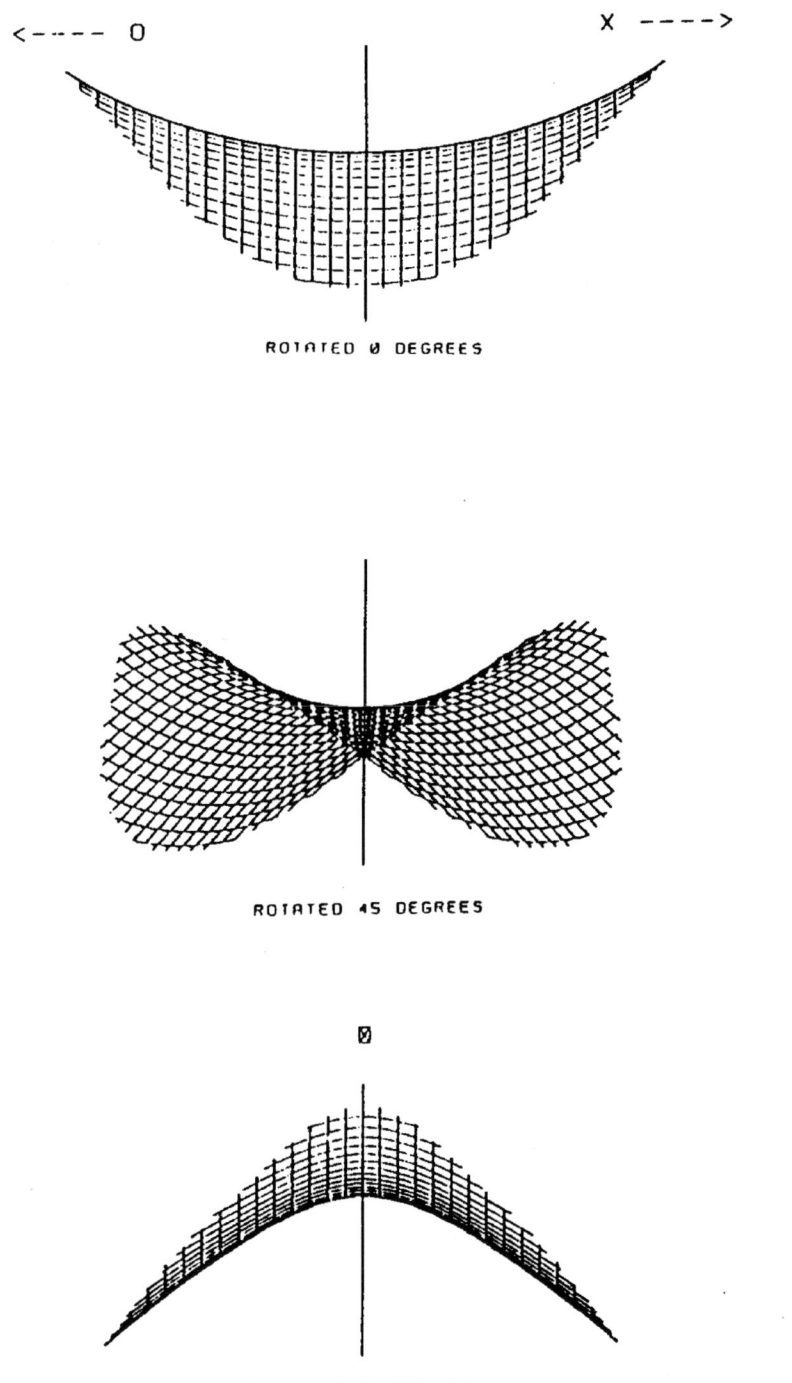

Figure 9.57 Various views of the saddle shape generated by MZO from [153].

9.18 Cross vs Down Dip, and Azimuth in Acquisition Design

There is still a great controversy about acquiring data in areas with dipping reflectors.

Should the project be acquired in the dip or strike direction? (Yes see [792])

- Will recording up (or down ???) dip allow migration to correctly position the data.
- Recording cross dip produces horizontal reflections that may be more easily migrated, but will have timing miss-ties. 3-D marine ???
- Should all 2-D recordings be center spreads. Why ??? (Reciprocity ?)

Does it matter how a 3-D project is laid out with respect to the dip?

- Is any 3-D orientation of the source and or receiver lines valid?
- Is it better to lay the project out in the dip or strike direction even though it may require significant additional expense?
- Should sources be up dip? Should receivers be up dip?
- Is an even spread of orientations required in each bin?
- Are there differences in stacked section migrations and prestack migrated sections?
- What about statics and datums?
- Should the geometry be aligned with azimuthal anisotropy?
- (See 3D marine data [790] Figure 6).

Some strong and differing opinions exist to all of the above.

One consideration is <u>trace interpolation</u>, where the source interval is twice the receiver interval. Interpolating data (before migration) in the strike direction would be easier as the reflections are horizontal, i.e. <u>shoot in strike direction</u>.

Marine data close to the shore may have a dipping water bottom that extends away from the shore. It may be desirable to acquire data in the dipping direction, but practice may require acquisition parallel to the shore to enable the streamer to turn around.
Typically the <u>longest dimension</u> of the project determines the sailing direction.
Marine <u>AVO</u> may be only valid in the sailing direction.
Marine projects with multi sources and streamers will require <u>single trace DMO</u> because of increased variability in source-receiver azimuths.

Consider the specula prestack illumination of a hemisphere by a marine 3D.

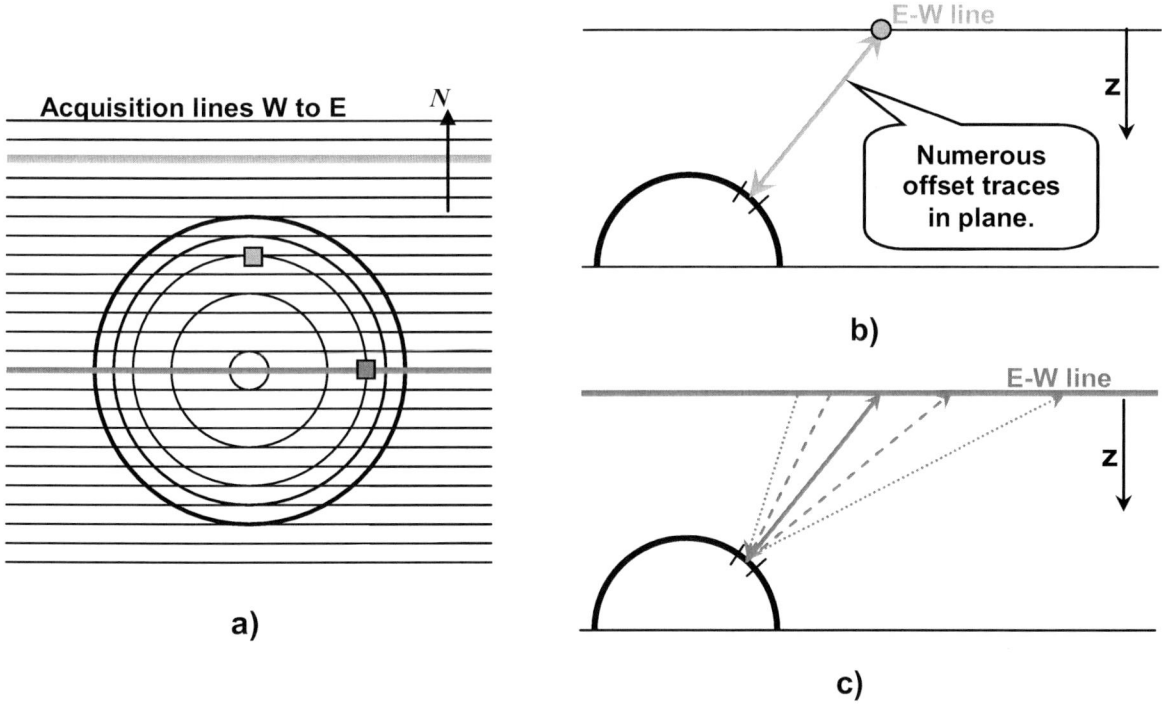

Figure 9.58 Illumination of a hemisphere by a marine 3D, a) the plan view showing north and east reflection point locations, b) multiple offset rays in a plane from a source line to the north, and c) the multiple offset rays for the east reflection point.

Consider the prestack illumination of the two points identified on the north and east side of the hemisphere in Figure 9.58a.

- When viewed from the east in (b), the north specula point contains offset traces in a plane from a line to the north of the specula reflection point. This energy may appear as sideswipe and is not in the correct time location for typical velocity analysis.

- In contrast, the east specula point (c) will have specula reflection energy on the line through the reflection point that and may appear as, and be processed as, a good reflector.

Good prestack processing (such as EOM) should eliminate these type of problems.

Commercial packages and courses are available such as Galbraith [439] to aid in the design of 3-D projects that allow selection of acquisition parameters based on the target objectives, equipment available, surface conditions, permitting costs etc.

> Perhaps the most important consideration may be the offset and azimuthal coverage of <u>all the prestack migration traces</u> that contribute to a single output bin.

9.19 General Comments

- Prestack migration <u>removes</u> the problems of dip-dependent MO and of overlapping reflections.

- As the DMO energy lies between the source and receiver, it may be applied to data at the floating datum stage; i.e. we may assume the surface is smooth or locally flat around the CMP gathers.

- Migration requires a much <u>larger aperture</u> than DMO or the formation of CMP gather.

- A poststack migration may sometimes appear to be superior to a prestack migration. If we assume that the pre and poststack migration algorithms are adequate, then the difference is usually due to uneven coverage of the acquisition geometry. <u>Stacking and amplitude equalization are very powerful tools</u> that smooth the wavefield for the poststack migration. An algorithm that balances the illumination of the reflectors may be required.

9.20 Summary of Points to Note in Chapter 9

- **Prestack migration by:**
    - **Direct Kirchhoff migration**
    - **Source records**
        - **Eikonal traveltimes and ray tracing**
        - **Full waveform propagation**
    - **Source/receiver (sg) method**
    - **Inversion**
    - **Constant offset sections**
    - **Constant offset angles**
    - **Stolt 2-D method**
    - **GDMO-PSI**
    - **EOM.**
- **3-D DMO and prestack migration for constant velocities.**
- **DMO for linearly increasing velocities (MZO).**
- **Cross dip, down dip, and azimuth should be considered in acquisition.**
- **Migration requires a much larger aperture than DMO**
- **Extensions to 3D**

Chapter Ten

Examples of DMO and Prestack Migrations

Objectives

- Recognize properties of the DMO operator.
- Know how to test a DMO operator.
- Identify areas where DMO benefits data.
- Identify areas where DMO may harm data.
- Know that test results only evaluate the implementation of an algorithm, not the algorithm.
- Identify the differences of time and depth migrations when they are applied to structured data.

10.1 Description of Figures

This chapter contains a collection of examples on DMO and prestack migration. A short explanation is provided for each section.

10.1.1 Comparison of DMO and Prestack Migration Ellipses

Figure 10.1

Kinematic family plots of DMO and prestack migration ellipse, both having the same source-receiver offset.

Note the small lateral extent of the DMO ellipses and the large lateral extent of the prestack migration ellipse. The prestack migration ellipses were truncated for the display.

10.1.2 3-D Model Test for DMO

Figure 10.2 to Figure 10.4

Figure 10.2 shows a constant velocity model with two horizontal layers and two dipping layers with dips of 10 and 20 degrees. The 10-degree event dips in the x direction and the 20-degree dips in the y direction.

Figure 10.3 and Figure 10.4 show three inline and three crossline sections taken from the center of the volume. In each figure, the first three sections (a) are stacked with no DMO and the second three sections (b) are stacked with DMO.

Note the improvement of the DMO processed sections in the area indicated by the arrow.

10.1.3 Example of the DMO Operator

Figure 10.5 – Figure 10.8

Figure 10.5 and Figure 10.6 are <u>theoretically correct examples</u> of DMO traces. The input trace contained spikes at 200 ms intervals that were zero-phase bandpass filtered. Note the increase in amplitude with offset at shorter times, and the decrease with amplitude at longer times. Also note the phase shift on the DMO'd traces in 45 degrees.

Figure 10.7 and Figure 10.8are example of DMO using time domain approach described in sections 8.4.10, and 8.4.12. Figure 10.7 used a boxcar AAF but has no phase correction filter ($rj\omega$). The $rj\omega$ filter is included in Figure 10.8.

Figure 10.6 to Figure 10.8 include a summed trace on the left side. This trace is a spatial (horizontal) sum of the DMO'd traces. This trace represents the result of DMO processing horizontal events made from the input trace. DMO should have no effect on horizontal events and the summed trace should be the same as the input traces as in Figure 10.6. Not the improvement in the summed trace with the inclusion of the $rj\omega$ filter.

10.1.4 Testing and Evaluating DMO

Figure 10.9 to Figure 10.12

These figures show an input trace (or traces), the time domain DMO result, and a number of summed traces at various stages of processing. The summed traces in Figure 10.9 and Figure 10.10 in order from the right contain no $rj\omega$ filter, a differential filter, only a 45 degree phase shift, and the $rj\omega$ filter. The last two traces are the difference between the $rj\omega$ filter trace and the input trace; the first is scaled by four (×4) to see the noise with the second at normal scale (×1).

Figure 10.9 is a small offset DMO. Small offsets are difficult to DMO and require the most testing to ensure acceptable amplitude and phase characteristics; this example is acceptable. A large offset is shown in Figure 10.10 with virtually no energy on the difference traces.

Figure 10.11 and Figure 10.12 simulate evaluation of the DMO applied to dipping events. Rather than a horizontal sum (as in the previous examples) the data is summed along a dipping trajectory as indicated by the arrow in Figure 10.12 to a trace located at the center. Energy tangent to the dip will sum coherently at the time delay where the arrow intersects the central trace. This time delay varies with time and dip as illustrated in the summed traces to the left.

All these traces are fully processed and represent the dips as indicated. Note as the dip increases the bandwidth of the wavelet decreases to prevent aliasing. As the input trace becomes aliased with increasing dip, a number of high-cut filters were applied to the input trace to prevent aliasing at the indicated dips. Now, the dipping sum of the traces may be compared with an unaliased-input trace. Note the input trace for the DMO was the original 0° dip; the other input traces were only provided for a comparison.

The amplitude of the dipping events should tend to zero for dips over 45-degrees.

These results show excellent phase characteristics that match the input trace, especially the small offset example in Figure 10.12. The residual amplitudes on traces above 45-degrees of dip are due to a boxcar AAF filter. This aliasing noise is evident in the following examples that use the same algorithm.

Practical approximation allow the $rj\omega$ filter to be applied after DMO and stacking, however, the AAF filter must be applied during the formation of the DMO'd traces.

10.1.5 DMO of an Aliased Dipping Event

Figure 10.13 to Figure 10.16

These figures contain two horizontal and one dipping event of a <u>constant offset section</u> in which MO correction has been applied. The intent of this test is to evaluate the amplitudes of horizontal events at different times, and to evaluate the DMO of a steeply dipping event ($\alpha \approx 40°$) that extends over a large time range (0 to 1700 ms). The zero offset position of the dipping event is illustrated by a thin line in Figure 10.13.

The DMO'd result in Figure 10.14 shows excellent amplitude and positioning results. A white line through the dipping wavelets indicated the zero offset position. This figure also shows aliasing noise above the dipping event that is due to the use of a boxcar AAF in the DMO algorithm. (The input dipping event contained the maximum frequencies before aliasing and tends to over emphasize the aliasing noise).

The input data of Figure 10.15 has the amplitudes in every second trace set to zero for an interpolation test, with the results in Figure 10.16. Note the <u>excellent interpolation</u> of the data by the DMO. The input data is severely aliased and the aliased noise is evident after DMO in Figure 10.16. Note that this is <u>only one</u> of <u>many constant offset sections</u>, which will reduce the aliased noise when stacked.

The input data, in essence, only contained every second trace of the input data or had double the trace interval. If the DMO algorithm had assumed this wider trace spacing, the aliasing noise would have been reduced substantially.

10.1.6 DMO Processing of Real Data

Figure 10.17 to Figure 10.20

The improvement, by including DMO in a processing sequence, is shown in this sequence of figures. Both sets of data (A and B) compare a best processing section with a DMO'd section. The data is taken from a marine line and illustrate the improved resolution of diffraction energy.

10.1.7 DMO of a Source Record

Figure 10.21 to Figure 10.22

A source record (with MO correction applied) is shown in Figure 10.21 with its DMO'd source record shown in Figure 10.22. Note the slight reduction in noise and the improved coherence of the reflectors. Also note the DMO'd energy is contained within the original source geometry.

10.1.8 Prestack Migration of a Source Record

Figure 10.23 to Figure 10.24

These figures show an input source record (with scaling decon etc.) and the corresponding prestack migration. The outline of the source record is shown on the prestack migration. Note on the input record that events, midway down, have their apex shifted away from zero offset. This shift is indicative of dipping events, which are evident after prestack migration. There is enough energy around these events for a partial reconstruction of the structure, however, there is much energy spread beyond the boundary of the input record that will require many other source records to reconstruct an image of the geological structure.

10.1.9 Comparisons of DMO and Migrations on Real Data

Figure 10.25 to Figure 10.34

Two examples (A and B) showing:

- a stacked section,
- a DMO'd stacked section,
- a poststack migration without DMO,
- a poststack migration with DMO, and a prestack migration.

In general the DMO'd sections have an improved image and contain more coherent dipping energy. However, the DMO'd section in Figure 10.26 does loose resolution in the area indicated by the white arrow relative to Figure 10.25. This could be due to a large velocity contrast above these events.

The DMO and poststack migration of A in Figure 10.28 has more steeply dipping energy than the poststack migration or the prestack migration as indicated by the white arrows. This indicates to me that a poor prestack migration was used. Note however, that the black arrow in the DMO migration of Figure 10.28 shows an under migrated diffraction that is better resolved in the prestack migration. The gray arrows also indicated improved resolution in the prestack migration.

The examples of the B data set also show DMO produces better coherence in the steeper dipping events as indicated by the black arrows in Figure 10.30, Figure 10.31, Figure 10.33, and Figure 10.34. The prestack migration of this area in Figure 10.34 is poor. However, the faulted area indicated by the white arrows in Figure 10.33 and Figure 10.34 show possible improvement in the prestack migration.

The prestack migration should produce an improved section in all areas, however it may have had a dip limit parameter that restricted the migration of the steeper dips.

10.1.10 Prestack Migration of a Structurally Complex Model

Figure 10.35 to Figure 10.39

The migrated images in this section were produced from prestack modelling of a structurally complex geological structure that represents a cross section through the Canadian Rockies as shown in Figure 10.35.

A prestack depth migration and a prestack time migration using the exact velocities are shown in Figure 10.36 and Figure 10.37.

Blind processing of the data (i.e. not knowing the velocities or structure) was attempted with terrible results. Four iteration of velocity analysis using DMO produced the results in Figure 10.39b (which is compared to the optimum prestack depth migration in (c).

An example of equivalent offset migration (EOM) is shown in Figure 10.38. This model started with a rough estimate of the RMS velocities to produce common scatter point (SCP) gathers, which were then analyzed for a refined velocity model and stacked. The results were surprisingly good when compared with other attempts at processing as illustrated in Figure 10.39.

10.1.11 Examples of 3-D Salt Imaging

Figure 10.40 to Figure 10.42

Examples of 3-D salt and subsalt imaging are included using:

- **prestack time migration** as an intermediate step, which is then
- **inverse poststack time migrated**, (modelled) and then
- **poststack depth migrated**.

The data was processed with MOVES, developed by Veritas DGC.

The results are very encouraging and provide an economical but accurate depth migration that is produced with prestack migration runtimes.

10.2 Comparison of DMO and Prestack Migration Ellipses

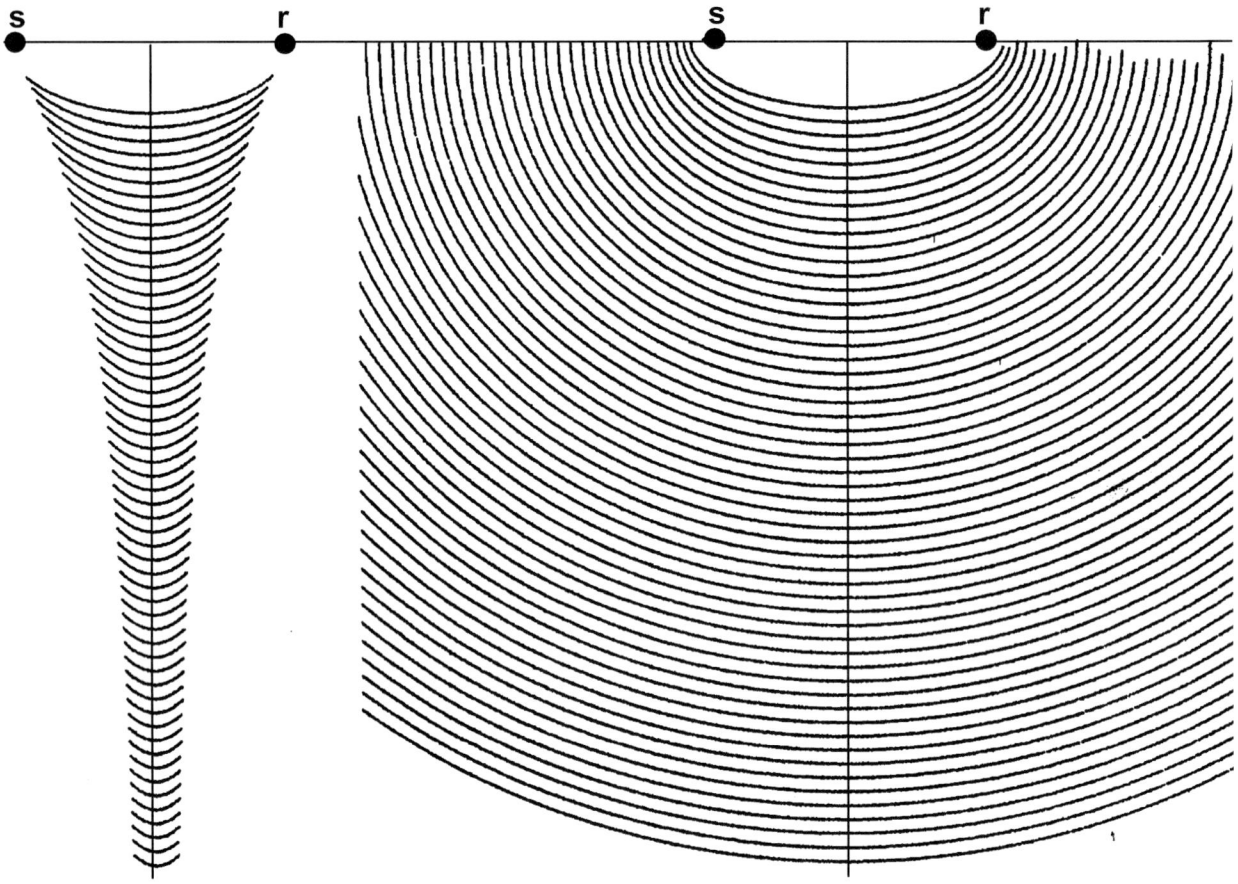

Figure 10.1 Kinematic comparison of DMO and truncated PSM ellipses.

Note:
- **Small extent of DMO ellipse**
- **Large extent of the prestack migration ellipse**
- **Prestack migration ellipse tends to a semicircle at large offsets**

10.3 3-D Model Test for DMO

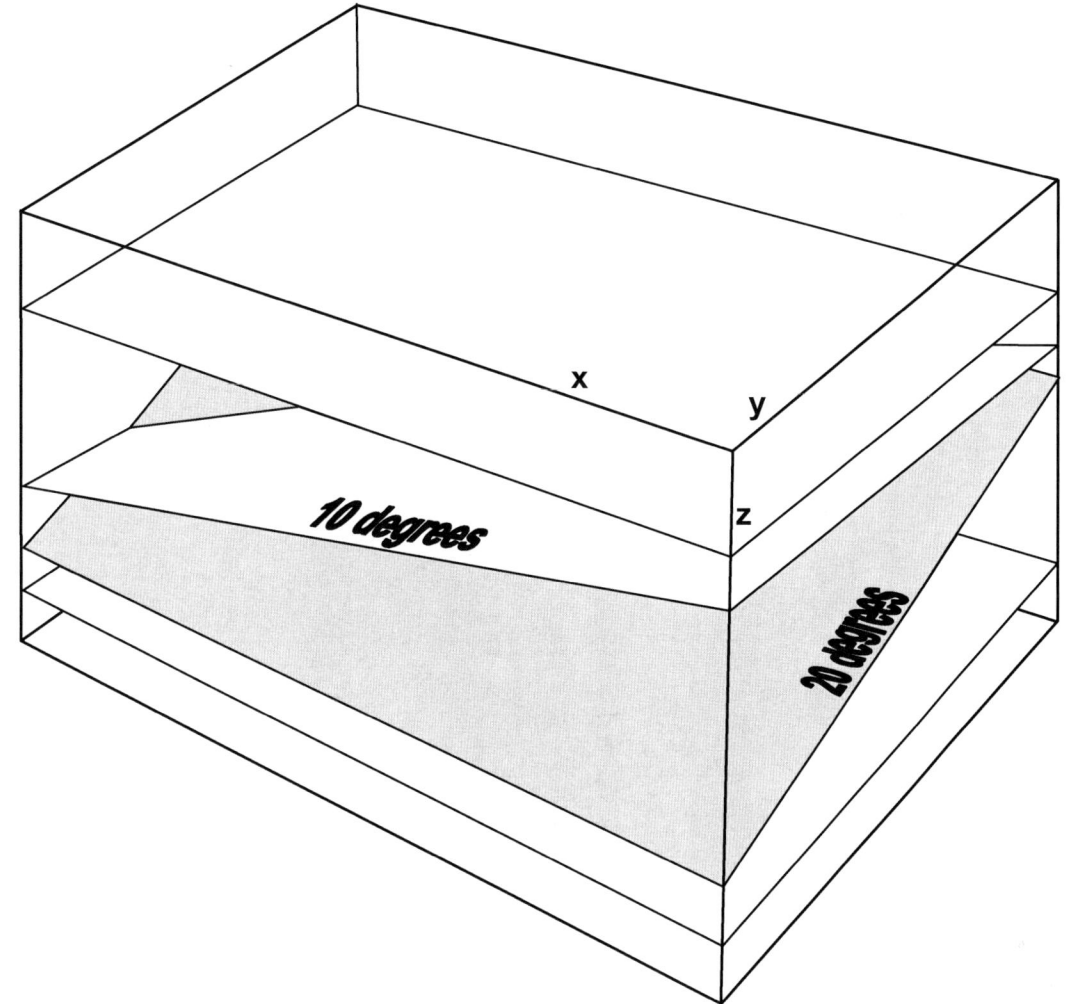

Figure 10.2 3-D DMO test structure for data in Figures 10.3 and 10.4.

3-D model for testing DMO:
- **Constant velocity.**
- **Two dipping events of 10 and 20 degrees with different dip directions.**
- **Two horizontal events.**

Figure 10.3 contains three inline sections that are parallel to the strike of the 10-degree reflector. They are taken from the center of the volume, (a) without and then (b) with DMO. Figure 10.4 shows three equivalent crosslines. Note:

- **DMO shows very slight improvement in the 10 degrees event.**
- **There is significant improvement with DMO in the 20-degree event.**

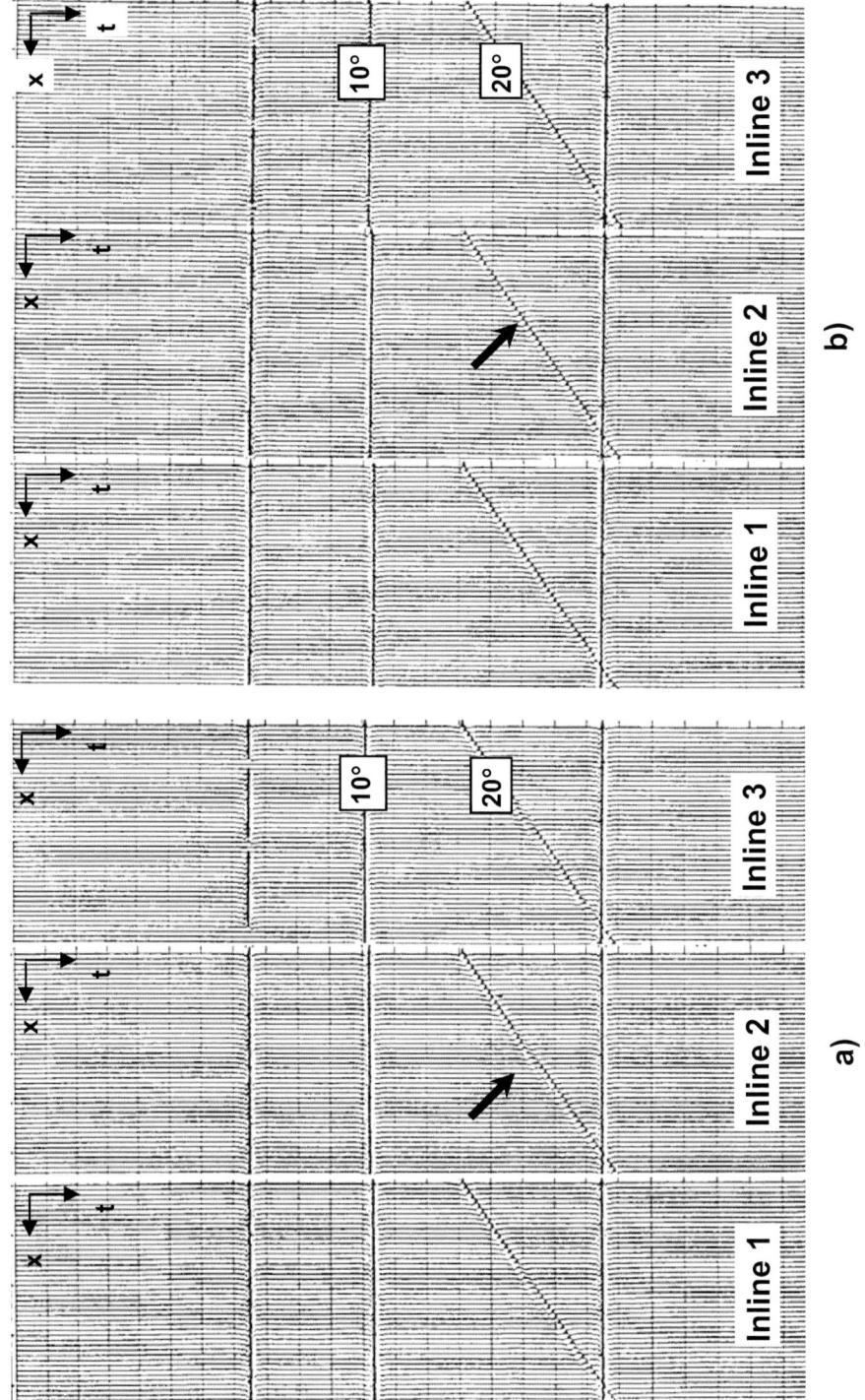

Figure 10.3 Three inline sections from center of 3-D model, a) normal processing, and b) including DMO.

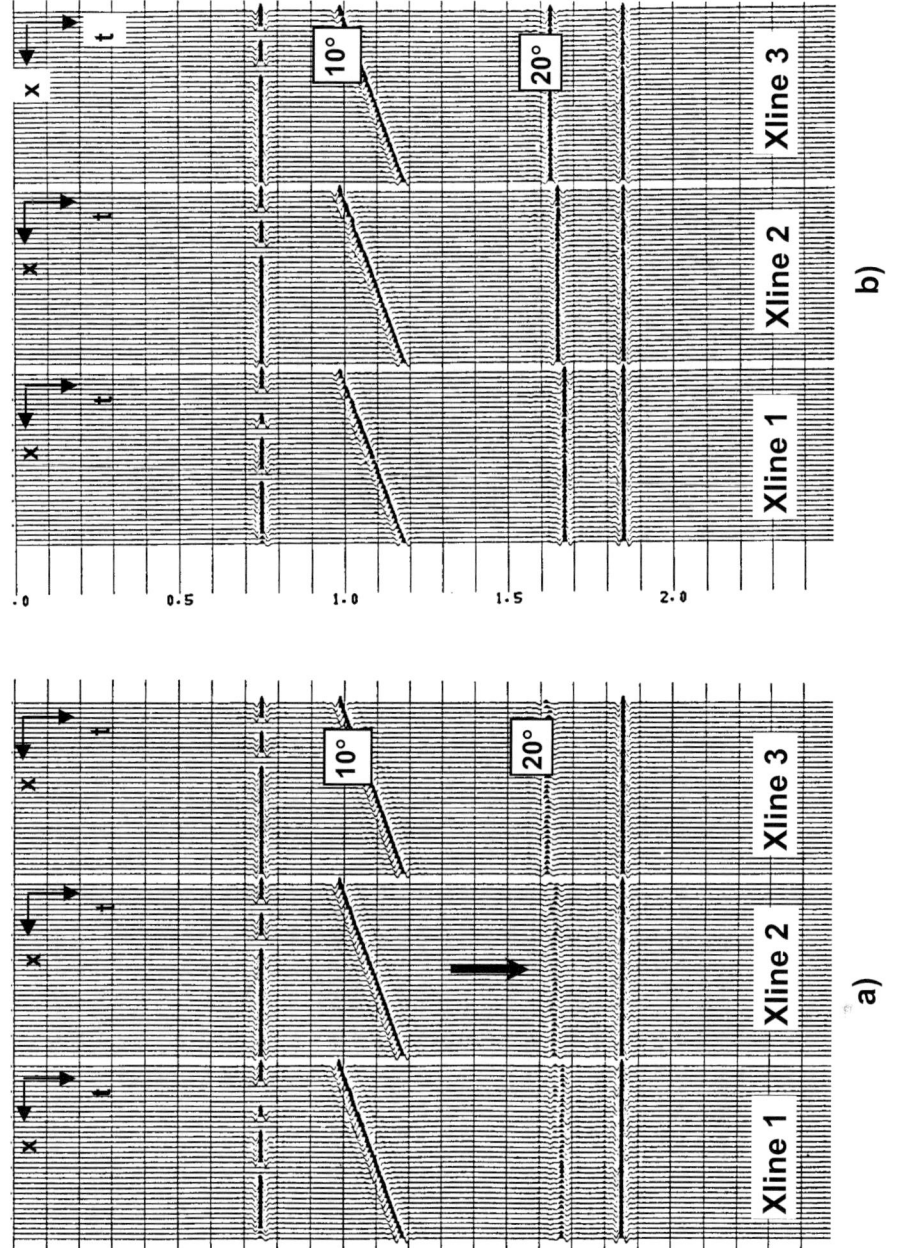

Figure 10.4 Three crossline sections from center of 3-D model, a) normal processing, and b) including DMO.

Note the degradation of the 20 degree event at area indicated by the arrow.

10.4 Examples of the DMO Operator

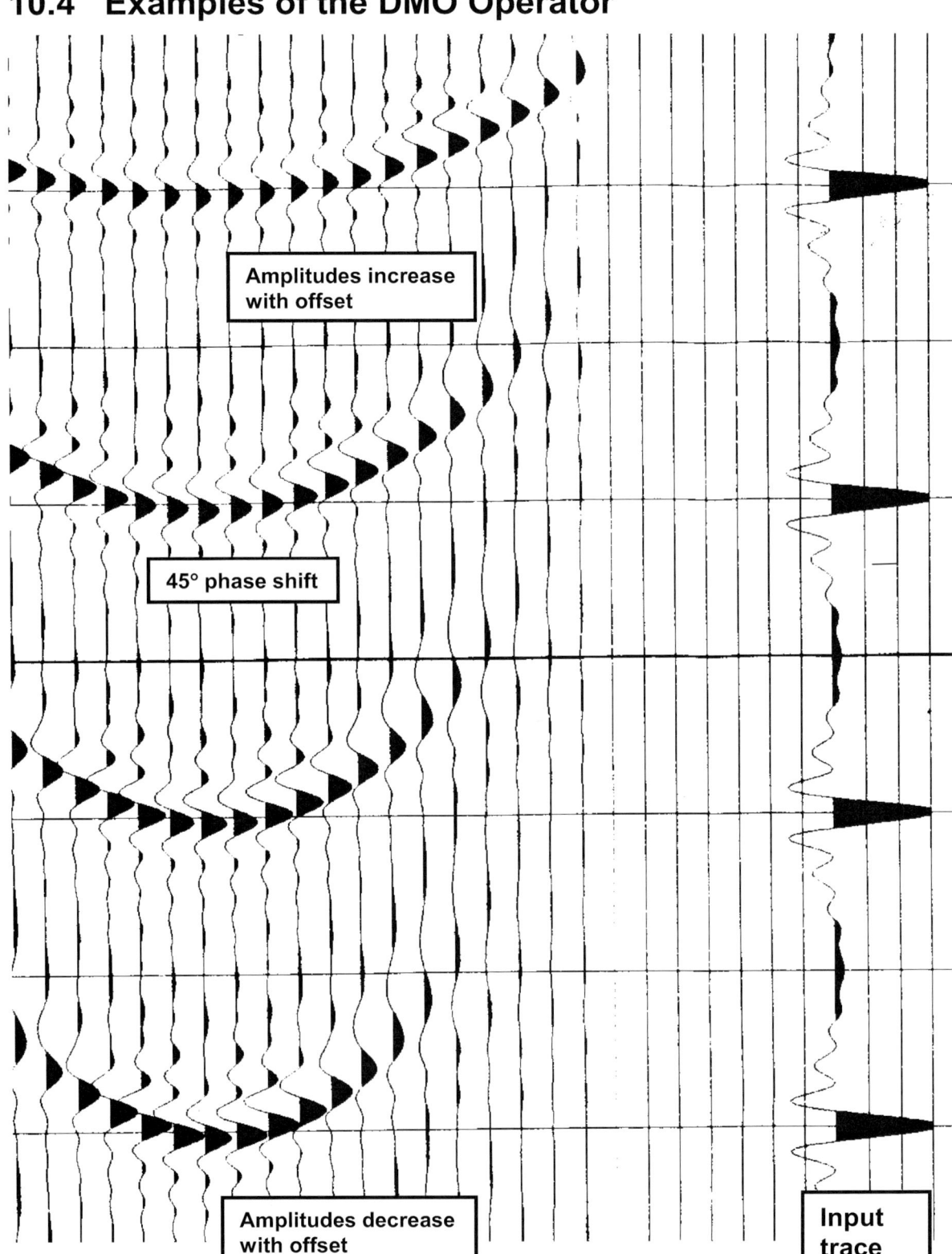

Figure 10.5 <u>Theoretically correct</u> DMO with input trace on the right. Note the variation of amplitude with offset, and the 45 degree phase change.

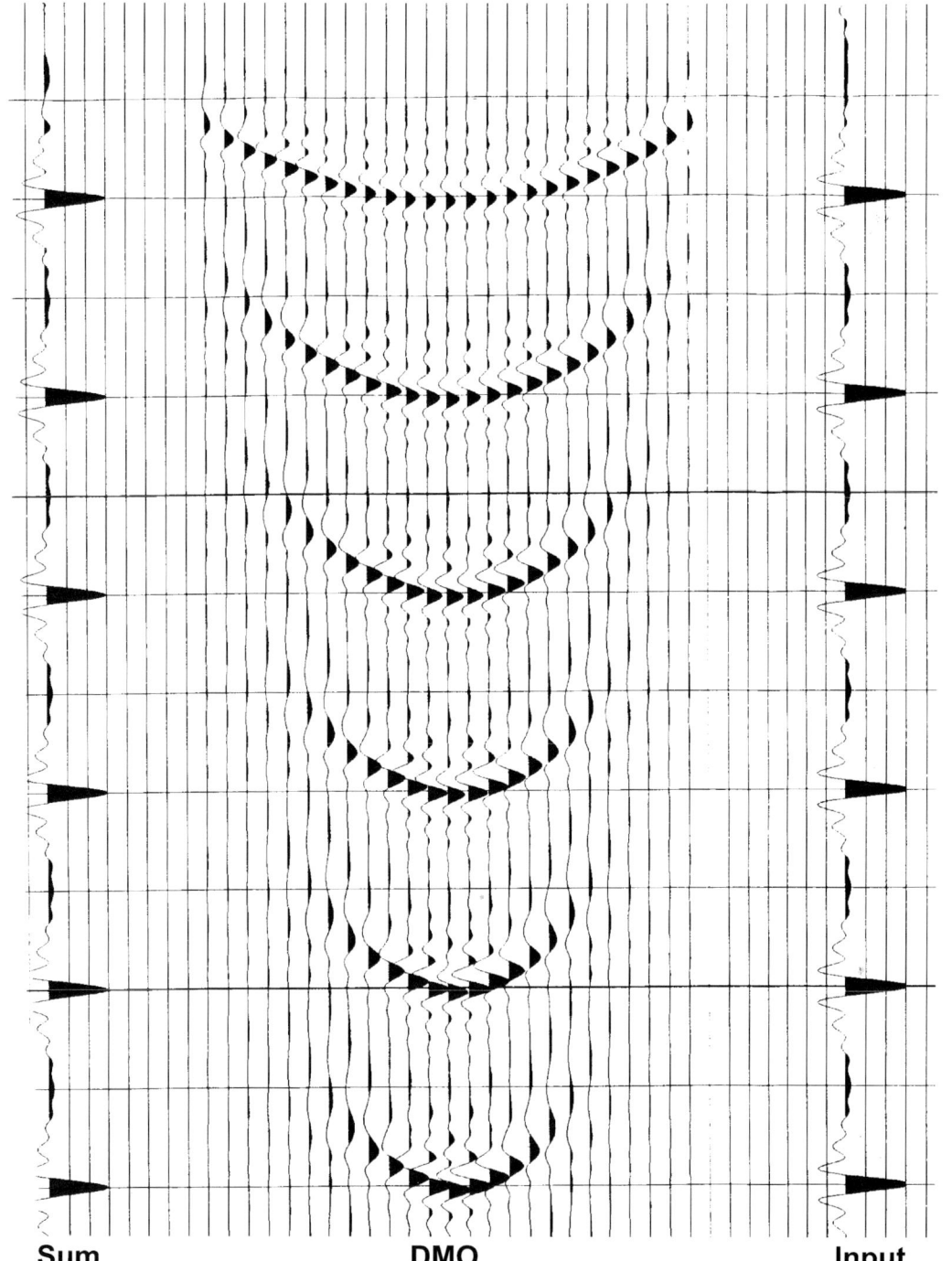

Sum　　　　　　　　　　DMO　　　　　　　　　　Input

Figure 10.6 <u>Theoretically correct</u> DMO curves by Liner method.

The trace on the left is formed by laterally summing all the DMO traces. It <u>represents</u> a single DMO'd trace taken from a <u>horizontal reflection</u>. The amplitudes and phase of the summed trace should be similar to the input trace.

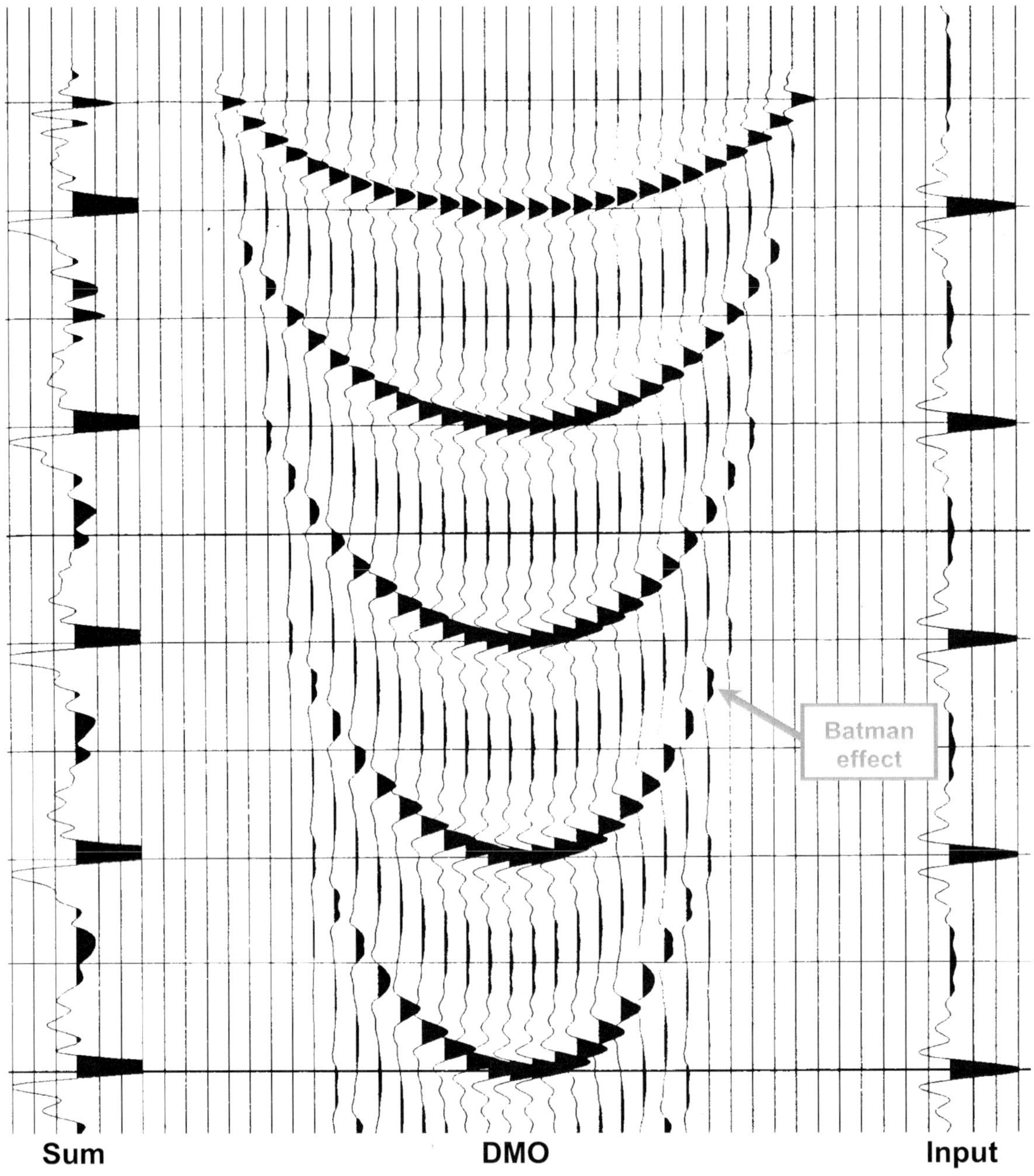

Figure 10.7 Time domain DMO with amplitude scaling and antialiasing filter.
<u>No rjω filter</u> was used.

A box car shape was used for the antialiasing filter. When the width of the box-car is larger than the wavelet period, the <u>"Batman" effect</u> occurs.

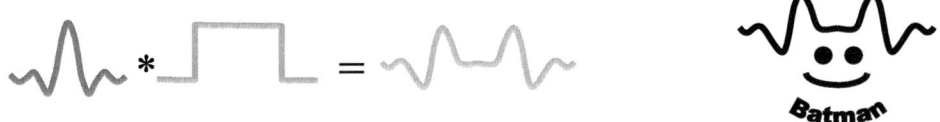

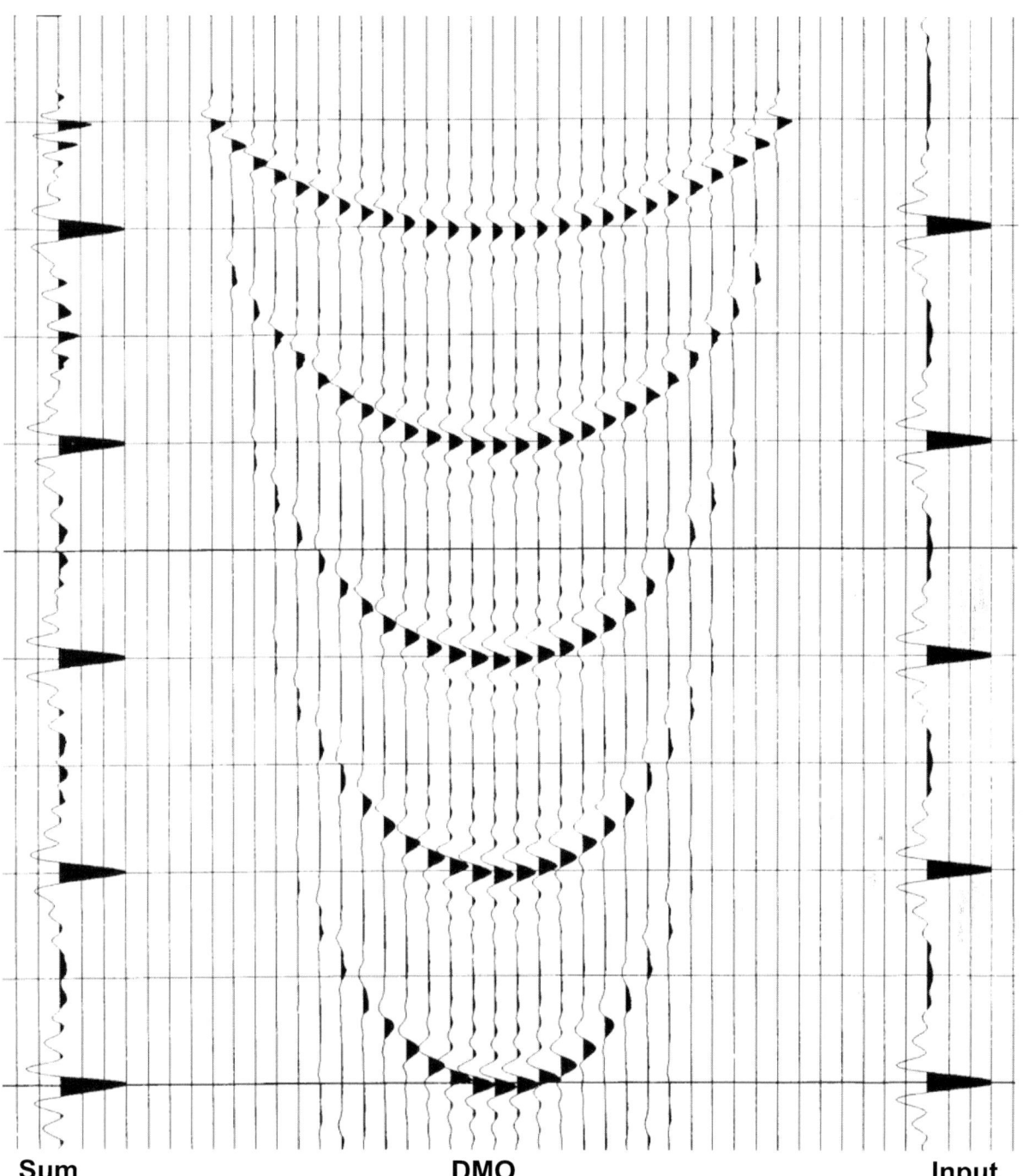

Figure 10.8 Time domain DMO, the same as previous figure, but now including the *rjw* filter.

The summed trace is very similar in amplitude and phase to the input trace. Compare with Figure 10.6

10.5 Testing and Evaluating DMO

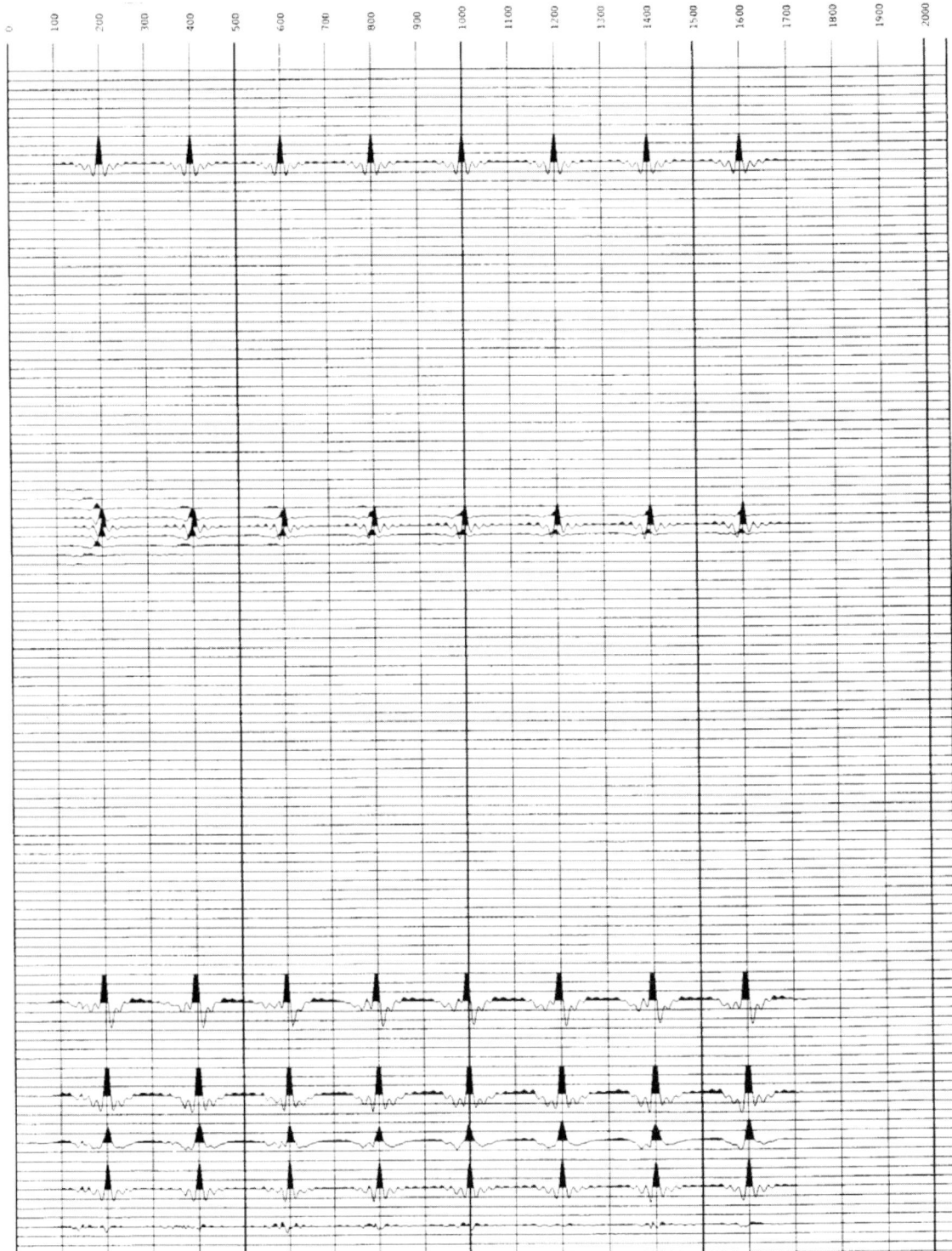

Figure 10.9 DMO sum and difference test for small offset.

Accurate amplitude and phase are difficult for small offsets.

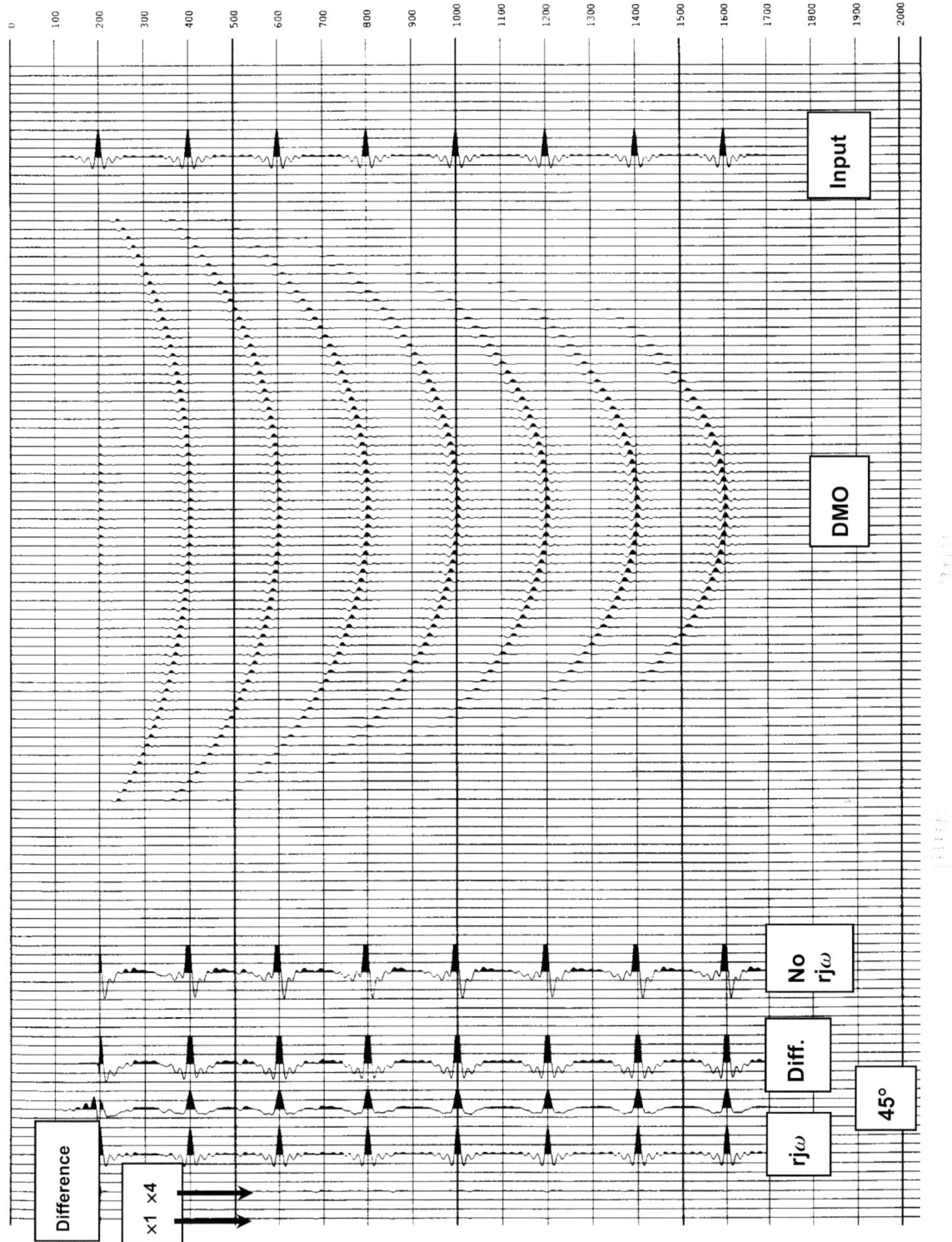

Figure 10.10 DMO, sum, and difference test for large offset.

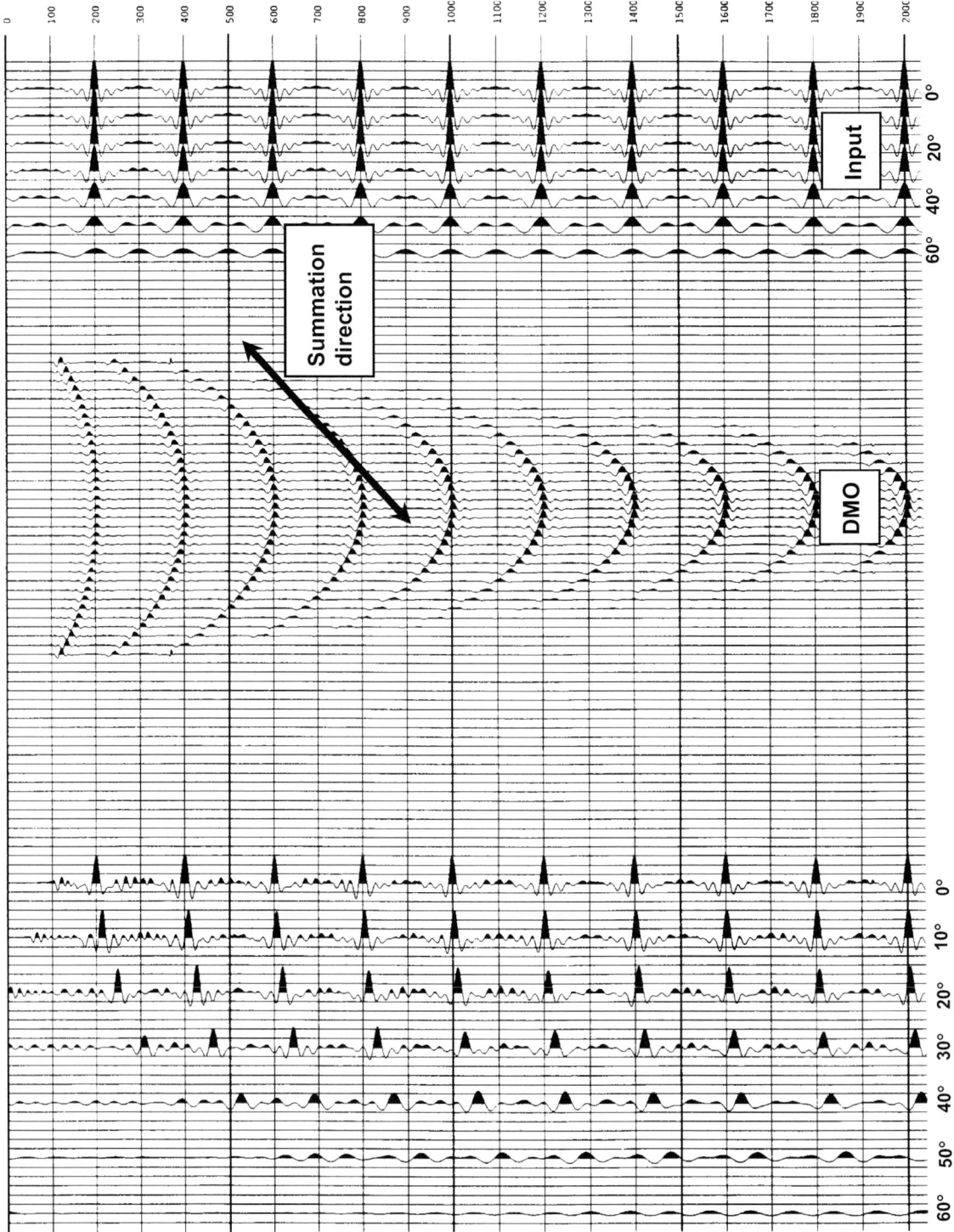

Figure 10.11 DMO dip summation test with medium offset.

The summation point intersects the zero displacement at deeper times resulting in an apparent time shift of the summed trace.

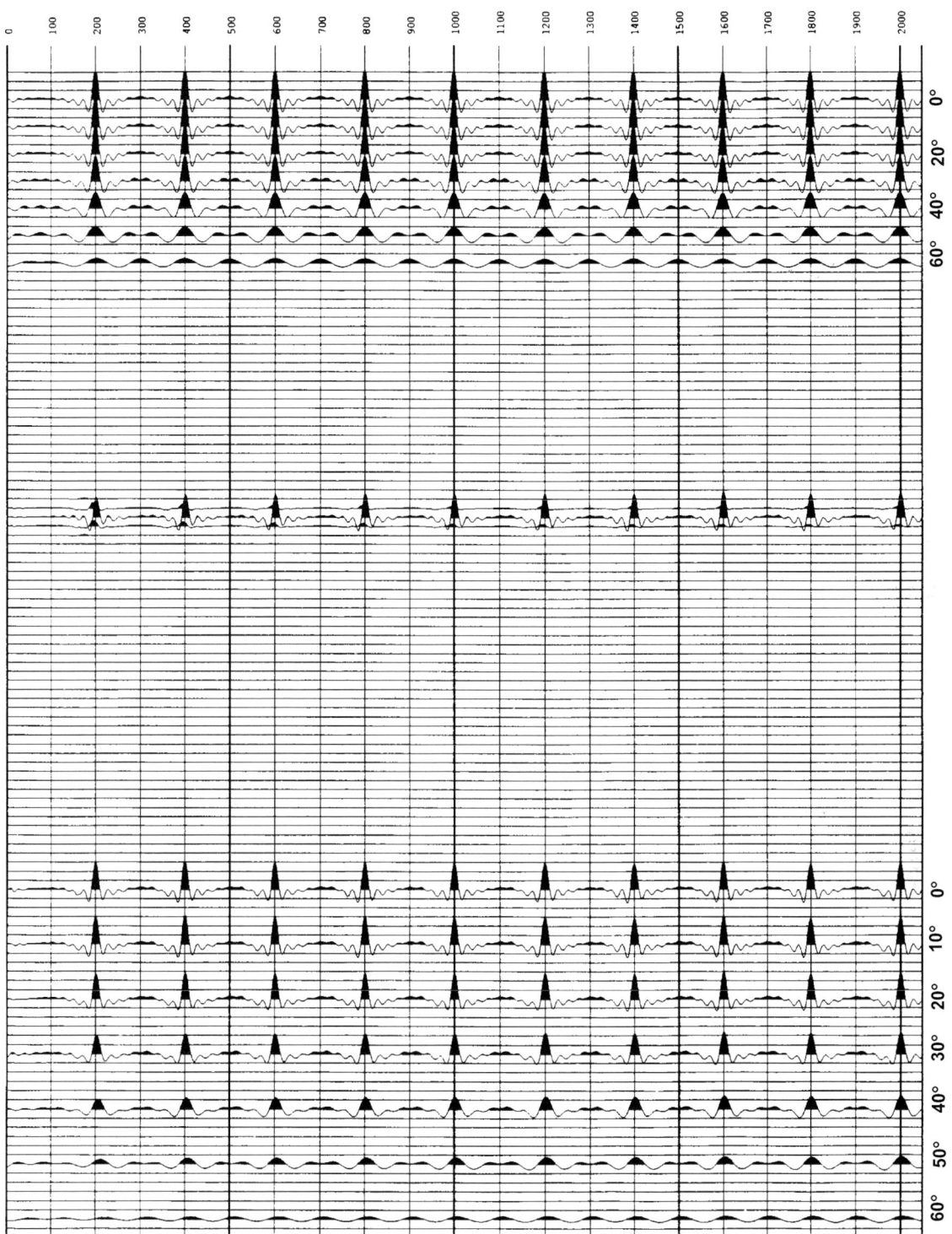

Figure 10.12 DMO dip summation test with small offset.

To prevent aliasing, the input trace was band limited for the appropriate dips as indicated above. The summation was along the appropriate dip. The resulting summed wavelets are still close to zero-phase.

10.6 DMO of an Aliased Dipping Event

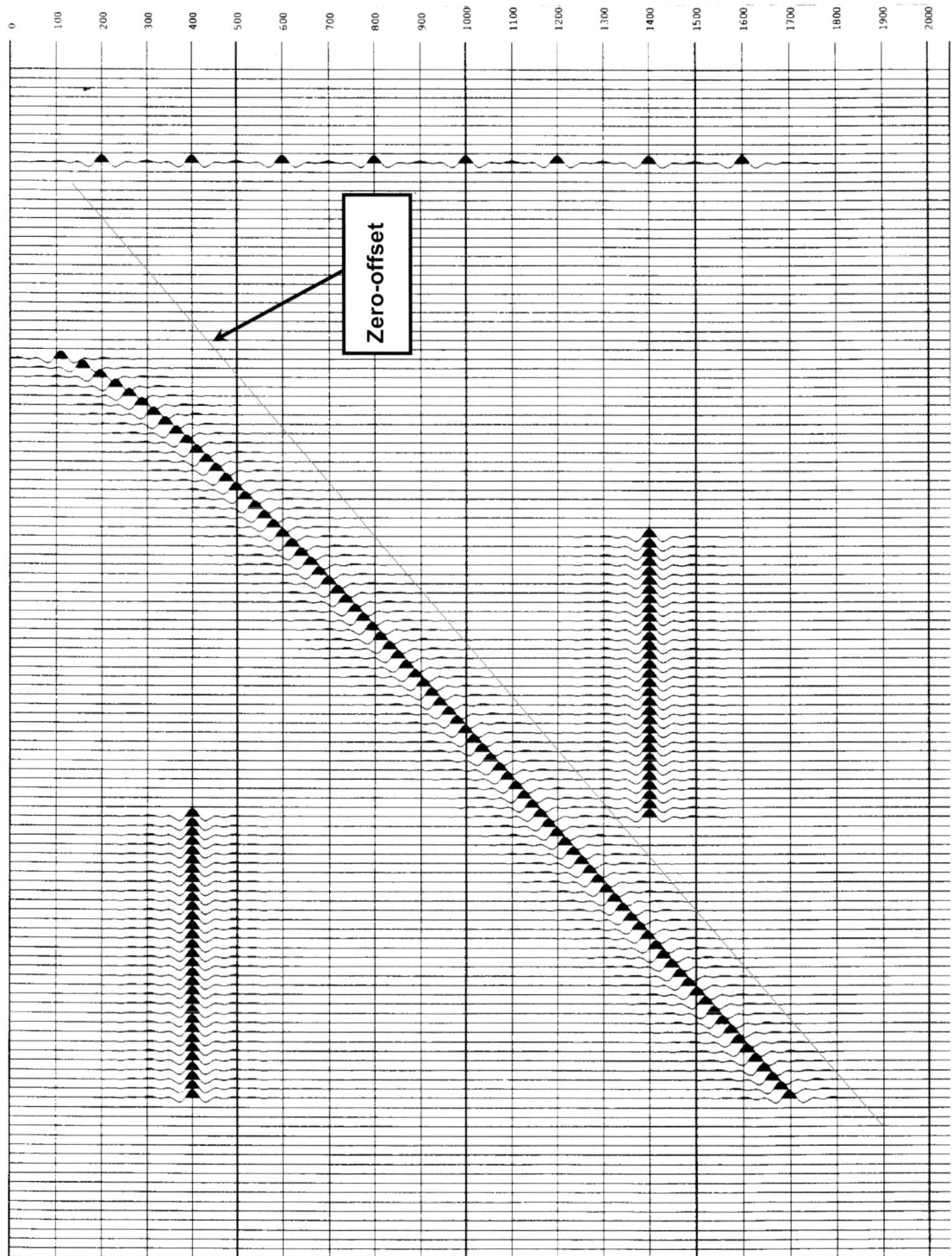

Figure 10.13 Constant offset (4000 ft) model, MO for zero dip applied. The thin black line shows the zero offset position.

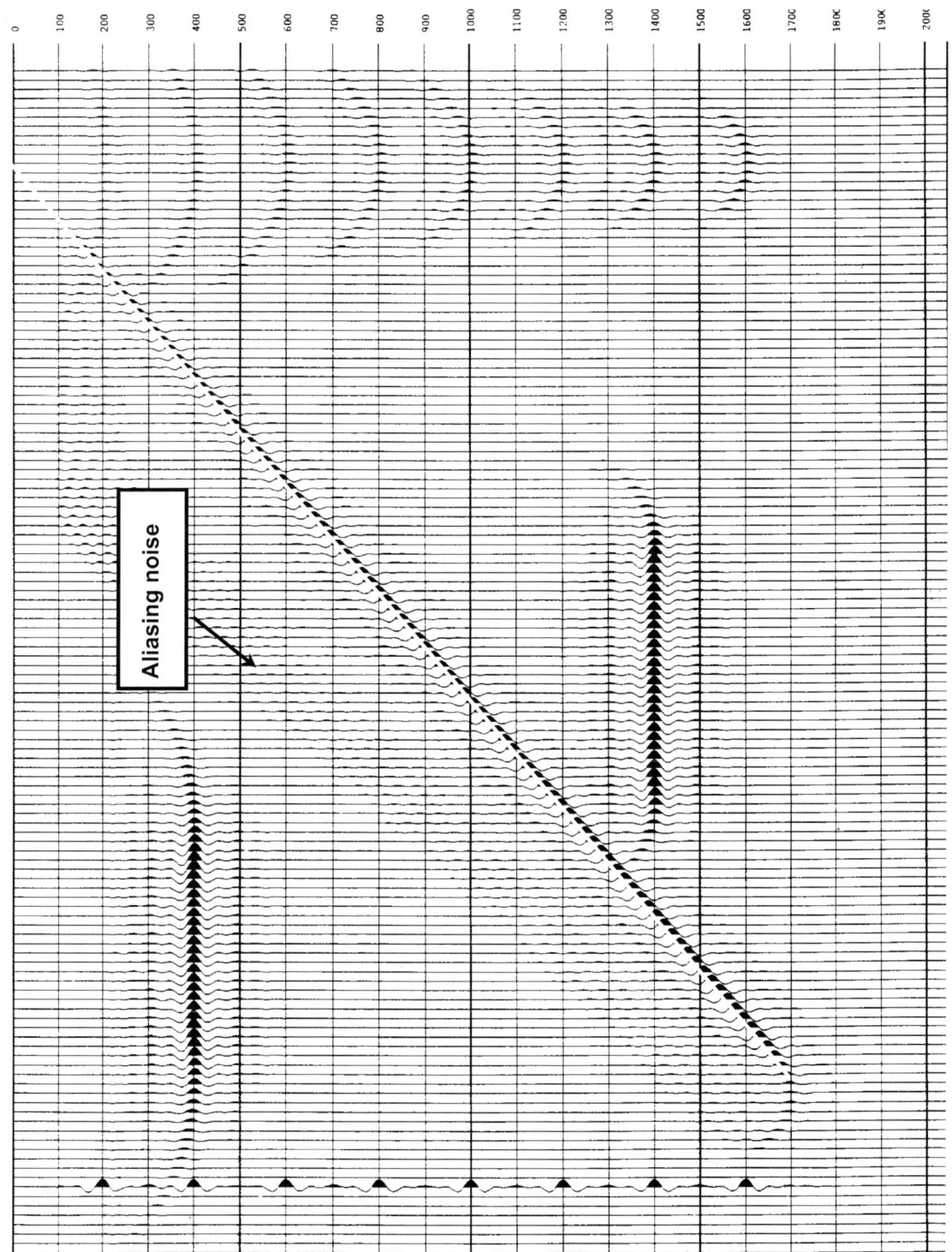

Figure 10.14 DMO of constant offset test model. The thin white line shows the zero offset position.

Note the aliasing noise that results from an imprecise box-car AAF filter

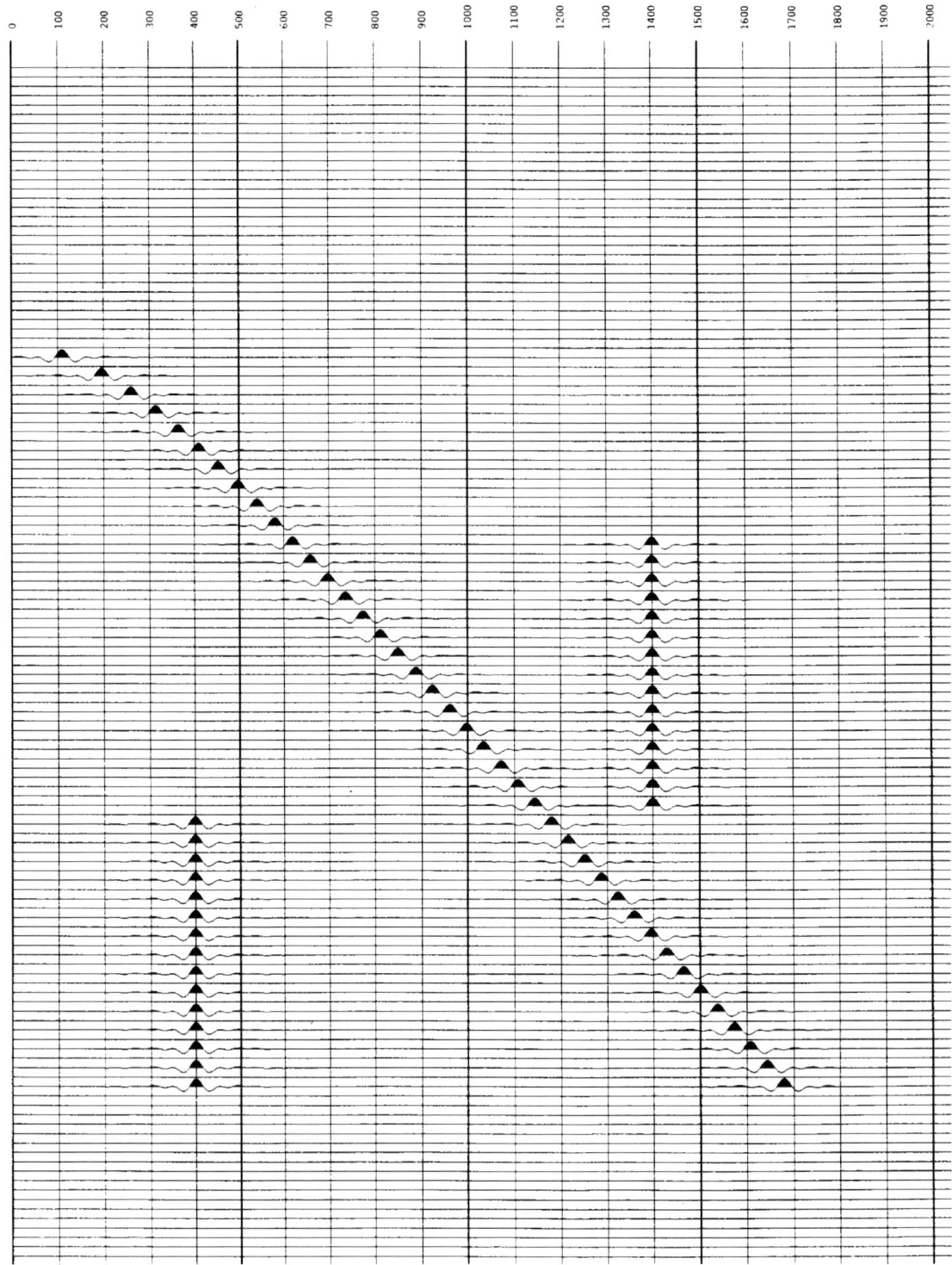

Figure 10.15 Model for trace interpolation; every alternate trace has zero amplitude.

Chapter 10 Example of DMO and prestack migrations

Figure 10.16 DMO interpolation of previous model.

Interpolation excellent, aliasing noise excessive. The aliasing noise could be reduced substantially by designing the AAF using the live input trace spacing.

10.7 DMO Processing of Real Data

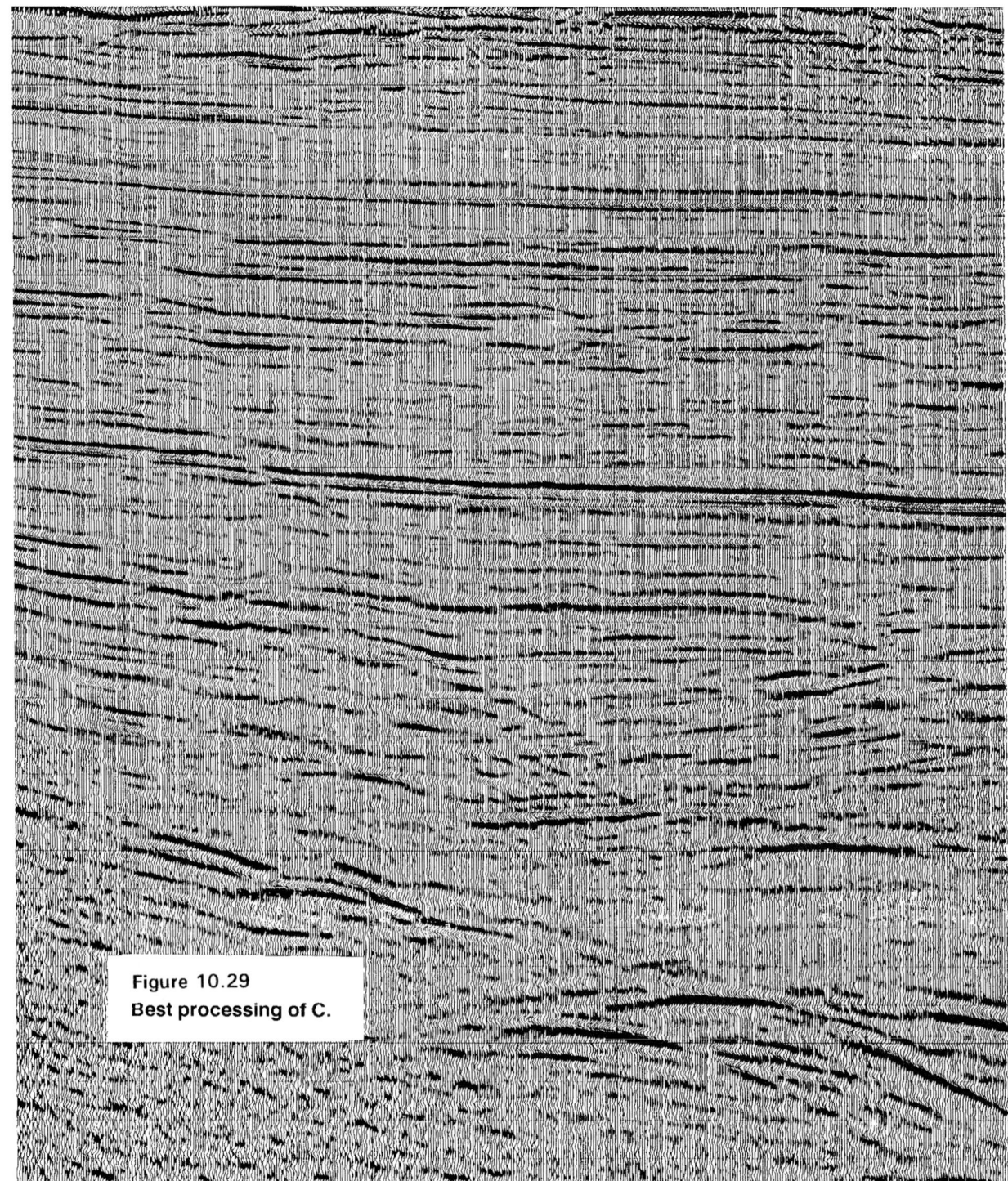

Figure 10.29 Best processing of C.

Figure 10.17 Best processing of example A.

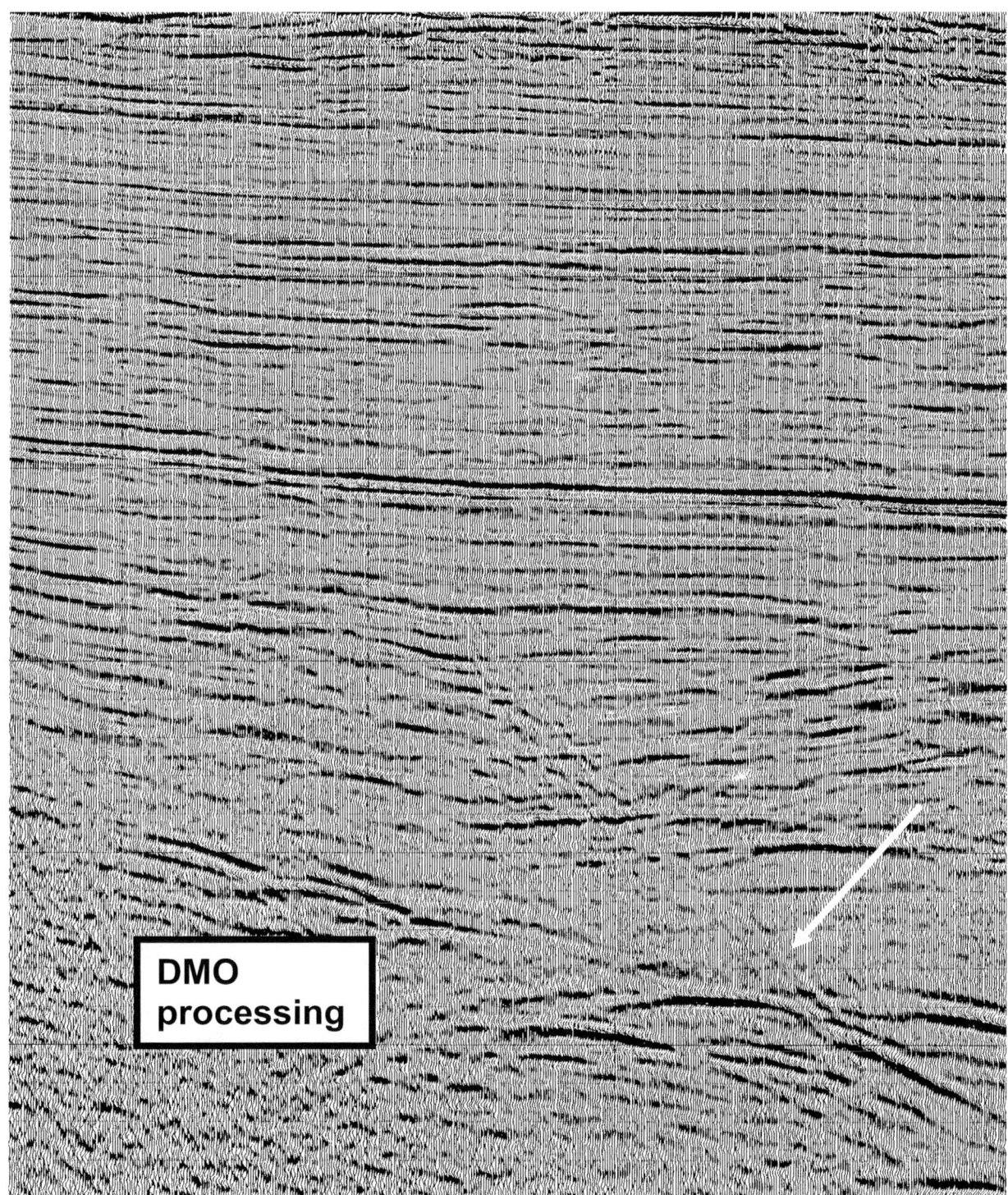

Figure 10.18 DMO of example A.

Compare the diffraction energy at the arrow with Figure 10.17.

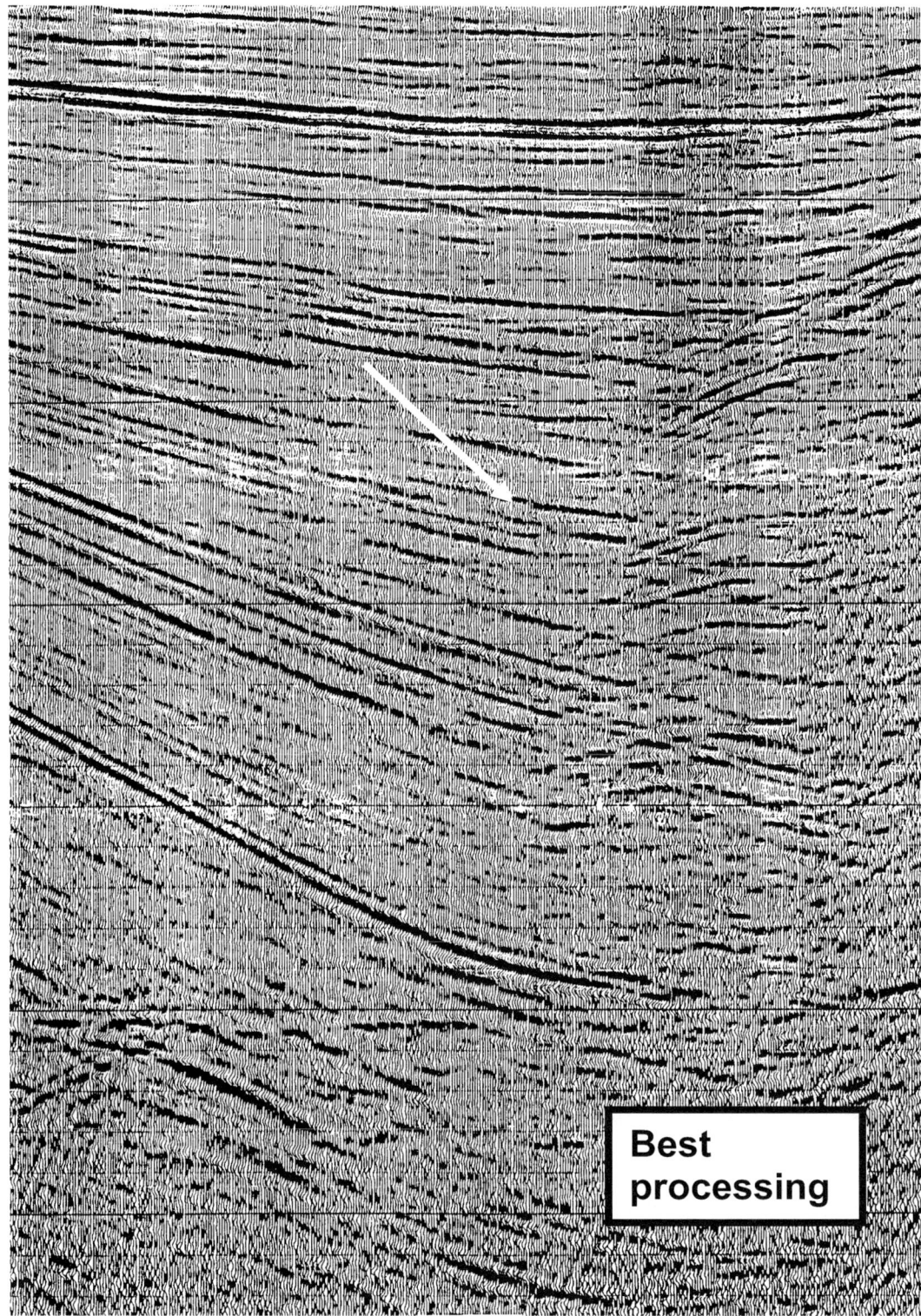

Figure 10.19 Best processing of example B.

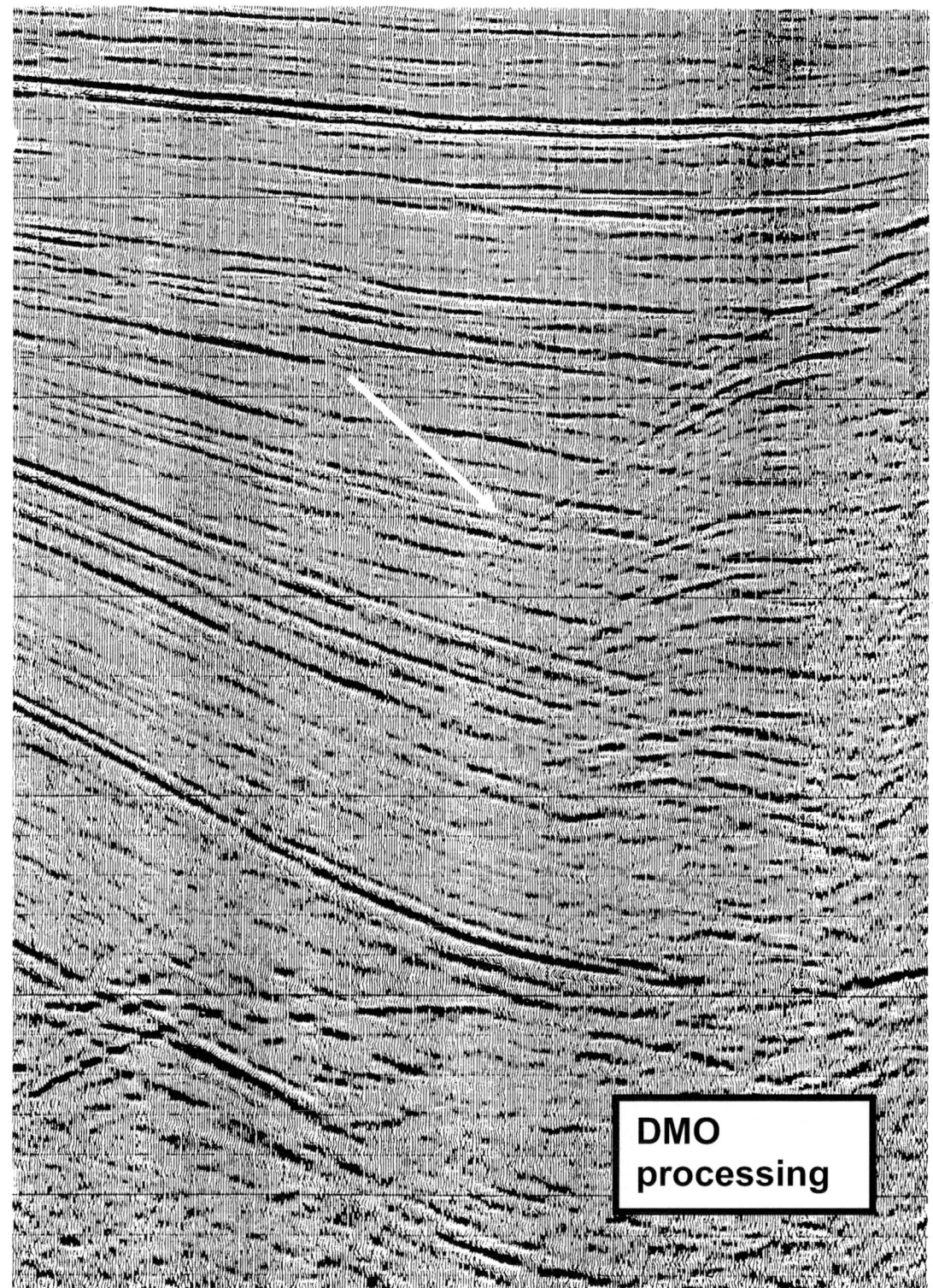

Figure 10.20 DMO of example B.

Compare the diffraction energy at the arrow with Figure 10.19.

10.8 DMO of a Source Record

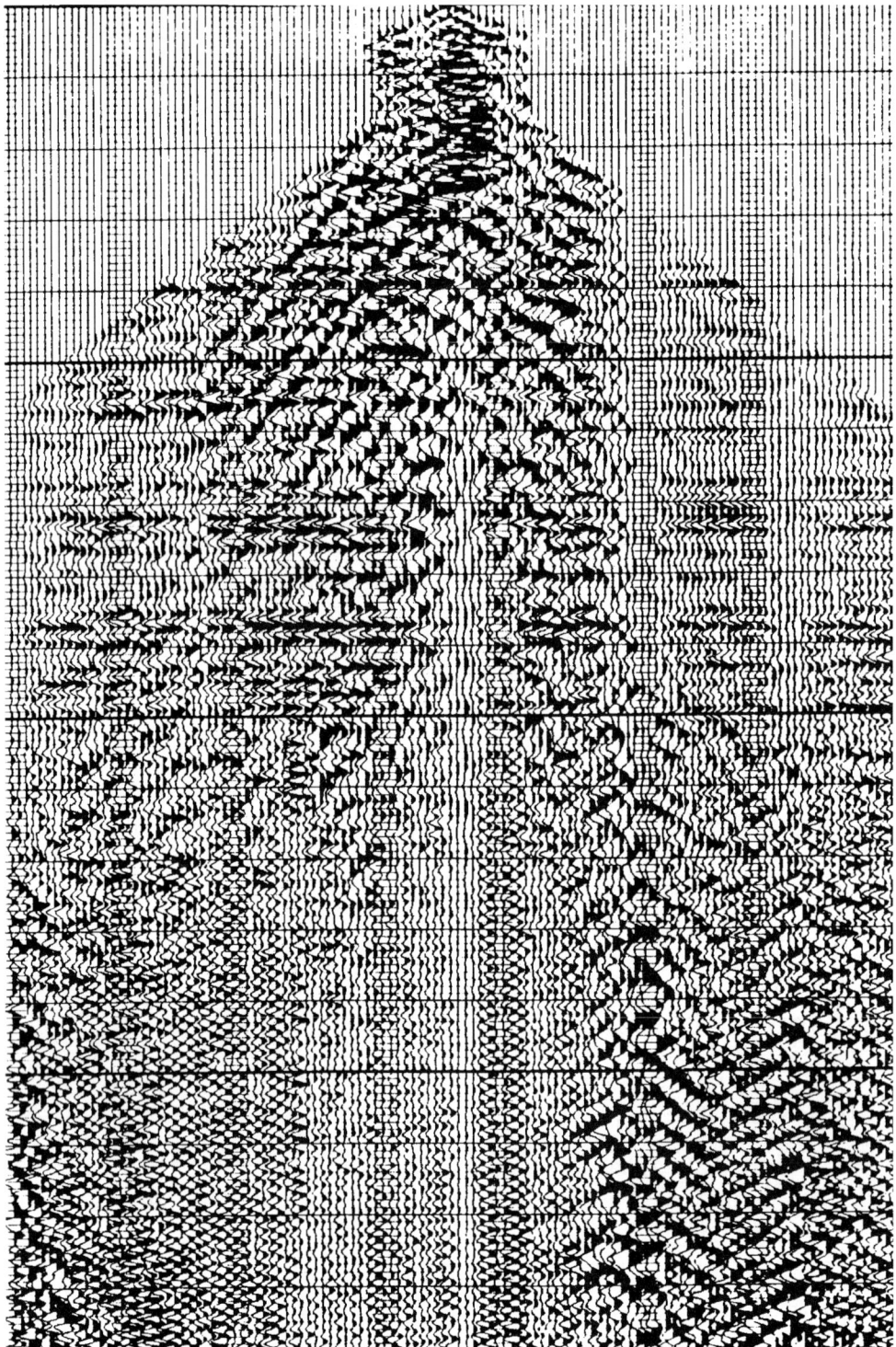

Figure 10.21 Source record for DMO.

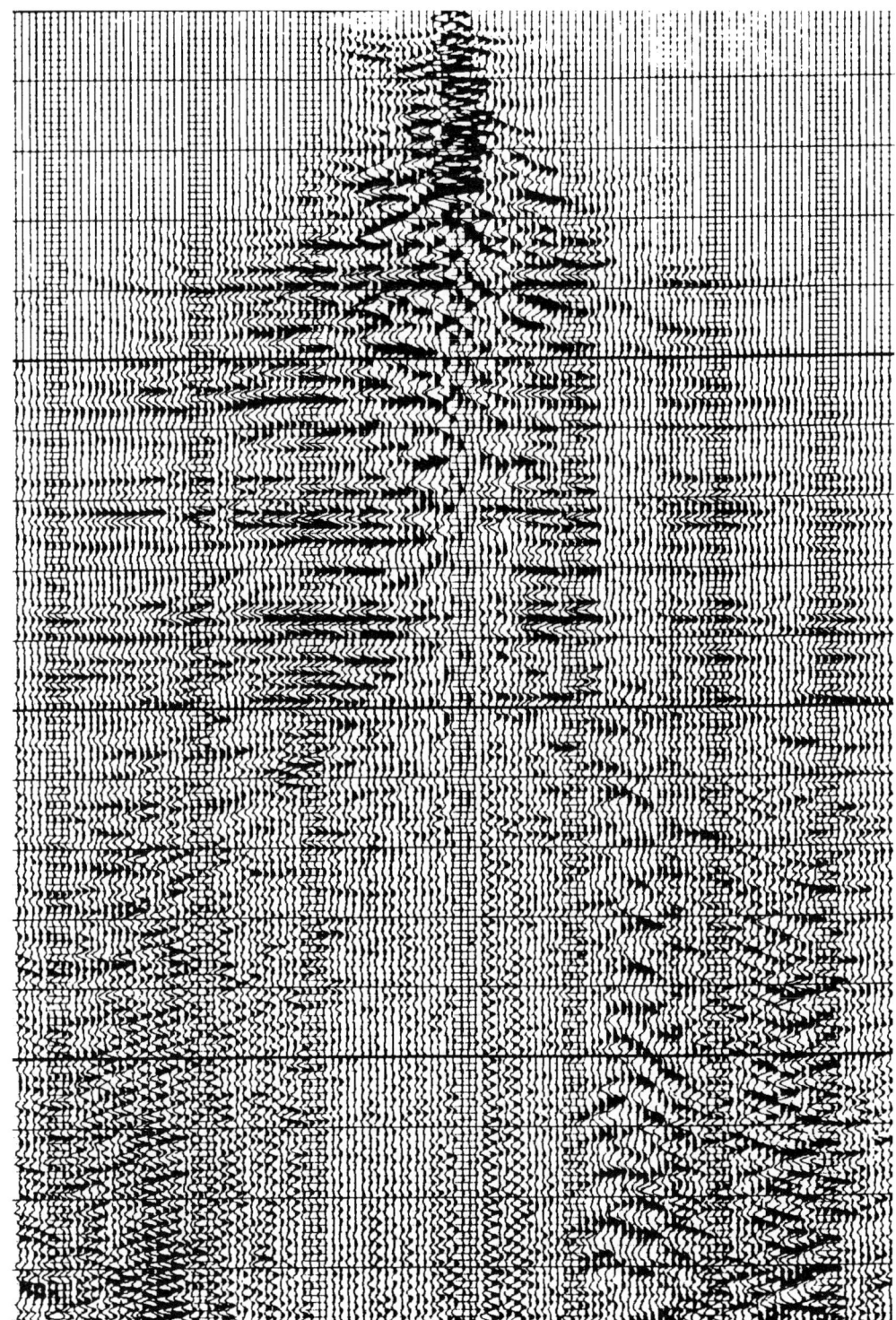

Figure 10.22 DMO of source record in Figure 10.21.

Note the reduced noise (especially dipping noise) and improved coherent events.

10.9 Prestack Migration of a Source Record

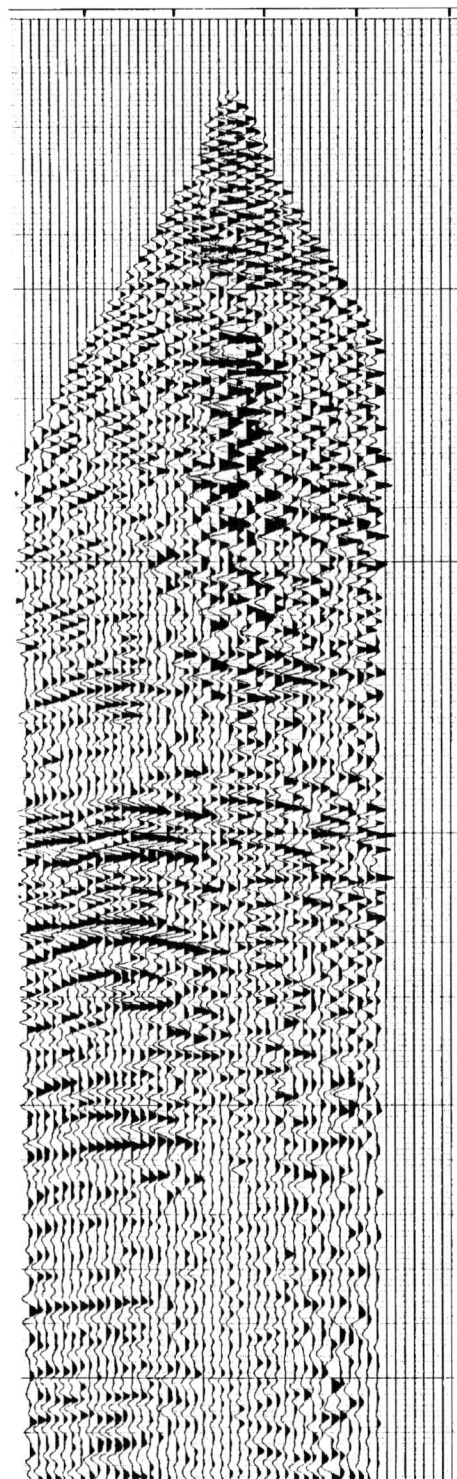

Figure 10.23 Source record for prestack migration test.

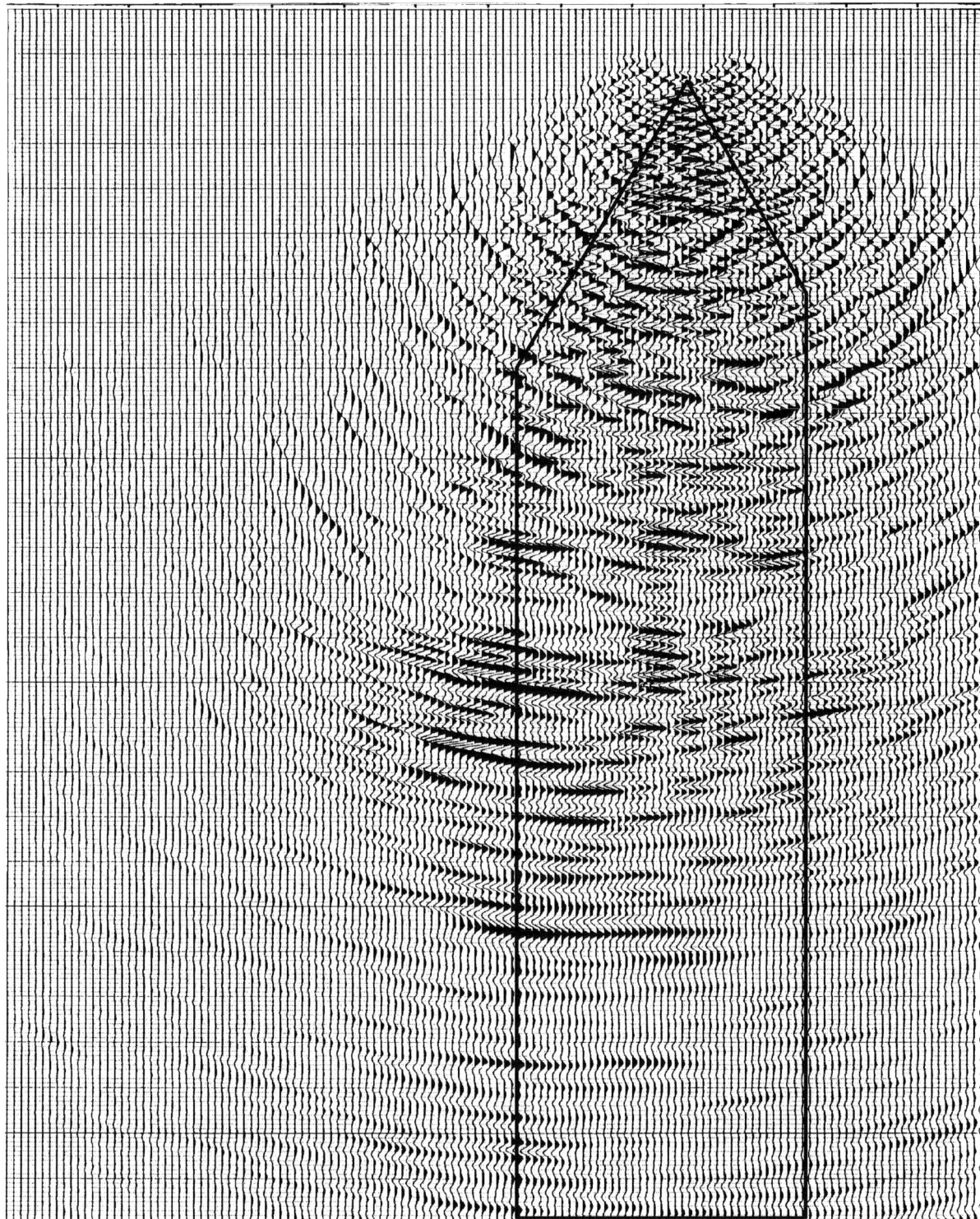

Figure 10.24 Prestack migration of source record in Figure 10.23.

Note the change in the shape of dipping energy, and amount of energy spread beyond the geometry of the input source record.

10.10 Comparison of DMO and Migrations on Real Data

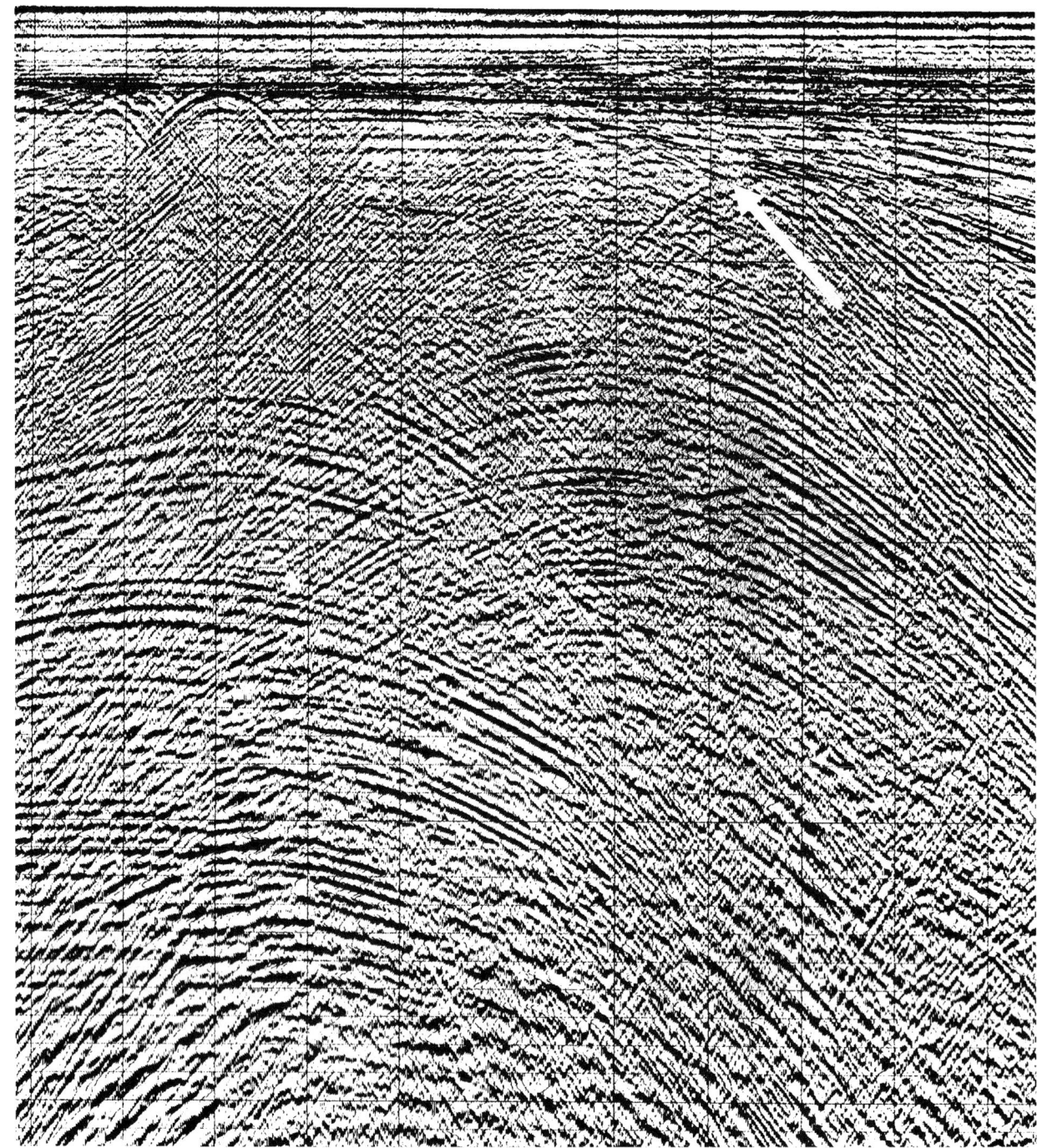

Figure 10.25 Stack example A.

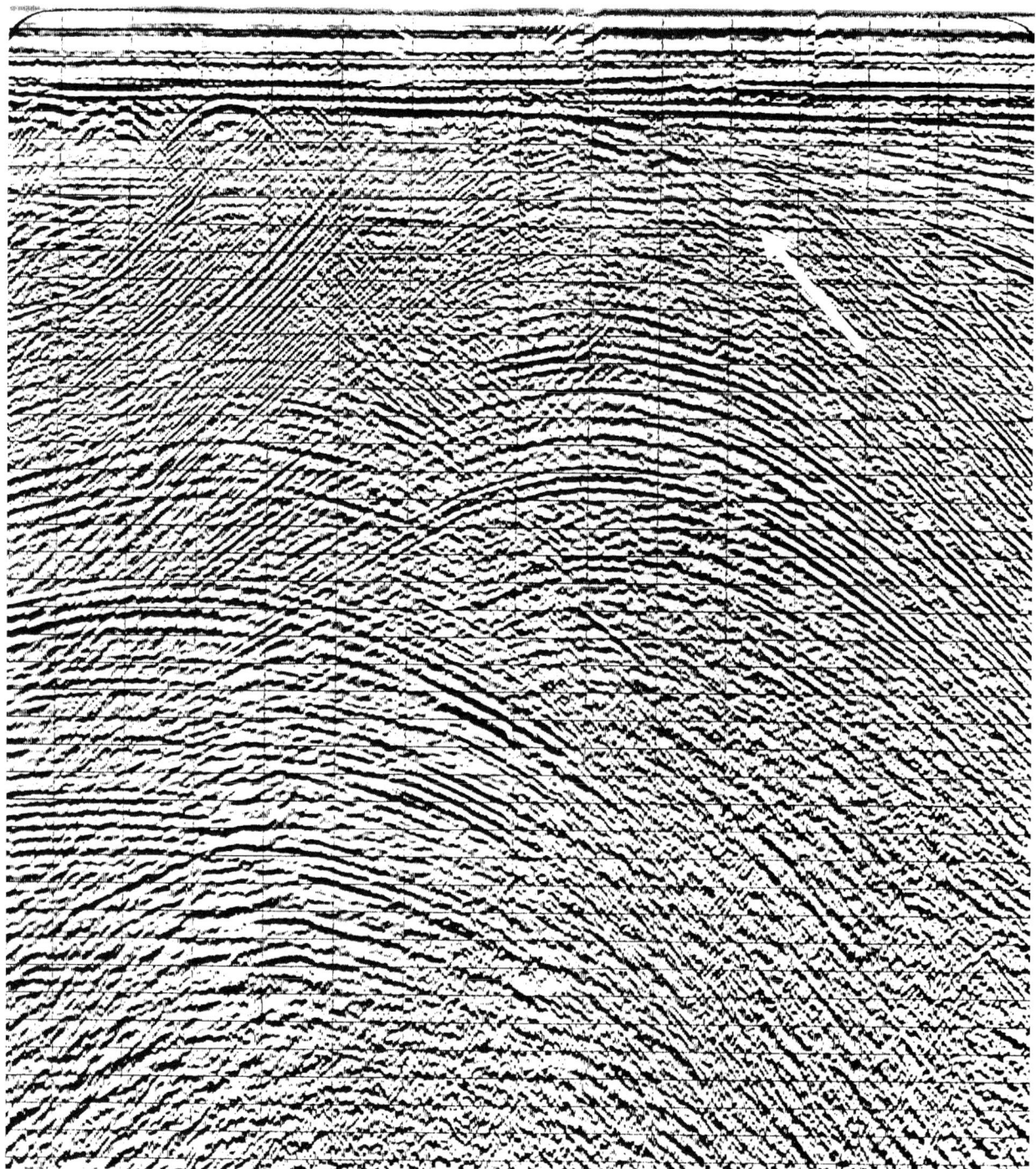

Figure 10.26 DMO of example A.

What has happened to the diffraction at the arrow?

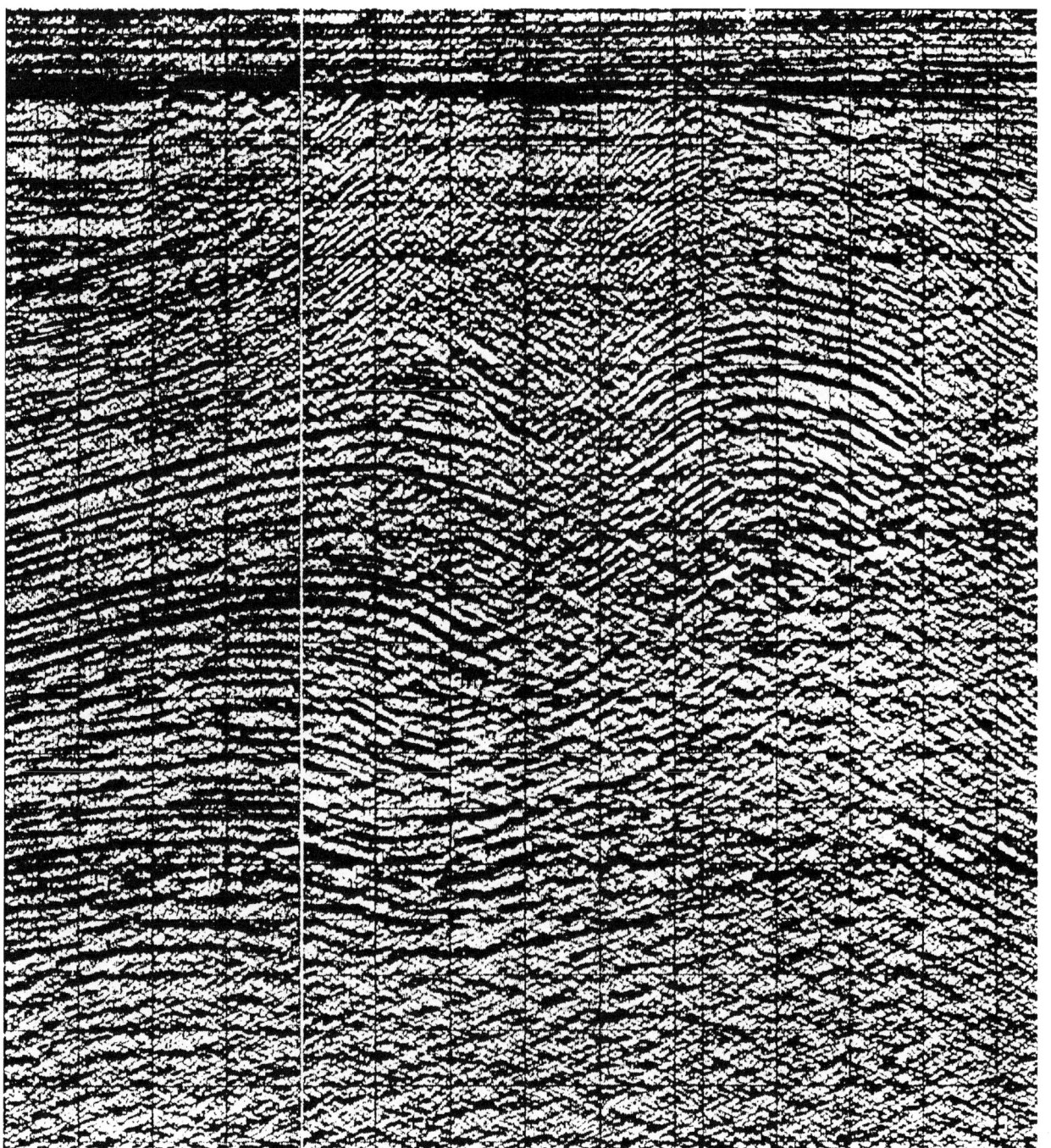

Figure 10.27 Post stack migration of A.

Figure 10.28 DMO and post stack migration of A.

Steeper dips have been preserved. Poor collapsing of energy at the black arrow.

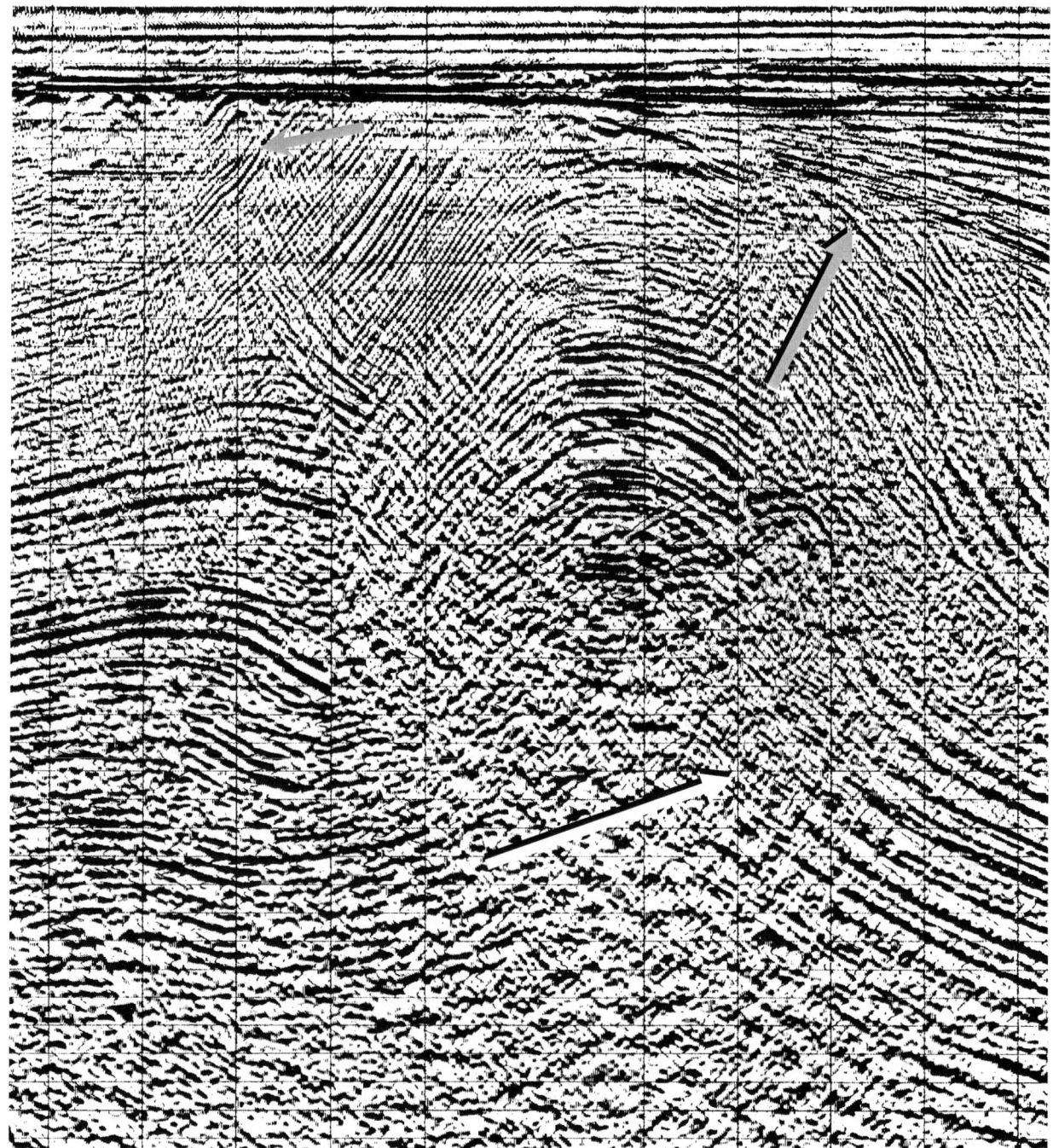

Figure 10.29 Prestack migration of A.

Better imaging below the erosional surface at the gray arrow but poor imaging of the steep dips at the white arrow.

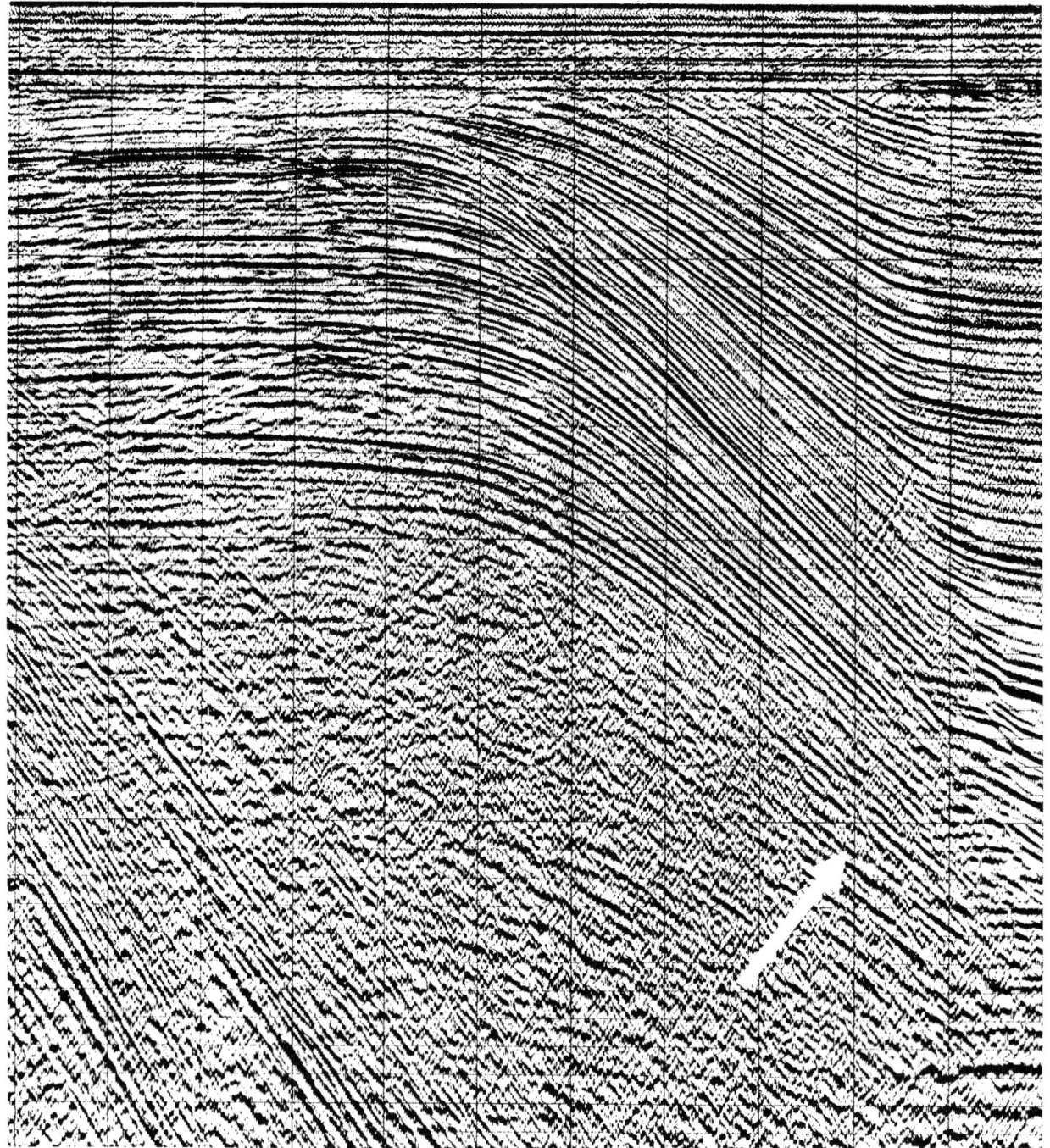

Figure 10.30 Stack example B.

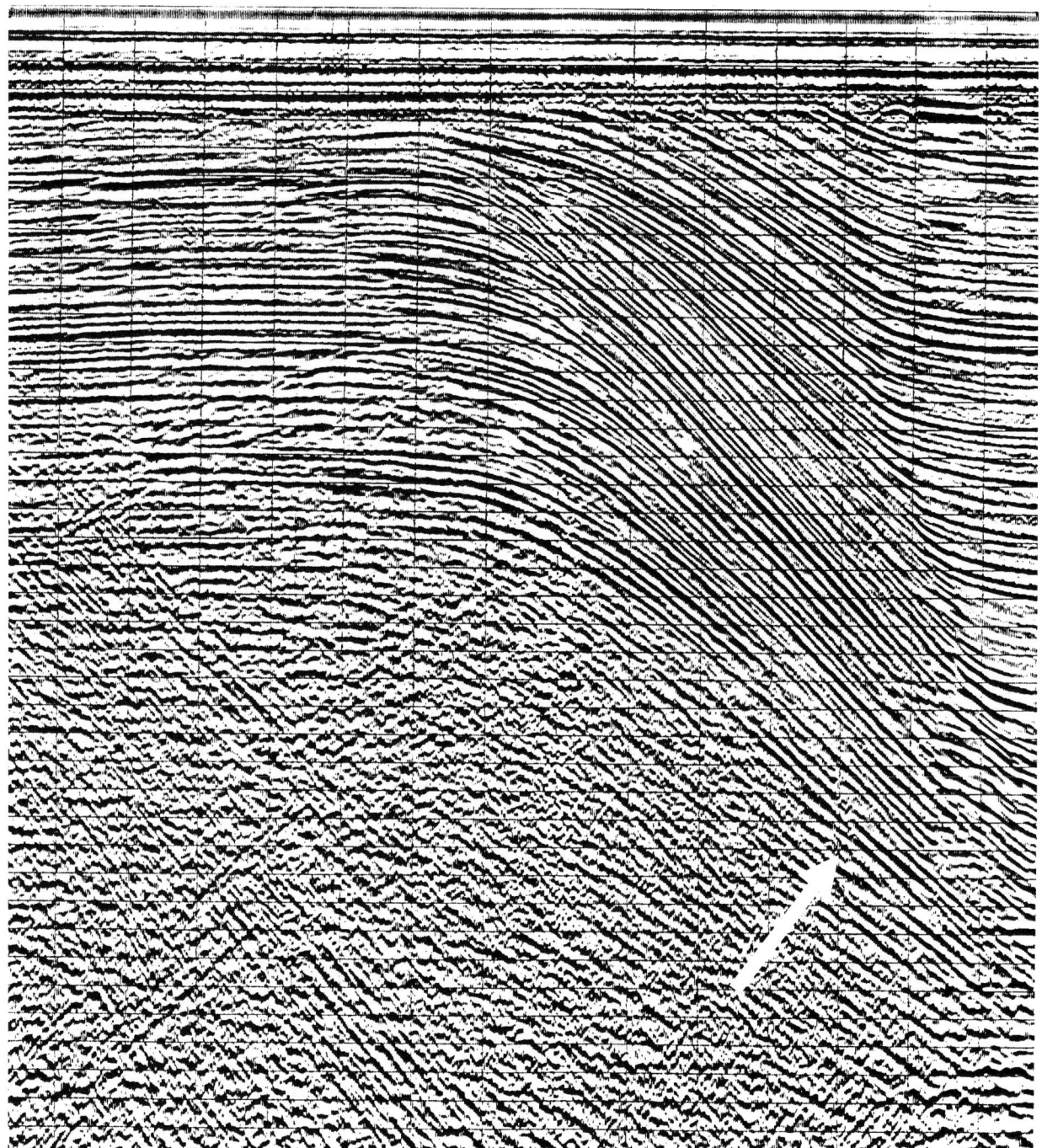

Figure 10.31 DMO of example B.

Note the better imaging of the steeper dips at the arrow.

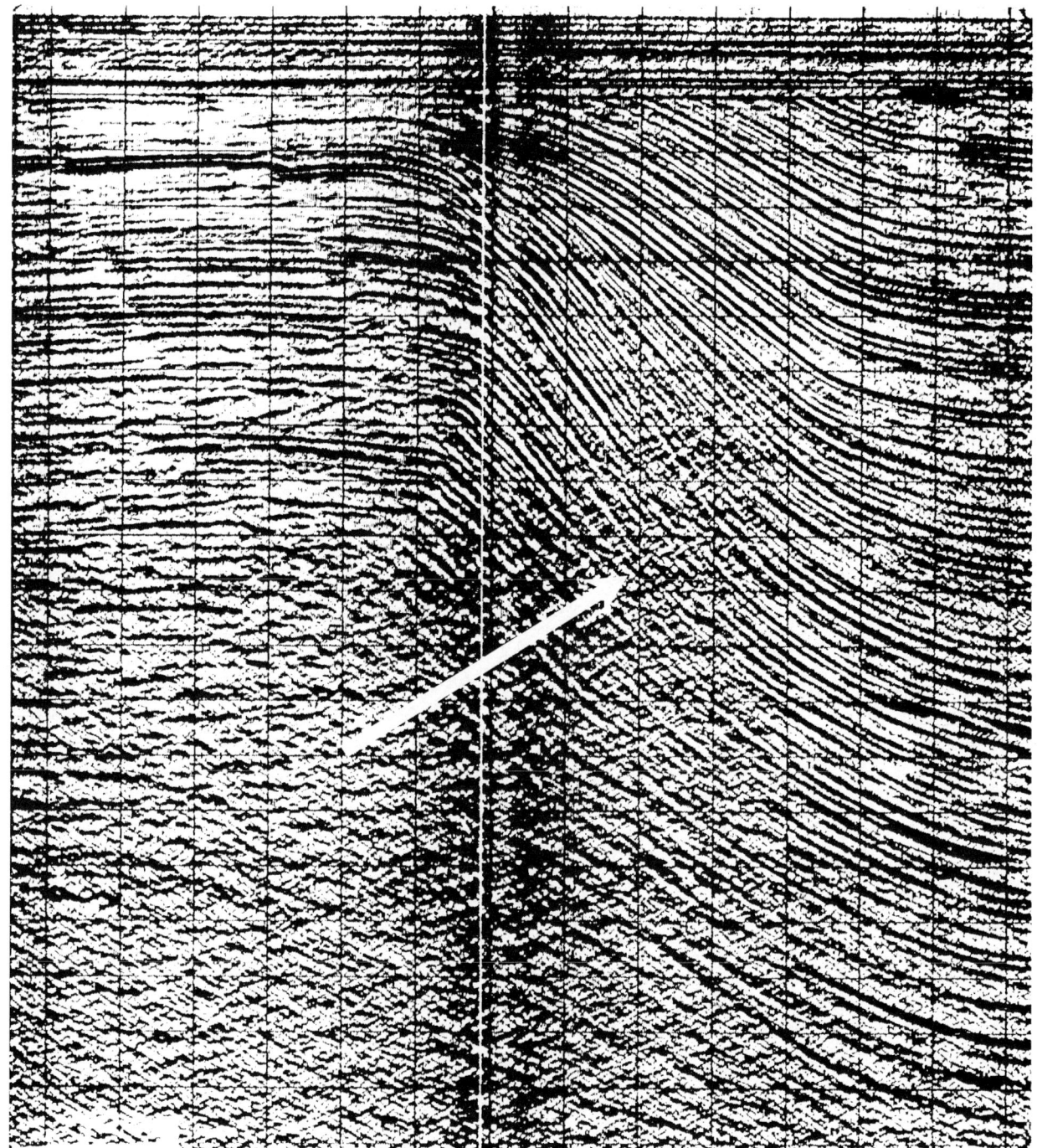

Figure 10.32 Post stack migration of B.

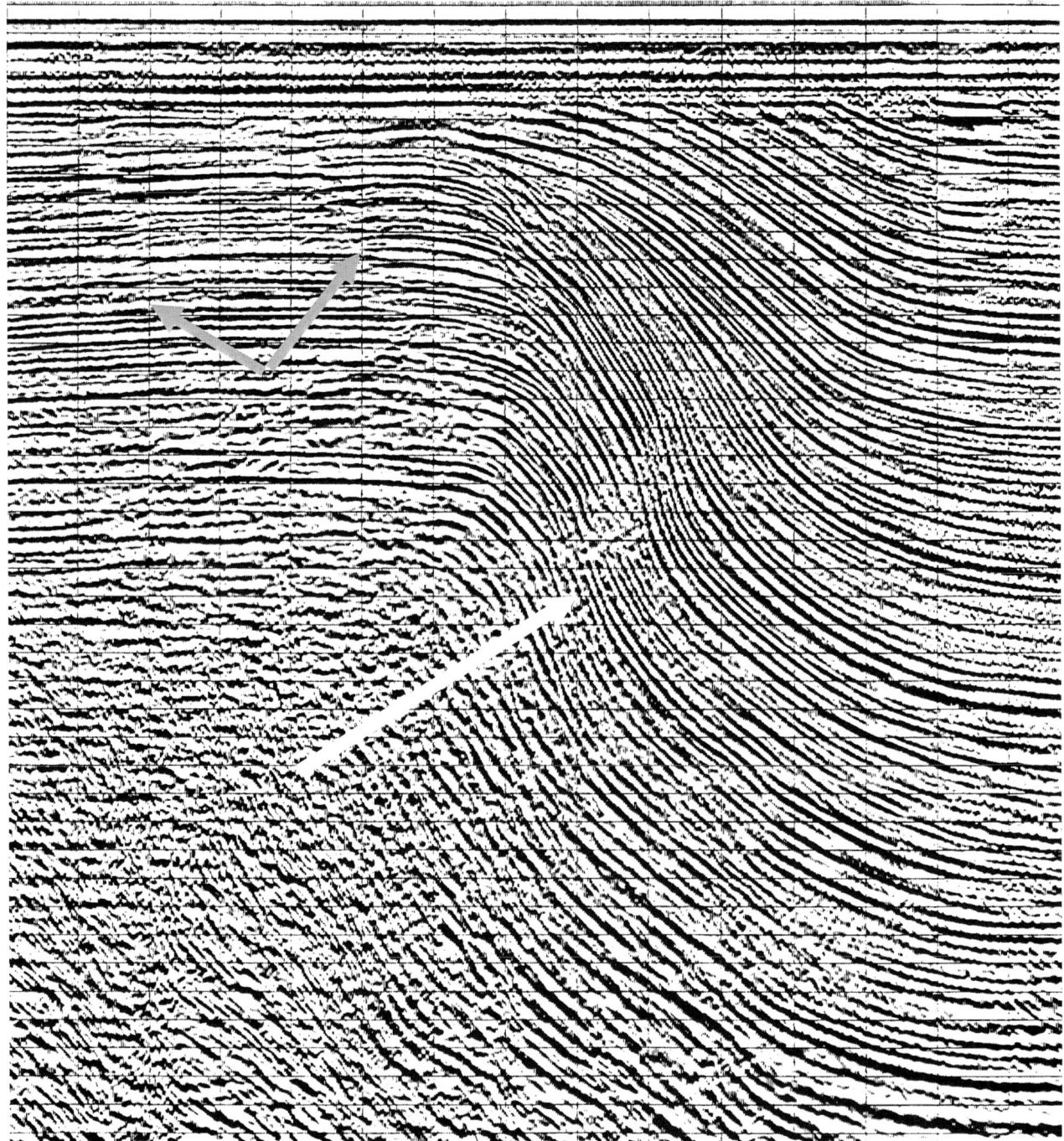

Figure 10.33 DMO and post stack migration of B.

Note that there are more coherent dipping events at the light gray arrow but poor resolution of the faults by the dark arrows.

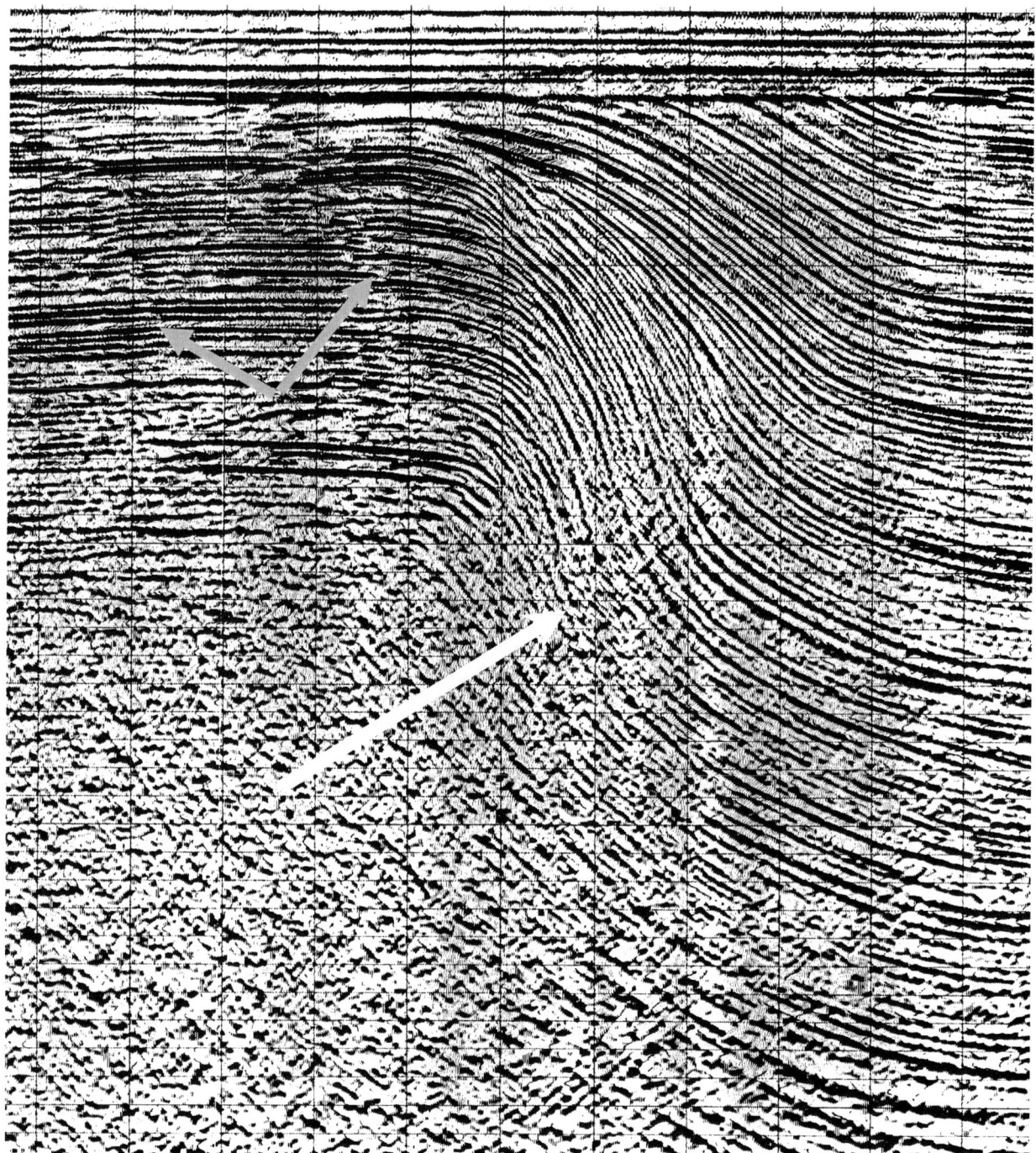

Figure 10.34 Prestack migration of B.

Possibly better resolution of faults by the dark gray arrows.

10.11 Prestack Migration of a structurally complex model.

The following figure represents a <u>structurally complex</u> geological cross-section through the <u>Canadian Rockies</u> that was used by a modelling package to create numerous source records. The model provides a standard for processing with known and unknown velocities.

The source records were produced by <u>raytracing</u> techniques.

Diffractions were <u>not</u> included in the modelling.

The surface of the model is horizontal, and <u>no static</u> corrections are required.

The velocities in m/s of the layers are 1 – 2900, 2 – 4250, 3 – 6100, 4 – 5500, 5 – 6400, 6 – 6250, and 7 – 6500.

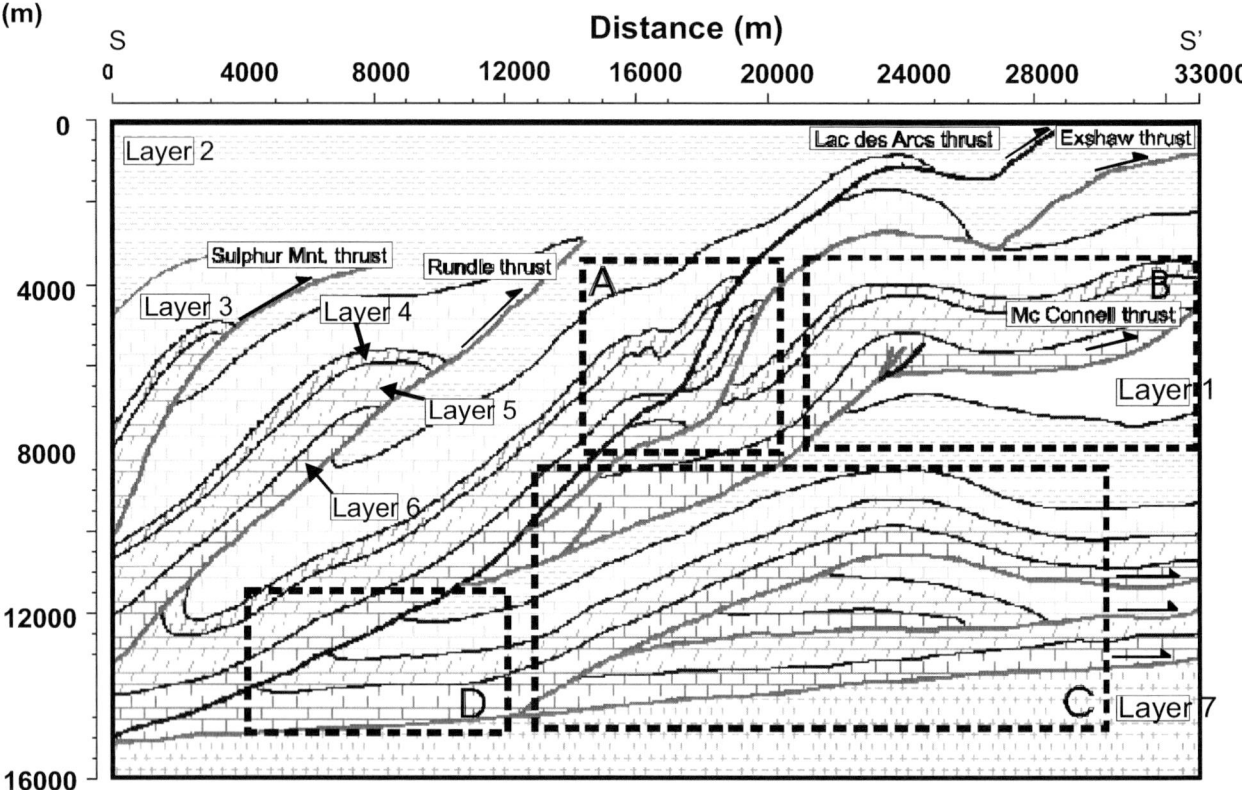

Figure 10.35 Complex geological model.

The overlay areas labeled A, B, C, and D in Figure 10.35 are discussed in more detail in the source paper for this material by G. K. Grech [641].

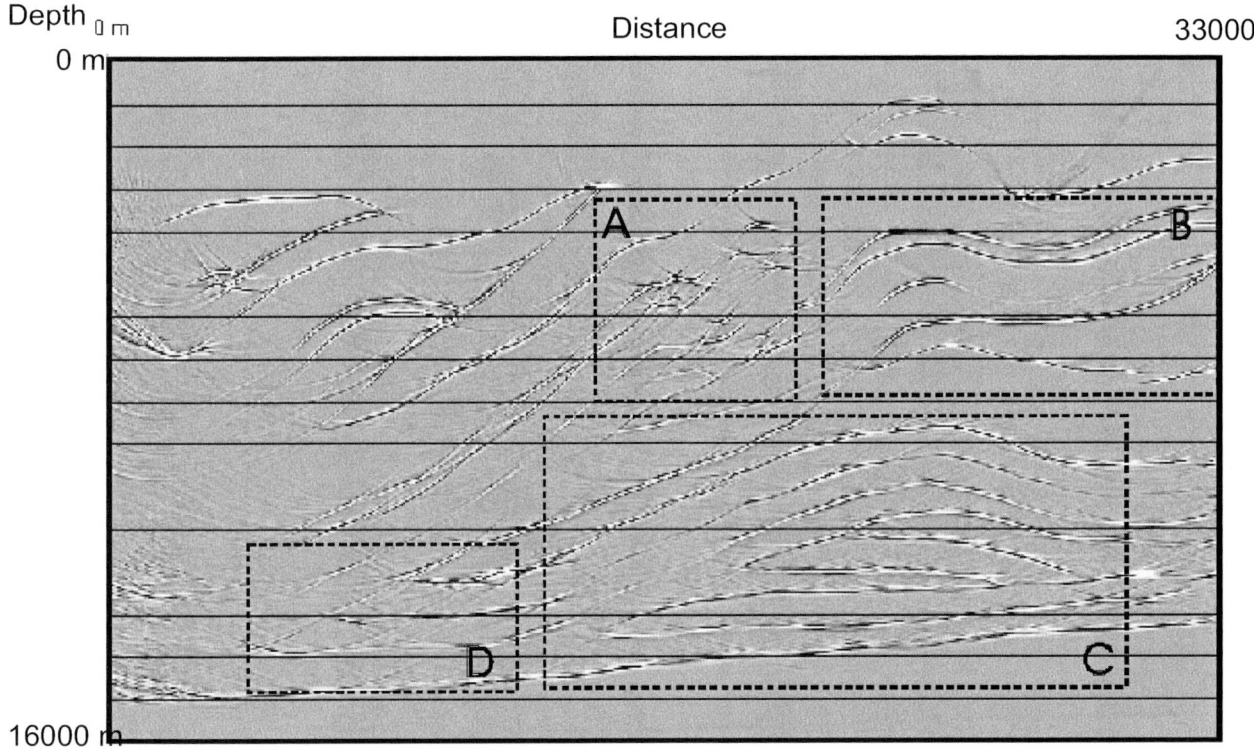

Figure 10.36 Prestack depth migration using the exact velocity model.

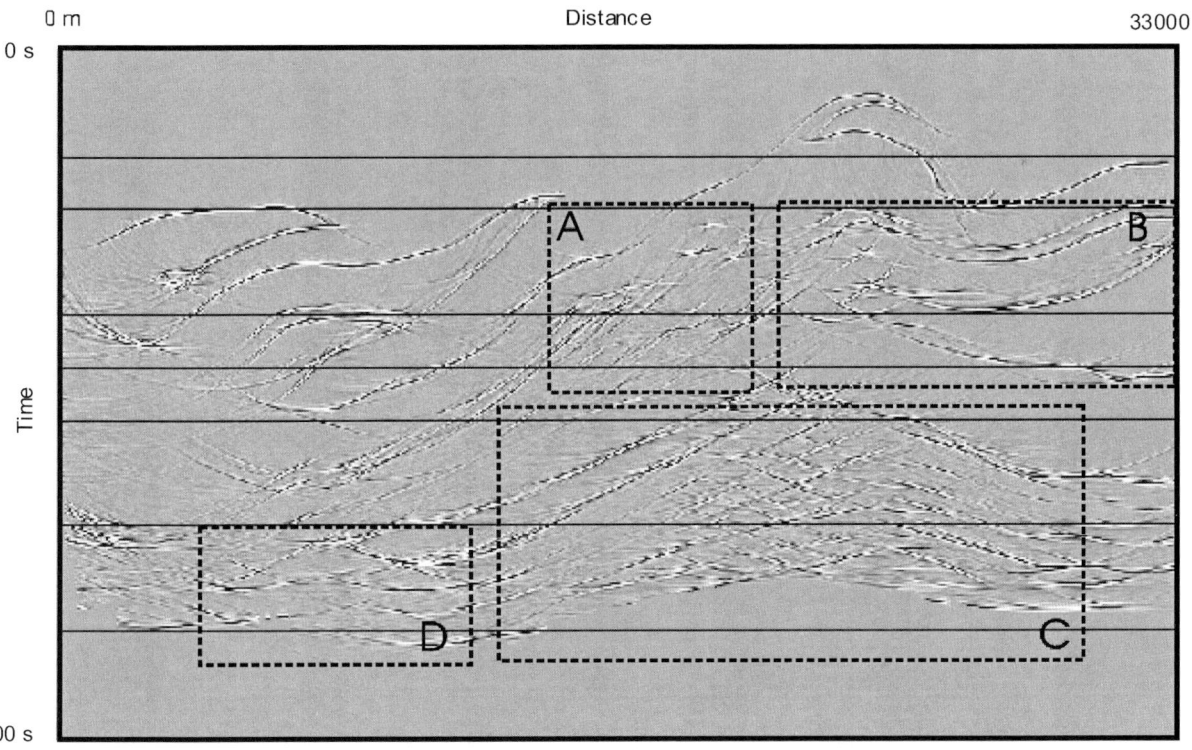

Figure 10.37 Poststack time migration using exact velocity model

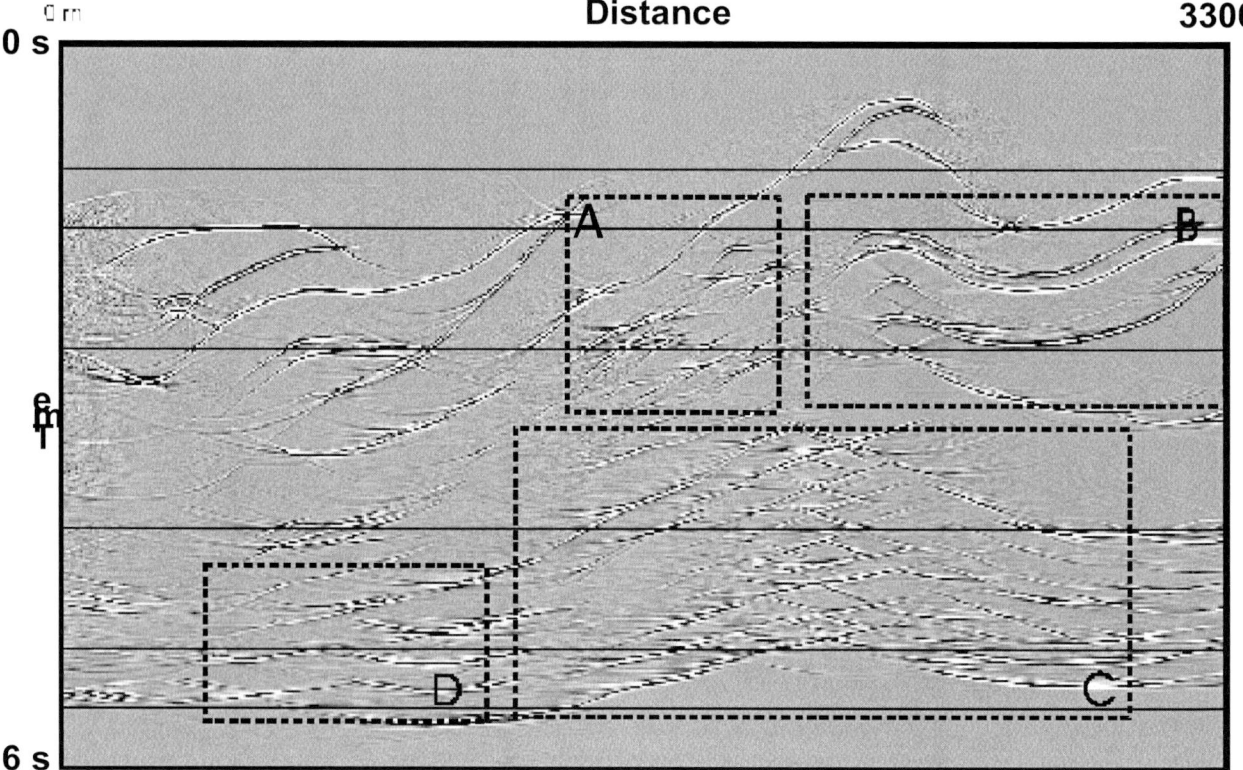

Figure 10.38 EOM migration with one input velocity function (approximately 10 km) to create the CSP gathers, followed by velocity analysis every 3 km for Kirchhoff MO.

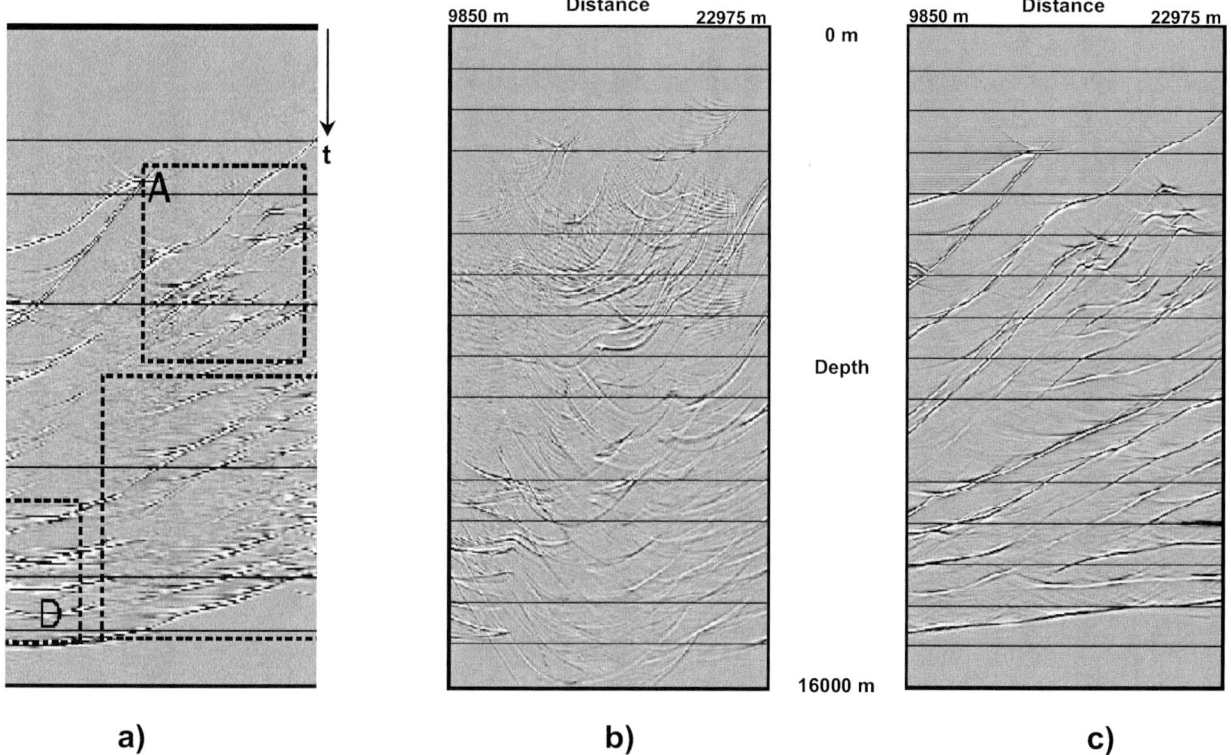

Figure 10.39 Comparison of a) EOM, b) four iterations of velocity analysis using MO-DMO-IMO followed by poststack depth migration, and c) prestack depth migration using the exact velocity model.

10.12 Salt Imaging Examples

The following examples are taken from a marine 3-D data set over a salt structure of Vermilion, Gulf of Mexico.

Figure 10.40 compares <u>poststack</u> time migration with <u>prestack</u> time migration.

Note the improved coherence of the reflectors in the prestack migration.

Figure 10.41 and Figure 10.42 compare prestack time migration with a poststack depth migration.

The prestack time migration was processed with MOVES, from Veritas DGC. MOVES is a prestack time migration that initially sorts data into binned offset groups (BOG's). Each BOG is processed separately with NMO + DMO + Zero-offset migration (ZOM). Additional proprietary processing is used to extract velocity information.

After the prestack time migration, the data is <u>inverse time migrated</u> to create a <u>zero offset stack</u> that is optimally focused.

This zero offset stack is then <u>poststack migrated</u>.

After the depth migration, the depth section is vertically stretched to a time section, enabling a convenient comparison with the prestack time migration.

The following results show that the base of salt (and possibly subsalt imaging) can be efficiently accomplished with:

- prestack time migration,
- inverse poststack time migration, and
- poststack depth migration.

Comparable data was shown by Ratcliff et al in [642]. Other examples of subsalt imaging may be found in [204], [146], [172], [174], [310], [332], [502], [598], [601-612] or The Leading Edge, August 1994, and [638], the SEG/EAGE Salt-Overthrust Model.

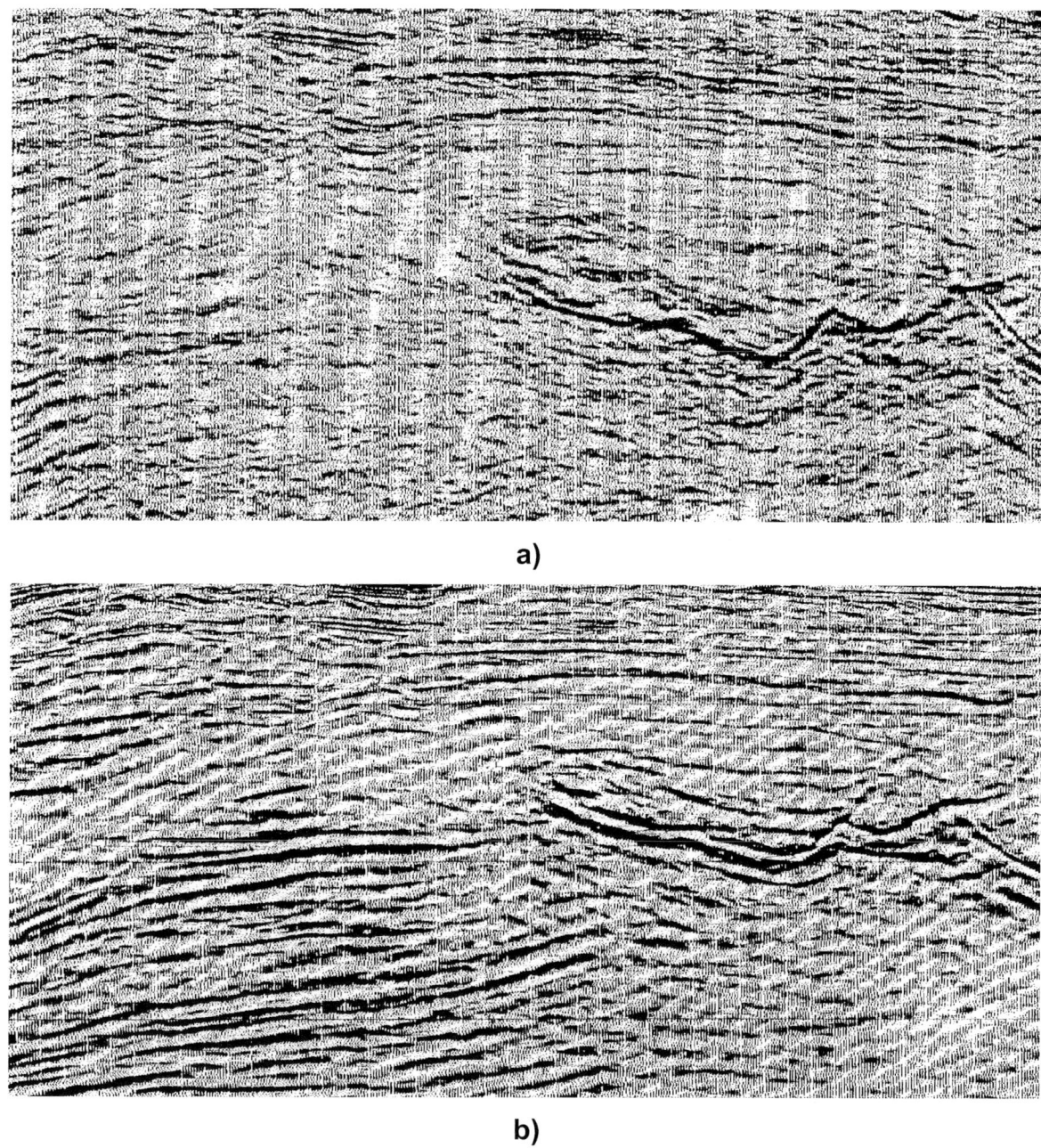

Figure 10.40 Comparison of processing a) a conventional poststack time migration, and b) a prestack time migration using MOVES (Veritas DGC). Only a portion of the section is shown

Note the improved coherence of the prestack migrated events at the center and lower left.

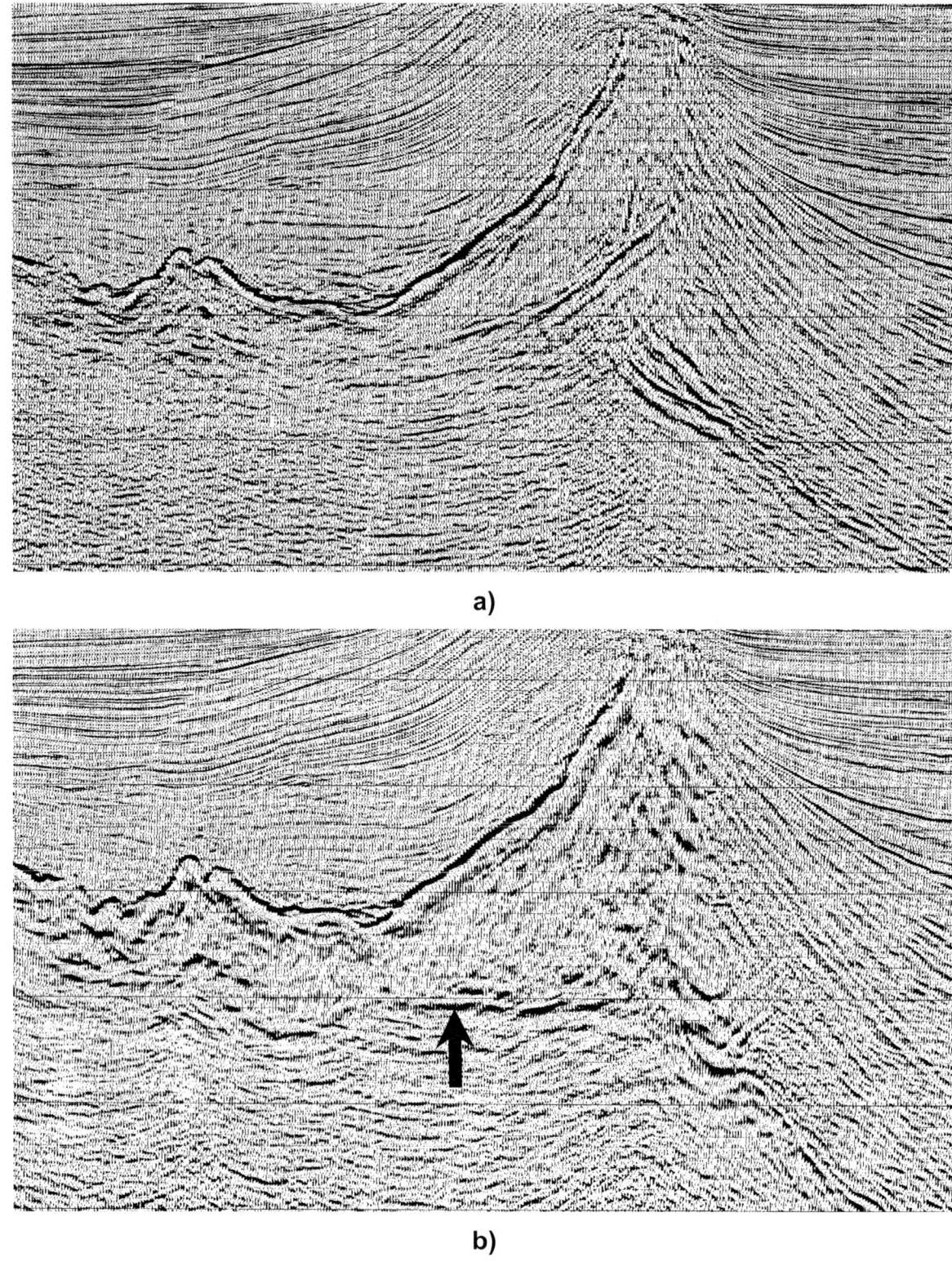

Figure 10.41 XL 1397 comparison of a) a prestack time migration using MOVES, and b) a poststack depth migration that used an inverse time migration of (a) as input; the depth has been vertically stretched to time for comparison with (a). Note the improved imaging below the base of the salt (indicted by the arrow).

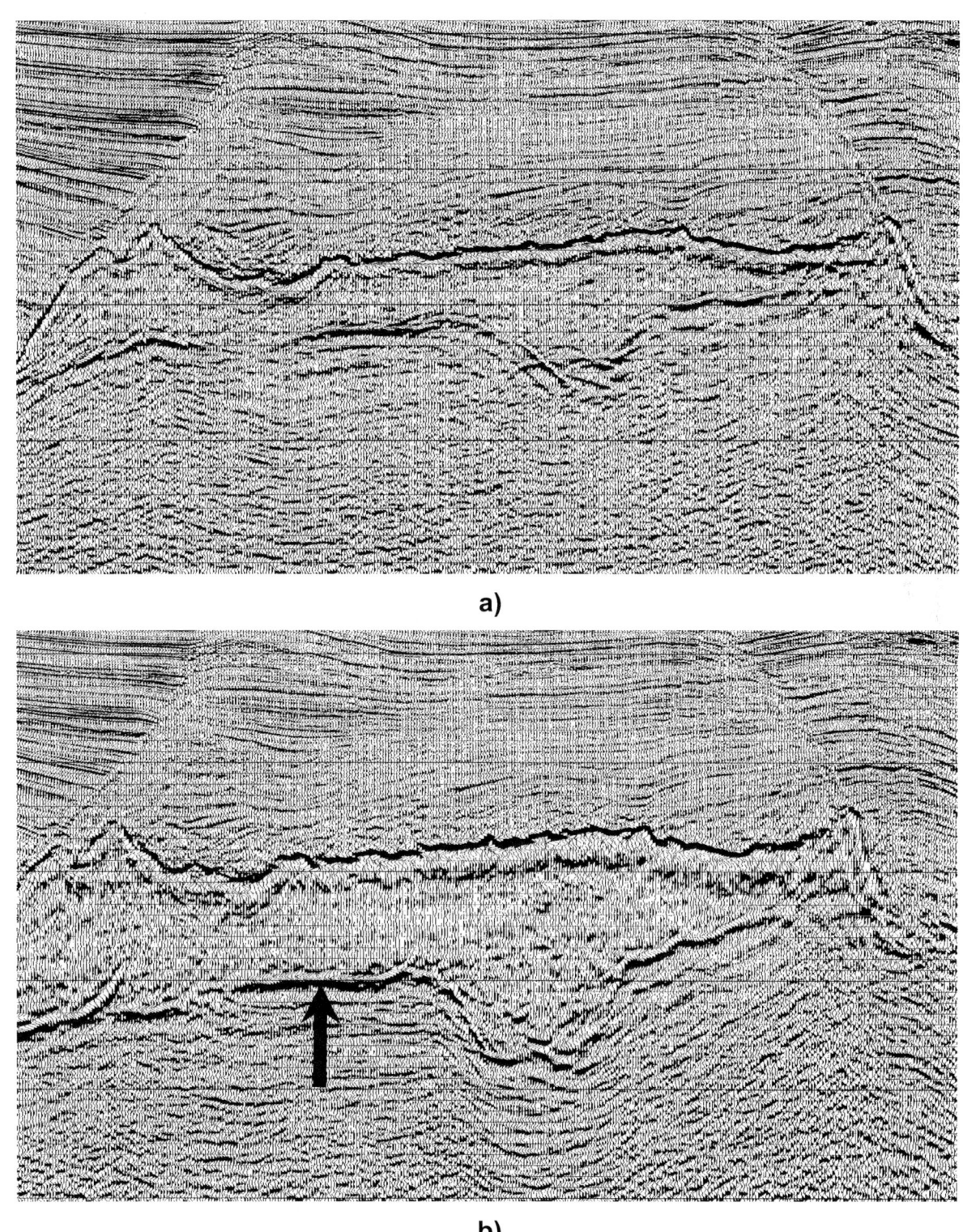

Figure 10.42 IL 19935 comparison of a) a prestack time migration using MOVES, and b) a poststack depth migration that used an inverse time migration of (a) as input; the depth has been vertically stretch to time for comparison with (a). Note the improved imaging below the base of the salt (indicated by the arrow).

10.13 Summary of Points to note in Chapter 10

- Time domain DMO can match the quality of transform methods.
- Tests to verify amplitude and phase of DMO.
- DMO may be used for interpolation.
- Prestack migration is not necessarily the best solution.
- The algorithm used is important (i.e. it works optimally).
- The velocity model is critical for depth migrations.
- The velocity model is not critical when forming CSP gathers for EOM.
- Prestack time migration/inverse poststack time migration/poststack depth migration may produce economical images comparable to prestack depth migration.

Chapter Eleven

Equivalent Offset Migration (EOM)

Introduction

Objectives
- Understand the principles of EOM.
- Know that EOM is based on prestack Kirchhoff migration.
- Know that EOM is a time and depth migration.
- Know the definition of equivalent offset.
- Understand how CSP gathers are formed.
- Know that CSP gathers are formed with no time shifting of the input data.
- Know that CSP gathers may be formed at random locations:
- Understand the principles of Kirchhoff MO.
- Know that CSP gathers are formed with slight dependence on velocity, but provide accurate velocities after they are formed.
- Know that principles of EOM may be extended to rugged topography, converted-wave processing, vertical array data, and may be used to extract residual statics.

11.1 Introduction

Equivalent offset migration (EOM) is a prestack time or depth migration that is based on Kirchhoff migration principles.

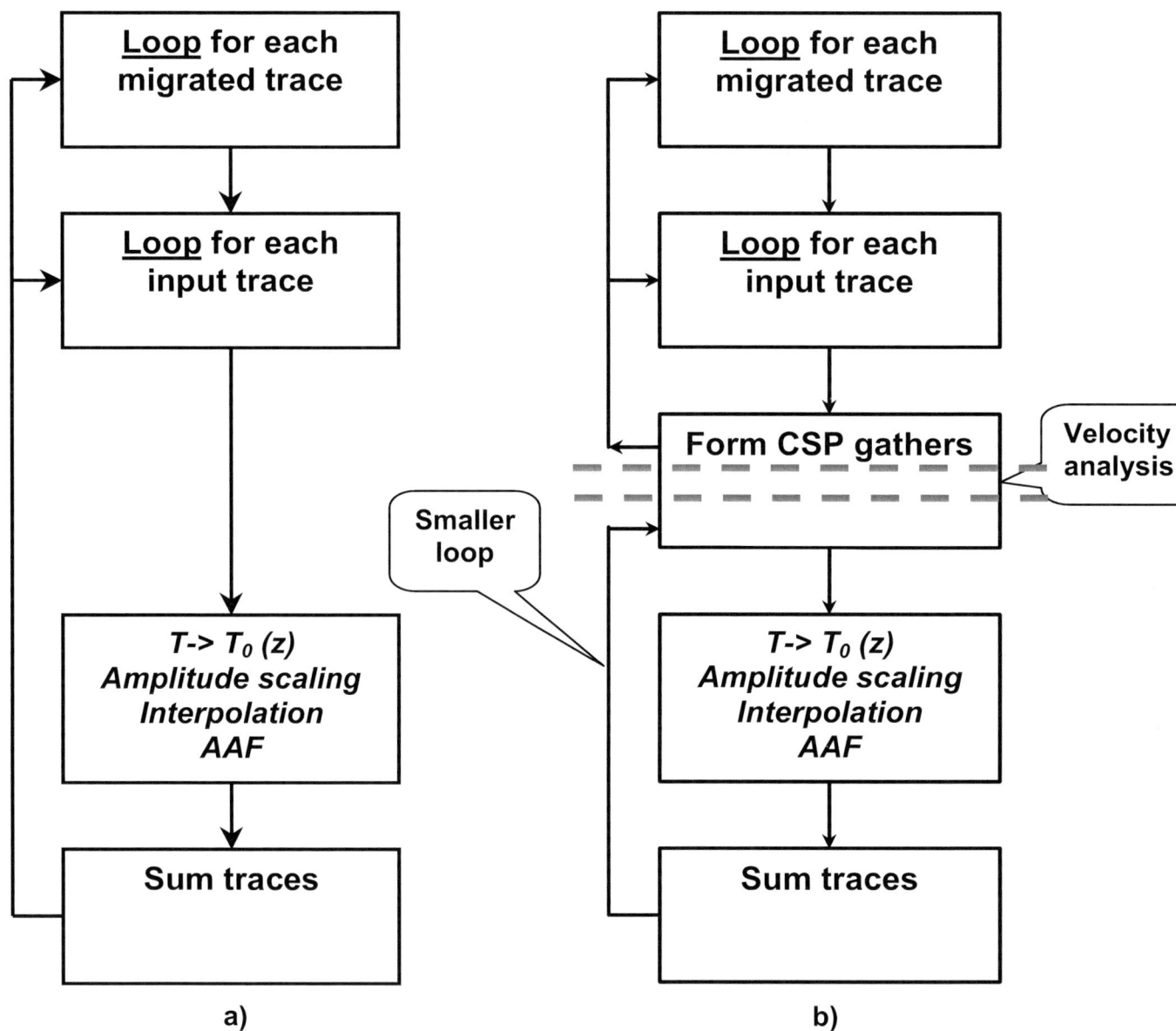

Figure 11.1 Block diagrams of a) Kirchhoff migration and b) EOM.

EOM:

- Includes an intermediate step that forms common scatterpoint (CSP) gathers.
- Allows a more accurate velocity analysis to be performed on the CSP gathers prior to Kirchhoff MO.
- Delays the computations of time shifting, amplitude scaling, interpolation, and anti-alias filtering until after the CSP gathers have been formed (Kirchhoff MO).
- Is fast. Reduces the number of Kirchhoff MO computation.
- Uses conventional software for velocity analysis.
- Prestack time or depth migrations.
- Converted wave processing.
- Vertical array processing.
- Provides model traces for estimating residual statics prior to MO correction.
- Can create a zero-offset section without MO correction or time shifting.
- Similar to DMO-PSI but does not require DMO.
- Includes all the benefits of Kirchhoff migrations:
    1. Irregular input geometry
    2. Arbitrary output locations
        - Single locations for refined velocity analysis
        - 2-D data output from 3-D input data
        - simplified crooked line processing (3-D)
    3. Migration from surface (rugged topography)
    4. Selective apertures (dips, offsets, etc)
    5. Combine data with different acquisition geometries

Fast ???

See section 4.7.7 for a fast poststack Kirchhoff 3-D migration.

For prestack, include an approximate fold in each bin of the 3-D.

What would be the potential reduction in computations?

EOM in time assumes:

- **RMS velocities.**
- Energy from a scatterpoint is spread over **Cheops pyramid** as defined by the double-square root equation.
- All reflectors can be formed from appropriately ordered scatterpoints.
- Prestack migration gathers are referred to as **equivalent offset** (EO) gathers when the gathered energy can be moved directly to the scatterpoint.
- EO gathers are **slightly dependent on velocity**.
- **Velocity analysis** is performed **after the EO gathers have been formed**.
- **Kirchhoff moveout** (KMO) (which involves filtering, scaling and stacking) is applied to the EO gathers to complete the prestack migration.
- A time migrated trace follows the path of its **image ray** in the depth section.
- Fast.

EOM in depth:

- **Uses <u>conventional</u> source and receiver traveltime maps.**
- **Can form EO gathers before or after "MO correction".**
- **Inclusion of <u>anisotropy</u> in the traveltime maps produces an anisotropic prestack migration.**
- **The EO gather may be formed with a <u>smoothed</u> (more than normal) velocity field that allows larger ray or wavefront increments.**
- **May aid in building the velocity model.**
- **Not fast.**

11.1.1 Pseudo code for EO migration

The Kirchhoff migration code is broken into two parts,
1. a process that forms the EO gathers,
2. then moveout correction and stack to complete the prestack migration.

Accurate velocity analysis may be performed after the first part as the data is hyperbolic with RMS type velocities.

Form EO gathers (simplified).

Loop for a <u>migrated trace</u>	`for i = 1 : I`
Migrated trace location	x_m
Loop for each <u>input trace</u>	`for k = 1 : K`
Geom., srs and rec. loc	$x_s, x_r, x_{cmp}, h, x, t_{alfa}, j_{alfa}$
Estimate EO,	$h_e = \sqrt{x^2 + h^2}$
Check migration aperture	`if he < aperture`
Loop for <u>input samples</u>	`for j = `j_{alfa}` : J`
Define the EO bin number	$i_b = \sqrt{x^2 + h^2 - (2xh/TV)^2}/d_{off}$
Sum input trace into EO gather	`EOgath(i,`i_b`,j) += Indata(k,j)`
End	`end`
End	`end`
End	`end`

MO and stack the EO gathers to complete the prestack migration, (similar to conventional NMO and stack)

Loop for a <u>migrated trace</u>	`for i = 1 : I`
Loop for each <u>offset bin</u>	`for k = 1 : K`
Define the (equivalent) offset	$h_e = k \cdot d_{bin}$
Loop for <u>migrated samples</u>	`for j = 1 : J`
Get the velocity at that location	`V = V(i, j)`
Get time from surface	$T_o = 2j \cdot dt$
Define traveltime	$T = 2\sqrt{(T_o/2)^2 + (h_e/V)^2}$
Define the sample number	$m_k = T/dt$
Sum sample into migrated trace	`EOmig(i,j) += EOgath(i,k,`m_k`)`
End	`end`
End	`end`
End	`end`

Simplifications:

No amplitude scaling or dip filtering, computed EO for each sample, don't show how to compute $V(T_0)$.

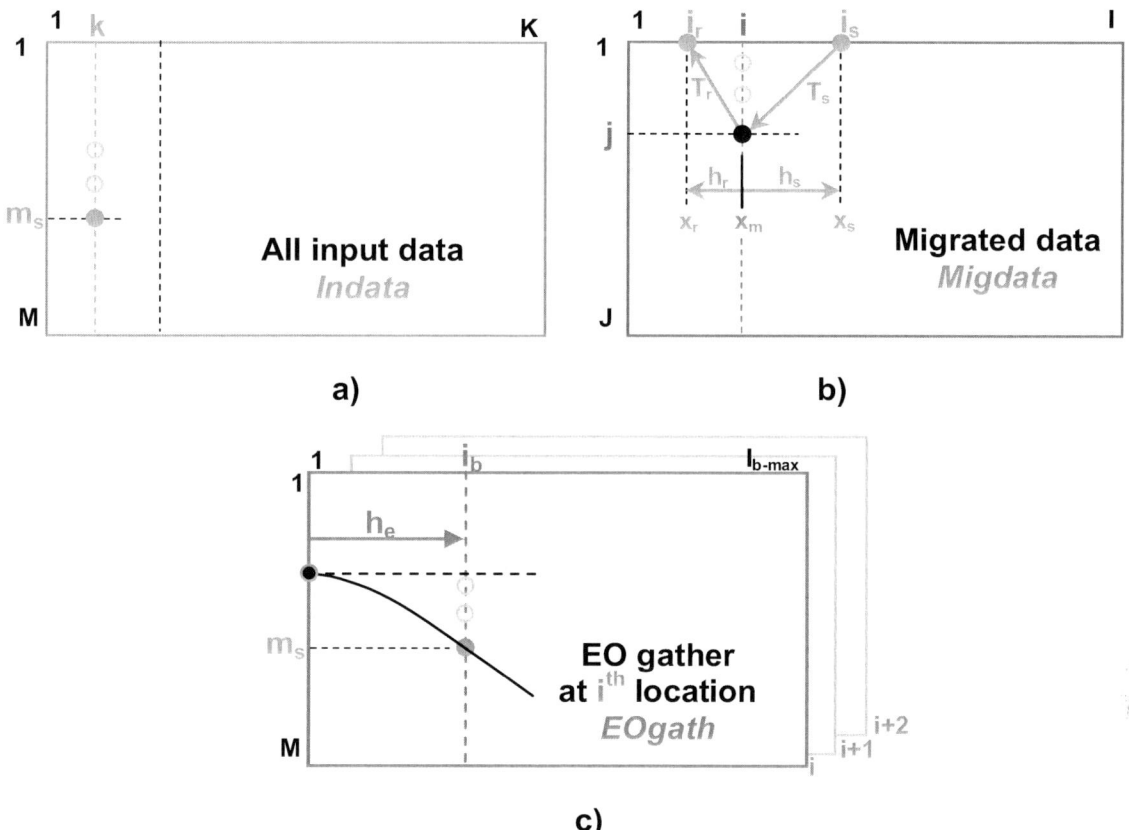

Figure 11.2 Formation of EO gathers, a) the input record, b) the geometry of a migrated trace and c) EO gathers.

The scatterpoint in Figure 11.2b produces energy in all input traces of (a), that reconstructs on the hyperbola in (c). This same energy will destructively cancel on neighbouring gathers.

Number of MO, scaling, AAF, and interpolations in prestack migration N_{opp}:

Basic prestack Kirchhoff

N_{mig} = 100

N_{CMP} = 100

N_{fold} = 20

N_{samp} = 1,000

$N_{opp} = N_{mig} * N_{CMP} * N_{fold} * N_{samp}$ = 200,000,000

EO migration

N_{bins} = 50

$N_{opp-EO} = N_{mig} * N_{bins} * N_{bins}$ = 5,000,000

11.1.2 Review of the RMS velocity and the hyperbolic equations

RMS velocities:
- Defined at the scatterpoint, which is the migrated (or modelled) sample.
- T_0 is the vertical zero-offset, two-way traveltime above the scatterpoint.
- Assumes <u>horizontal</u> layers above the scatterpoint.
- MO correction is assumed to be hyperbolic.
- Diffractions are assumed to be hyperbolic.
- Rays to and from the scatterpoint are <u>above</u> the depth of the scatterpoint Z_0.
- The diffraction extends <u>below</u> T_0.
- Simplify the computation of MO correction and Kirchhoff time migrations by assuming <u>a constant velocity</u> at a specified time a T_0.

For convenience, when working with a single scatterpoint, we define it to be at the spatial origin $x = 0$. That is also the location for the apex of the corresponding Cheops pyramid.

✳✳✳

We <u>define T_n</u> to be the zero-offset, two-way traveltime, displace from the surface location of the scatterpoint ($x \neq 0$). This parameter was referred to as T_0 when working with horizontally layered post-stack data.

We now use T_0 specifically as the vertical two-way zero-offset time to a scatterpoint, i.e. at the apex of Cheops pyramid.

The two-way time <u>MO equation</u> at a CMP location approximated by,

$$T^2 = T_n^2 + \frac{4h^2 \cos^2 \beta}{V_{RMS}^2}, \tag{11.1}$$

where T is the recorded time, h half the source-receiver offset, β the geological dip, and V the RMS type velocity.

The kinematic two-way time for zero offset <u>Kirchhoff migration</u> is,

$$T_n^2 = T_0^2 + \frac{4x^2}{V_{RMS}^2}, \tag{11.2}$$

where x is the distance from the migrated trace to the CMP location of the input trace and T_0 is the vertical two-way time from a scatterpoint to the surface.

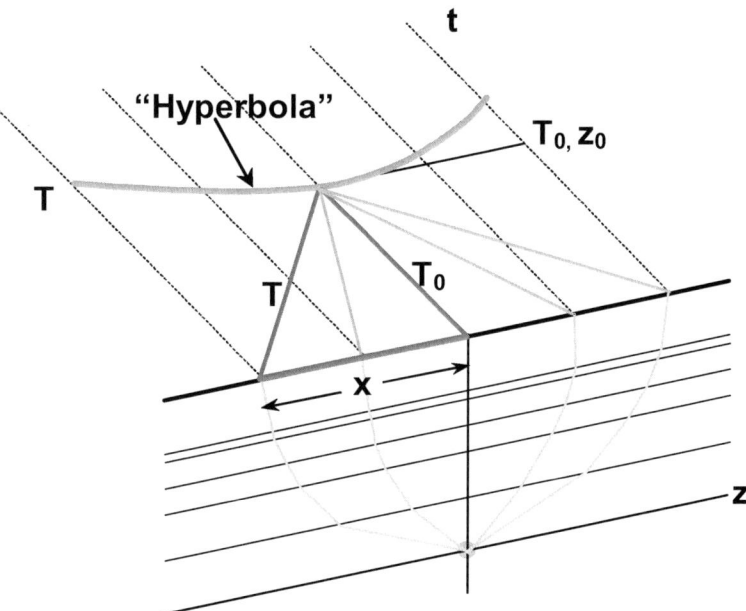

Figure 11.3 Figure reviewing the power of the RMS velocity assumption. The traveltimes of rays between a scatterpoint and the surface can be approximated by straight rays on a time section.

11.1.3 Zero-offset Kirchhoff migration as MO correction

Each output point in Kirchhoff migration is typically found by summing data within a diffraction shape as illustrated in Figure 11.4a.

This migration could also be accomplished by using MO correction by:

- Rotating (or folding) the zero offset data into a CMP gather at the scatterpoint location as illustrated in Figure 11.41b.
- MO correction would flatten the diffraction energy in this migration gather.
- Stacking the migration gather would complete the migration for this trace.

We only considered the energy from scatterpoints on the migrated trace.

- This same zero-offset energy, when folded at a different location, will not be hyperbolic, and will not stack, as illustrated in Figure 11.4c.

A zero-offset section with a continuum of diffractions can therefore be accomplished by:

- Rotating the zero-offset data about each migrated trace, to form CMP gathers, and
- MO correction and stacking will complete the post-stack migration.

The only difference between this method and a conventional Kirchhoff method is that there will be additional weighting applied in the Kirchhoff method.

> The main reason for introducing this concept is that the zero offset data can be moved to a corresponding offset to help form a prestack migration gather.
> Can we do the same the offset data?

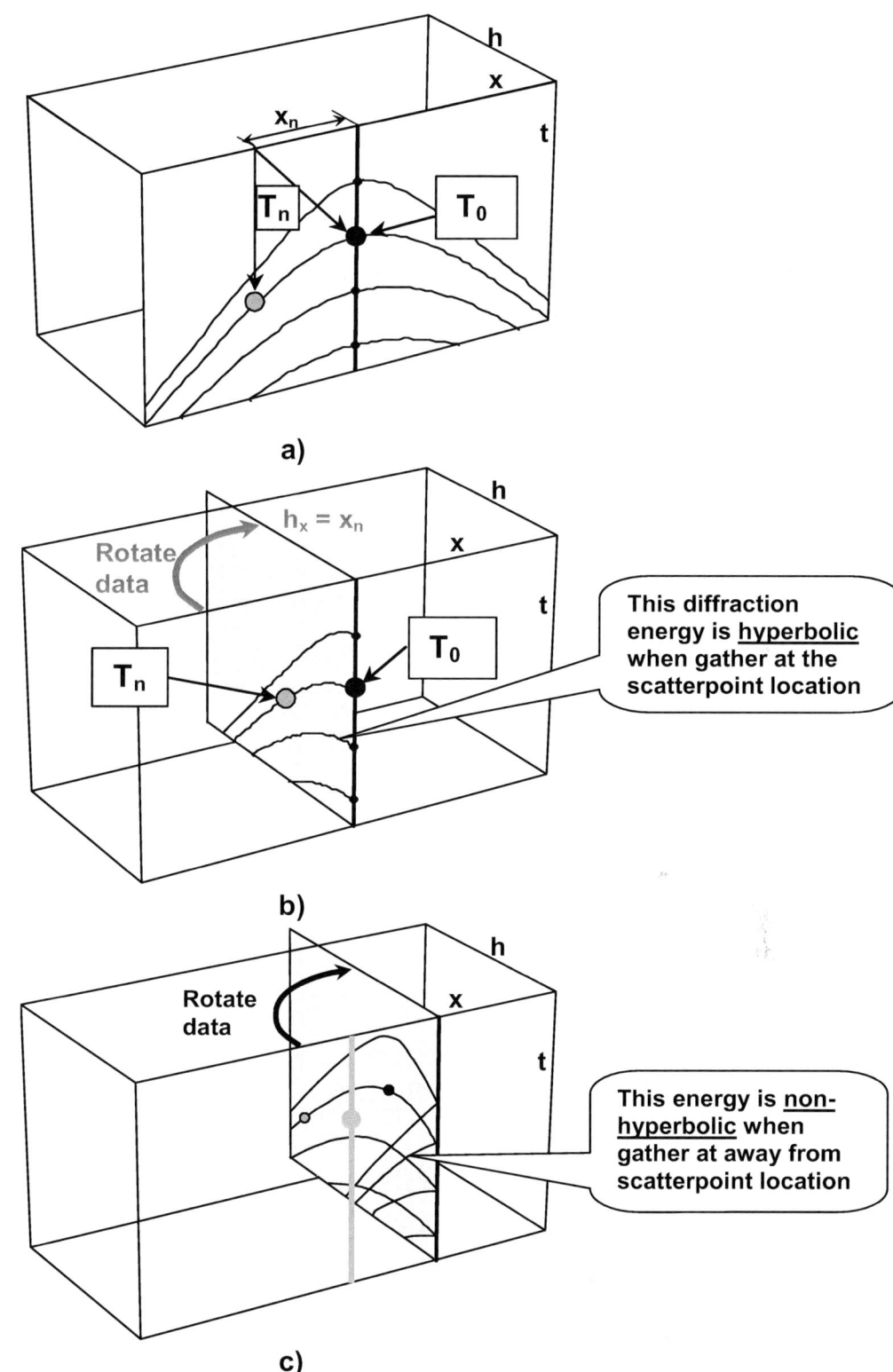

Figure 11.4 Resorting the data from a) the input zero-offset section into b) a one-sided migration gather located at the scatterpoint, and c) a one-sided migration gather that is removed from the scatterpoint.

11.1.4 Combination of MO and poststack migration

Each input trace contributes energy to each output migrated trace. In conventional processing this is accomplished by the two kinematic steps of MO and poststack migration as illustrated in Figure 11.5a.

We now combine these two processes to prestack migrate energy that lies on a Cheops pyramid.

Recall our two-way time for a scatterpoint located at *x = 0*.:

- *T* is still the time of the sample on an input trace.
- T_n is the zero-offset time after MO correction but before migration. (This
- T_0 is the vertical two-way time for a scatterpoint, and,
- β is now the dip at the *x* location on the zero-offset diffraction.

MO of equation (11.1) and poststack migration of equation (11.2) can be combined into one process as illustrated in Figure 11.5b and by the following;

Substituting $\quad T_n^2 = T_0^2 + \dfrac{4x^2}{V^2} \quad$ into $\quad T^2 = T_n^2 + \dfrac{4h^2 \cos^2 \beta}{V^2} \quad$ (11.1 and 2)

gives $\quad T^2 = T_0^2 + \dfrac{4x^2}{V^2} + \dfrac{4h^2 \cos^2 \beta}{V^2} = T_0^2 + \dfrac{4\left(x^2 + h^2 \cos^2 \beta\right)}{V^2} = T_0^2 + \dfrac{4h_n^2}{V^2},$

where $\quad h_n^2 = x^2 + h^2 \cos^2 \beta = x^2 + h^2 - h^2 \sin^2 \beta,$ (11.3)

or, when β is small, $\quad h_n^2 \approx x^2 + h^2$ (11.4)

Energy from an input trace can now be <u>stacked directly</u> to the <u>migrated trace</u> using an offset h_n and conventional MO correction as illustrated in Figure 11.5b.

However, rather than MO correcting and stacking the data at the migrated trace,

- simply move it to the <u>CMP (now a prestack migration) gather</u> at an offset defined by h_n with
- no time shifting as illustrated in Figure 11.5c, where a sample from Cheops pyramid will now lie on an <u>approximate hyperbolic path</u>.
- The path is approximate because the moveout equation is only an approximation to the double-square root equation (see section 7.8).
- An exact solution can be found that is independent of β as described in section 11.2.

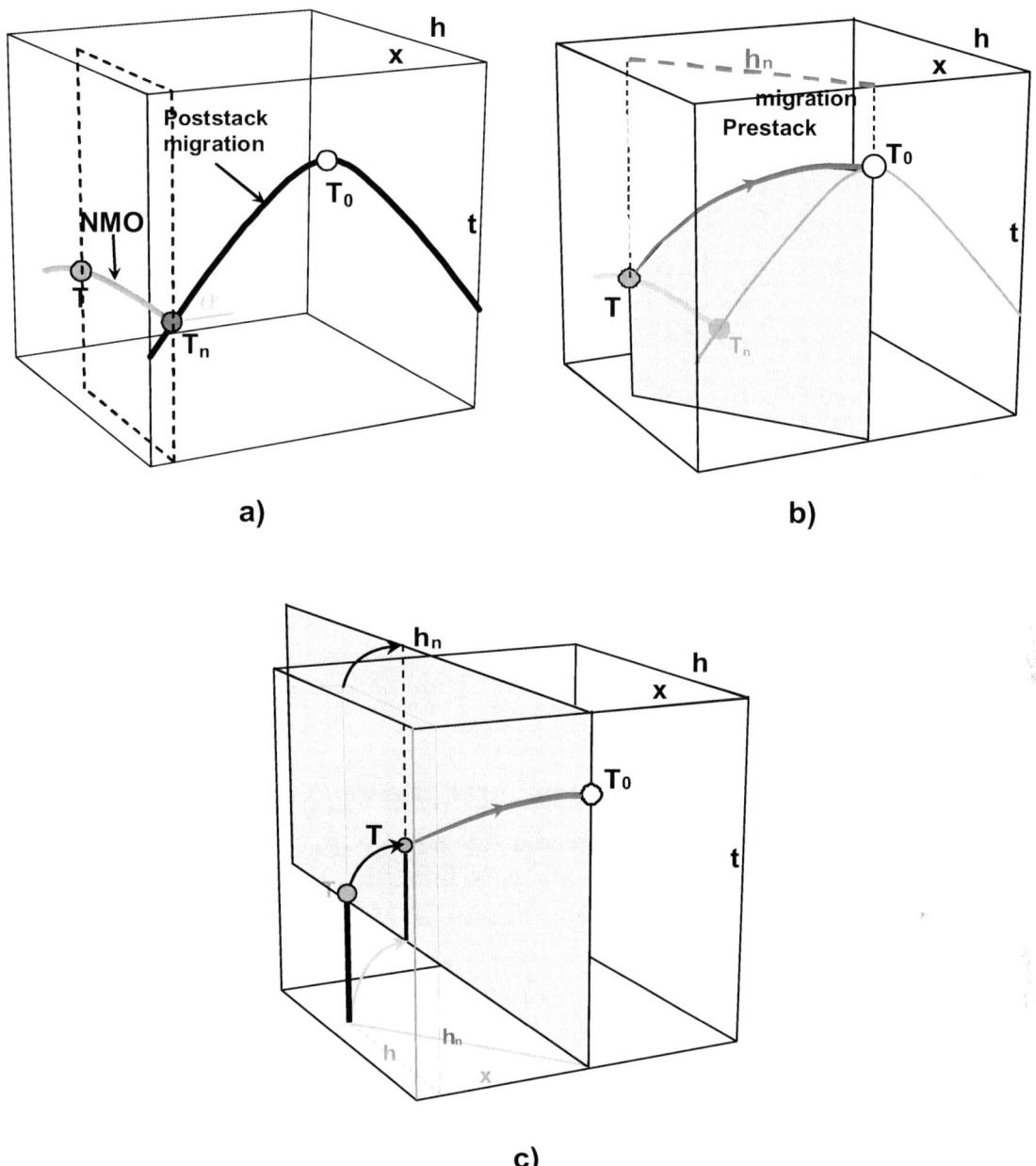

Figure 11.5 Comparison of data movement between a) MO and poststack migration, b) the direct movement of energy from the input trace to a migrated trace, and c) the input sample moved to the corresponding migration gather. The ,two blue hyperbolic paths are the same.

11.1.5 A prestack migration gather

All input traces, within the prestack migration aperture, can be gathered into a prestack migration gather.

- The CMP traces are already there.
- Section 11.0.2 showed how to include the zero-offset traces with offset x.
- Section 11.0.3 showed we could include arbitrary input samples with offset h_n,

where
$$h_n^2 \approx x^2 + h^2.$$ (11.6)

All samples in the input trace will have an offset h_n as illustrated in Figure 11.6a. This input trace can be summed into the prestack migration gather at offset h_n.

In a similar manner, all input traces could be moved to the prestack migration gather as illustrated in Figure 11.6b, where each point on the grid represents an input trace.

> The prestack migration of one output trace is then formed by MO correction and stacking of the prestack migration gather.

This process can then be repeated for all input traces.

Note that the same energy is summed into all the prestack migration gathers. However, the new offsets, h_n, will be slightly different in the different gathers and cause the reconstruction of reflectors at that trace, and the cancellation of all other energy.

The prestack migration also requires phase filtering ($rj\omega$), antialiasing filters, and amplitude scaling that are typical of Kirchhoff migration. When combined with MO correction, we refer to these processing steps on the prestack migration gather as Kirchhoff MO or KMO.

> Note there is no time shifting of the input trace samples when they are added to the prestack migration gather.
>
> The data movement is prior to MO correction.
>
> Different migration locations (x) will have different radius or offsets h_n.

Very important

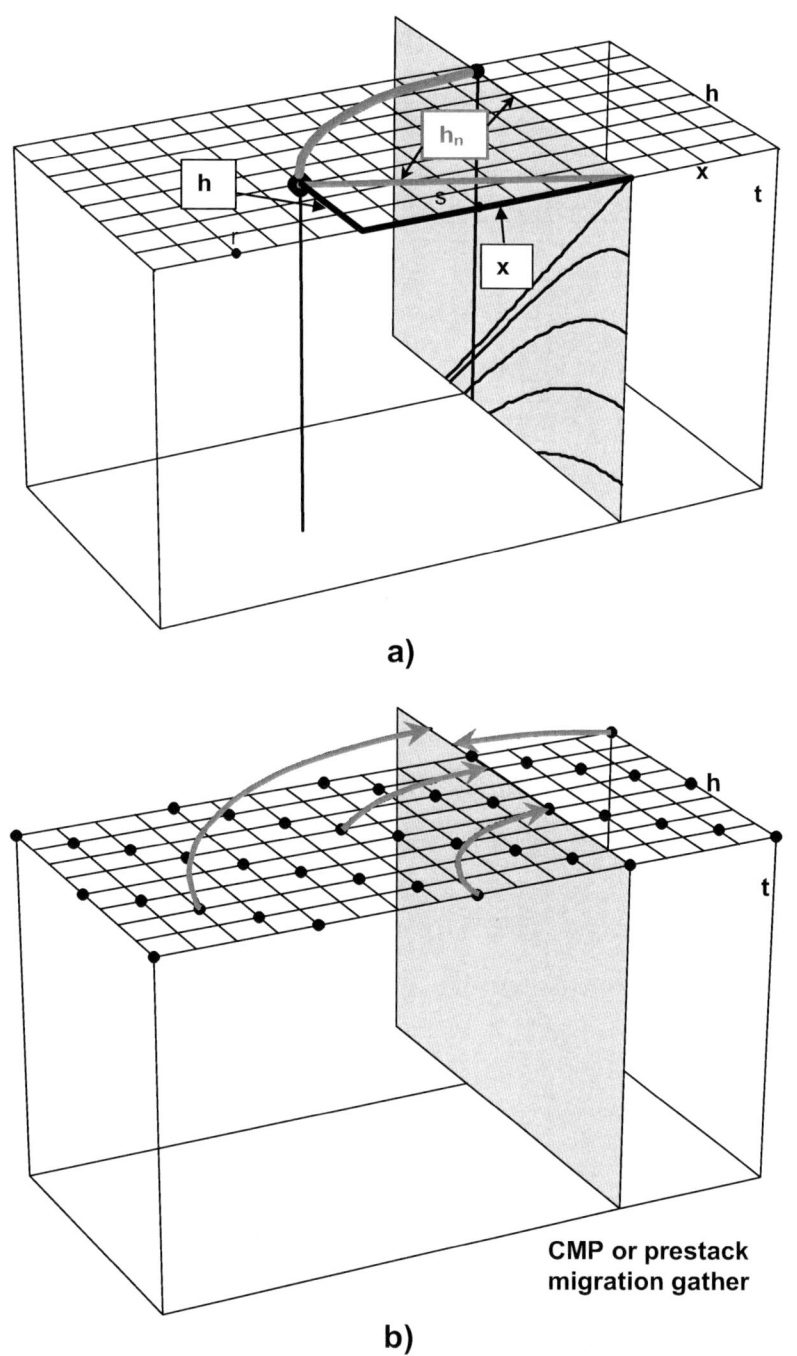

Figure 11.6 Illustration of energy movement for a) one input trace to the prestack migration gather and b) many traces contributing to the prestack migration gather.

11.1.6 Raypaths for a Scatterpoint

The previous sections formed prestack migration gathers that were based on the two processing steps of MO correction and poststack migration. The formation of prestack migration gathers may be optimized by defining the prestack migration offset on <u>scatterpoint principles</u>.

Consider the scatterpoint in Figure 11.7a.

- Energy is "reflected" from all sources to all receivers.
- Each input trace contains energy from the scatterpoint.
- CMP reflections are a small subset of all possible reflections.
- When including all the input traces into a prestack migration gather, <u>what offset should be used</u>?

All input prestack traces could be included in a prestack migration gather if we could assign an offset for each sample in the trace.

<u>Prestack migration gathers</u> that are based on the scatterpoint principles are referred to as <u>common scatterpoint (CSP) gathers</u>.

The scatterpoints are not restricted to the 2-D case, but also apply to 3-D data, as illustrated in Figure 11.7b.

All rays in Figure 11.7a-b have the <u>same RMS velocity</u> defined at the scatterpoint.

<u>Travel times</u> are based on this RMS velocity and the <u>surface distance</u> between the source/receiver and the CSP location.

<u>Velocity analysis</u> should be based on these surface distances.

<u>Conventional processing</u> is based on the half source-receiver offset h.

How will conventional velocity analysis compare with scatterpoint velocity analysis?

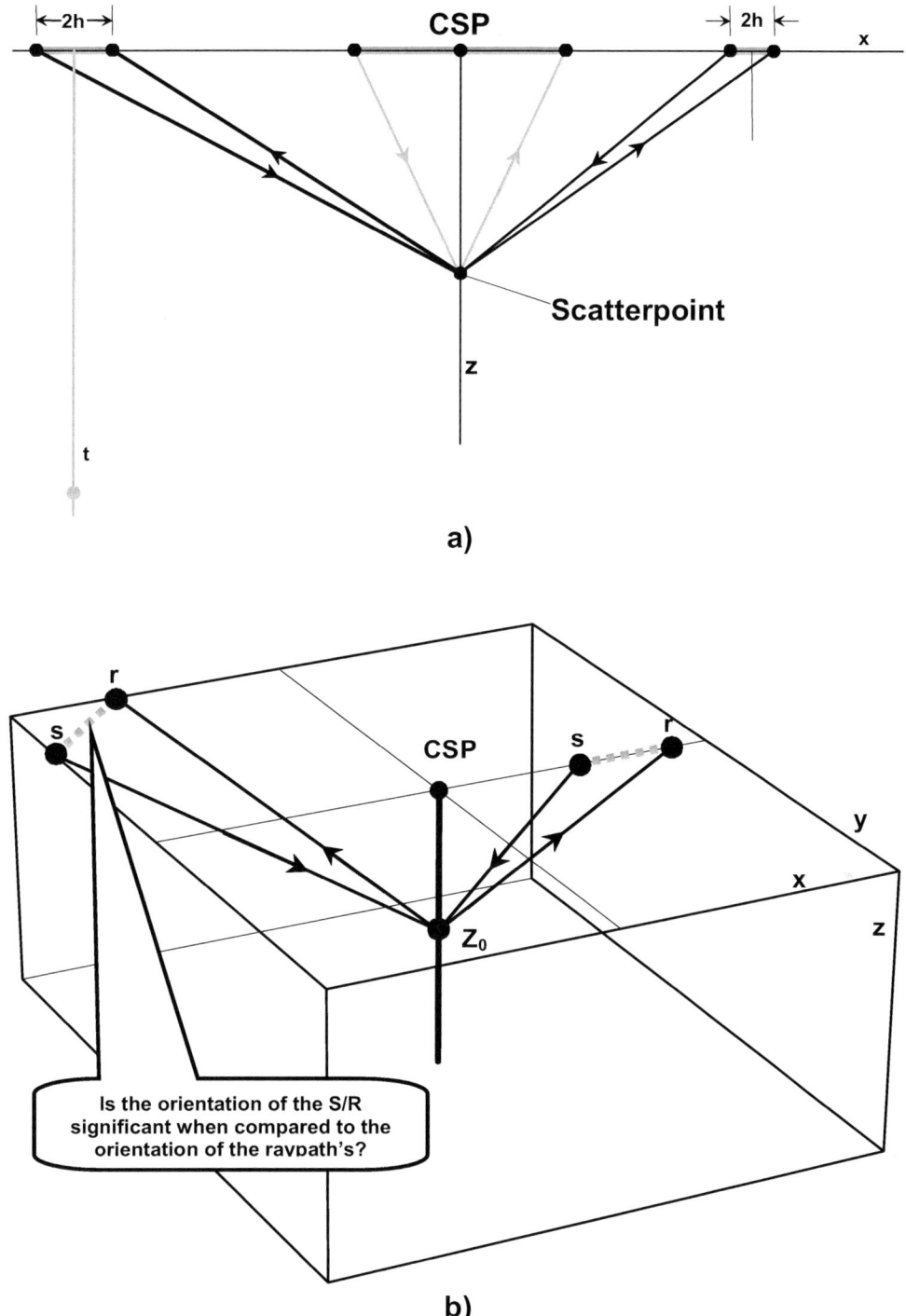

Figure 11.7 Each input trace contains energy from a scatterpoint, as illustrated in a) for 2-D, and b) 3-D data.

11.2 The Equivalent Offset h_e

11.2.1 Exact prestack time migration definition

The equivalent offset h_e is defined using the raypaths *to* and *from* a scatterpoint.

- The source and receiver are <u>colocated</u>.
- The two-way traveltime T remains the <u>same</u>.

Figure 11.8a illustrates the time-migration raypaths for a given source and receiver location, while (b) shows the raypaths for a colocated source and receiver with the same traveltime.

The traveltimes of the Figure 11.8a are defined by the double square root equation (DSR)

$$T = \left[\frac{T_0^2}{4} + \frac{(x+h)^2}{V^2}\right]^{1/2} + \left[\frac{T_0^2}{4} + \frac{(x-h)^2}{V^2}\right]^{1/2}. \tag{11.7}$$

The collocated source and receiver traveltime is given by

$$T = 2\left[\frac{T_0^2}{4} + \frac{h_e^2}{V^2}\right]^{1/2}. \tag{11.8}$$

By forcing the traveltimes in equations (11.7) and (11.8) to be equal, we have forced the double square root equation into a single hyperbolic form, i.e.

$$2\left[\frac{T_0^2}{4} + \frac{h_e^2}{V^2}\right]^{1/2} = \left[\frac{T_0^2}{4} + \frac{(x+h)^2}{V^2}\right]^{1/2} + \left[\frac{T_0^2}{4} + \frac{(x-h)^2}{V^2}\right]^{1/2}. \tag{11.9}$$

The value of h_e can be solved exactly (see Appendix 4.7) from equation (11.9) (Bancroft et al 1998) to be

$$\boxed{h_e^2 = x^2 + h^2 - \frac{4x^2 h^2}{T^2 V^2}.} \tag{11.10}$$

This equation enables all input points to be <u>mapped one-to-one</u> in a CSP gather.

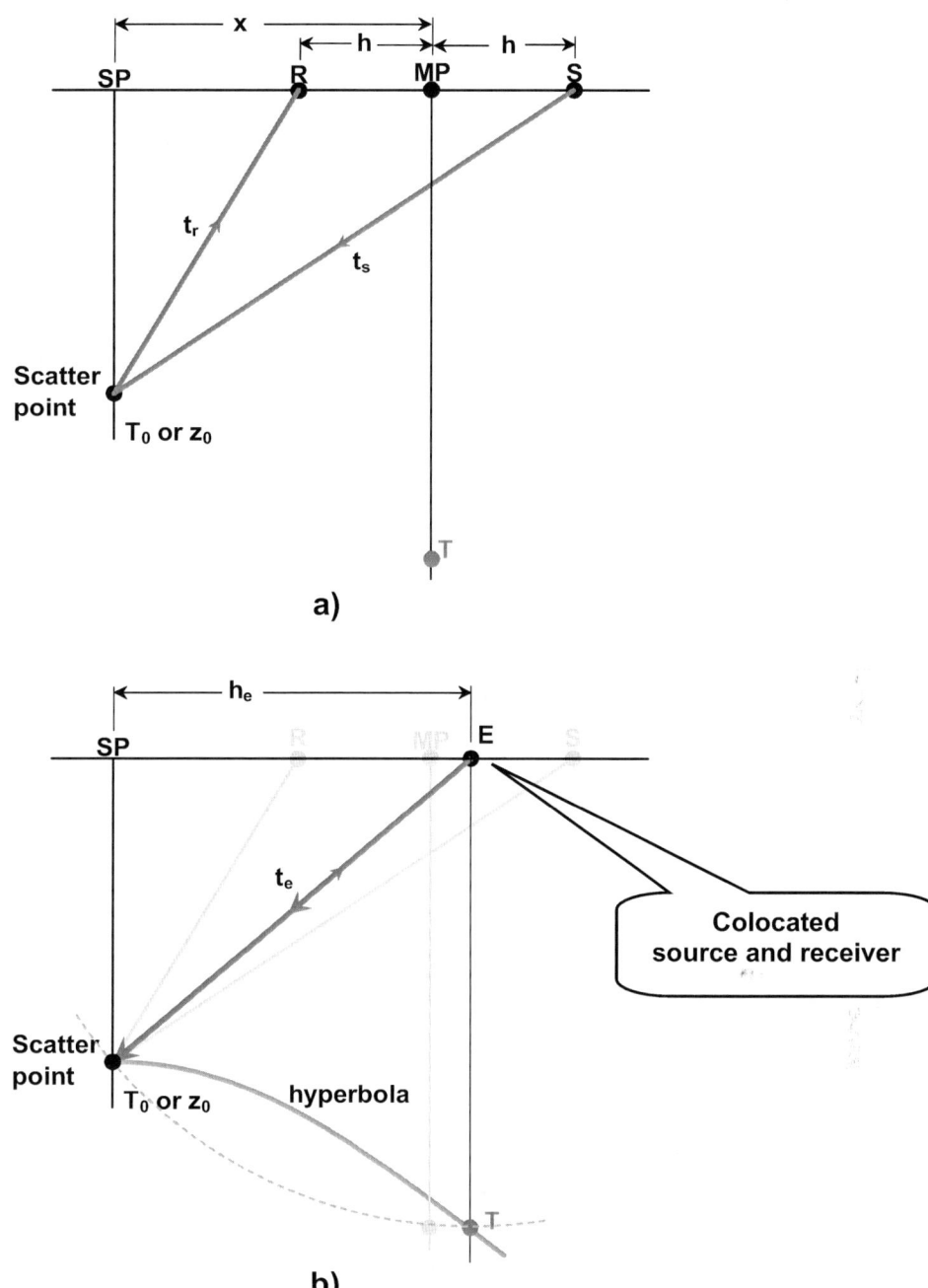

Figure 11.8 Raypaths to scatterpoint a) from a distant CMP, and b) the equivalent offset.

Note the energy at the equivalent offset is on a <u>hyperbolic path</u> (velocity V) from the scatterpoint.

The time sample in (a) lies on <u>Cheops pyramid</u>, while the time sample in (b) lies on a <u>hyperboloid</u>.

11.2.2 Construction of the equivalent offset

Mapping input energy to an EO gather is illustrated in Figure 11.9. An <u>input trace</u> with offset 2h, source S, receiver R, and midpoint MP, is shown relative to an arbitrary location of a <u>vertical array of scatterpoints</u> SP, which represent a migrated trace The displacement between SP and MP is x. We will assume the velocity to be one, and that time can be equated to distance. For convenience, we will consider the times on the trace and gather to be one-way times.

When the time T on the input trace tends to infinity, the cross term in equation (11.10) tends to zero defining a maximum equivalent offset $h_{e\text{-}max}$ as

$$h^2_{e-\max} = x^2 + h^2. \qquad (11.11)$$

The construction to get $h_{e\text{-}max}$ is a simple right angle triangle with sides equal to x and h as illustrated in (a). The maximum equivalent offset for this trace, relative to the scatterpoint is plotted on the EO gather in (b).

An interesting question: What happens to h_e when T tend to zero?

It turns out that there will be a minimum time T_{min} that occurs when T_0, in equations (11.7) and (11.8), goes to zero, giving a minimum equivalent offset $h_{e\text{-}min}$

$$h_{e-\min} = x. \qquad (11.12)$$

This minimum offset corresponds to a scatterpoint located at the surface and is the displacement between SP and MP, which is also plotted on the EO gather in (b).

Note that the minimum time T_{min} and the minimum and maximum equivalent offsets $h_{e\text{-}min}$ and $h_{e\text{-}max}$ will vary with the relative location of the migrated trace.

The <u>equivalent offset construction</u> for a scatterpoint labeled SP_4 is illustrated in (c). We start by projecting the t_r onto t_s, using SP4 as a common point, to define the location u. A point v in then constructed midway between u and S to define the average or zero-offset time t_e. The point at v, is rotated about SP_4 to the surface to define the equivalent zero-offset location h_e.

The time t_e is located on the input trace represented by (d), and then mapped onto the EO gather in (e) with the same time t_e and offset h_e.

Chapter 11 Equivalent offset migration (EOM)

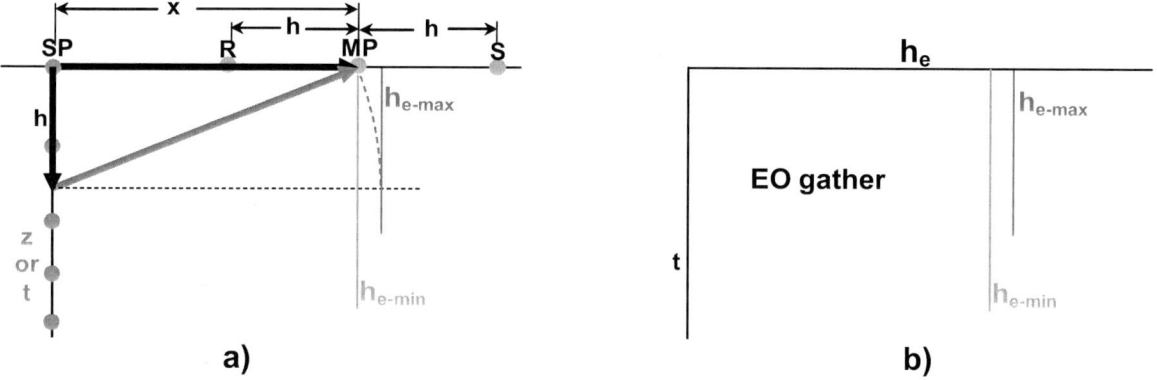

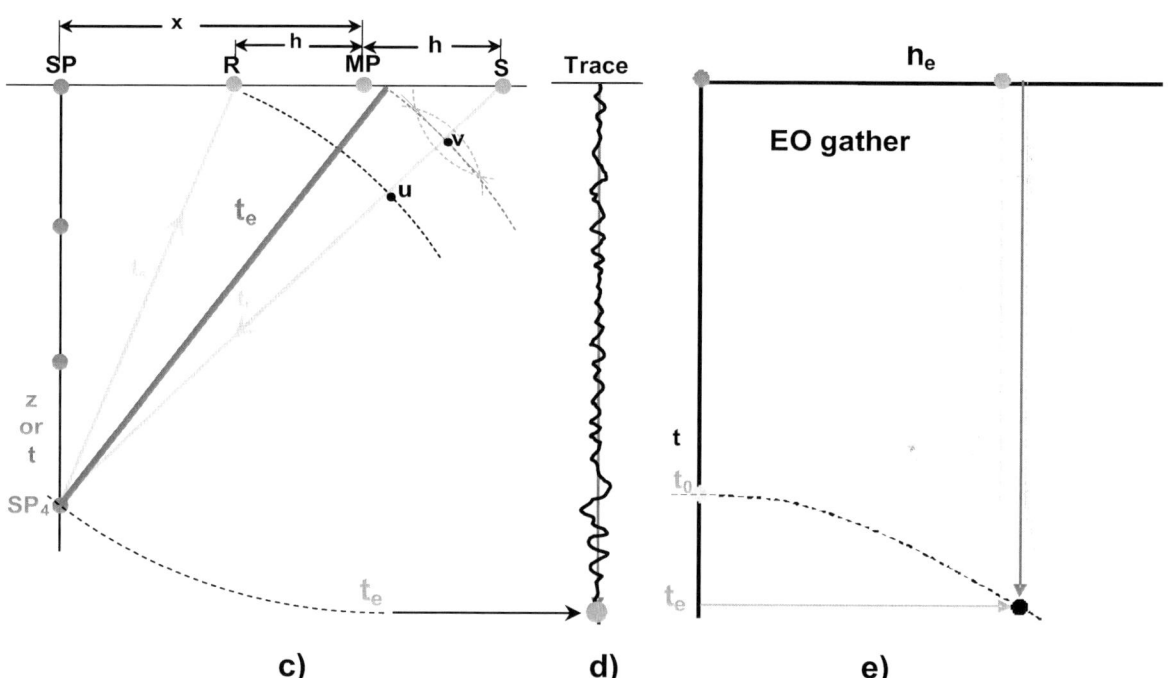

Figure 11.9 Construction of equivalent offset for one scatterpoint.

Construction exercise:

Repeat the construction of Figure 11.9 in Figure 11.10 for SP_2. Map the corresponding input time samples from the trace in (b) to the EO gather in (c).

What is the minimum time on the input trace that will contribute to the EO gather in (c).

A new trace with source S_2 and receiver R_2 that has the same offset is recorded closer to the array of scatterpoints as illustrated in Figure 11.11a. Repeat the construction for locating the energy for SP_2 from the new input trace to the EO gather.

What is the significance of the dashed curve in Figure 11.11c?

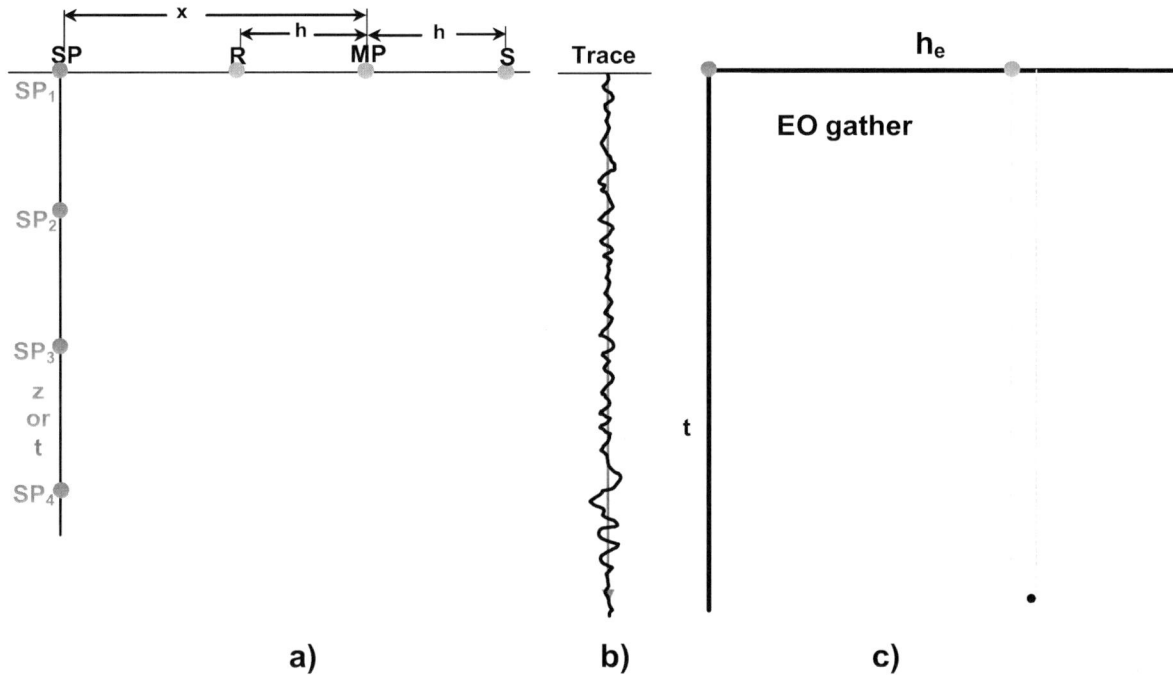

Figure 11.10 Construction of equivalent offset for three additional scatterpoints, a) the cross-section, b) the input trace, and c) the EO gather.

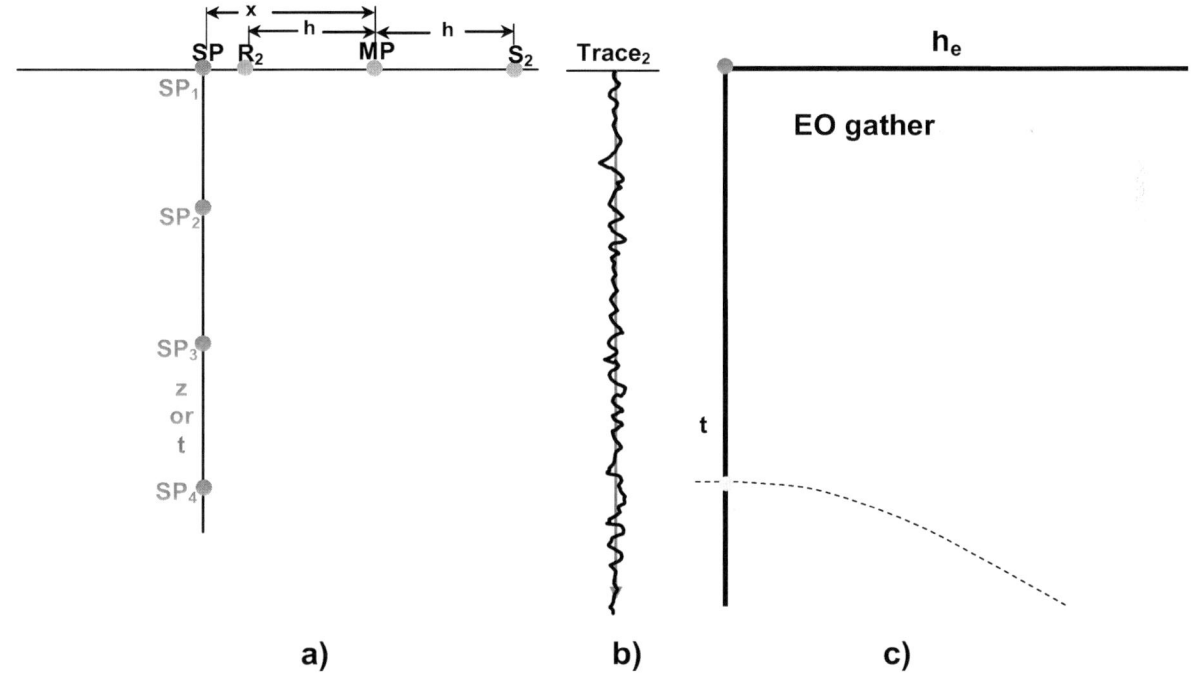

Figure 11.11 Second construction of equivalent offset with a) new source-receiver locations on the cross-section, b) the new input trace, and c) the EO gather..

11.2.3 Construction for various traces and one scatterpoint

Construction:

Figure 11.12 contains a geological cross-section with one scatterpoint and numerous source receiver pairs identified by S_i and R_i. Part (b) of the figure contains an EO gather that is located at the scatterpoint. Assume the traveltimes are equivalent to the distances

Construct the equivalent one-way time t_e at the equivalent offset h_e for each source-receiver pair, and plot them on the EO gather.

Also plot the one-way time t_e with the source receiver offset on the EO gather.

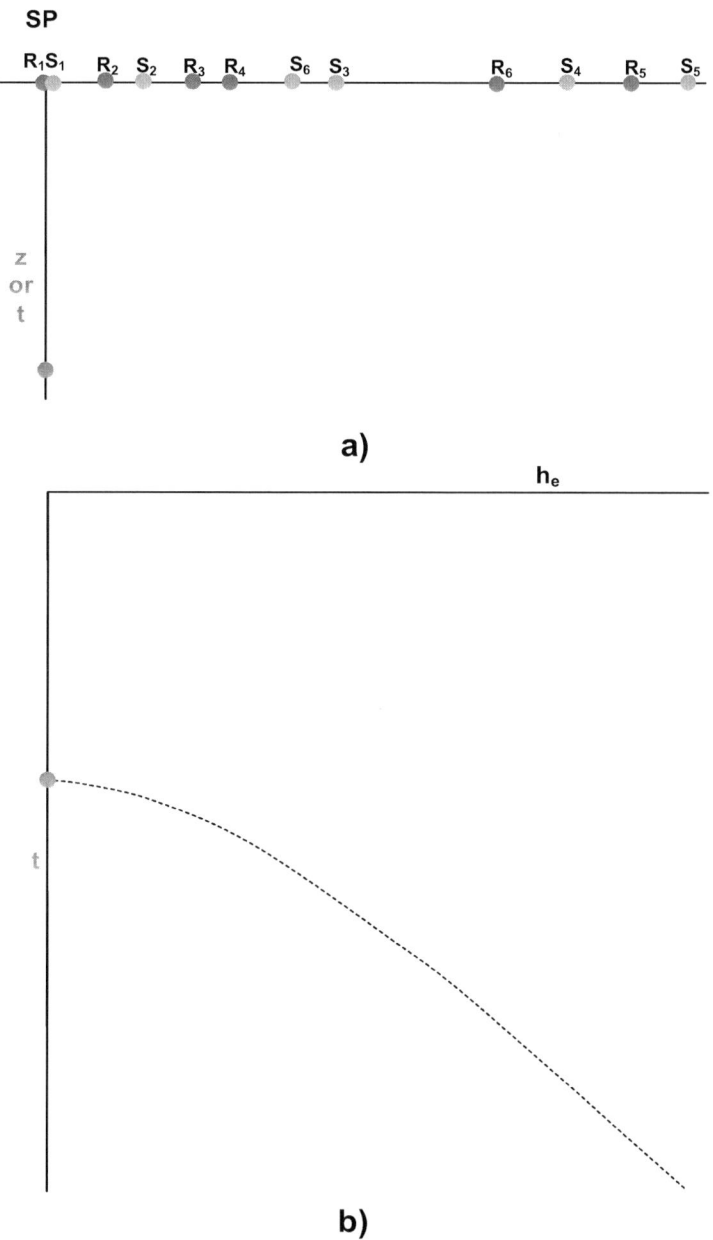

Figure 11.12   Construction for the contribution of energy from a) a number of input traces that have reflections from one scatterpoint to b) the EO gather.

Observations:

- **Reflection energy from a scatterpoint will be recorded on all input traces within the migration aperture.**
- **An EO gather at the scatterpoint will align this energy along a hyperbola with a velocity defined at the scatterpoint.**
- **True reflection energy will construct while other energy will cancel.**

11.3 Mapping one input trace to one CSP gather

The mapping of one input trace to a CSP gather with relative distances $x = 3$ and $h=1$ is shown in Figure 11.13.

The CSP gather (h_e, t) is <u>orthogonal</u> to the ray geometry (x, z), but for convenience they are overlaid with the geometry shown in gray.

- Times are plotted to allow distance to match two-way times.
- The <u>first sample $T\alpha$</u> is shown relative to the surface scatterpoint.
- The offset of $T\alpha$ is $h_{e\alpha} = x$, the distance between the CMP and CSP locations.
- The <u>maximum offset</u> $h_{ew} = \sqrt{x^2 + h^2}$ is shown (dashed) relative to the CMP offset x.
- The source/receiver offset is identified by the surface symbols with $h = 1.0$.
- Raypaths are shown for one scatterpoint at a depth of 4 units.
- Times relative to the scatterpoints are shown by the semicircles of migration.
- These times T lie on hyperbolic paths from the scatterpoints. The hyperbolic path for $T\alpha$ is the asymptote at an angle of 45 degrees. It will migrate to $t = 0$ with a dip of *90 degrees*.

In 1997, Fowler [619] showed that the DSR equation could be simplified into <u>many</u> different hyperbolic forms.

From these possibilities, the equivalent offset method is the <u>only formulation</u> that moves data to all CSP gathers with no time shifting.

The migration time shift is accomplished with Kirchhoff MO <u>after</u> the CSP gather has been formed.

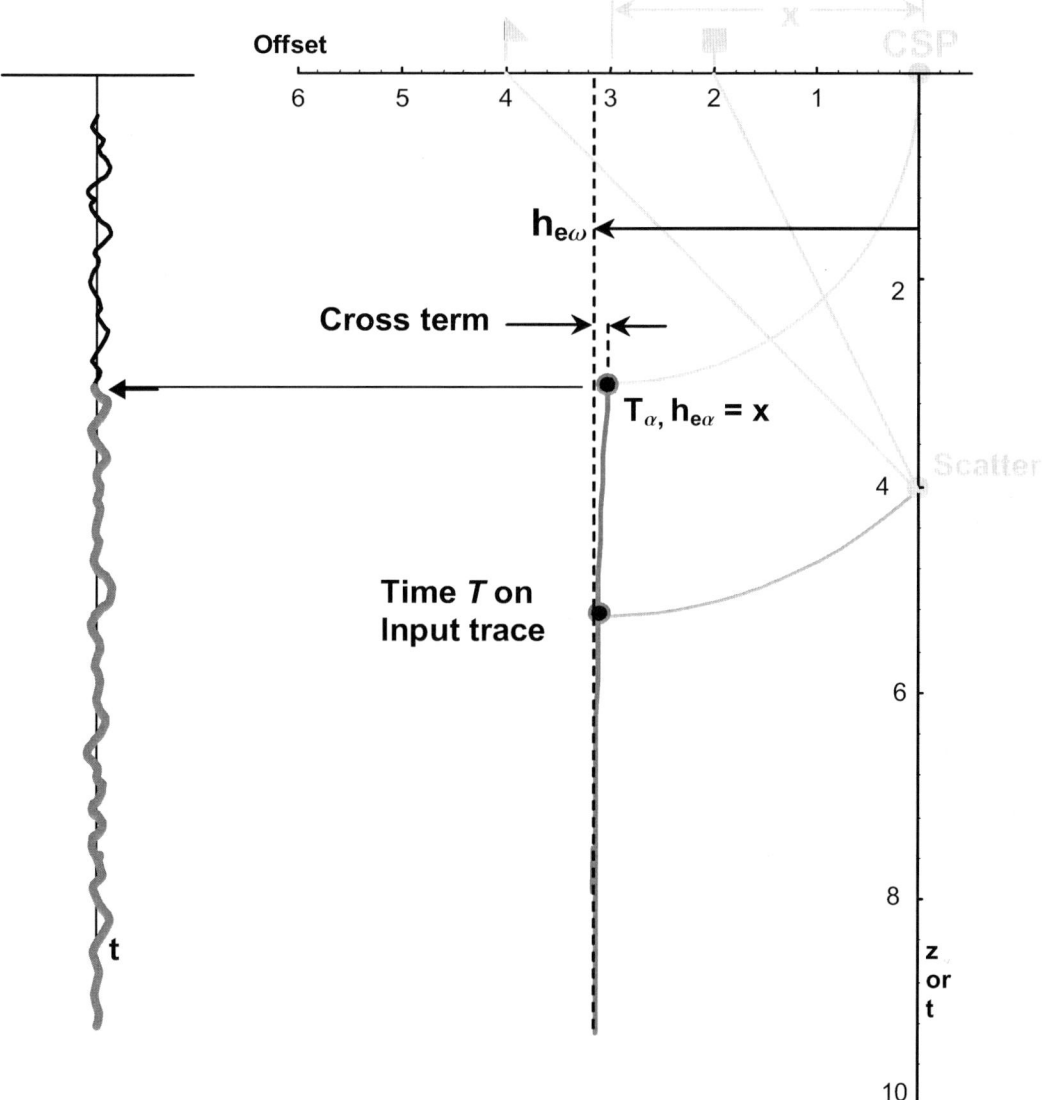

Figure 11.13 Plot of an input trace mapped to a CSP gather.

11.3.1 Mapping one input trace to a number of CSP gathers

Figure 11.14 contains four families of curves, each with source/receiver offsets of $2h = 0.5, 1.0, 2.0,$ and 4.0. The family of curves in each figure is for one input trace with an offset $2h$ with each curve showing the equivalent offset location as the CMP to CSP distance x ranges from 0.0 to 5.0 in increments of 0.5.

Note:

- The equivalent offsets tend to an asymptote at longer times.
- The top of each equivalent offset curve forms a 45-degree slope. This point will move to $T_0 = 0$ and represents a 90 degree migration.
- A sloping curve from ($t = 0$, $h_e = 0$) represents a dip limit α for the poststack migration as shown on (c).
- The dipping line has the angle α when measured from the vertical, and using the poststack migrators equation $\tan \alpha = \sin \beta$, an <u>approximate dip limit</u> β may be imposed.
- The equivalent offset range of a trace below the dip limit is <u>smaller</u> and may:
    1. Increase the <u>speed</u> of the gathering process.
    2. <u>Reduce noise</u>.
- In some circumstances, such as marine data, the range of dip limited offsets may be assumed to fall within <u>one bin</u>, increasing the speed of the gathering process.

The CSP gathers in Figure 11.14 have been formed with a constant velocity. When the velocity varies with depth, (at the CSP location), then more complex equivalent offset curves will be observed. In some circumstances, the curve may become <u>multi-valued</u> in time as with conventional MO correction.

Chapter 11 Equivalent offset migration (EOM)

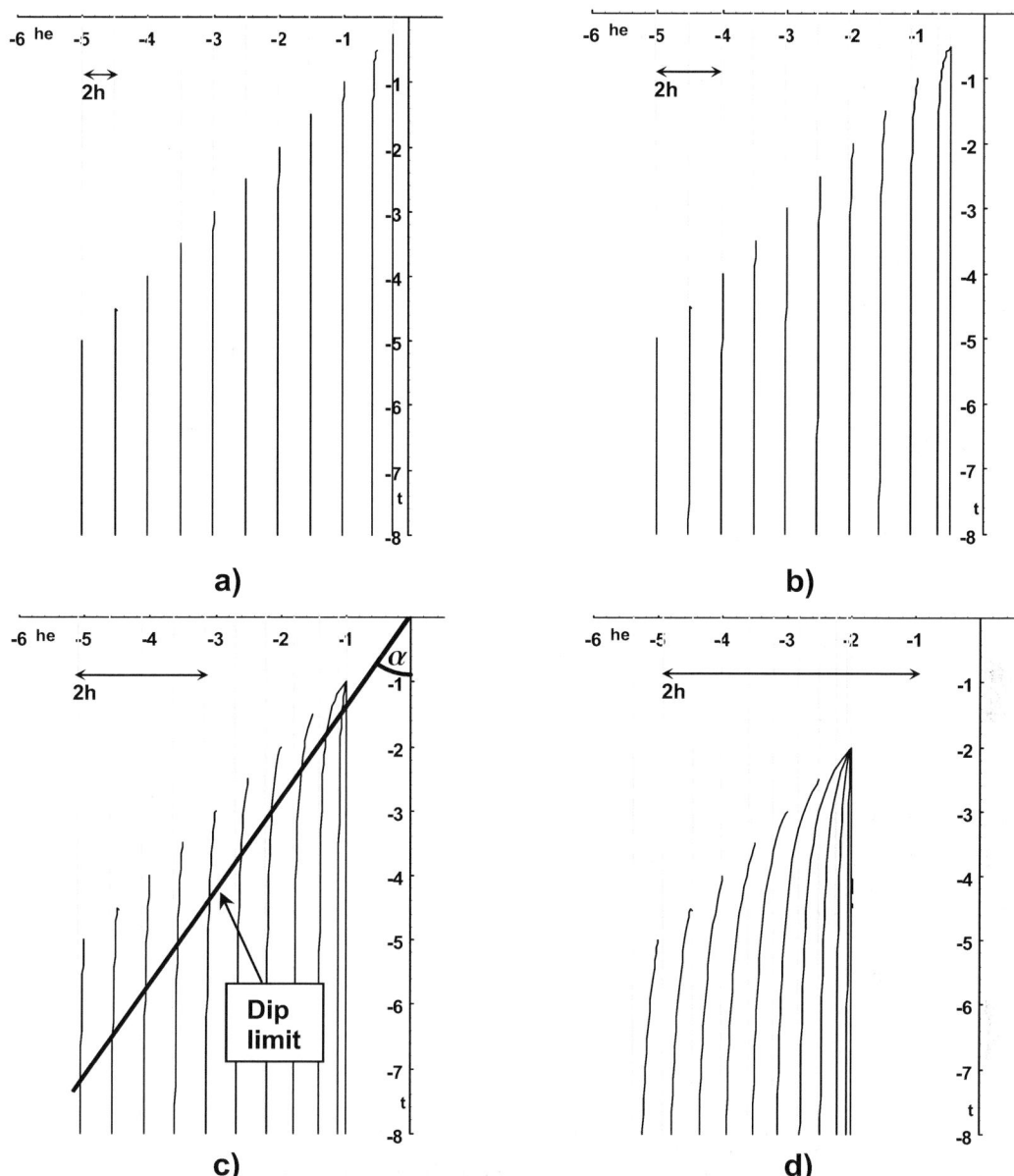

Figure 11.14 Families of equivalent offsets, each with different source/receiver offsets of a) 0.5, b) 1.0, c) 2.0, and d) 4.0. A normalized scale for offset and time (*V*=0.5) was used.

11.3.2 Mapping one input trace in the prestack volume

The mapping of one input trace to the neighbouring CSP gathers may also be viewed in a perspective view of the prestack volume. Figure 11.15 shows one input trace with a geometry similar to that of Figure 11.14d.

We refer to the surface produced by the equivalent offset as the bow (of a ship) effect.

The shape of the bow becomes sharper as the source/receiver offset is reduced.

One input sample

The mapping of <u>one input sample</u> to all <u>neighbouring CSP gathers</u> was illustrated in Section 9.7. Figure 9.29 is repeated as Figure 11.16.

Figure 11.16 to show the energy from one input sample is first mapped to an equivalent offset hyperbola at constant time, after which Kirchhoff MO moves the energy directly to the prestack migration ellipse (assuming constant velocities).

The <u>extent</u> (or amplitude) of the equivalent offset hyperbola is limited by the migration time (T_0) that goes to zero at the sides of the prestack migration ellipse.

Exercise:

Use equation 11.10 to prove that an input sample, when mapped to the neighbouring CSP gathers, lies on a hyperbolic path.

(Hint: h, T, and V are all constant.)

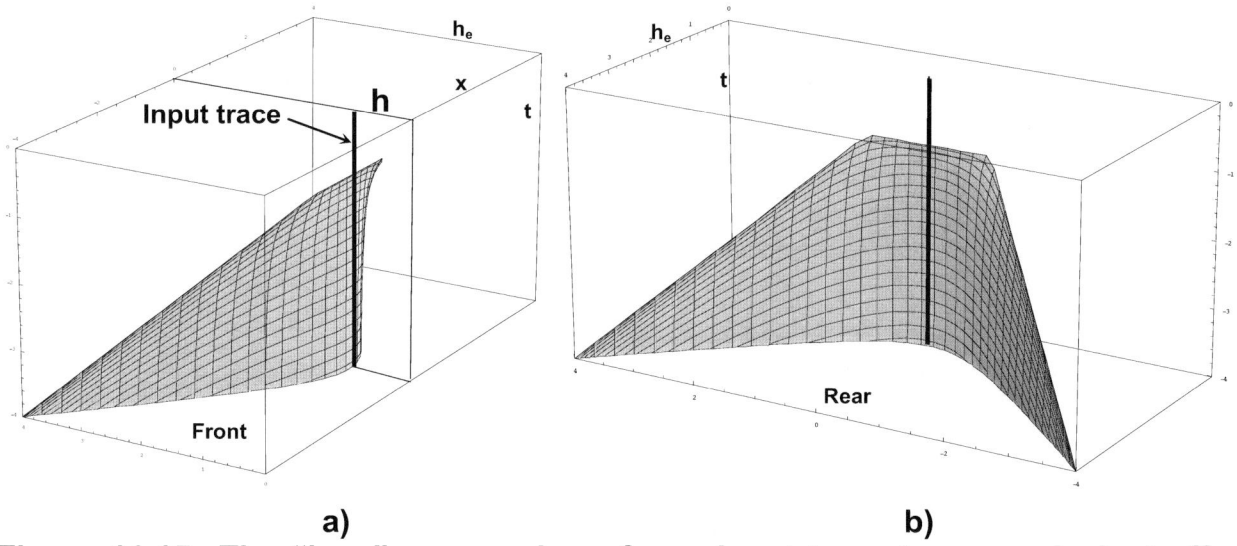

Figure 11.15 The "bow" or mapping of one input trace to an equivalent offset surface in the 3-D prestack volume with a) the front and b) the rear view.

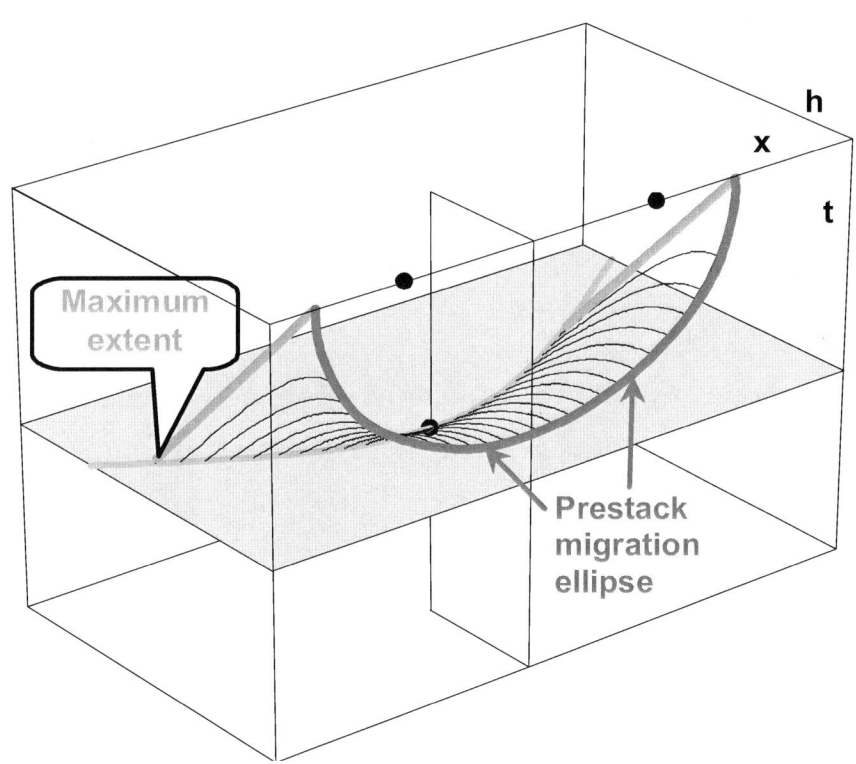

Figure 11.16 A repeat of Figure 9.29 showing the mapping of one input sample at a constant time to the equivalent offset hyperbola and then with Kirchhoff MO to the prestack migration ellipse.

11.3.3 Efficient computation of the equivalent offset

We now consider the time migration contribution of <u>one input *trace*</u> to a CSP gather that is located at the scatterpoint position *x = 0*.

The values of *x* and *h* are now constant in equation (11.10). The equivalent offset h_e will, however, increase with increasing time *T*. The velocity *V* is defined at the scatterpoint time T_0 (not *T*) and has a slight effect on the computation of h_e.

As with any offset trace, only the <u>lower portion</u> will contribute to the CSP gather. The <u>first useful time</u> t_α defines the two-way time for a scatterpoint located at the surface (*x = 0, T_0 = 0*) and is computed from

$$T_\alpha = \frac{2x}{V_0}, \tag{11.11}$$

where V_0 is the velocity at the surface. The <u>first equivalent offset</u> $h_{e\alpha}$ for T_α is given by

$$h_{e\alpha} = x, \tag{11.12}$$

and is rounded to the bin center h_{e0}. Subsequent bin centers will have offsets $h_{ei} = h_{e0} + i\, \delta h$ where *i* is the incremental bin number, and δh the bin width.

The maximum equivalent offset $h_{e\omega}$ is given by

$$h_{e\omega}^2 = x^2 + h^2, \tag{11.13}$$

which is the asymptotic limit of the equivalent offset as *T* tends to a large value.

These points are shown in Figure 11.17, which shows the mapping of an input trace to a few bins of the CSP gather.

It may appear from equation (11.10) that the <u>equivalent offset</u> needs to be computed for <u>each sample</u> in an input trace. <u>That is not the case.</u>

Since the CSP gather is composed of offset bins,

- the only occasions for which calculations are computed are when the input samples shift to a new offset bin h_{ei}.

Chapter 11 Equivalent offset migration (EOM)

Movement of energy for a typical case follows:

- **The <u>initial equivalent offset</u> $h_{e\alpha}$ is computed using equation (11.12) and assigned to an appropriate (closest) offset bin.**

- **The samples following T_α are <u>added to this bin</u> until the equivalent offset increases to the <u>next bin boundary</u>, at which point the input samples are added to the new bin.**

- **The <u>transition time</u> t_i at the equivalent offset bin h_{ei} is found by rearranging equation (11.10) to give**

$$t_i = \frac{(2xh)}{V_{mig}\left(x^2 + h^2 - h_{ei}^2\right)^{1/2}}. \qquad (11.14)$$

- **These bins and <u>transition times are pre-computed</u> for each input trace to allow efficient copying of the samples into the respective bins.**

For the example trace shown in Figure 11.17, <u>only two transition times</u> are computed, then three "DO LOOPS" add the input trace to the CSP gather.

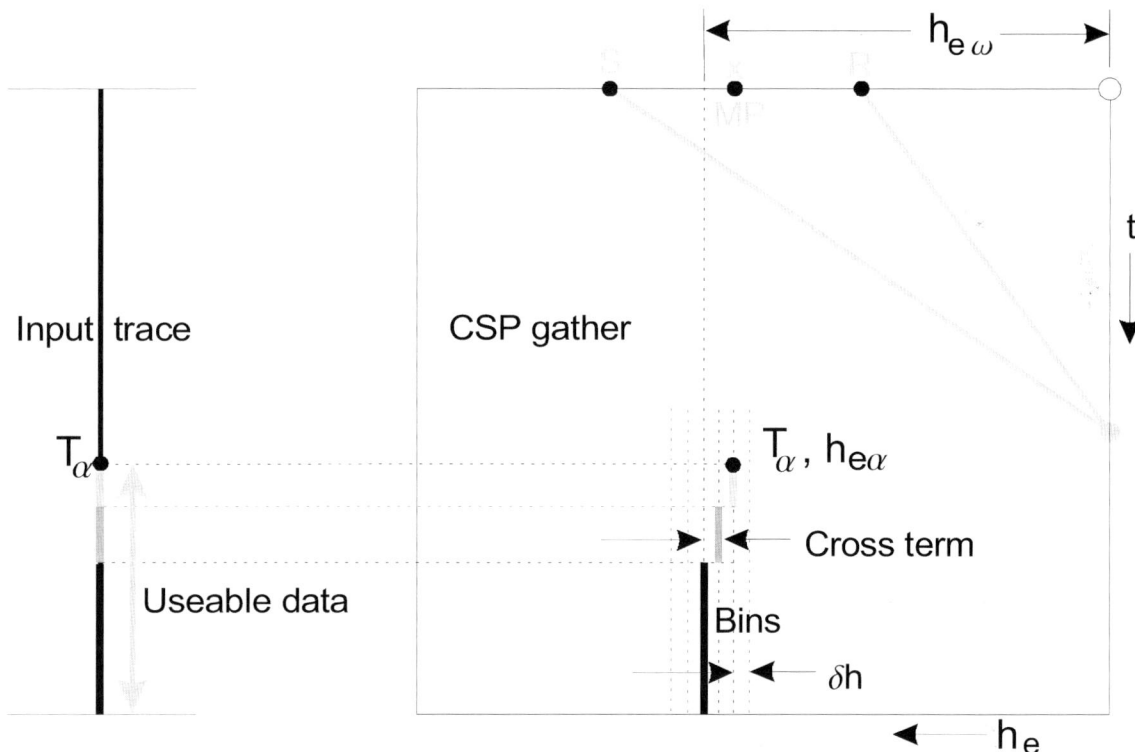

Figure 11.17 Mapping an input trace to a CSP gather.

11.3.4 Computation of the equivalent offset for depth migration

Chernis [707] extended the equivalent offset method to prestack depth migration. The two-way traveltime T (that was approximated from the RMS velocities and the double square root equation) is computed from <u>raytracing</u> or traveltime computations.

The traveltimes of the complex raypaths now define the two-way traveltime T to a scatter point at a depth z_0. A <u>hyperbolic equation</u> (similar to the time method) may be used to define equivalent offset h_e as

$$T^2 = T_0^2 + \frac{4h_e^2}{\hat{V}^2}, \quad (11.15)$$

where $\hat{V}$ is an arbitrary velocity. Choosing $\hat{V}$ to be the average velocity V_{ave}, enables the depth of the scatter point z_0 to be computed from T_0 giving:

$$T^2 V_{ave}^2 = z_0^2 + h_e^2, \quad (11.16)$$

or

$$h_e^2 = T^2 V_{ave}^2 - z_0^2. \quad (11.17)$$

Now all the energy from the scatter point in all input traces will align on the hyperbolic path at offsets that approximate the original geometry of the raypaths. Kirchhoff MO and stacking will complete the prestack depth migration for that scatter point.

Figure 11.18 shows an input trace (a) and its mapping to a depth section (b). The traveltime maps from the source and receiver are combined to produce the isochrons for the input trace. Note, for a defined output location, the input time T is mapped to a depth of z_0 or time T_0.

- Energy from the input traces is added (with no time shifting) to a CSP gather.
- After Kirchhoff MO and stacking, the vertical time trace may be converted directly to a depth trace using the average velocity V_{ave}.
- One advantage of the method is the reduced computations for moving each input sample to the depth location (see Chernis).
- Another advantage of the method is the scattered energy is placed at an equivalent offset that is similar to the geometry of the original offset trace.
- Errors in the initial velocity model are manifest by the location of energy about the ideal hyperbolic path.

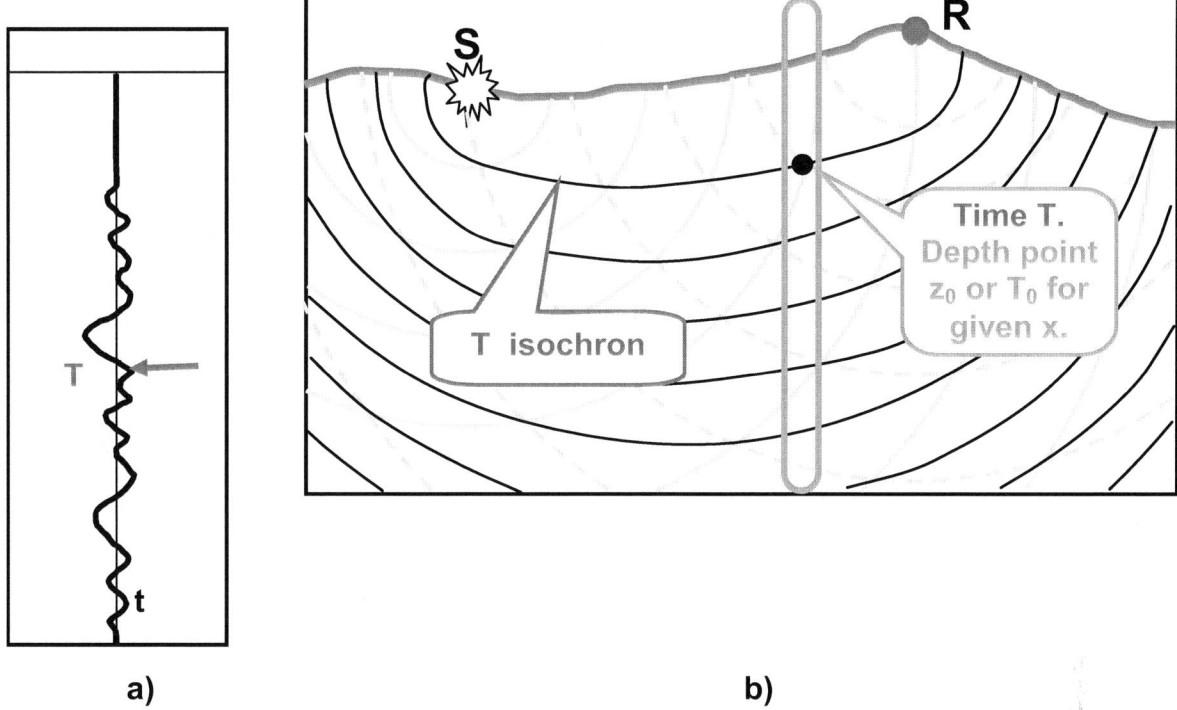

Figure 11.18 An input trace a) and its corresponding traveltime maps and isochrons on a depth section.

CSP gathers may also be formed that <u>include the moveout correction</u> by using the T to z_0 relationship. These gathers may also be evaluated by the flatness of energy from reflectors.

The velocity model used to form the CSP gathers may not have to be very accurate. Consequently it could be <u>smoothed</u> (more than normal) to allow larger ray increments or a larger wavefront grid to increase the speed of computing the traveltime maps.

After the CSP gathers have been formed the vertical traveltime T_0 and the distribution of the diffraction energy may assist in <u>refining the velocity model</u>.

11.4 The CSP Gather
11.4.1 Time migration examples

All the input traces within the migration aperture may be summed into a CSP gather at an offset defined by the equivalent offset.

- The CSP gather is similar to a CMP gather as:
    1. Both represent a subsurface location.
    2. Both have dimensions of offset and time.
- The offsets in the CSP gather are not limited by the source-receiver offset, but extend to the size of the migration aperture.
- The maximum equivalent offset in the CSP gather is limited by the recording time.
- Traces with larger equivalent offset (as with large migration window) will only contain contributing energy at larger times.
- Signals will constructively reinforce at the scatterpoint to form the correct image.
- Unwanted signal from other scatterpoints will be attenuated by destructive interference.

Figure 11.19 shows a CMP and CSP gather, taken at the same subsurface location on a marine line.

- The CMP required eight CMP's to fill all the offset positions with single fold.
- The fold on the two-sided CSP gather averages around twelve.
- Since the data is mainly below 3.0 seconds, the CSP gather was formed using one equivalent offset at $h_{e\omega}$ for each input trace.
- The single offset enabled the entire trace to be summed into the CSP gather.

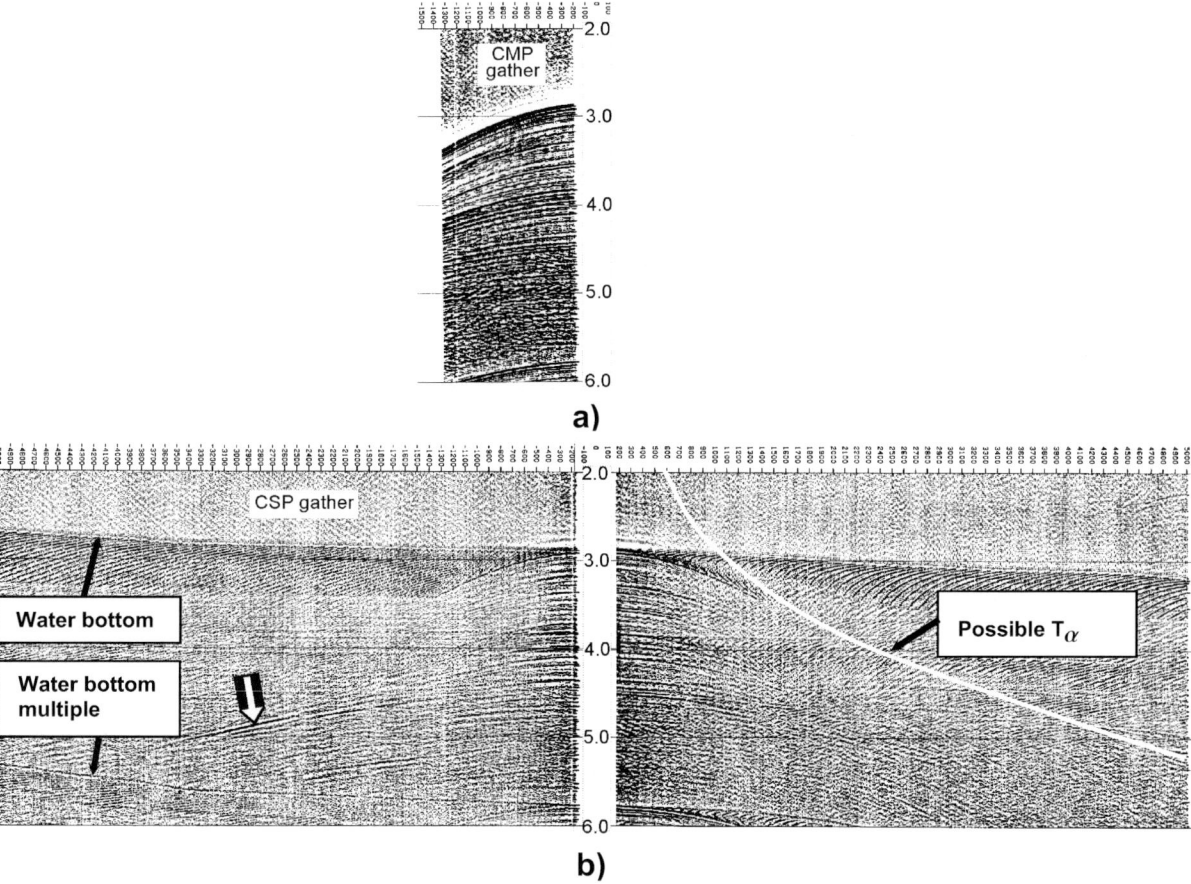

Figure 11.19 Comparison of a) singe fold CMP gather (8 CMP's wide), and b) a two sided CSP gather.

Note on the above figure:

- A white curve shows a possible $T\alpha$; information above this curve is usually not present in CSP gathers but illustrates the presence of zero-offset energy.
- The water bottom around 3.0 seconds.
- The first multiple of the water bottom around 6.0 seconds.
- Energy from a dipping event at a large offset, indicated by the white arrow.
- Much larger offsets on the CSP gather that contain coherent energy.

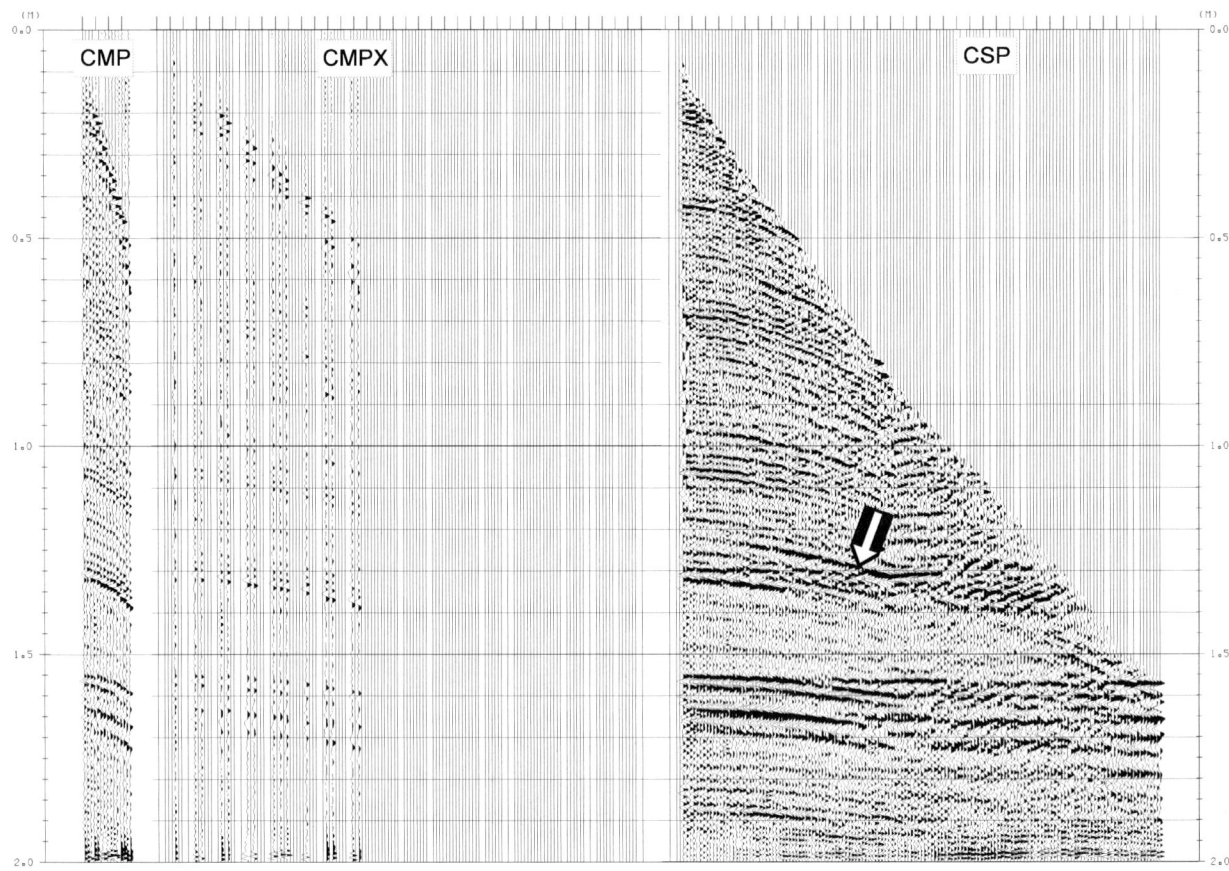

Figure 11.20 Plains data with a) a CMP gather of traces, b) the CMP gather with traces positioned by offset, and c) a CSP gather.

The dipping energy that is identified by the arrow in the CSP gather above comes from the flank of a channel (see Figure 11.16). This energy reconstructs at the appropriate channel location, but cancels at other CSP locations.

This energy is not present in the CMP gather.

The previous CSP gather in Figure 11.12 showed the zero-offset data of the water-bottom. The above CSP gather in Figure 11.20 also shows <u>zero-offset data</u>, especially at 1.570 s where horizontal and hyperbolic features tend to combine at zero-offset. We refer to this feature as the "<u>prow effect</u>" [643] and [646].

After Kirchhoff MO, the horizontal features curve upward and stack out as in conventional poststack migration of zero-offset data.

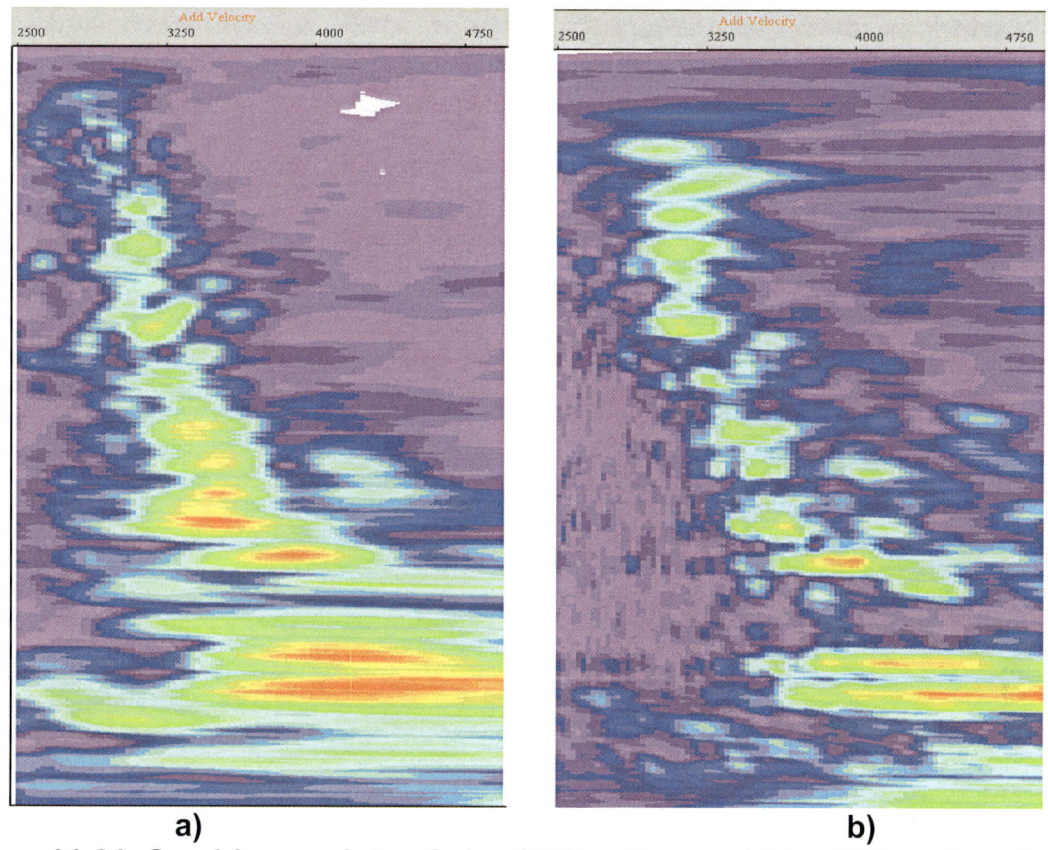

Figure 11.21 Semblance plots of a) a CMP gather and b) a CSP gather from a location close to the gathers in Figure 11.15

The following figure contain high resolution CSP gathers that show <u>tighter clustering of the energy</u>, which enables (and requires) a more accurate picking of the velocities.

The improved clustering is a benefit as <u>multiples</u>, and possibly mode <u>conversions</u>, form tight clusters of energy, which, when displaced from the stacking velocity, <u>will tend to "stack out"</u> or can be removed using Radon transforms. (The semblance plot is a colour display of a hyperbolic Radon transform.)

In contrast, the CMP semblance (above and below) shows that a large range of velocities will produce a reasonable stack, however multiple energy, etc. will also be stacked.

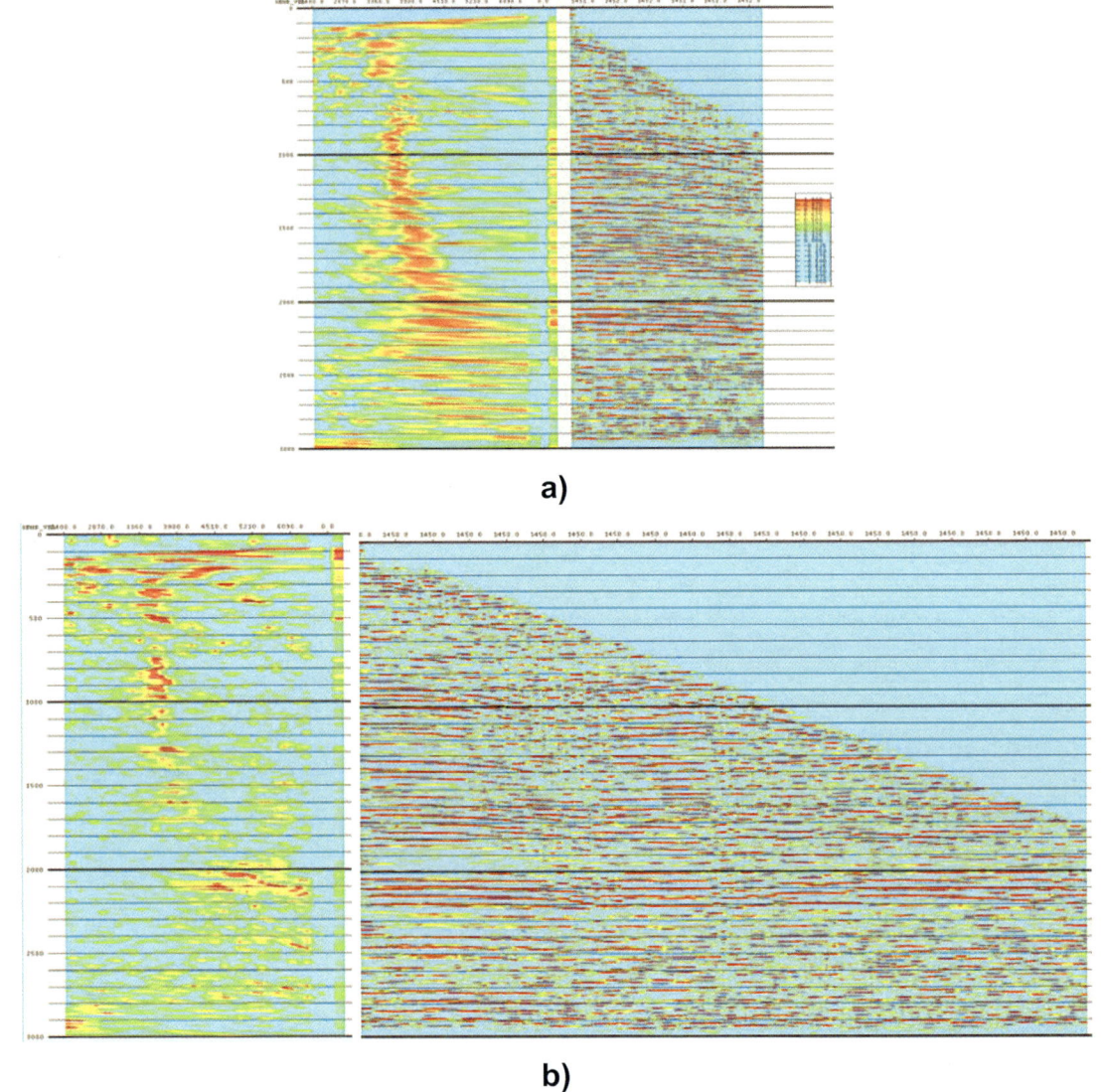

Figure 11.22 High resolution semblance plots of a) a CMP gather and b) a CSP gather

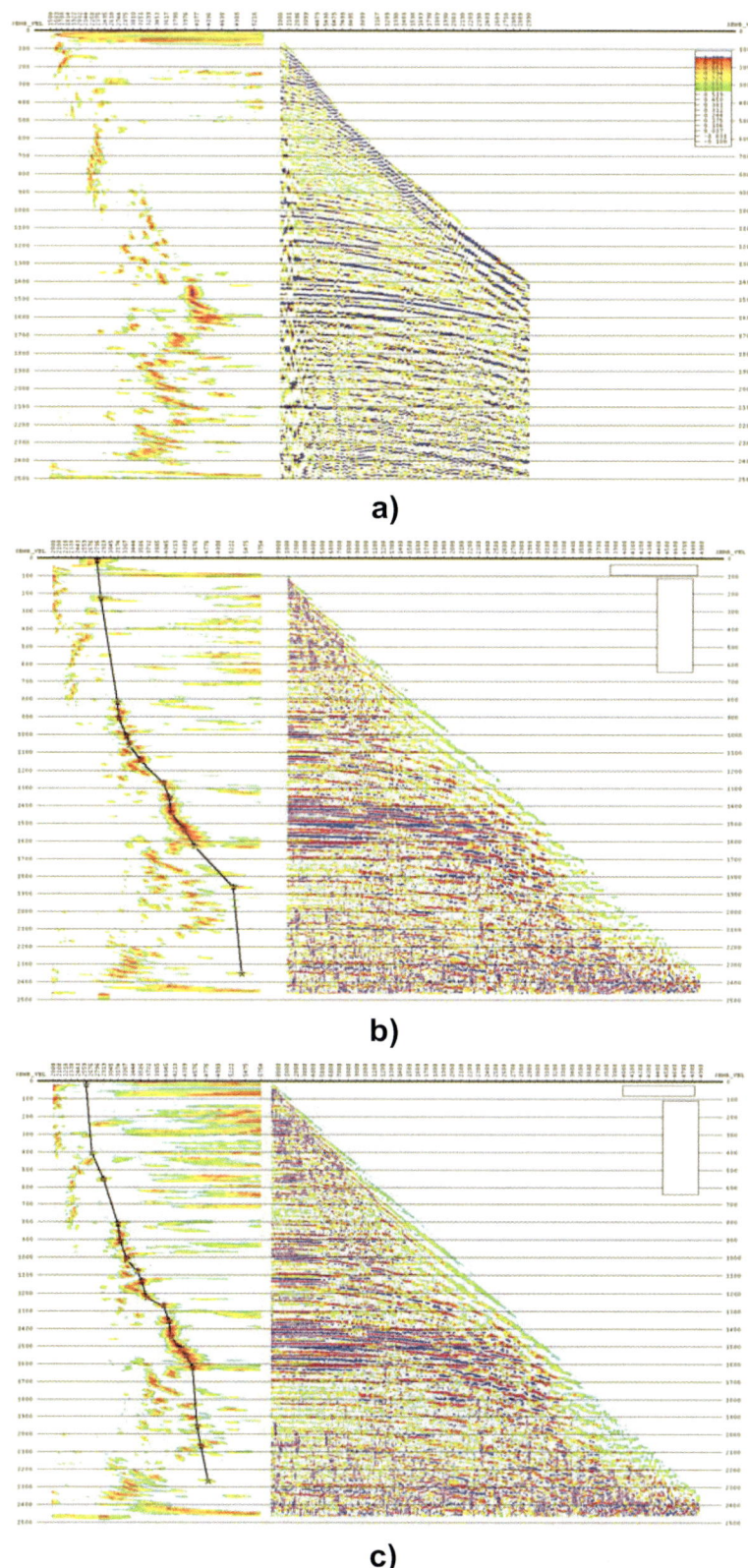

Figure 11.23 Semblance and a) a CMP gather, b) a SCP gather using infinite velocities, and c) the after one velocity iteration using b) as input.

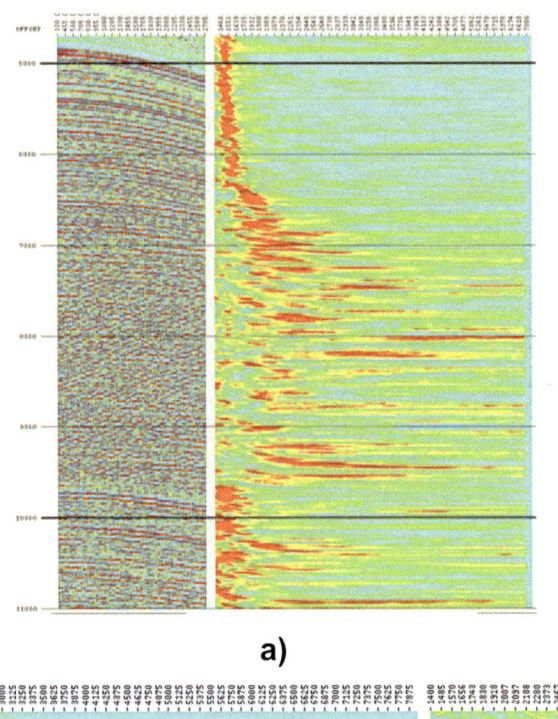

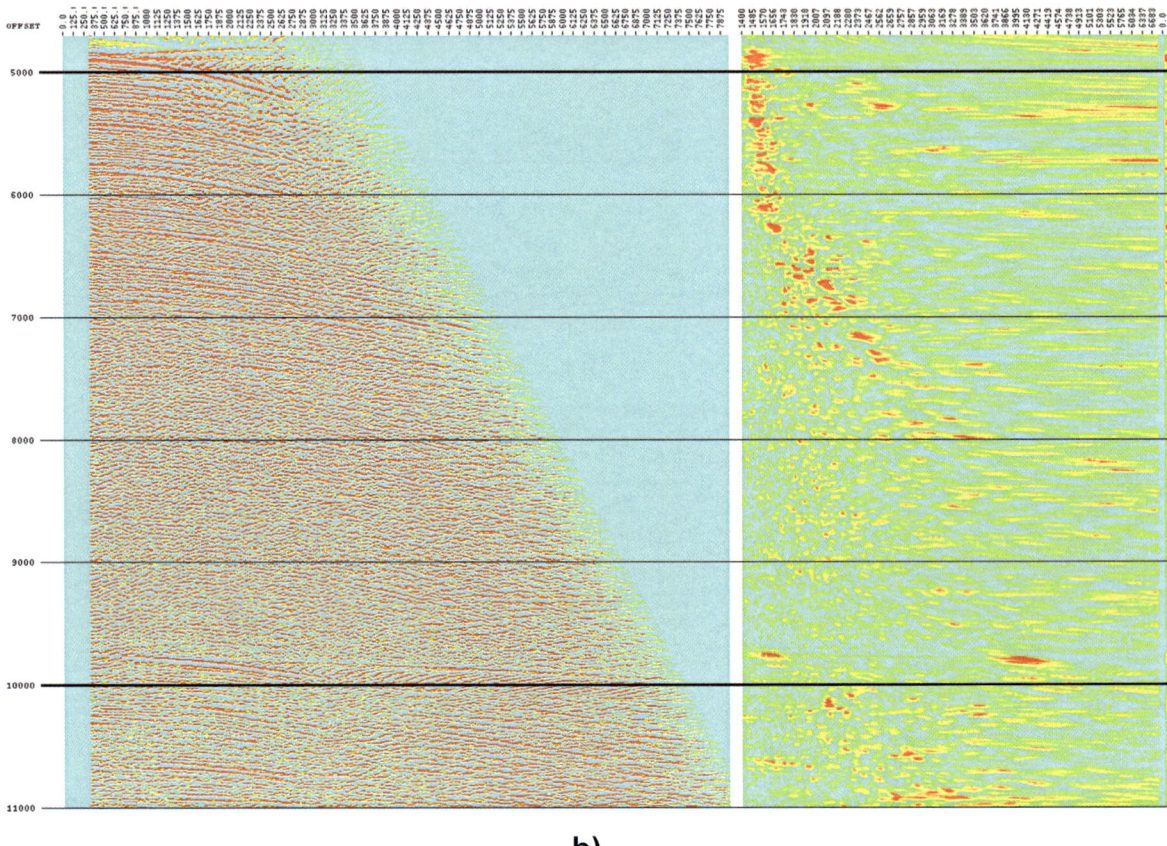

Figure 11.24 Crustal study data illustrating in a) a CMP gather and semblance, and b) the corresponding CSP gather and semblance. Note the possible moho reflection at a higher velocity than the multiple energy just above 10.000sec.

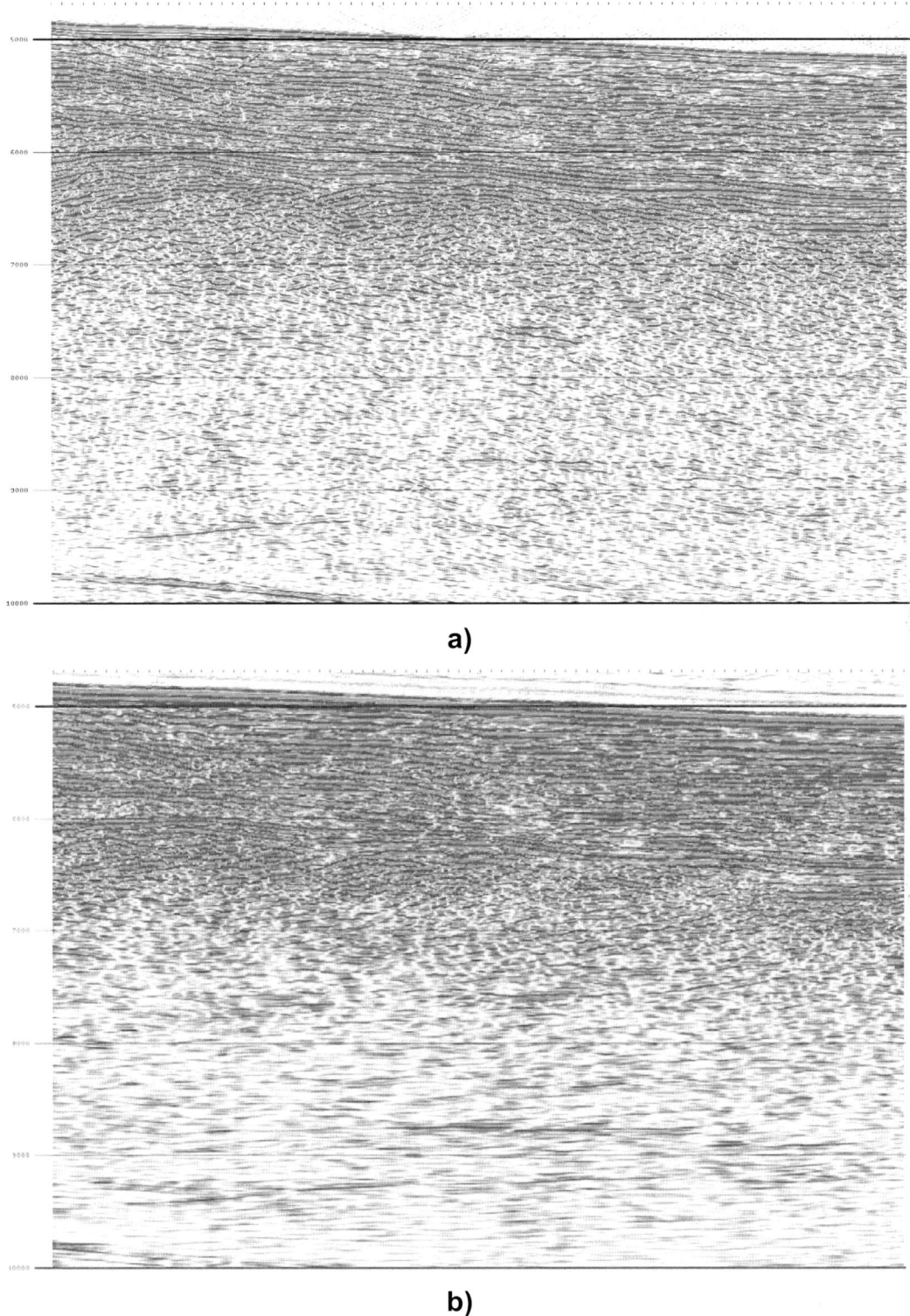

a)

b)

Figure 11.25 Stack and CSP stack (EO prestack migration) of the crustal study data

11.4.2 Data Examples and Comparisons with EOM

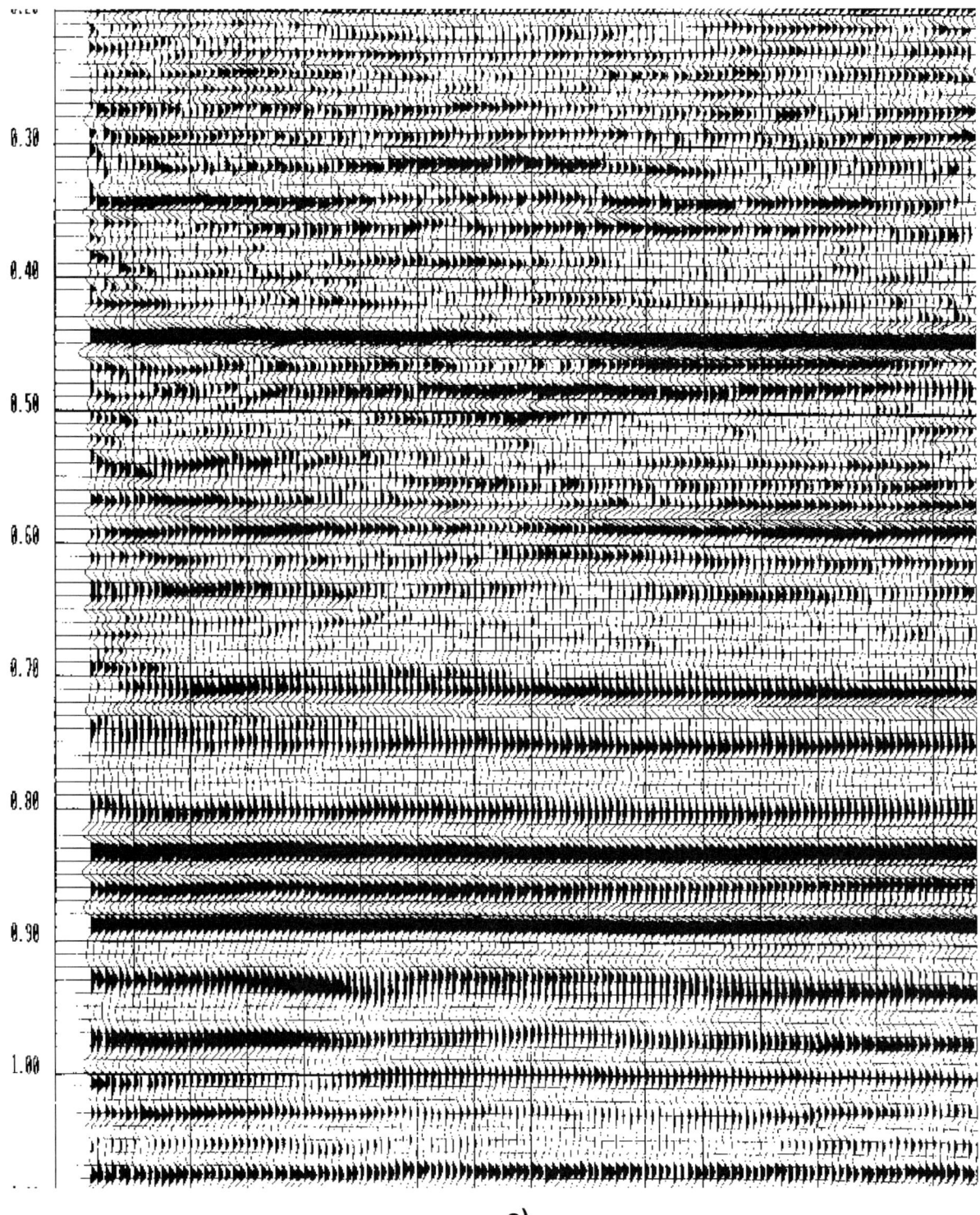

a)

Figure 11.26 Comparison of "railroad" data with a) a poststack migration, and b) an EOM prestack migration.

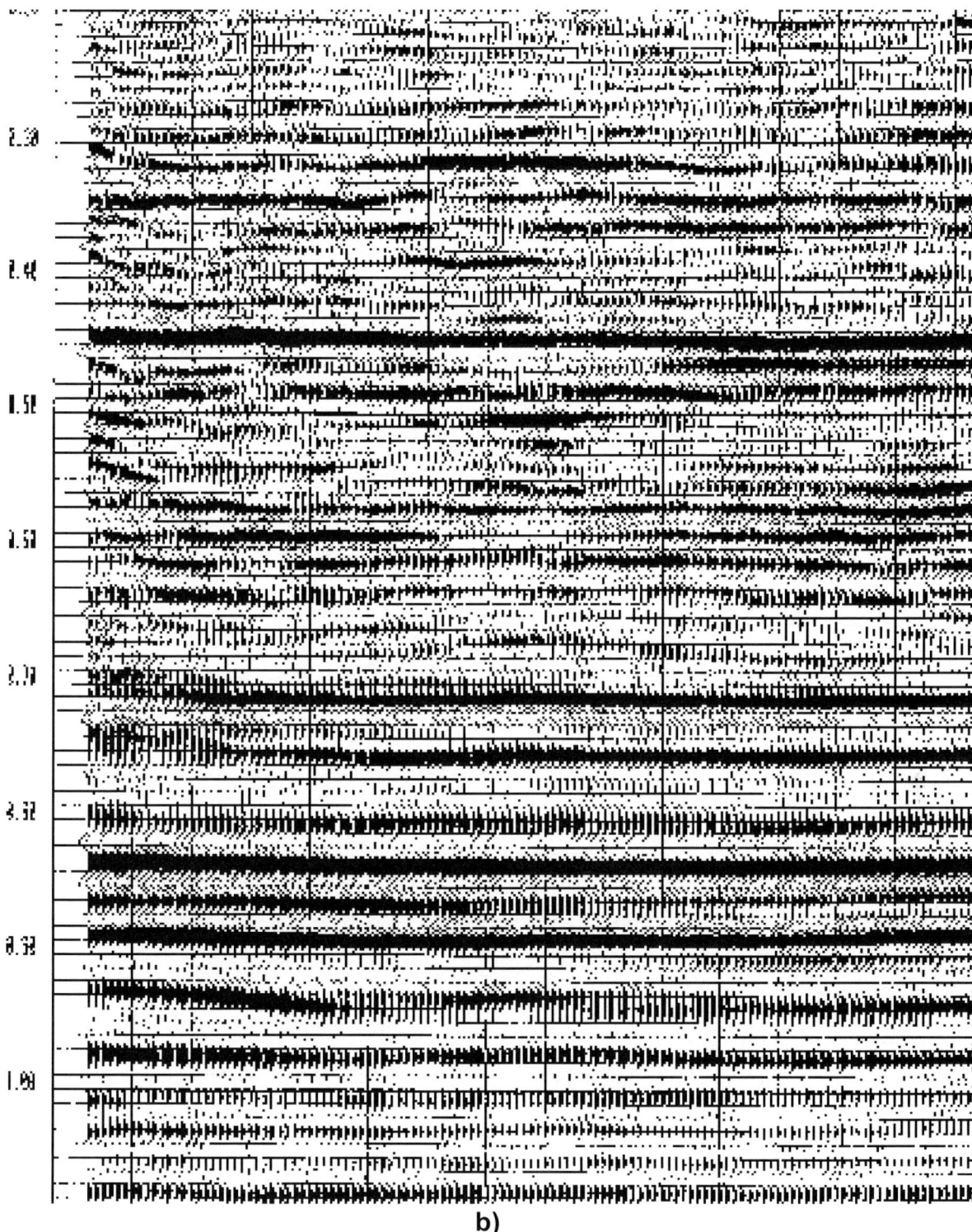

b)

Note the <u>improved resolution</u> of the events between times of 0.600 and 0.700 s identified by the arrows, and the improved signal-to-noise ratio.

Channel sand; previous examples of CSP gathers and their semblance were taken from this data set.

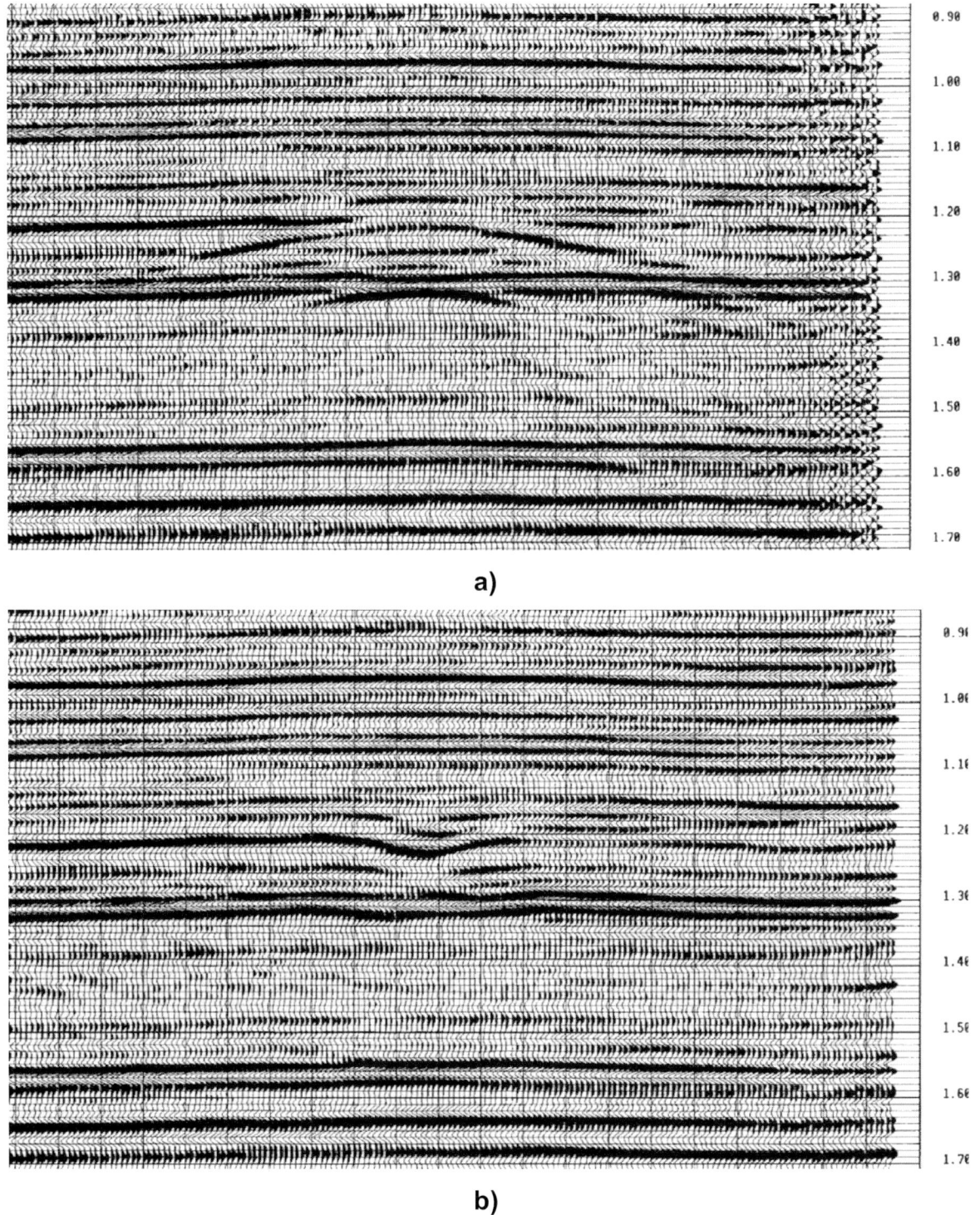

Figure 11.27 A channel sand example; a) showing the stacked section, b) conventional Kirchhoff prestack migration, c) EOM direct to stack simulating (b), and d) EOM prestack migration with fold division and Kirchhoff weighting.

Compare (b) and (c); noise reduction in (c) but similar character of the channel.

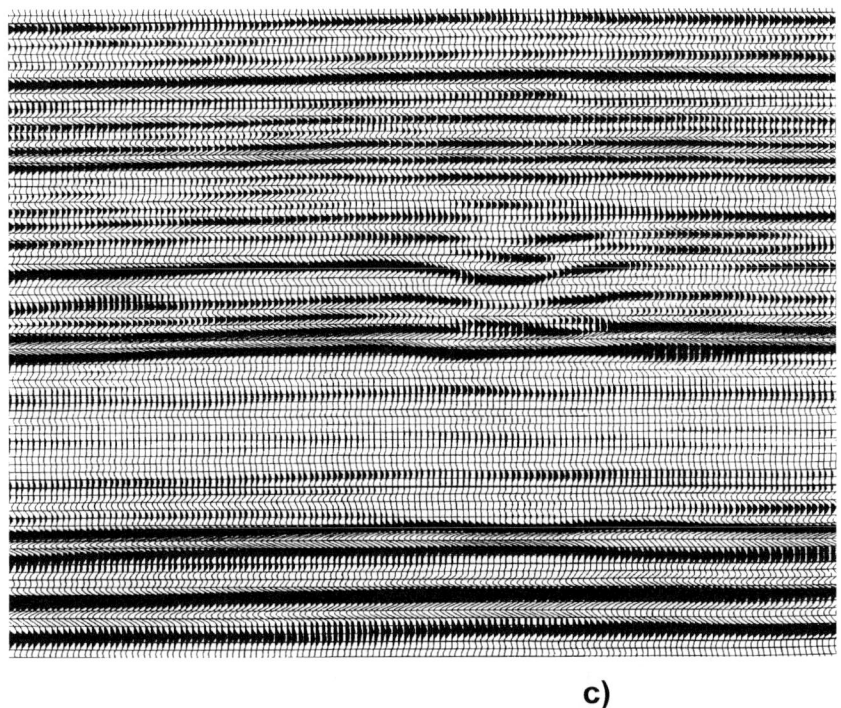

c)

d)

Compare (c) and (d); improved character of the channel in (d). This difference in these two EOM examples is due to the amplitude scaling of the CSP gathers.

11.5 3-D Processing

The previous discussion has assumed 2-D data. Figure 11.28 shows a plan view of a source *S*, receiver *R*, and scatterpoint *SP*.

Extension to 3-D assumes:

- Traveltimes are <u>independent of the azimuthal</u> direction of the raypaths.

- The source-scatterpoint-receiver traveltime *T* would be the same for any source on the circle C_s, or any receiver on circle C_r.

- The kinematics of 3-D geometry are made equivalent to 2-D geometry by using <u>radial distances</u> from the scatterpoint to the source d_s and scatterpoint to receiver d_r.

- New 3-D variables $\hat{x}$ and $\hat{h}$ are now computed from the scalars d_s and d_r by

$$\hat{x} = \frac{d_s + d_r}{2}, \qquad (11.18)$$

and

$$\hat{h} = \frac{|d_s - d_r|}{2}. \qquad (11.19)$$

The values of $\hat{x}$ and $\hat{h}$, computed from 3-D geometry, may be substituted for the 2-D values of *x* and *y* in equation (7) to compute the 3-D equivalent offset.

Azimuthal information of the original source and receiver raypaths is lost in the above method, but can be preserved by forming multiple CSP gathers that are ordered by azimuth.

The time domain formulation of the method allows any single CSP gather to be formed using the entire 3-D input data. An arbitrary 2-D line may be fully 3-D prestack migrated by computing only those CSP gathers that fall on the line.

Chapter 11 Equivalent offset migration (EOM)

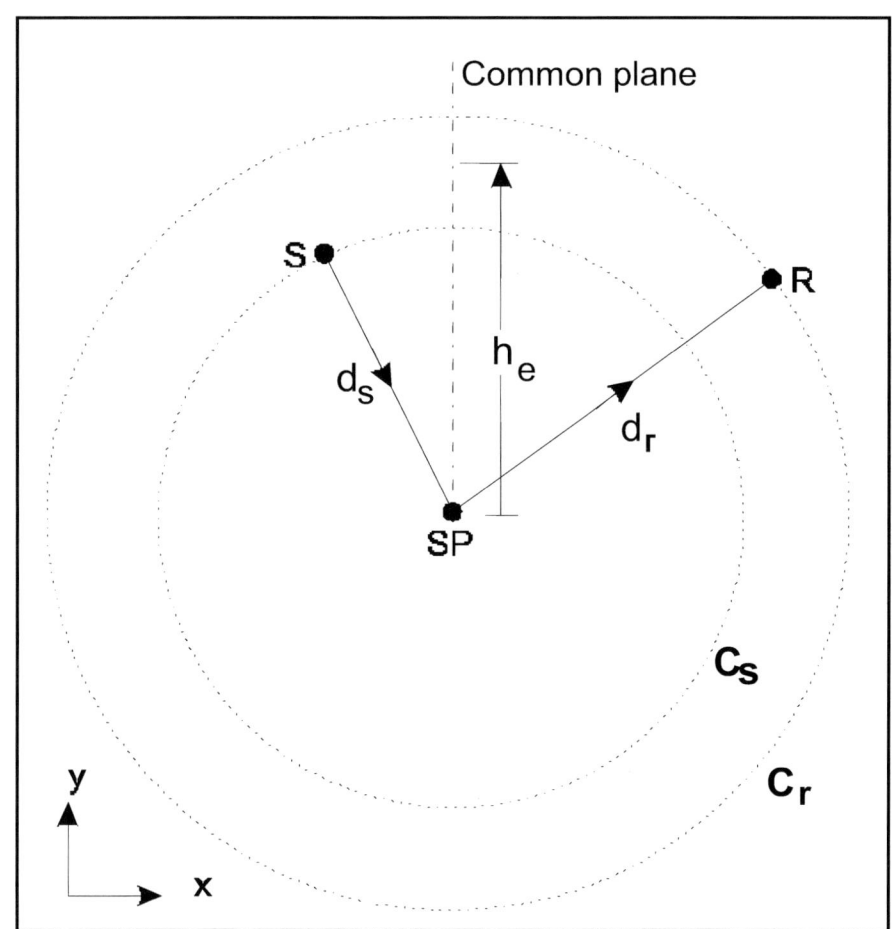

Figure 11.28 Plan view of 3-D data.

Example of 2-D lines extracted from a 3-D volume. The following images contain a cross-section of a reef.

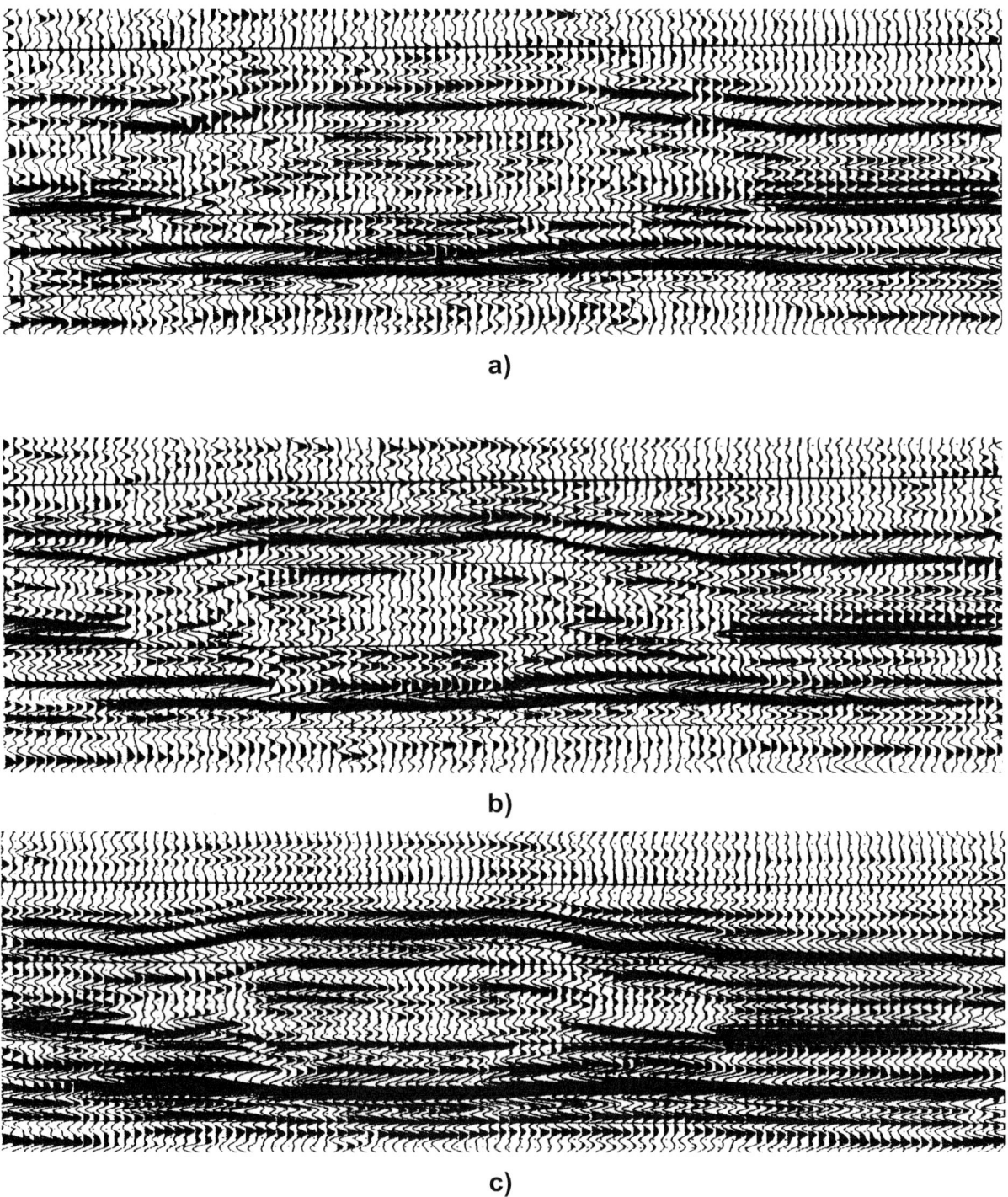

Figure 11.29 Reef example with a) the stacked section, b) a 3-D poststack migration, c) an EOM prestack migration.

11.6 Depth migration examples (including anisotropy)

The principles of equivalent-offset prestack time migration may be applied to depth migration (Chernis [707]) by using the source and receiver traveltime maps of traditional depth migration. The CSP gathers may be formed in time (t) for additional velocity analysis, or in depth (z) for direct stacking.

- Choose a scatterpoint location (vertical array of scatterpoints in depth).
- Estimate the two-way traveltimes T (t or z) for each scatterpoint from the travel-time maps.
- Estimate the vertical zero-offset traveltimes T_0 (t or z) using the average velocity.
- The equivalent offset is computed using the following equations:

Time CSP gather

$$T^2 = T_0^2 + \frac{4h_e^2}{V_{ave}^2(T_0)}, \tag{11.20}$$

The input traces are added to a CSP gather with no time shifting using equation 11.20 to define the equivalent offset. After the CSP gathers are formed, hyperbolic Kirchhoff MO and stacking forms a time trace that may be converted to depth directly using the average velocities.

Depth CSP gather

$$T^2 = \frac{z^2 + 4h_e^2}{V_{ave}^2(T_0)}, \tag{11.21}$$

The input traces may be mapped from the input timer T to the corresponding depth z at an equivalent offset given from equation 11.21. Weighting and stacking completes the prestack migration.

Inclusion of the anisotropic parameters into the raytracing or traveltime computations for each source and receiver allow the total traveltime T to include the effects of anisotropy.

The effect of including anisotropy in the computation of traveltimes is illustrated in Figure Figure 11.30 with data acquired from a <u>physical model</u> of dipping anisotropic material (Isaac and Lawton 1998 [709]). The model consisted of a block of transversely isotropic (TI) phenolic material with a slow velocity axis of symmetry dipping at an angle of 45 degrees. An anomaly in the base simulates a target reef or fault. The scaled thickness is 1500m and the anisotropic effects of the dipping beds produce a 300 m lateral displace the anomaly. The following two-sided CMP and CSP gathers have been normal moveout (MO) corrected, and are ready for stacking.

Figure 11.30a and (b) are <u>CMP</u> gathers, where (a) was formed with an <u>isotropic</u> algorithm, and (b) formed with an <u>anisotropic</u> algorithm. Similarly, figures (c) and (d) are <u>CSP</u> gathers that have been formed with isotropic and anisotropic algorithms. The offset of the CMP gathers range from –2000m to +2000m, while the CSP gathers range from –4000m to +4000m.

The effect of anisotropic migration can be observed by the difference in amplitudes on the left and right sides of the CMP gathers in (a) and (b). This results from the two-sided gathers being formed from the positive or negative migration offset rather than source-receiver offset. Traces on the left side of the gather migrated into that gather from the left and vice versa.

Note that energy in the isotropic CSP gather (c) is not flat. This is most probably due to velocity errors caused by the isotropic assumption. Also note the data appears to be symmetric about zero offset. The anisotropic CSP gather (d) has laterally displaced energy that is now flatter because the migration has moved the data to the left or in the down-dip direction of the anisotropic overburden.

Figure 11.31 shows the comparison of the effects on anisotropy on CMP and CSP gathers for real data. Figure 11.32 then shows a comparison of EOM prestack migrated depth sections with (a) showing the isotropic case, and (b) the anisotropic case.

Note that EOM depth migration requires the computation of traveltime maps similar to conventional prestack depth migrations and runs much slower that the EOM time migrations.

The formation of the <u>time CSP gathers</u> (for depth migration) does save computation time by delaying the AAF, scaling etc. until after the gathers are formed.

The inclusion of anisotropy in the velocity model has the potential for more accurate spatial and depth positioning.

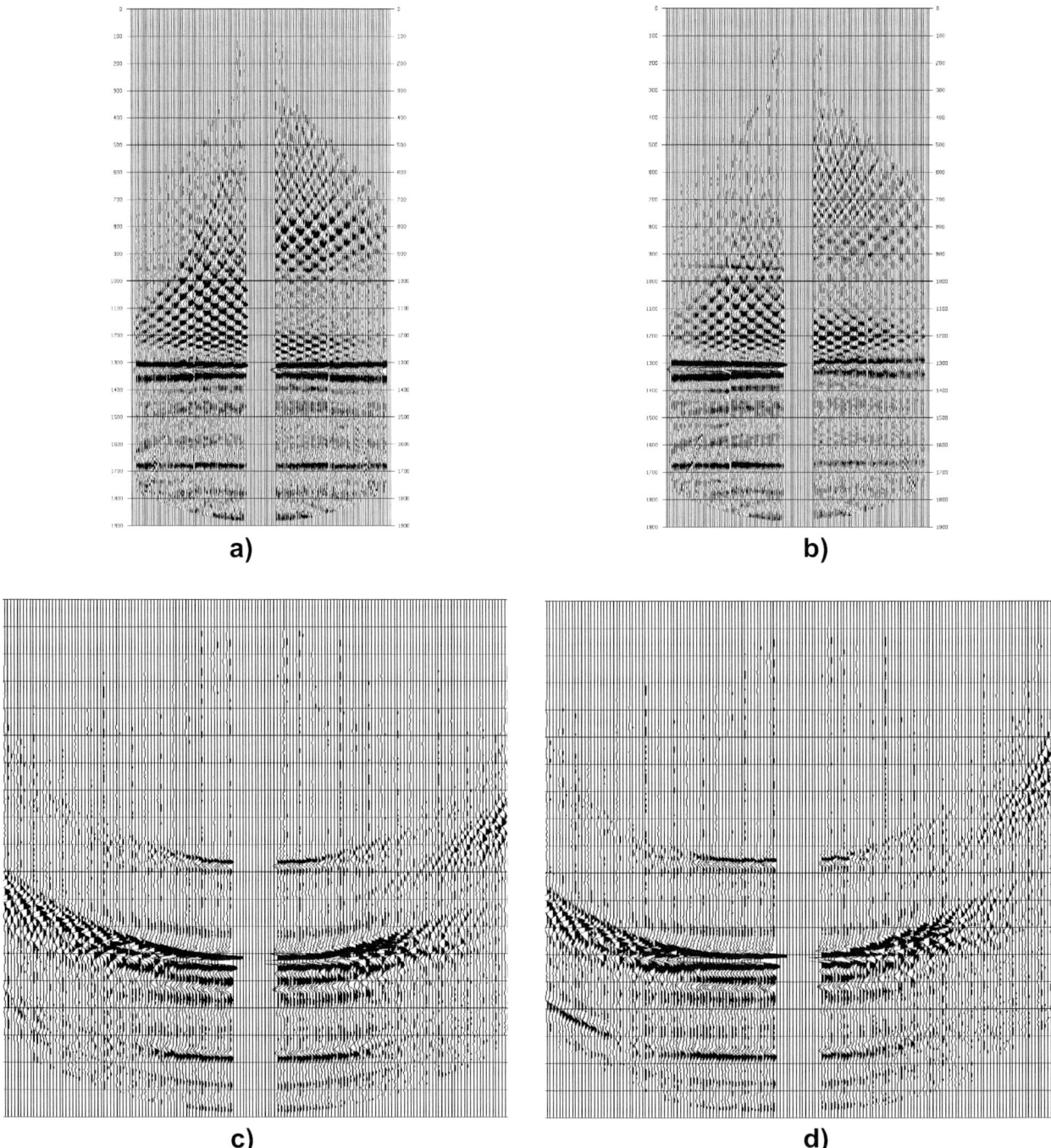

Figure 11.30 Comparison of CMP and CSP gathers that first ignore then include anisotropic traveltime computations, i.e. a) isotropic CMP gather, b) anisotropic CMP gather, c) isotropic CSP gather and d) anisotropic CSP gather.

Chapter 11 Equivalent offset migration (EOM)

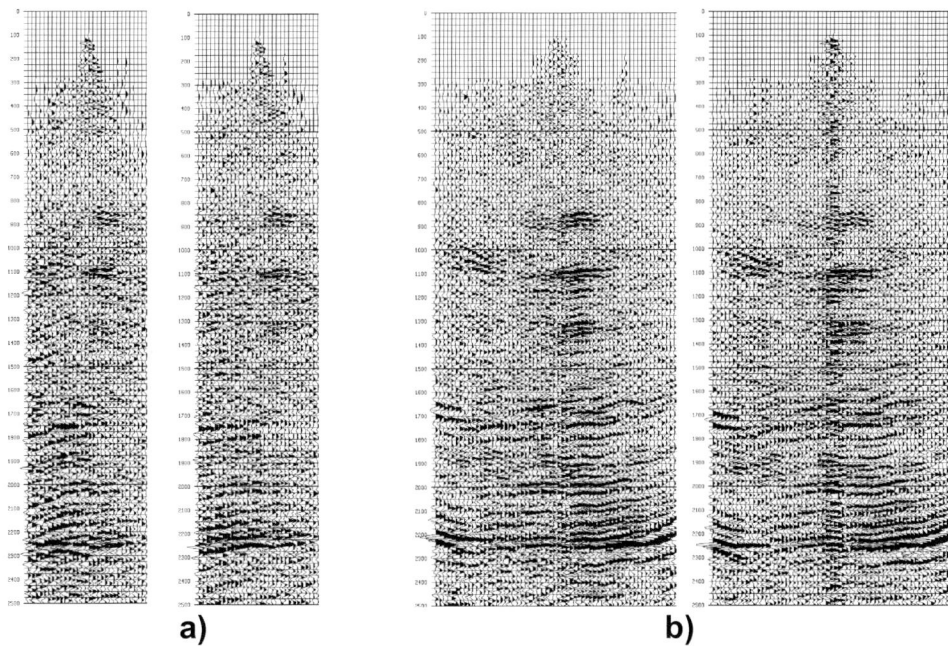

Figure 11.31 Real data comparison: a) an isotropic CMP gather is compared to an anisotropic CMP gather, and b) an isotropic CSP gather is compared to an anisotropic CSP gather

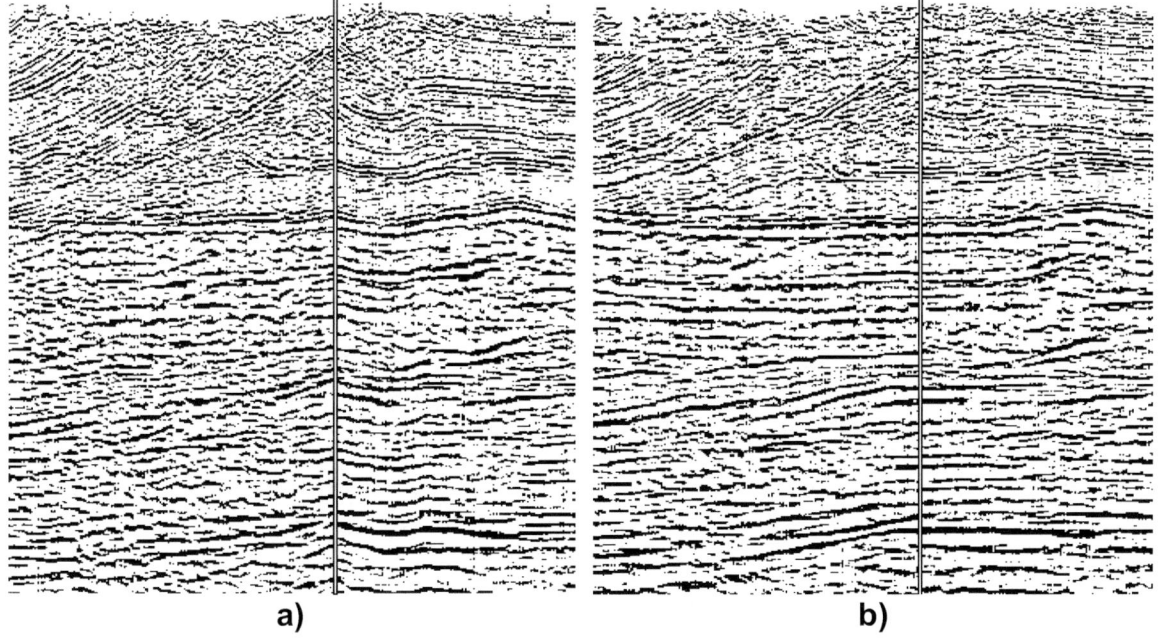

Figure 11.32 EOM sections using a) isotropic velocities, and b) anisotropic velocities. Note the relative continuity in the area indicated.

11.7 Rugged Topography Processing

The equivalent offset method of migration may also be performed from surface by computing the equivalent offset at a datum set for each CSP gather. The equivalent offset is found by equating the original source-receiver traveltimes with a colocated source and receiver on a datum, as illustrated in Figure 11.33.

The total traveltimes for the original raypaths are computed using the appropriate offsets h_s and h_r. The zero-offset time for the source $T_0 + t_s$, the zero-offset time for the receiver $T_0 + t_r$, and appropriate velocities for the source V_{srs} and receiver V_{rec} defined from the surface enable T to be computed from

$$T = \left[\left(\frac{T_0}{2}+t_s\right)^2 + \frac{h_s^2}{V_{srs}^2(T_0+t_s)}\right]^{1/2} + \left[\left(\frac{T_0}{2}+t_r\right)^2 + \frac{h_r^2}{V_{rec}^2(T_0+t_r)}\right]^{1/2}, \quad (11.22)$$

where t_s and t_r are the vertical traveltimes from source and receiver to the datum. The equivalent offset h_e is computed on the datum with velocity V_{sc} by solving,

$$T = 2\left(\frac{T_0^2}{4} + \frac{h_e^2}{V_{sc}^2(T_0)}\right)^{1/2}, \quad (11.23)$$

giving

$$h_e = \frac{V_{sc}(T_0)}{2}\left(T^2 - T_0^2\right)^{1/2}. \quad (11.24)$$

The definition of surface may be defined as:
1. The actual surface elevation.
2. The elevation of a floating datum.

The first method would assume an accurate model of the shallow subsurface such as that given by refraction statics.

The second method assumes the data was processed to a floating datum in which the near-surface geology was replaced with a constant velocity. Migration may then be performed from the floating datum, or with an elevation correction from the floating datum to the actual surface.

An example of rugged topography EOM is shown in Figure 11.34.

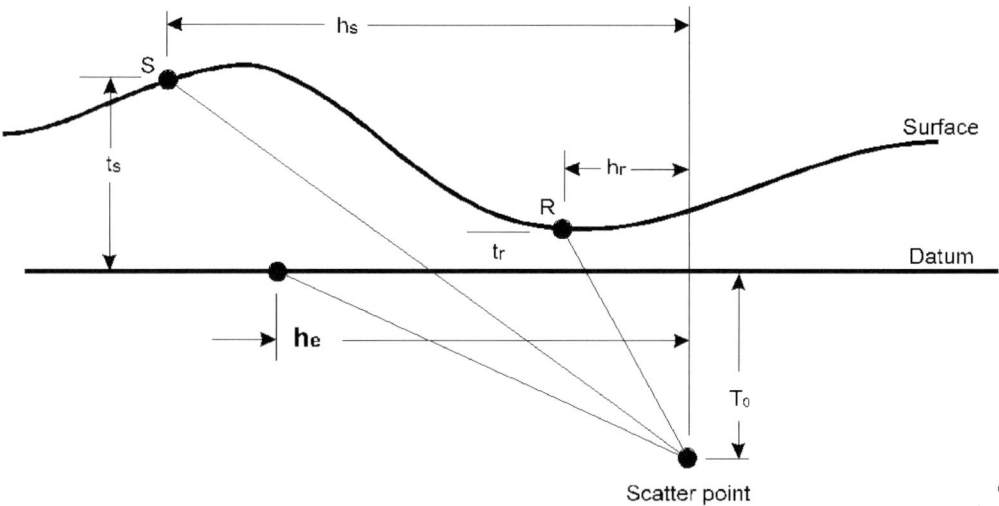

Figure 11.33 Raypaths, offsets, and traveltimes for computing the equivalent offset with rugged topography.

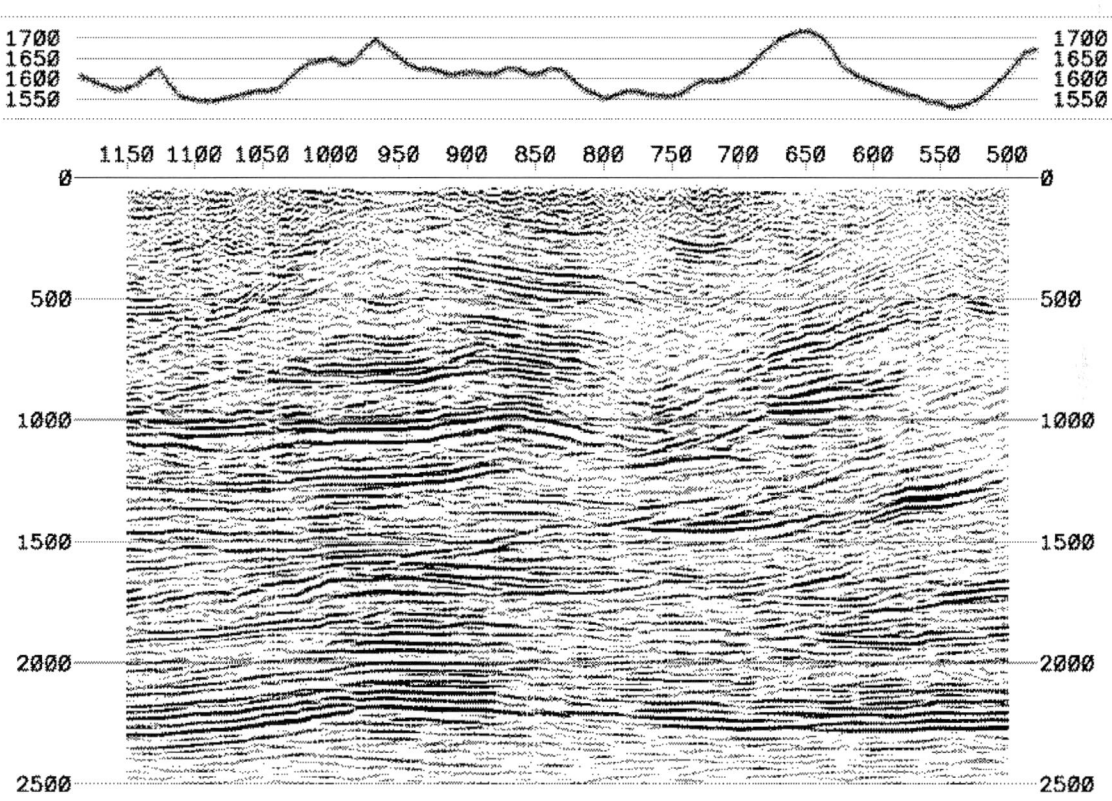

Figure 11.34 Rugged topography EOM of an Alberta foothills 2-D data set.

Additional information may be found in Geiger and Bancroft [557] and [624].

11.8 Residual Statics Before MO
11.8.1 Modelled data

Conventional residual statics methods require the application of MO to the input traces followed by correlation with a model trace that has usually been formed from a filtered brute stack.

Li and Bancroft [618]) have demonstrated an algorithm for the computation of residual static corrections *before* the application of MO by using model traces from CSP gathers.

Source records were created from a linear reflector model. Each source and each receiver were shifted by statics with a linear distribution between ± 20 ms to give a maximum static range of 80 ms as shown in the stacked section in Figure 11.35a. The result using EOM statics is shown in (b).

The migration aperture may be much larger than the CMP aperture and will tend to give a better low frequency estimation. However, it is recommended that elevation and refraction statics be applied as with conventional processing to aid in the estimation process.

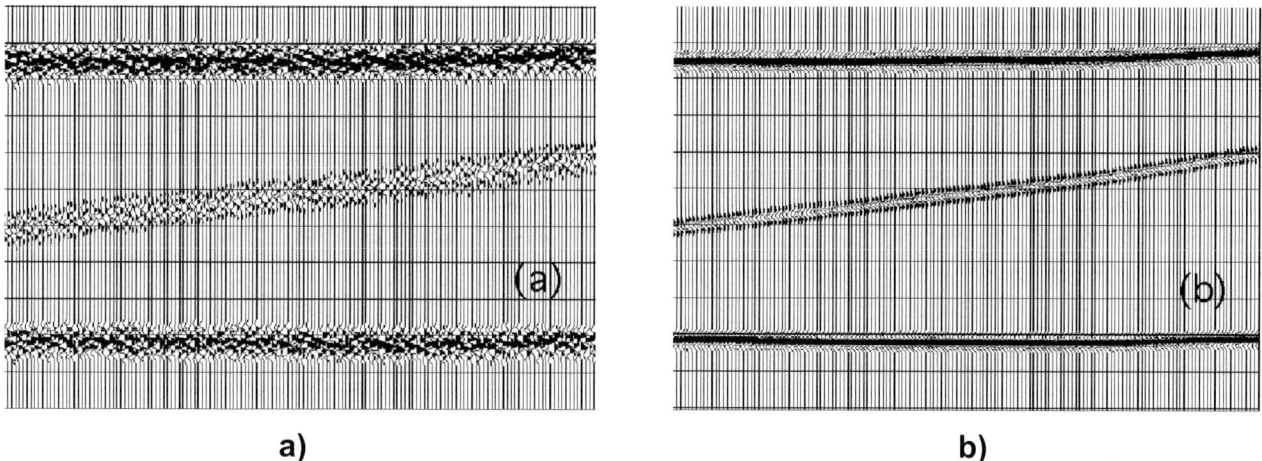

Figure 11.35 Example of residual statics removal for numerical model with a) showing the modelled section with the statics, and b) the section after the statics have been estimated and removed.

11.8.2 Real data

The following pages show one real data source record at various stages of EOM statics processing. The data is taken from a 2-D line that has an elevation change of 40 m and is located in the Blackfoot area of Alberta, Canada. The data quality is very good. Because the line was relatively flat, no elevation or first break statics were applied to the data before EOM statics were estimated.

Every input trace has a <u>unique model trace</u>. Cross correlation yields the total static time of each trace before MO correction.

Figure 11.36 on the following page show:

(a) raw source record with only gain recovery applied,

(b) decon, filtered and AGC'd source record,

(c) model source record formed by combining traces after CSP gathers have been formed,

(d) application of estimated statics to decon, filtered and AGC'd source record with a correlation window from 800 to 2000 ms.

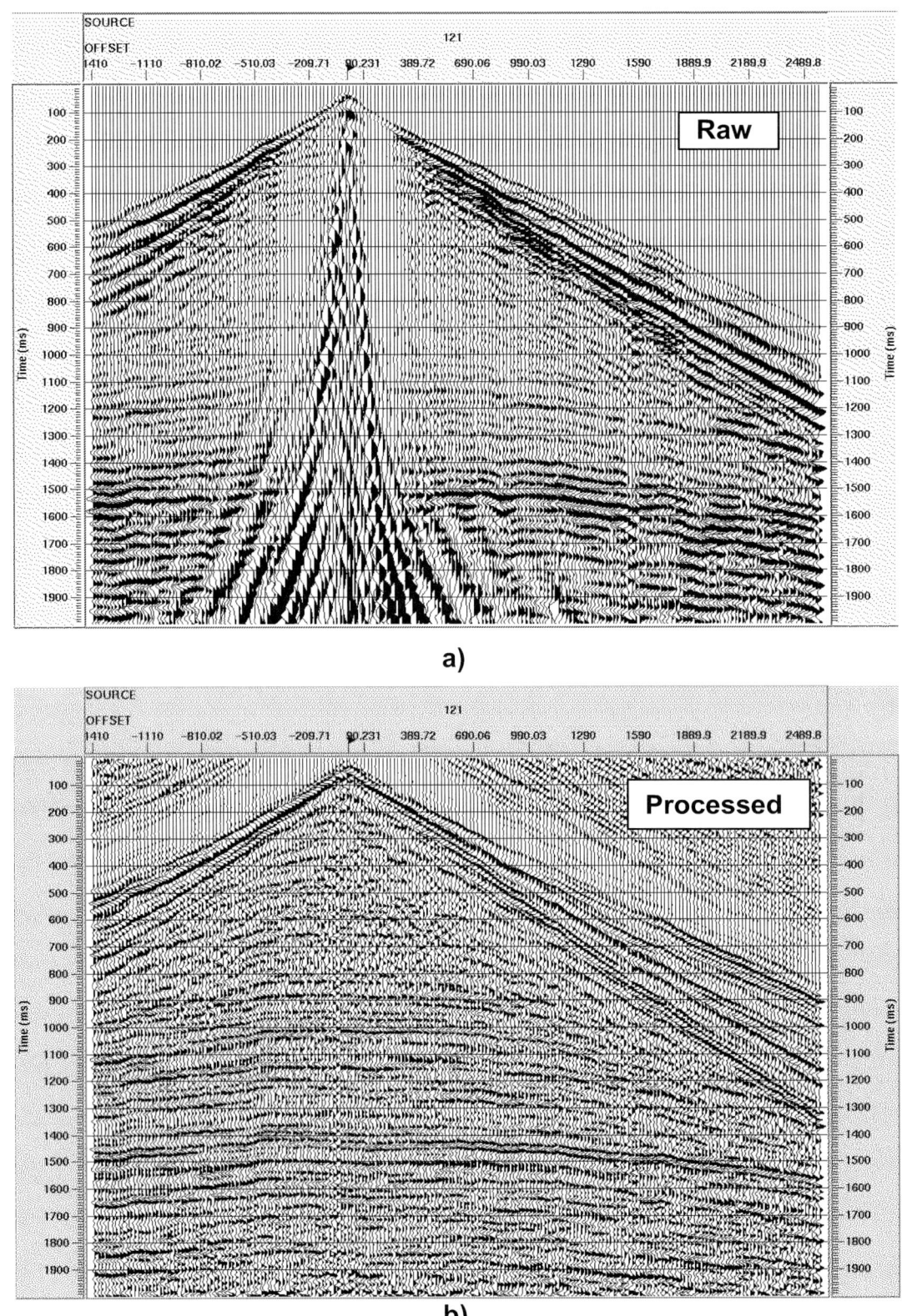

Figure 11.36 Real data source record; a) original record, b) deconvolution, filtered, and AGC'd, c) EOM model, and d) with statics applied to (b).

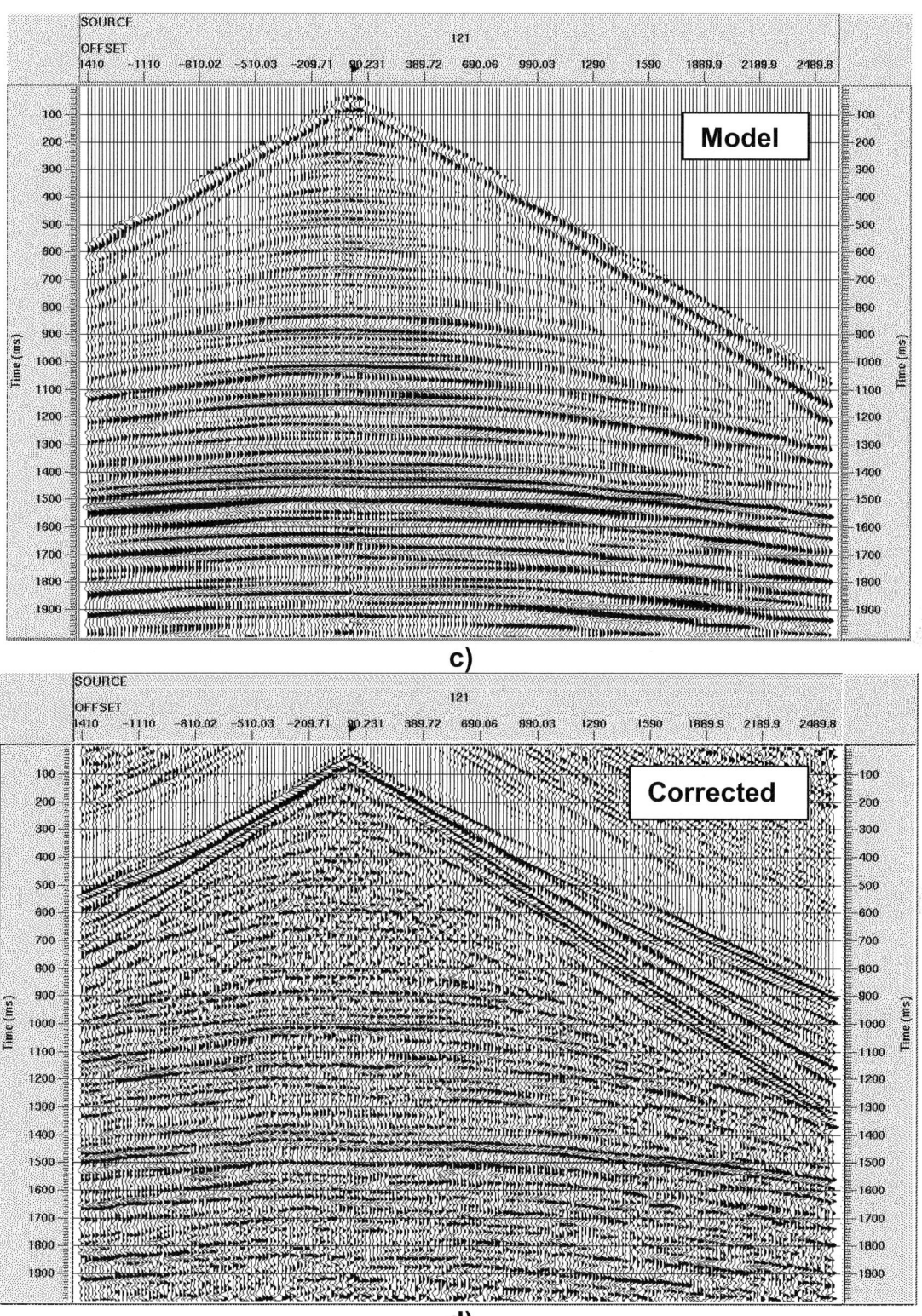

c)

d)

Conventional residual statics estimation is performed **after MO correction** and usually requires:

- a time-variant estimation
- a velocity term, and
- iterative solutions between statics and velocities

EOM statics are computed prior to MO correction and are

- valid over a larger time window
- independent of velocity, and
- require no iterative solutions.

Note in the preceding example the improved static corrections from the **first breaks** to **maximum time**.

Static results from above were applied to input traces, which were then conventionally processed and compared with data processed with elevation, first break, and trim statics. There were no perceivable differences in the stacked sections.

On the opposite page, Figure 11.37 compares the statics estimated using a correlation time window from 800 to 2000 ms with one using 1200 to 1700 ms.

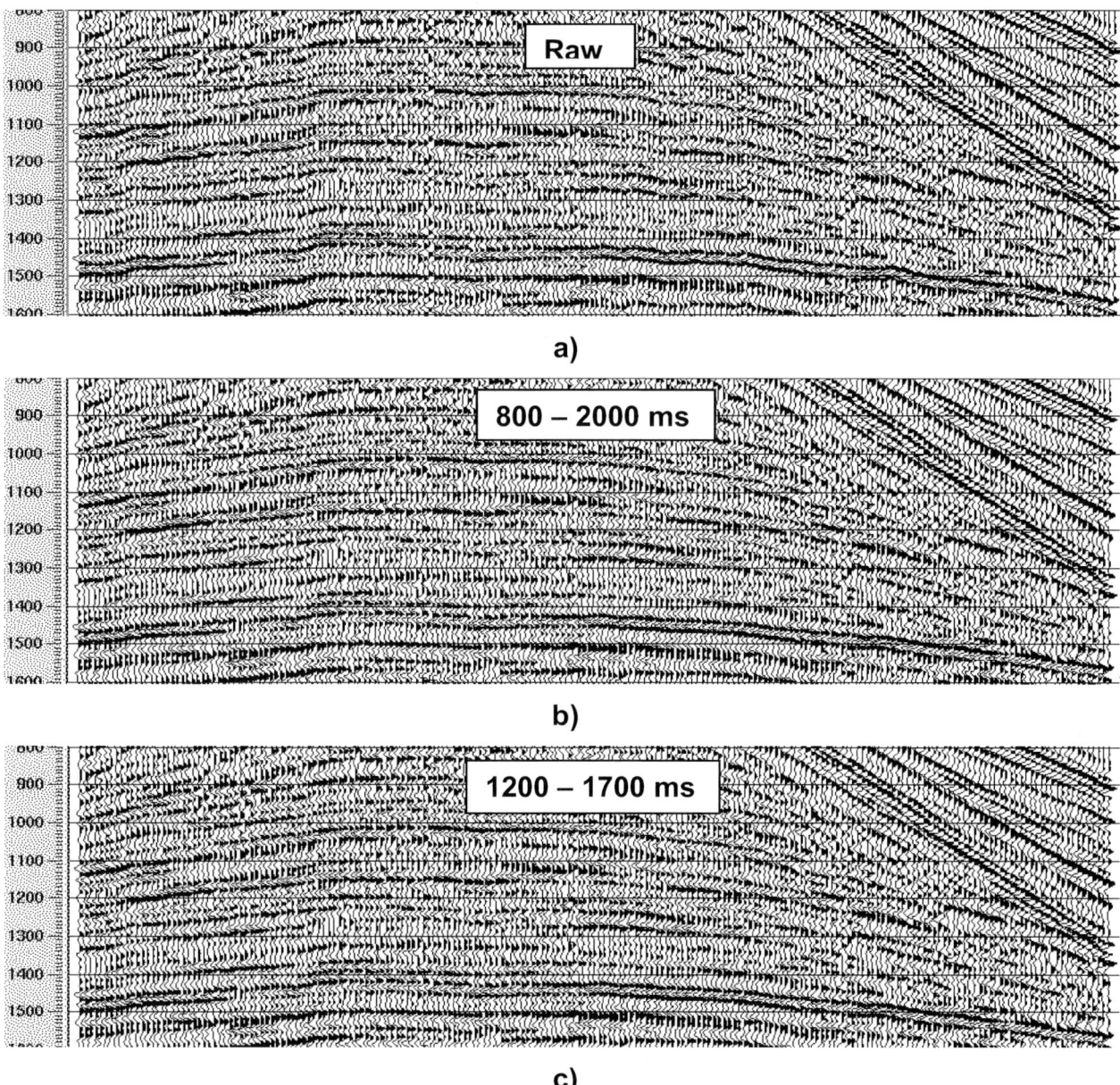

Figure 11.37 Comparison of a) the input source record with b) EOM statics estimated with an 800 to 2000 ms correlation window, and c) with a 1200 to 1700 ms correlation window.

11.8.3 Statics for structured data (Marmousi data)

The Marmousi data was created with no static errors to test migration algorithms and velocity analysis techniques. The data is quite structured and conventional CMP gathers are difficult to analyze for velocity information. Forming model traces for statics analysis is also difficult, especially when the velocity is not known.

Conventional residual statics analysis packages were applied to the Marmousi data (no synthetic statics). All attempts to estimate the zero statics failed, and instead gave estimated statics that ranges to the maximum allowed by the defined maximum static window.

EOM statics analysis was performed with an infinite-velocity assumption. The estimated statics tended to less than 1 ms, however, a few values approached 3 ms.

Synthetic statics ±20 ms were applied to the sources and receivers. Once again the conventional methods failed to yield any useful results. The EOM results that assumed infinite input velocities were reasonably successful as illustrated in Figure 11.38.

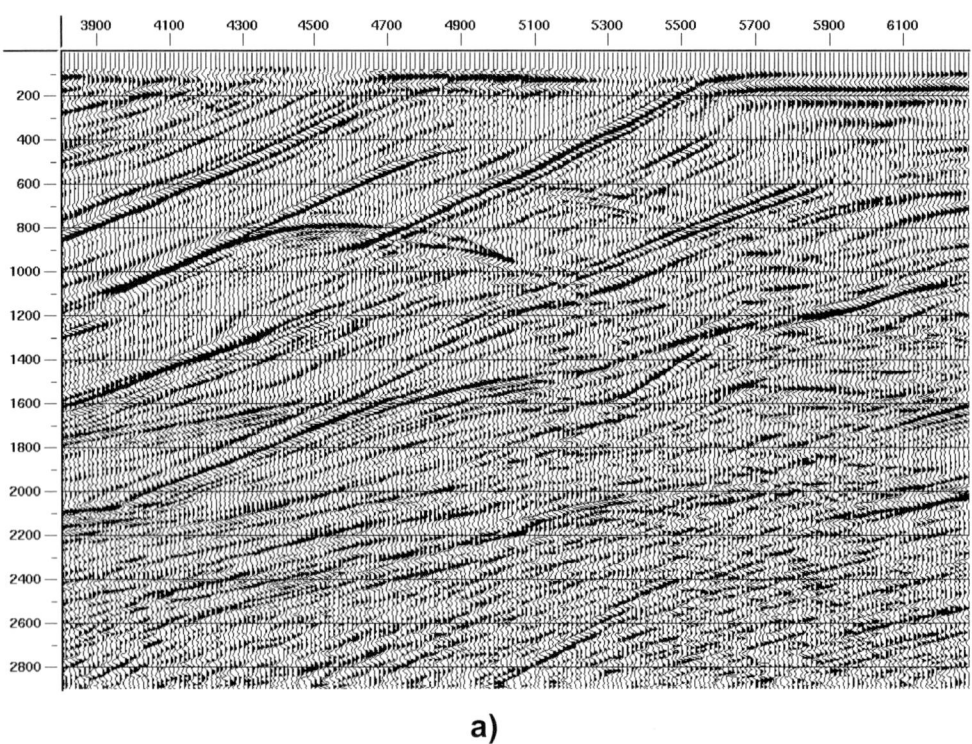

a)

Figure 11.38 The Marmousi data showing a) the stacked section, b) the stacked section with synthetic statics, and c) the stacked section with EOM statics.

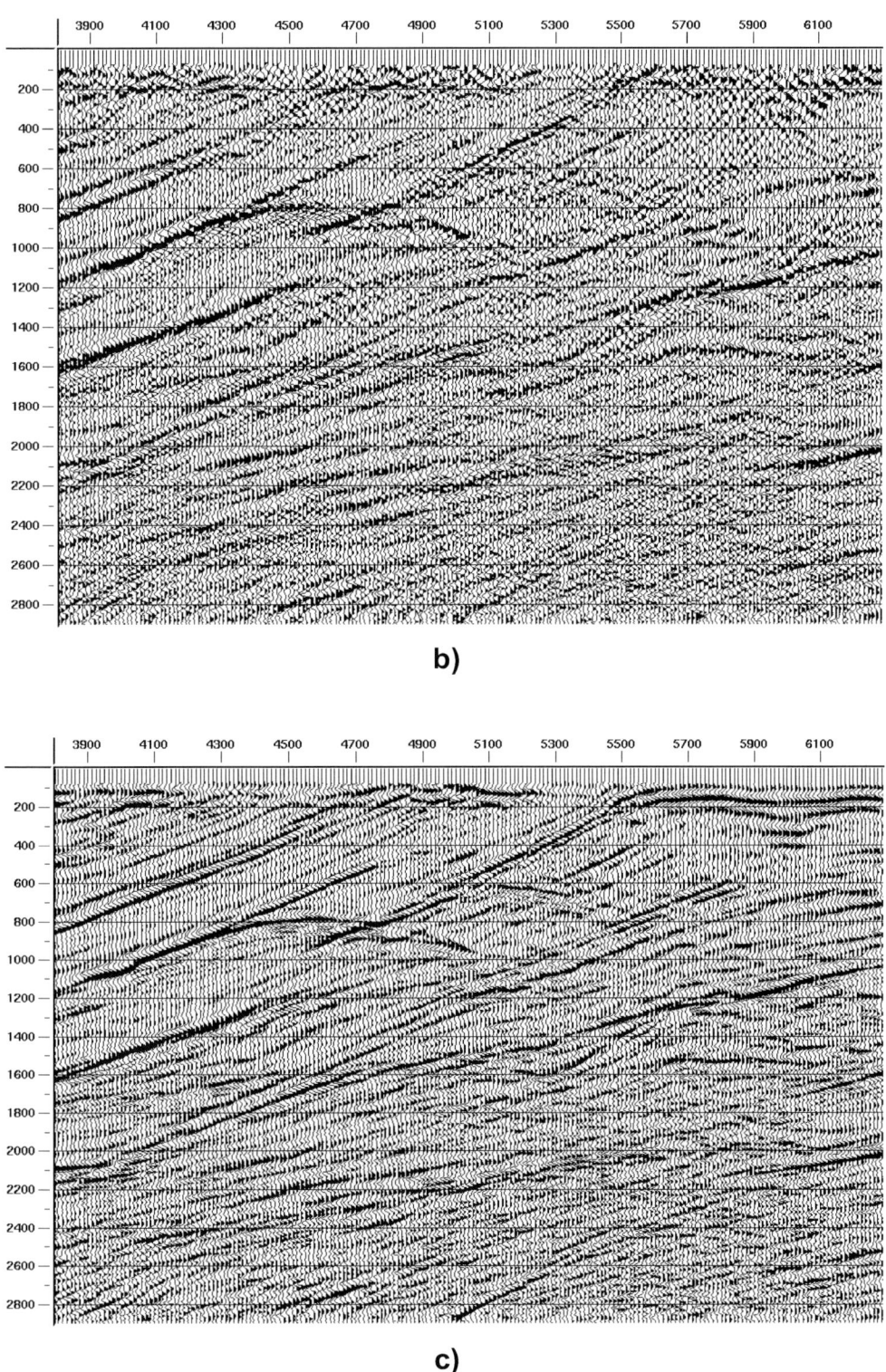

b)

c)

Some low frequency statics remain, but much of the high frequency information was recovered.

11.9 Converted-wave Processing

11.9.1 Converted-wave processing using EOM

The principle of equivalent offsets may be applied to <u>converted-waves</u> where the down going energy from the source is a P-wave and the reflected or scattered energy is a shear or S-wave. We now refer to the scatterpoint as a <u>conversion scatterpoint</u>.

The assumption of linear raypaths and RMS velocities is maintained, with RMS velocities defined as V_p for the P-wave and V_s for the S-wave. A ratio of these velocities γ is defined as $\gamma = V_p/V_s$.

The <u>law of reciprocity</u> does not apply to converted-waves, i.e., the locations of the source and receiver cannot be interchanged as the respective raypaths are different. In addition, the reflection or conversion point for horizontally layered media is not at the midpoint creating significant difficulty in processing the data and estimating the velocities Wang [633].

The concepts of EOM greatly simplify the processing of converted-wave data.

The total traveltime T is computed from the double square root equation using the P- and S-wave velocities. The time T is also equated with a colocated source and receiver at an equivalent offset h_{ec}, i.e.,

$$T = \frac{(\widetilde{Z}_0^2 + h_{ec}^2)^{1/2}}{V_p} + \frac{(\widetilde{Z}_0^2 + h_{ec}^2)^{1/2}}{V_s} = \frac{(\widetilde{Z}_0^2 + h_s^2)^{1/2}}{V_p} + \frac{(\widetilde{Z}_0^2 + h_r^2)^{1/2}}{V_s}, \quad (11.25)$$

or

$$T = \left(\frac{1+\gamma}{V_p}\right)(\widetilde{Z}_0^2 + h_{ec}^2)^{1/2} = \frac{(\widetilde{Z}_0^2 + h_s^2)^{1/2}}{V_p} + \frac{\gamma(\widetilde{Z}_0^2 + h_r^2)^{1/2}}{V_p}, \quad (11.26)$$

where Z_0 is the approximate pseudo depth of the P- and S-wave data on a time section. Eliminating the velocity and solving for h_{ec} gives

$$h_{ec} = \left\{\left[\frac{\gamma}{1+\gamma}(\widetilde{Z}_0^2 + h_s^2)^{1/2} + \frac{1}{1+\gamma}(\widetilde{Z}_0^2 + h_r^2)^{1/2}\right]^2 - \widetilde{Z}_0^2\right\}^{1/2} \quad (11.27)$$

The left side of equation (11.24) may be written in hyperbolic form

$$T^2 = T_{0-ps}^2 + \frac{4h_{ec}^2}{V_{ps}^2}. \tag{11.28}$$

where

$$T_{0-ps} = \frac{(1+\gamma)\widetilde{Z}_0}{V_p}, \tag{11.29}$$

and

$$V_{ps} = \frac{2V_p}{1+\gamma}. \tag{11.30}$$

Once again the <u>double square root equation has been reduced to a hyperbolic form</u>. The common-conversion-scatterpoint (CCSP) gather is created with no time shifting, however the scattered converted-wave energy lies on hyperbolic paths. After the CCSP gathers have been formed, the values of T_{0-ps} and V_{ps} are estimated using conventional velocity analysis techniques. When combined with conventional P-P processing, where T_0 and V_p are known, the RMS values of γ or V_s may be estimated from equation (11.30), i.e.,

where

$$\gamma = \frac{2V_p}{V_{ps}} - 1, \tag{11.31}$$

and

$$V_s = \frac{V_p V_{ps}}{2V_p - V_{ps}}. \tag{11.32}$$

The velocities V_p, V_s, and V_{ps} are all RMS-type velocities and could be used to estimate <u>interval velocities</u>.

In addition, the zero-offset times T_0 from P-P processing can be combined with the zero-offset times T_{0-ps} from the converted-wave processing to also estimate <u>interval values</u> for γ and the velocity.

The processing of converted-waves is simplified to a gathering process that provides velocity information for conventional MO type processing. In addition, the results are prestack migrated.

Additional information and data examples may be found in [526], [556], [623], [626], [633], and [621].

11.9.2 Valhall converted-wave data

Converted-wave data is illustrated with 4-C data from Valhall offshore Norway [649] and [650]. The data was acquired with an ocean-bottom dragged array that had three-component geophones and one hydrophone at each receiver location.

The hydrophone data was processed with conventional P-P velocities and is shown in Figure 11.39. This data compared well with the vertical component data. Data in the center of the section is distorted when the P-waves pass through a gas chimney.

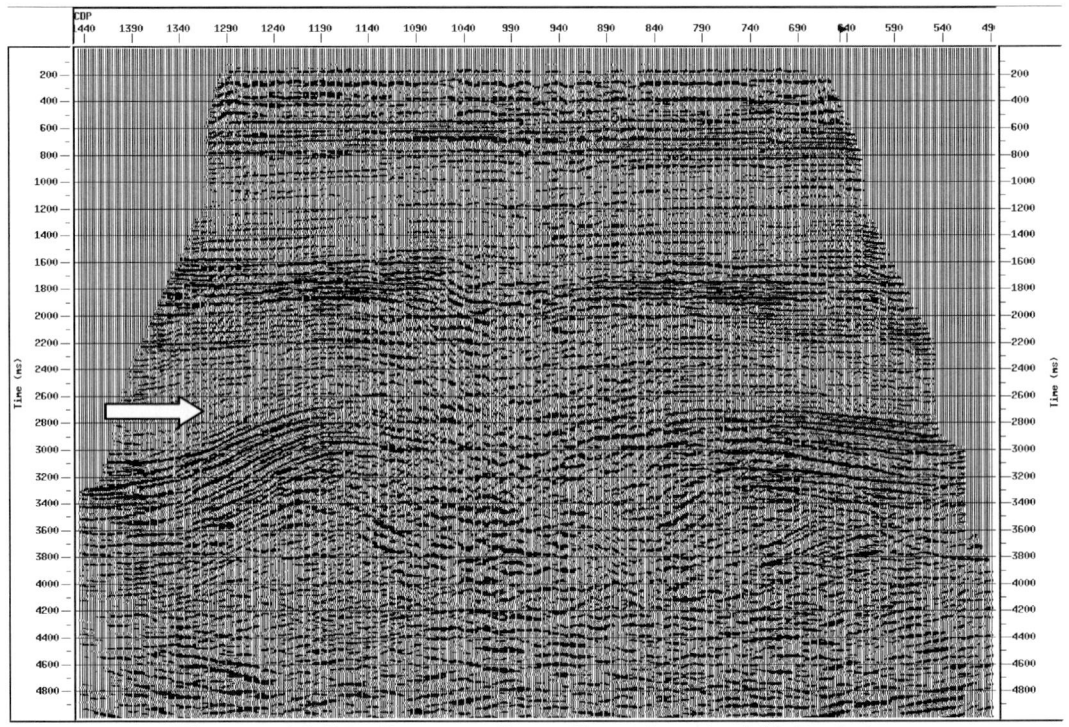

Figure 11.39 Conventionally migrated P-P data.

<u>Ocean-bottom acquisition</u> is necessary to <u>acquire shear wave reflections</u> (as fluids do not propagate shear waves). The main shear-waves are generated at the primary reflector and the gas chimney does not affect their velocities.

Figure 11.40 shows processing of the radial component by:

(a) asymptotic binning and poststack migration,

(b) converted-wave DMO, conventional stack, and poststack migration,

(c) equivalent-offset migration for converted-waves.

Compare the same event at P-P time of 2.7 with the P-S time of 5.5 s as indicated by the arrow.

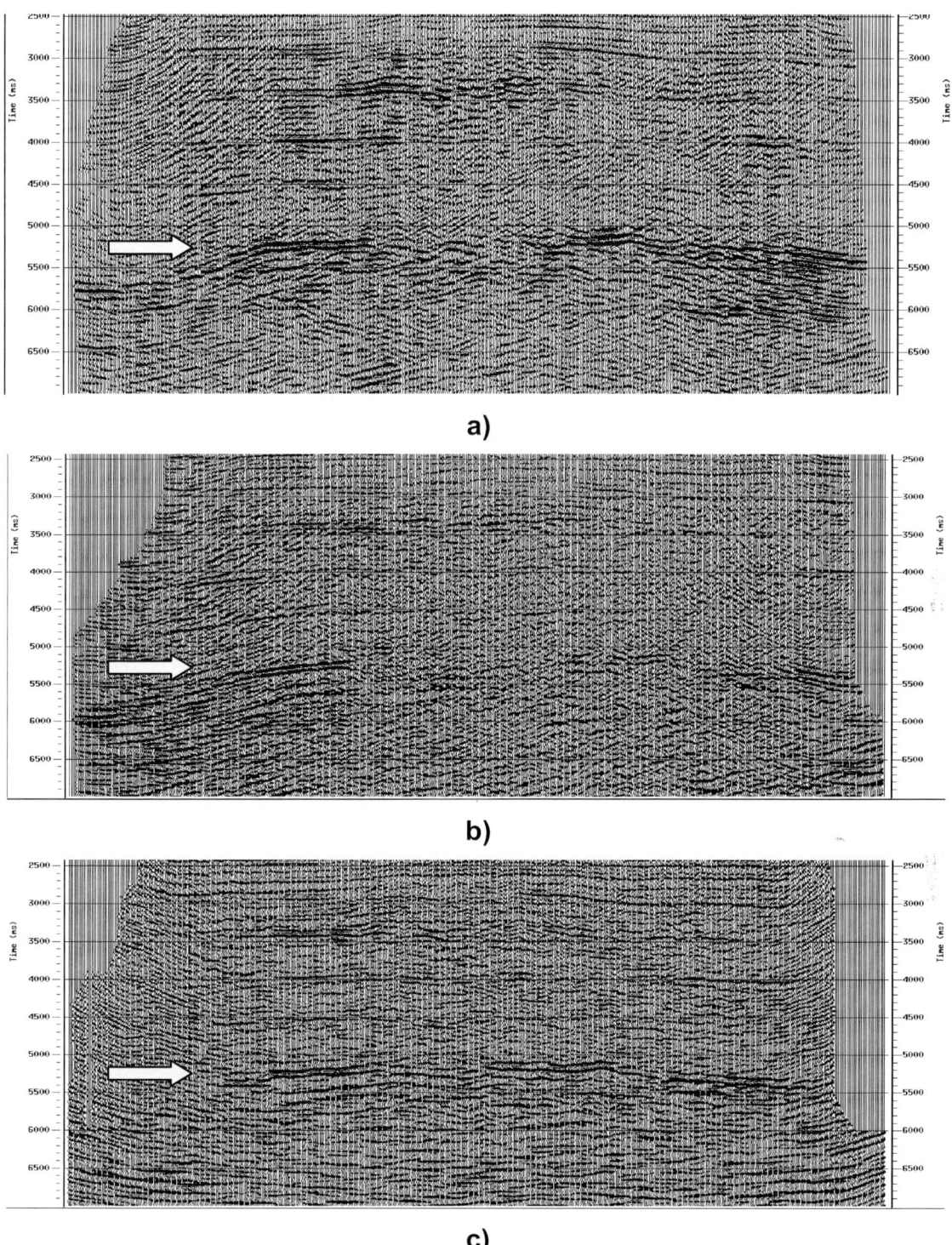

Figure 11.40 Converted-wave P-S processing by a) asymptotic binning and poststack migration, b) converted-wave DMO, conventional stack, and poststack migration, and c) equivalent offset migration for converted-waves. The arrow identifies the area of interest.

11.10 Other Features of EOM

11.10.1 Crooked line processing

2-D acquisition often requires the line to follow existing roadways or riverbanks in rugged terrain. Processing this data is difficult because of the distribution of CMP bins. As with any Kirchhoff migration <u>3-D EOM</u> can be <u>prestack migrate</u> directly to an <u>arbitrary line</u> or swath (Geiger et al [635]).

11.10.2 Fold count and amplitude scaling

A benefit of EOM is the ability to maintain an accurate fold count for <u>each sample</u> in the CSP gathers. The fold is used to balance the amplitudes of each sample in the CSP gather to improve the focussing in areas with irregular geometries (Geiger et al [635]).

11.10.3 Vertical array prestack migrations

VSP data that is recorded (or shot) in deviated wells, or marine data that has been recorded with vertical arrays of hypdrophones may be prestack migrated using the EOM process as reported in [700], [714], and [715].

11.10.4 Forming a zero offset section with no MO correction

Rotation of the input data to the CSP gathers may also be rotated to zero offset where scaling and stacking can create a stacked "zero offset" section. Note that no MO correction is required. These stacked sections are in time (x, t) and are ready for poststack time or depth migration. Papers that discuss a similar process are [631] and [624].

When combined with velocities derived from CSP gathers, the stacked "zero offset" section may provide a <u>fast and accurate time or depth migration</u>.

11.10.5 DMO

If desired, the equivalent offset may be used to distribute input data that converts Cheops pyramids to hyperboloids: a DMO operation.

11.10.6 Velocity analysis

A few CSP gathers may be formed at selected locations using crude velocities, and these gathers used to estimate more accurate velocities. These new velocities could be used to reform the selected CSP gathers and then repeat the velocity analysis. In practice, velocity estimation converges rapidly (Bancroft and Geiger [553], [632]) and typically, only one iteration is required. After a suitable input velocity is found, the entire set of CSP gathers can be formed after which velocity analysis may again be applied to selected CSP gathers to determine the optimum velocity for Kirchhoff MO.

Velocity analysis of CSP gathers compares the recorded traveltime with the migrated output traveltime. In contrast, velocity analysis on CMP gathers uses only a portion of this traveltime differential. An alternative view is that CSP velocity analysis is based on the distances from the sources and receivers to the CSP location, and not on the source/receiver offset.

The increased offset range and high fold of the CSP gather may create higher resolution semblance plots than those from CMP methods. This improved resolution is a consequence of the increased resolving power of prestack migration and implies that velocities must be accurate to maximize the prestack migrated energy.

11.10.7 Natural antialiasing

Energy from a scatterpoint will map to a hyperbola on a CSP gather as illustrated in Figure 11.41.

- When many traces are binned, the equivalent offset jitter is averaged over a bin width δh.
- This is equivalent to convolving a bin-centered trace with a boxcar filter of width δt that is proportional to the slope on the hyperbola. This process attenuates higher frequencies at steeper dips.
- The bin size δh could be chosen to limit the attenuated frequencies to those above the signal bandwidth.

The loss of high frequency with dip may appear to be a detriment to the CSP gathering process, but it is actually a benefit (Bancroft [554], [625]). It has similar properties to a temporal filter designed to attenuate spatially aliased frequencies as required for Kirchhoff migration. In effect, it <u>provides a natural antialiasing filter</u> for the data in a CSP gather. In practice, the bin spacing may equal that of the 3-D subsurface grid or the 2-D CMP spacing.

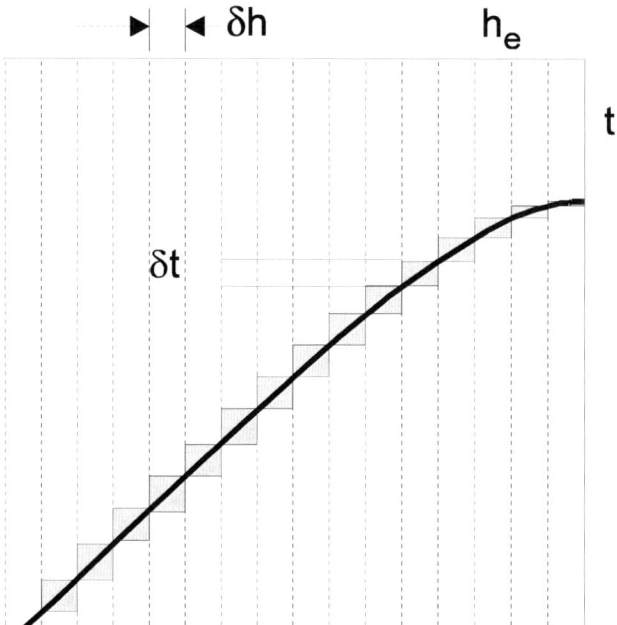

Figure 11.41 CSP gather showing the scatterpoint hyperbola and the time smear caused by binning.

11.11 Benefits of EOM Processing

The data in the CSP gather lie in hyperbolic paths as in a CMP gather.

Only hyperbolic NMO correction is required to move energy to zero-offset.

Prestack migration has been reduced to:
- a trace <u>gathering</u> process,
- followed by <u>Kirchhoff MO</u>.

Other benefits of the CSP gather include:

- longer offsets,
- accurate prestack migration velocity analysis,
- easier picking of velocities in structurally complex areas,
- incorporated rugged topography,
- use of <u>conventional processing tools</u> such as velocity analysis,
- direct application to <u>converted-wave (P-S) processing</u>,
- prestack migration <u>static</u> analysis,
- reduced run <u>time</u>.
- prestack migration <u>multiple attenuation.</u>

Computational speed evaluation

Input traces are summed directly into the CSP gather with no anti-aliasing filters, no scaling, no interpolation, and no time shifting. All these processes are delayed until after the CSP gather is completely formed. Differences in computing times with a full prestack Kirchhoff migration may be found by evaluating the fold in each bin of the CSP gather.

Fold retention

The fold of each sample in a CSP gather is recorded in a companion array. This fold is used to balance the amplitudes in a CSP gather, especially in areas with uneven geometry.

Noise reduction

EOM migration tends to reduce noise (Figures 11.14b-c, and 11.15a-b), which enables the bandwidth of the signal to noise ratio to be increased for improved event resolution.

11.12 Disadvantages of the Process

The equivalent offset is <u>slightly dependent on the velocity</u>, and may require the <u>preprocessing of a few CSP gathers</u> to define a velocity function.

It is precisely this property that properly focuses the data.

In practice, the initial velocity may be assumed to be infinite. The processing of a few CSP gathers at selected velocity analysis points will provide velocities that are accurate enough for processing the entire data set to final migrated section. If the CSP gathers are saved, additional velocity analysis and stacking could be performed.

The <u>amplitude computations of the individual traces</u> are lost by gathering traces into CSP bins. The loss of input trace identity may restrict the use of amplitude computations involving Zoeppritz's equations.

Not so, AVO effects are preserved.

The AVO effects in a CSP gather will not be the same as a CMP gather. AVO effects are also available for dipping data, and marine 3D data orthogonal to shooting direction.

Amplitude scaling before and after gathering also preserves trace weighting.

If the entire data set is processed to CSP gathers, the <u>volume of the gathered data</u> may exceed that of the original input data. True.

When there are memory restriction, in the CPU or disk, portions of the data set may be processed. An example, when processing 3-D data, would be to produce a few 2-D inlines at a time.

The great speed of EOM is limited to <u>prestack time migration</u>.

When combined with an inverse time migration, EOM can provide an optimally focused stacked section. A following <u>poststack depth migration</u> may be <u>comparable</u> to a <u>prestack depth migration</u>.

11.13 Summary of Points to Note in Chapter 11

- EOM prestack time migration is based on RMS velocity assumptions.
- Prestack migration gathers may be formed by a number of methods.
- CSP gathers collect energy from scatterpoints on an image ray.
- The equivalent offset assumes the source and receiver are collocated.
- The equivalent offset positions an input trace relative to the scatterpoint location.
- No time shifting of the input data is required in the formation of CSP gathers.
- Crude velocities may be used to form CSP gathers.
- Accurate velocities are derived from CSP gathers after they are formed.
- A natural antialiasing filter aids in the formation of the CSP gathers.
- EOM may migrate from surface.
- CSP gathers provide model traces for residual statics analysis.
- EOM simplifies converted-wave (P-S) processing.
- EOM simplifies vertical array processing. (vertical marine, borehole, logging tool)
- EOM time domain migration is fast.
- EOM depth domain migration has no speed advantage, but make velocity analysis much easier.

Chapter Twelve

Comparisons and Evaluations

"Unanswered Questions".

12.1 Which Algorithm is Best for a Given Application?

Kirchhoff:

- This method can be a simple time migration, or a very complex prestack depth migration.
- Its application therefore depends on the complexity of the algorithm used.

FK or KK:

- This method is ideal when the velocities are constant.
- It works very well when the velocities vary smoothly.
- It gives excellent results on many marine lines.
- It can give an accurate time migration when many constant velocity passes are made.

Downward continuation:

- This method has the potential to accurately migrate any complex structure.

However:

- Its accuracy is dependent on the algorithm used.
- Finite difference algorithms are typically time migrations.
- Phaseshift and ωx migrations are typically depth migrations.

Evaluate the following

	Kirchhoff	**FK**	**Downward continuation**
Wave eqn, migration			
Robust			
Accuracy			
Ease of use			
Noise			
Focusing			
Positioning			
Carbonate/shale velocity variations			
Speed			
Constant velocity			
Vertical velocities $V(z)$			
Lateral velocities $V(x, y, z)$			
Structured geology			
Aliasing			
Processing datums			
Others ???			

12.2 Differences Between Time and Depth Migrations?

	Time migration	Depth Migration
Wave eqn, migration		
Robust		
Accuracy		
Ease of use		
Noise		
Focusing		
Positioning		
Carbonate/shale velocity variations		
Speed		
Constant velocity		
Vertical velocities $V(z)$		
Lateral velocities $V(x, y, z)$		
Structured geology		
Aliasing		
Processing datums		
Others ???		

12.3 What Factors Are Considered when Choosing a Migration Algorithm?

Speed	
Sideswipe	
Obliquity	
Vertical velocity	
Lateral velocity	
Structured geology	
Statics	
Crooked lines	
Positioning	
Relative positioning	
Focusing	
Resolution	
Cost	
Mode conversions	

Appendix 4: Kinematic derivations for DMO-PSI and EOM

A4.1 Definitions

V_{rms} generalization term used for NMO and migration of horizontal data

V_{stk} dip dependent velocity for NMO of dipping data $V_{stk} = V_{rms}/cos(dip)$

V local velocity of interest (typically RMS)

All time variables are two-way time.

T input time

T_n after NMO using RMS velocities

T_{dn} dip dependent NMO using V_{stk}

T_d tangential time after NMO-DMO

T_0 vertical zero-offset time to scatterpoint (image ray)

T_g tangential time (relative to T_0) after Gardner's DMO

$t=2h/V$ radius (in time) of DMO unit circle (x, t) from $-h$ to h

T_c time at offset b in DMO unit circle.

The following pages derive kinematic equations for DMO-PSI in the time domain. PSI is a migration-like process that deals exclusively with spatial parameters (x, h) at constant time layers. This process is independent of velocities and is computed efficiently in the Fourier transform domain (K_x, K_h).

DMO-PSI is virtually velocity independent. The derived point of tangency (T_g, k) is defined for a specified velocity to enable comparison with EOM.

References

ID	Authors	Date	Title	Journal or Publisher	Vol	Pages
111	Forel D., and Gardner,	May. 1988	A Three-Dimensional Perspective on Two-	Geophysics	53	604-610
322	Gardner, G. H. F.,	May. 1986	Dip moveout and prestack imaging	Exp. Abs., Offshore		75-84
373	Ottolini, R. A.	Nov. 1982	Migration of reflection seismic data in	Dissertation,		
452	Ottolini, R. A., and	Mar. 1984	The migration of common midpoint slant	Geophysics	49	237-249
544	Bancroft, J. C.,	Jun. 1998	A comparison of prestack migration	Exp. Abs. EAGE		1-18
619	Fowler, P. J.	Nov. 1997	A comparative overview of prestack time	Exp. Abs. SEG 1997		1571-1574
620	Bancroft, J. C.,	Nov. 1997	A kinematic comparison of DMO-PSI and	Exp. Abs. SEG 1997		1575-1578
658	Canning, A., and	Mar. 1996	A two-pass approximation to 3-D prestack	Geophysics	61	409-421
690	Fowler, P. J.	Jun. 1998	A comparative overview of prestack time	Exp. Abs. EAGE		1-17
697	Fowler, P. J.	Sep. 1998	A comparative overview of dip moveout	Exp. Abs. SEG		1744-1747

A4.2 Derivation of GDMO offset *k* (See [111])

For this section, the times T_g, T_d, and T_c are arbitrary times on the DMO ellipse and unit circle, as shown in Figure A4.1. (Later they will be specific times relative to the scatterpoint at T_0.)

The conventional NMO-DMO path in Figure A4.1a gives the NMO time *Tn* from,

$$T_n^2 = T^2 - \frac{4h^2}{V^2}. \tag{A4.1}$$

The time T_d (after DMO) may be found using T_c on the unit circle, i.e.,

$$\frac{T_d}{T_c} = \frac{T_n}{2h/V}, \quad \text{where} \quad T_c = \frac{2\sqrt{h^2 - b^2}}{V}, \quad \text{giving} \quad T_d = T_n \frac{\sqrt{h^2 - b^2}}{h}. \tag{A4.2}$$

Combining NMO then DMO we get;

$$T_n^2 = \left(\frac{h^2 - b^2}{h^2}\right)\left(T^2 - \frac{4h^2}{V^2}\right). \tag{A4.3}$$

Alternatively using Gardner's method in Figure A4.1b, apply DMO first,

$$T_g = T\frac{\sqrt{h^2 - b^2}}{h}, \quad \text{and NMO with new offset } k, \quad T_k^2 = T_g^2 - \frac{4k^2}{V^2}, \tag{A4.4}$$

Combining these equations we get:

$$T_k^2 = \left(\frac{h^2 - b^2}{h^2}\right)\left(T^2 - \frac{4k^2}{V^2}\frac{h^2}{h^2 - b^2}\right). \tag{A4.5}$$

Equating both methods of DMO, i.e. equations (A4.3) and (A4.5), yields the definition of the new offset *k*:

$$\boxed{k^2 = h^2 - b^2.} \tag{A4.6}$$

Appendix 4 Kinematic derivations for DMO-PSI and EOM

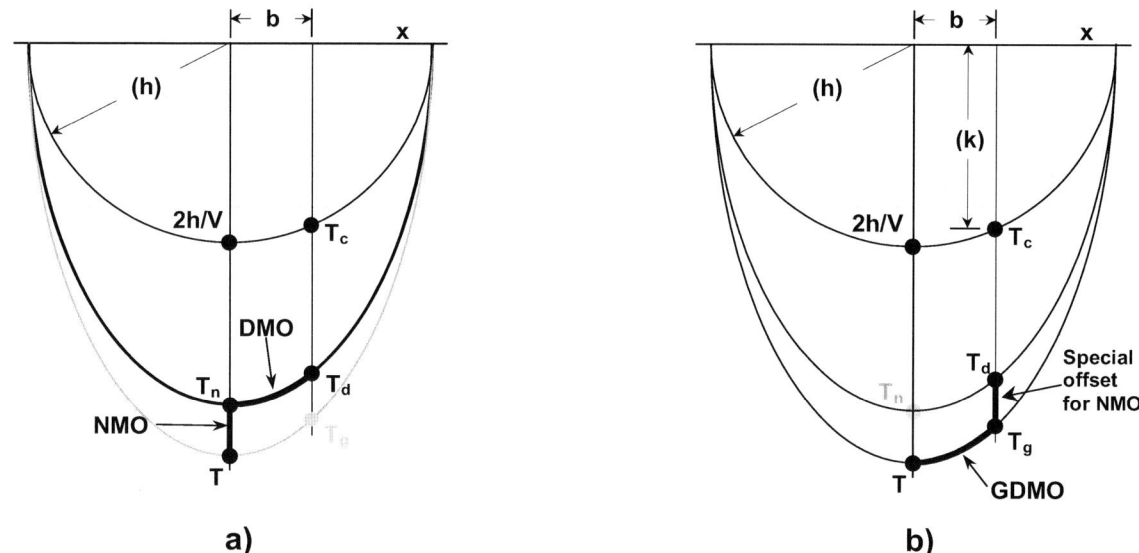

Figure A4.1   DMO ellipses for defining a) conventional NMO and DMO, and b) GDMO.

A4.3 Define the location *b* for a given scatterpoint

An input sample at *T* will prestack migrate to the prestack migration ellipse (constant velocity *V*). A specific scatterpoint may be defined on the ellipse at depth Z_0 or time T_0 as illustrated in Figure A4.2.

We define the following equations,

prestack migration ellipse
$$\frac{t^2}{T_n^2} + \frac{4x^2}{T^2 V^2} = 1,\qquad (A4.7)$$

NMO removal
$$T^2 = T_n^2 + \frac{4h^2}{V^2},\qquad (A4.8)$$

NMO removal after GDMO
$$T_g^2 = T_d^2 + \frac{4k^2}{V^2},\qquad (A4.9)$$

radius of poststack migration
$$T_d^2 = T_0^2 + \frac{4(x_0 - b)^2}{V^2},\qquad (A4.10)$$

poststack migration semicircle
$$\frac{4(x - b)^2}{V^2} + t^2 = T_d^2,\qquad (A4.11)$$

offset on radial plane
$$h_g = h \frac{T_g}{T}.\qquad (A4.12)$$

Solve for *b* in terms of input parameters *T*, x_0, *h*, and *V*.

The post-stack migration semicircle and the prestack migration ellipse are tangential at (x_0, t_0). The zero-offset point of tangency at *b* is found by equating the derivative of these curves. Starting with the equation (A4.7) for the ellipse:

$$\frac{2t}{T_n^2}\frac{dt}{dx} + \frac{8x}{T^2 V^2} = 0,\qquad (A4.13)$$

and evaluated at (x_0, T_0)
$$\frac{dt}{dx} = -\frac{4x_0 T_n^2}{T_0 T^2 V^2}.\qquad (A4.14)$$

The slope for the circle is found from equation (A4.11)
giving

$$\frac{8(x - b)}{V^2} + 2t\frac{dt}{dx} = 0,\qquad (A4.15)$$

and when evaluated at (x_0, T_0),
$$\frac{dt}{dx} = -\frac{4(x_0 - b)}{T_0 V^2}.\qquad (A4.16)$$

Appendix 4 Kinematic derivations for DMO-PSI and EOM

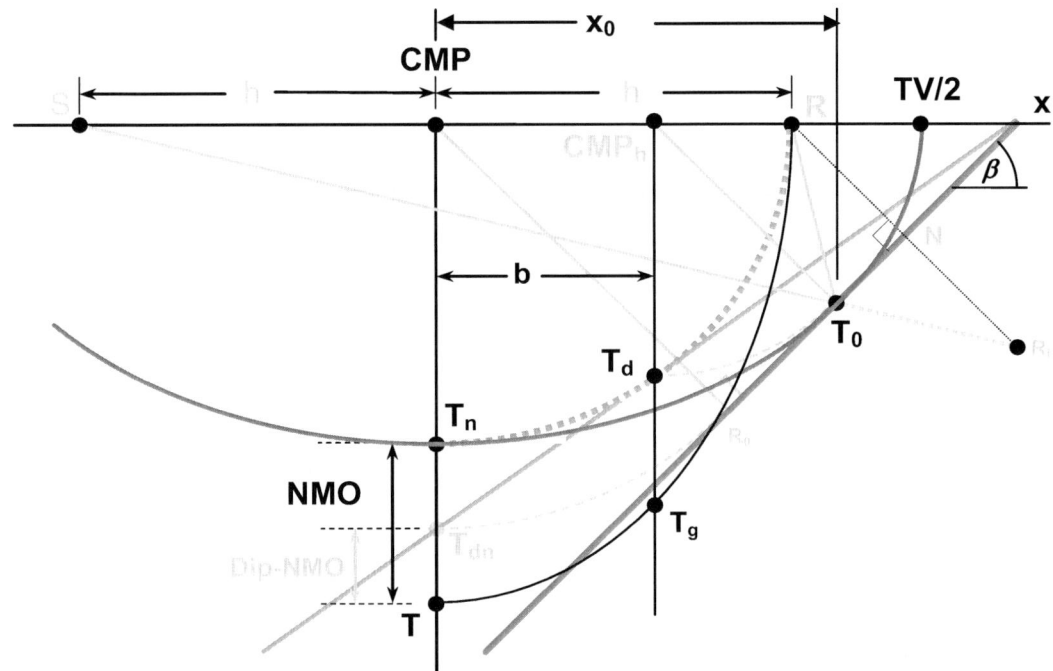

Figure A4.2

Equating the two slopes from equations (A4.14) and (A4.16),

$$\frac{4x_0 T_n^2}{T_0 T^2 V^2} = \frac{4(x_0 - b)}{T_0 V^2}, \tag{A4.17}$$

solving for *b* gives

$$b = x_0 \left(1 - \frac{T_n^2}{T^2}\right), \tag{A4.18}$$

when substituting T_n using equation (A4.8) we get

$$b = x_0 \left(1 - \frac{T^2 - \frac{4h^2}{V^2}}{T^2}\right), \tag{A4.19}$$

which becomes

$$\boxed{b = \frac{4x_0 h^2}{T^2 V^2}.} \tag{A4.20}$$

A4.4 Define T_0 in terms of T, x_0, h, and V

At the point (x_0, T_0), equation (A4.7) becomes

$$T_0^2 = T_n^2\left(1 - \frac{4x_0^2}{T^2 V^2}\right) \tag{A4.21}$$

Solving the NMO equation (A4.8) for T_n, and substituting into (A4.21) we get

$$T_0^2 = \left(T^2 - \frac{4h^2}{V^2}\right)\left(1 - \frac{4x_0^2}{T^2 V^2}\right) \tag{A4.22}$$

which becomes

$$\boxed{T_0^2 = T^2 - \frac{4}{V^2}\left(x_0^2 + h^2 - \frac{4x_0^2 h^2}{T^2 V^2}\right)} \tag{A4.23}$$

Note that the bracketed term is the equivalent offset h_e that will be derived in section A4.7.

The above derivation for h_e is based on the prestack migration ellipse that assumes a constant velocity. A more general derivation that only assumes RMS velocities at the scatterpoint is presented in A4.7.

A4.5 Define T_g and k in terms of T, x_0, h, and V

Start with NMO for GDMO in equation (A4.9) then substitute the value of T_d given by poststack migration in equation (A4.10) to get

$$T_g^2 = T_0^2 + \frac{4(x_0 - b)^2}{V^2} + \frac{4k^2}{V^2}, \qquad (A4.24)$$

Substituting the value of T_0 defined in equation (A4.23) we get

$$T_g^2 = T^2 - \frac{4}{V^2}\left(x_0^2 + h^2 - \frac{4x_0^2 h^2}{T^2 V^2} - (x_0 - b)^2 - k^2\right), \qquad (A4.25)$$

which simplifies to

$$T_g^2 = T^2 - \frac{4}{V^2}\left(2x_0 b - \frac{4x_0^2 h^2}{T^2 V^2}\right). \qquad (A4.26)$$

Substituting for b from equation (A4.20) we get

$$T_g^2 = T^2 - \frac{4}{V^2}\left(\frac{8x_0^2 h^2}{T^2 V^2} - \frac{4x_0^2 h^2}{T^2 V^2}\right). \qquad (A4.27)$$

or

$$T_g^2 = T^2\left(1 - \frac{16 x_0^2 h^2}{T^4 V^4}\right), \qquad (A4.28)$$

which is the desired result for defining T_g in terms of input parameters.

The value of the GDMO offset k is found by substituting the value of b in equation (A4.20) into equation (A4.6) giving

$$k^2 = h^2\left(1 - \frac{16 x_0^2 h^2}{T^4 V^4}\right). \qquad (A4.29)$$

Note also that $T_g^2 = T^2\left(1 - \frac{b^2}{h^2}\right)$.

A4.6 Define the PSI offset $h_{g\text{-}PSI}$ for the point of tangency T_g

After GDMO, PSI is applied to the prestack volume. Energy at the point of tangency T_g will be rotated to the prestack migration gather at an offset $h_{g\text{-}PSI}$.

The offset $h_{g\text{-}PSI}$ is the radial distance from the scatterpoint location ($x = x_s$, $h = 0$) to the GDMO'd location ($x = b$, $h = k$), i.e.,

$$h_{g-PSI}^2 = (x_0 - b)^2 + k^2 , \qquad (A4.30)$$

expanding and substituting the value of b,

$$h_{g-PSI}^2 = x_0^2 - 2x_0 b + b^2 + h^2 - b^2 , \qquad (A4.31)$$

$$\boxed{h_{g-PSI}^2 = x_0^2 + h^2 - \frac{8x_0^2 h^2}{T^2 V^2} .} \qquad (A4.32)$$

A4.7 Derivation of the equivalent offset equation (11.11)

Starting with the DSR equation (7.6, 11.7), and equating it with hyperbolic form of the equivalent offset (11.8) we have,

$$T = \left[\left(\frac{T_0}{2}\right)^2 + \frac{(x+h)^2}{V^2}\right]^{1/2} + \left[\left(\frac{T_0}{2}\right)^2 + \frac{(x-h)^2}{V^2}\right]^{1/2} = 2\left[\left(\frac{T_0}{2}\right)^2 + \frac{h_e^2}{V^2}\right]^{1/2}, \quad \text{(A4.33)}$$

rearranging, we get

$$\left[\left(\frac{T_0 V}{2}\right)^2 + (x+h)^2\right]^{1/2} = 2\left[\left(\frac{T_0 V}{2}\right)^2 + h_e^2\right]^{1/2} - \left[\left(\frac{T_0 V}{2}\right)^2 + (x-h)^2\right]^{1/2}, \quad \text{(A4.34)}$$

then by squaring both sides and simplifying,

$$\left(\frac{T_0 V}{2}\right)^2 + (x+h)^2 = \left\{2\left[\left(\frac{T_0 V}{2}\right)^2 + h_e^2\right]^{1/2} - \left[\left(\frac{T_0 V}{2}\right)^2 + (x-h)^2\right]^{1/2}\right\}^2, \quad \text{(A4.35)}$$

$$\left(\frac{T_0 V}{2}\right)^2 + (x+h)^2 = 4\left[\left(\frac{T_0 V}{2}\right)^2 + h_e^2\right] + \left[\left(\frac{T_0 V}{2}\right)^2 + (x-h)^2\right] \\ - 4\left[\left(\frac{T_0 V}{2}\right)^2 + h_e^2\right]^{1/2}\left[\left(\frac{T_0 V}{2}\right)^2 + (x-h)^2\right]^{1/2} \quad \text{(A4.36)}$$

$$\left(\frac{T_0 V}{2}\right)^2 + x^2 + 2xh + h^2 = 4\left(\frac{T_0 V}{2}\right)^2 + 4h_e^2 + \left(\frac{T_0 V}{2}\right)^2 + x^2 - 2xh + h^2 \\ - 4\left[\left(\frac{T_0 V}{2}\right)^2 + h_e^2\right]^{1/2}\left[\left(\frac{T_0 V}{2}\right)^2 + (x-h)^2\right]^{1/2} \quad \text{(A4.37)}$$

$$4xh = 4\left[\left(\frac{T_0 V}{2}\right)^2 + h_e^2\right] - 4\left[\left(\frac{T_0 V}{2}\right)^2 + h_e^2\right]^{1/2}\left[\left(\frac{T_0 V}{2}\right)^2 + (x-h)^2\right]^{1/2}, \quad \text{(A4.38)}$$

rearranging gives

$$\left[\left(\frac{T_0V}{2}\right)^2+(x-h)^2\right]^{1/2}=\left[\left(\frac{T_0V}{2}\right)^2+h_e^2\right]^{1/2}-\frac{xh}{\left[\left(\frac{T_0V}{2}\right)^2+h_e^2\right]^{1/2}},\qquad(A4.39)$$

Squaring each side, eliminating terms, then resorting gives

$$\left(\frac{T_0V}{2}\right)^2+(x-h)^2=\left(\frac{T_0V}{2}\right)^2+h_e^2+\frac{x^2h^2}{\left(\frac{T_0V}{2}\right)^2+h_e^2}-2\left[\left(\frac{T_0V}{2}\right)^2+h_e^2\right]^{1/2}\frac{xh}{\left[\left(\frac{T_0V}{2}\right)^2+h_e^2\right]^{1/2}},\qquad(A4.40)$$

$$x^2-2xh+h^2=+h_e^2+\frac{(xh)^2}{\left(\frac{T_0V}{2}\right)^2+h_e^2}-2xh,\qquad(A4.41)$$

$$h_e^2=x^2+h^2-\frac{(xh)^2}{\left[\left(\frac{T_0V}{2}\right)^2+h_e^2\right]},\qquad(A4.42)$$

where the denominator term is simplified using equation (A4.33) to get the total two way time *T* giving

$$\boxed{h_e^2=x^2+h^2-\frac{4x^2h^2}{T^2V^2}.}\qquad(A4.43)$$

A4.8 Approximate but similar solutions

Approximate Solution #1

We will make use of the power series expansion of the square-root;

$$(1+x)^{1/2} = 1 + \frac{x}{2} - \frac{x^2}{8} + \frac{x^3}{16} \mp \dots . \tag{A4.44}$$

in the DSR equation and in the hyperbolic form of the equivalent offset.

Starting with the exact traveltimes of equation (A4.33) we have

$$T = \left[\left(\frac{T_0}{2}\right)^2 + \frac{(x+h)^2}{V^2}\right]^{1/2} + \left[\left(\frac{T_0}{2}\right)^2 + \frac{(x-h)^2}{V^2}\right]^{1/2} = 2\left[\left(\frac{T_0}{2}\right)^2 + \frac{h_e^2}{V^2}\right]^{1/2}.$$

Expressed in the normalized form we have

$$T = \frac{T_0}{2}\left[1 + \frac{4(x+h)^2}{T_0^2 V^2}\right]^{1/2} + \frac{T_0}{2}\left[1 + \frac{4(x-h)^2}{T_0^2 V^2}\right]^{1/2} = 2\frac{T_0}{2}\left[1 + \frac{4h_e^2}{T_0^2 V^2}\right]^{1/2}.$$

or

$$\left[1 + \frac{4(x+h)^2}{T_0^2 V^2}\right]^{1/2} + \left[1 + \frac{4(x-h)^2}{T_0^2 V^2}\right]^{1/2} = 2\left[1 + \frac{4h_e^2}{T_0^2 V^2}\right]^{1/2}.$$

Using only the first two terms of the power series expansion we get

$$\left[1 + \frac{2(x+h)^2}{T_0^2 V^2}\right] + \left[1 + \frac{2(x-h)^2}{T_0^2 V^2}\right] = 2\left[1 + \frac{2h_e^2}{T_0^2 V^2}\right].$$

Solving for h_e,

$$\frac{4h_e^2}{T_0^2 V^2} = \frac{2(x+h)^2 + 2(x-h)^2}{T_0^2 V^2},$$

we get our solution

$$\boxed{h_e^2 = x^2 + h^2}.$$

Note that this form has no cross term and is an asymptotic solution for the exact solution.

Approximate Solution #2

This following solution takes the dip-dependent moveout DD-MO correction and Post-stack migration equations and then uses a simple geometry substitution for the cosine term. The geometry is defined in Figure 4A.2, where we know that this method will stack a dipping event, but will smear energy along the dip.

Note that T_{dn} is the moveout corrected data that lies on the zero-offset hyperbola.

Substituting $T_{dn}^2 = T_0^2 + \dfrac{4x^2}{V^2}$ **into** $T^2 = T_{dn}^2 + \dfrac{4h^2 \cos^2 \beta}{V^2}$ (11.1 and 2)

gives $T^2 = T_0^2 + \dfrac{4(x^2 + h^2 \cos^2 \beta)}{V^2} = T_0^2 + \dfrac{4(x^2 + h^2 - h^2 \sin^2 \beta)}{V^2} = T_0^2 + \dfrac{4\tilde{h}_e^2}{V}.$

Recall the geometry of zero-offset Kirchhoff migration from Figure 4.8 in the Migration course notes as;

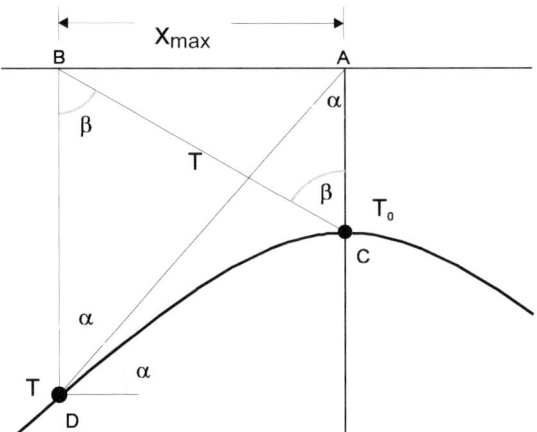

Figure 4.8. Angle relationships with hyperbola.

In this figure, the times represented by T in the post-stack now represent T_n in the prestack considerations. From this figure

$$\sin \beta = \frac{2x}{VT_n}.$$

giving a similar form for the equivalent offset

$$\boxed{h_e^2 = x^2 + h^2 - \frac{4x^2 h^2}{T_n^2 V^2}}.$$

Note that this solution uses T_n instead of the theoretically correct T.

Approximate Solution #3

This solution will approximate the square-roots in the DSR with the first three terms of the power series, and then equate them with the hyperbolic form of the equivalent offset moveout equation.

Starting again with the DSR equation (7.6, or 11.7),

$$T = \left[\left(\frac{T_0}{2}\right)^2 + \frac{(x+h)^2}{V^2}\right]^{1/2} + \left[\left(\frac{T_0}{2}\right)^2 + \frac{(x-h)^2}{V^2}\right]^{1/2}, \quad \text{(A4.34)}$$

we will express the square-roots in the above power series form and then only consider the first three terms. We now start with

$$T = \frac{T_0}{2}\left\{\left[1 + \frac{4(x+h)^2}{T_0^2 V^2}\right]^{1/2} + \left[1 + \frac{4(x-h)^2}{T_0^2 V^2}\right]^{1/2}\right\}, \quad \text{(A4.34)}$$

Expanding,

$$T = \frac{T_0}{2}\left\{\left[1 + \frac{4(x+h)^2}{2T_0^2 V^2} - \frac{16(x+h)^4}{8T_0^4 V^4}...\right] + \left[1 + \frac{4(x-h)^2}{2T_0^2 V^2} - \frac{16(x-h)^4}{8T_0^4 V^4}...\right]\right\}, \quad \text{(A4.34)}$$

$$T = \frac{T_0}{2}\left\{2 + \frac{2\left[(x+h)^2 + (x-h)^2\right]}{T_0^2 V^2} - \frac{2\left[(x+h)^4 + (x-h)^4\right]}{T_0^4 V^4}...\right\}, \quad \text{(A4.34)}$$

$$T = T_0\left\{1 + \frac{\left[(x+h)^2 + (x-h)^2\right]}{T_0^2 V^2} - \frac{\left[(x+h)^4 + (x-h)^4\right]}{T_0^4 V^4}...\right\}, \quad \text{(A4.34)}$$

Squaring both sides, and only considering terms up to a power of four,

$$T^2 = T_0^2\left\{1 + \frac{2\left[(x+h)^2 + (x-h)^2\right]}{T_0^2 V^2} + \frac{\left[(x+h)^2 + (x-h)^2\right]^2}{T_0^4 V^4} - \frac{2\left[(x+h)^4 + (x-h)^4\right]}{T_0^4 V^4}...\right\}, \quad \text{(A4.34)}$$

$$T^2 = T_0^2 + \frac{4(x^2 + h^2)}{V^2} + \frac{4(x^2 + h^2)^2}{T_0^2 V^4} - \frac{2\left[(x+h)^4 + (x-h)^4\right]}{T_0^2 V^4} \pm ..., \quad \text{(A4.34)}$$

$$T^2 = T_0^2 + \frac{4}{V^2}\left\{(x^2+h^2) + \frac{(x^2+h^2)^2}{T_0^2 V^2} - \frac{\left[(x+h)^4 + (x-h)^4\right]}{2T_0^2 V^2} \pm ...\right\},$$ (A4.34)

This is now equated to the equivalent offset hyperbola

$$T^2 = T_0^2 + \frac{4}{V^2} h_e^2,$$ (A4.34)

$$h_e^2 = x^2 + h^2 + \frac{(x^2+h^2)^2}{T_0^2 V^2} - \frac{\left[(x+h)^4 + (x-h)^4\right]}{2T_0^2 V^2} \pm ...,$$ (A4.34)

$$h_e^2 = x^2 + h^2 + \frac{2(x^2+h^2)^2 - \left[(x^2+h^2+2xh)^2 + (x^2+h^2-2xh)^2\right]}{2T_0^2 V^2} \pm ...,$$ (A4.34)

$$h_e^2 = x^2 + h^2 + \frac{2(x^2+h^2)^2 - \left[\left[(x^2+h^2)+2xh\right]^2 + \left[(x^2+h^2)-2xh\right]^2\right]}{2T_0^2 V^2} \pm ...,$$ (A4.34)

$$h_e^2 = x^2 + h^2 + \frac{2(x^2+h^2)^2 - \left[(x^2+h^2)^2 + 2xh(x^2+h^2) + 4x^2h^2 + (x^2+h^2)^2 - 2xh(x^2+h^2) + 4x^2h^2\right]}{2T_0^2 V^2} \pm ...$$

, (A4.34)

$$h_e^2 = x^2 + h^2 + \frac{-8x^2h^2}{2T_0^2 V^2} \pm ...,$$ (A4.34)

$$\boxed{h_e^2 = x^2 + h^2 - \frac{4x^2h^2}{T_0^2 V^2} \pm ...,}$$ (A4.34)

Note this solution has a T_0 in the cross term, not T as in the exact solution.

Approximate Solutions #4

We will now start with the exact solution for h_e, i.e. equation (A4.43)

$$h_e^2 = x^2 + h^2 - \frac{4x^2 h^2}{T^2 V^2}.$$

Note that the velocity is defined at T_0, not at the time of the input sample T. This expressed in the following form of the equation

$$h_e^2 = x^2 + h^2 - \frac{4x^2 h^2}{T^2 V_{T_0}^2}.$$

One possibility is to convert the time T into T0 with the hyperbolic form of the equivalent offset moveout correction to give

$$h_e^2 = x^2 + h^2 - \frac{4x^2 h^2}{T_0^2 V_{T_0}^2 + 4h_e^2}.$$

We now have an implicit equation for h_e, but since the cross term is quite small, this equation is quite good for getting an iterative solution of the form,

$$\boxed{h_{e,n+1}^2 = x^2 + h^2 - \frac{4x^2 h^2}{T_0^2 V^2 + 4h_{e,n}^2}.}$$

We can also replace the equivalent offset in the cross term with the approximate solution #1, i.e.,

$$\boxed{h_e^2 = x^2 + h^2 - \frac{4x^2 h^2}{T_0^2 V^2 + 4\left(x^2 + h^2\right)}.}$$

These forms are all OK, (just OK), as h_e is defined at T. We really want to define the velocity (which is still at T_0) but with a function of (x, h, T), the input sample location.

Comments

- The above equations may imply that the equivalent offset is computed each time one input sample is summed into all the prestack migration gathers. That is not the case.

    An efficient operation copies one input trace into all surrounding prestack migration gathers. Since one input trace may cover a number of offset bins in the same prestack migration gather, we only need to know the starting bin number, and the times at which the data moves to a new offset bin.

    At this point, the only way I know how to do this practically is to create <u>a table for T, h_e, and bin number</u> as a function of T_0. I then find the times of the bin transitions, which define the size of the loops that sum the data into the corresponding offset bins. The cosine of the dip can be included in this table to provide a dip limiting filter at the time the prestack migration gathers are formed.

- If the velocity function is spatially independent, then one set of tables can be generated for all displacements and offset.

- When the velocity varies with depth, one input sample can occupy a number of different moveout corrected times. Similarly, one input sample may have a number of different equivalent offsets in the same prestack migration gather. This data mapping can be defined by starting at $T_0 = 0$ and then defining T and h_e for the defined sample on the input trace. Building a table also helps to define these areas and aids in selecting the desirable data.

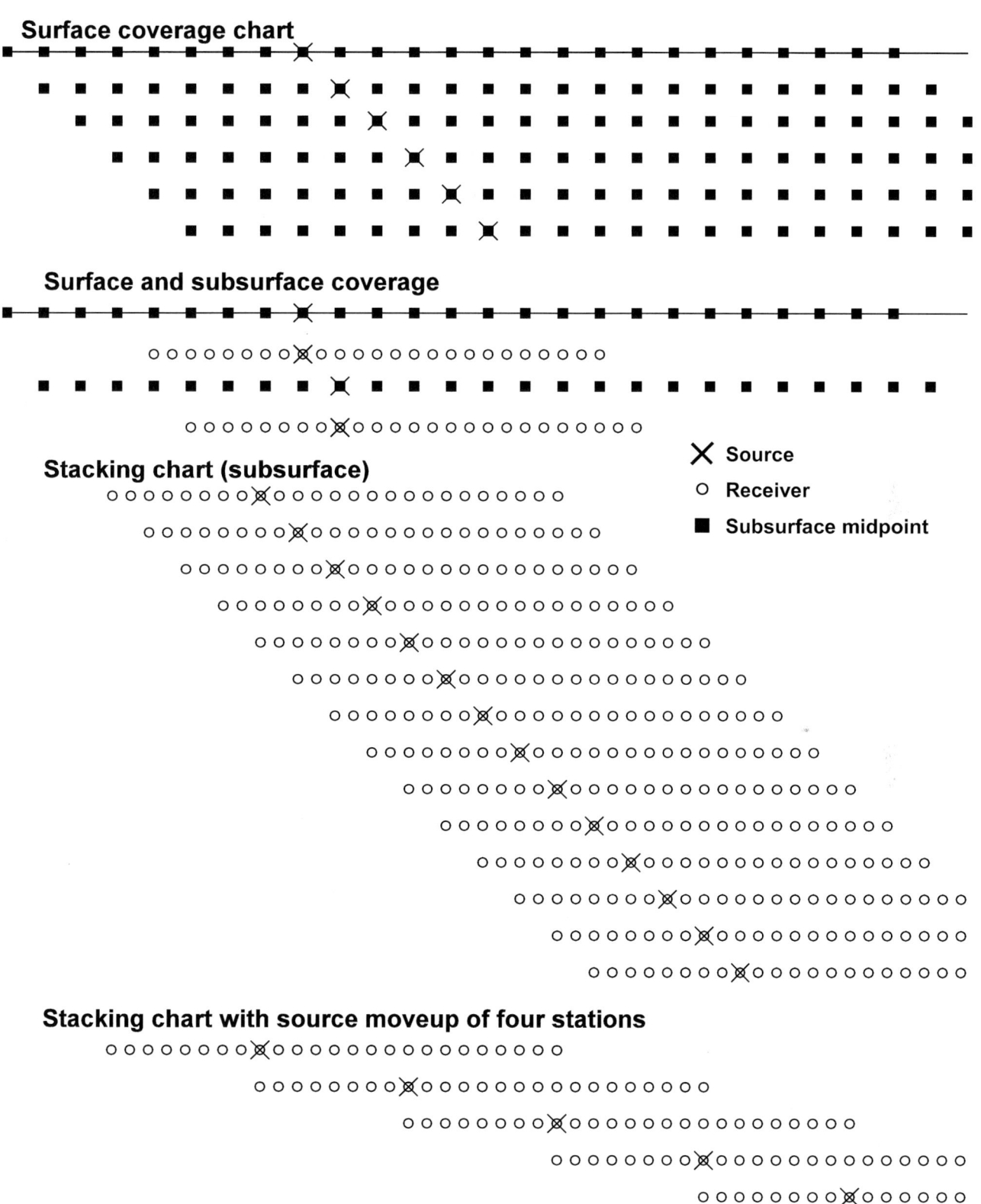

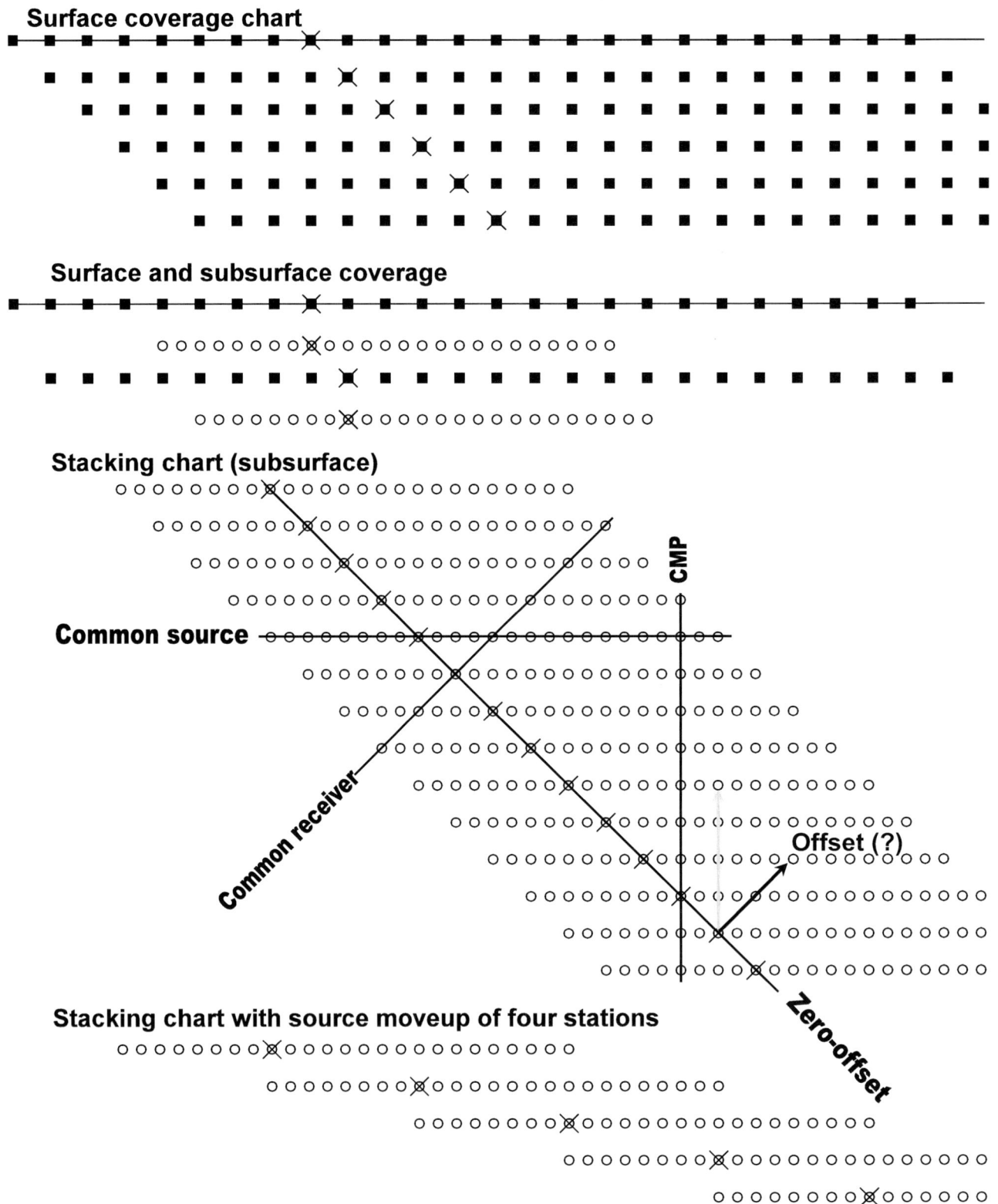

Appendix 5 Stacking Charts

Vertical movement of the CMP gathers produces the following stacking chart.

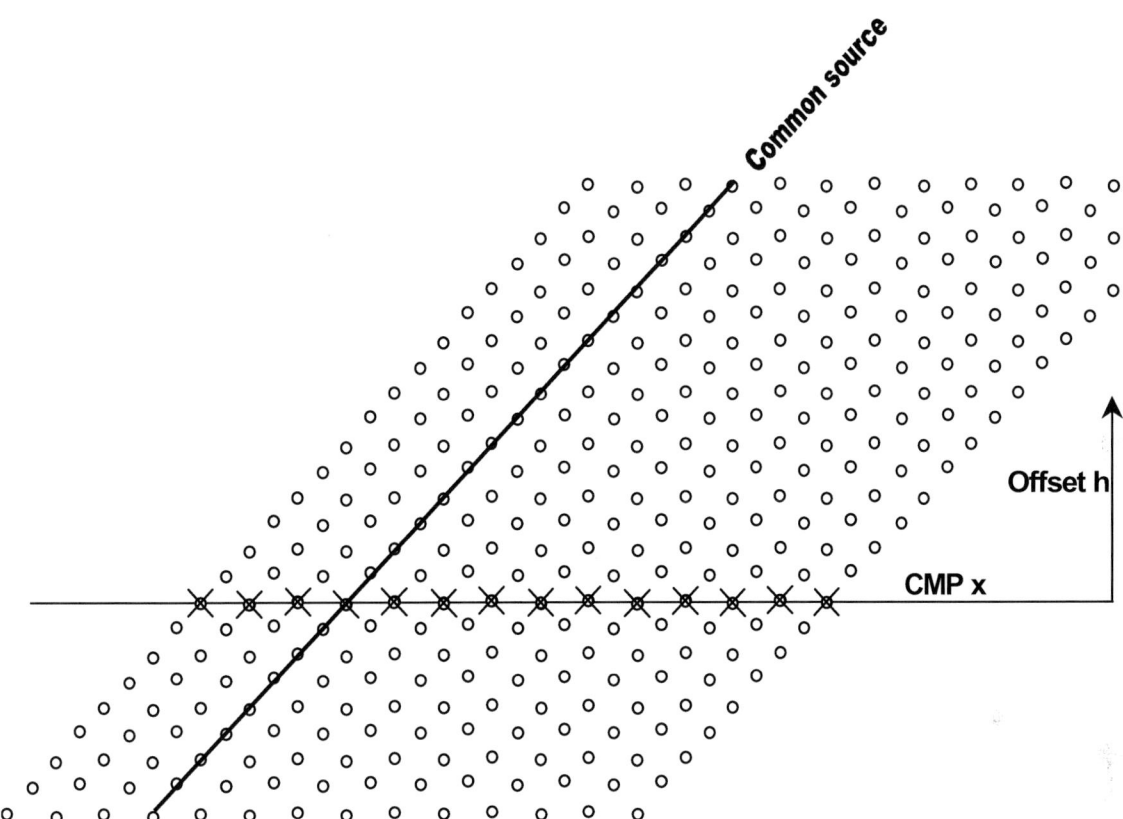

References: Alphabetical by authors' names

Author(s)   ID number   Date   Title   Journal/publisher   Pages

Abrial, W. L., and Wright, R. M. 608 Aug. 1994 The shape of Gulf Coast salt intrusions related to seismic imaging The Leading Edge 868-872

Agarwal, B. N. P., and Lal, T. 320 Jun. 1971 Application of rational approximation in calculation of the second derivative of the gravity field Geophysics 571-581

Akbar, F. E., Sen, M. K., and Stoffa, P. L. 518 Oct. 1994 Prestack plane-wave Kirchhoff depth migration using a cluster of workstations Exp. Abs., SEG Nat. Conv. 225-228

Akbar, F. E., Sen, M., and Stoffa, P. L. 616 Sep. 1996 Prestack plane-wave Kirchhoff migration in laterally varying media Geophysics 1068-1079

Al-Chalabi, M. 615 Jul. 1997 Time-depth relationships for multilayer depth conversions Geophys. Prosp. 715-720

Aldridge, D., and Oldenburg D. 144 Mar. 1992 Refractor Imaging Using an Automated Wavefront Reconstruction Method Geophysics 378-385

Alfaraj, M., and Larner, K. 335 Nov. 1991 Dip moveout for mode-converted waves Exp. Abs., SEG Nat. Conv. 1191-1193

Alfaraj, M. 682 Oct. 1992 Transformation to zero offset for mode-converted waves by Fourier transform Exp. Abs. SEG 974-978

Alfaraj, M., and Larner, K. 479 Oct. 1994 Anti-aliased Kirchhoff 3-D migration Exp. Abs. SEG Nat. Conv. 1282-1285

Alford, R., Kelly, K., and Boore, D., 13 Dec. 1974 Accuracy of Finite Difference Modelling of the Acoustic Wave Equation Geophysics 834-842

Alkhalifah, T. 272 Sep. 1993 Gaussian Beam Migration for Anisotropic Media Exp. Abs., SEG Nat. Conv. 847-851

Alkhalifah, T., and Larner, K. 354 Sep. 1994 Migration error in transversely isotropic media Geophysics 1405-1418

Alkhalifah, T., and Tsvankin, I. 772 Sep. 1995 Velocity analysis for transversly isotropic media Geophysics 1550-1566

Alkhalifah, T., Tsvankin, I., Larner, K., and Toldi, J 565 May. 1996 Velocity analysis and imaging in transverse isotropic media: Methology and a case study. The Leading Edge 371-378

Alkhalifah, T. 673 Jul. 1997 Kinematics of 3-D DMO operators in transversely isotropic media Geophysics 1214-1219

Allan D., and Bale, R. 561 Jun. 1998 Components of 3-D converted wave imaging Exp. Abs. EAGE 1-46

Al-Yahya, K. 192 Jun. 1989 Velocity analysis by iterative profile migration Geophysics 718-729

Ameely, L., Krey, T., Muhtadie, F., Rau, H., and Wrist, H. 118 Aug. 1983 Migration in the Presence of a Rugged Interface with High Velocity Contrast Geophys. Prosp. 561-573

Amestoy, P., Gardner, G., Leiss, E. 415 Oct. 1987 Phase shift-Based prestack depth migration for laterally varying velocities Exp. Abs., SEG Nat. Conv. 737-740

Aminzadeh, F., Brac, J., and Kunz, T. 638 Jan. 1997 SEG/EAGE 3-D Salt and Overthrust Models SEG/EAGE 2-CD's

Amundsen, L., and Reitan, A. 492 Oct. 1994 The relationship between 2-D and 3-D wave propagation Exp. Abs., SEG Nat. Conv. 1387-1381

Anderson, J., Alkhalifah, T., and Tsvankin, I 575 May. 1996 Fowler DMO and time migration for transversely isotropic media Geophysics 835-856

Anderson, J. E., and Tsvankin, I. 670 Jul. 1997 Dip-moveout processing by Fourier transform in anisotropic media Geophysics 1260-1269

Andrews, H. C., and Hunt, B. R. 92 Jan. 1977 Digital Image Restoration (Text) Prentice-Hall Inc.

Apostoiu-Marin, I., and Ehinger, A. 351 Sep. 1993 Kinematics of prestack depth migration Exp. Abs., SEG Nat. Conv. 873-875

Apostoiu-Marin, I., and Ehinger, A. 671 Jul. 1997 Kinematic interpretation in the prestack depth-migrated domain Geophysics 1226-1237

Armstrong, P. N., Chmela, W., and Leaney, W. S. 541 Aug. 1995 AVO calibration using borehole data First Break 319-328

Artley, C. T. 338 Nov. 1991 Dip-moveout processing for depth-variable velocity Exp. Abs., SEG Nat. Conv. 1204-1207

Asakawa, E., and Kawanaka, T. 218 Jan. 1993 Seismic ray tracing using linear traveltime interpolation Geophys. Prosp. 99-111

Ata, E., Michelena, R. J., Gonzales, M., Cerquone, H., and Carry, M 520 Oct. 1994 Exploiting P-S converted waves: Part 2, Application to a fracture reservoir Exp. Abs., SEG Nat. Conv. 240-243

Audebert, F., and Diet, J. P., 274 Sep. 1993 Migrated focus panels: focusing analysis reconciled with prestack depth migration Exp. Abs., SEG Nat. Conv. 961-964

Audebert, F., Nichols, D., Rekdal, T., Biondi, B., Lumley, D. E., and Urdaneta, H. 654 Sep. 1997 Imaging complex geologic structure with single-arrival Kirchhoff prestack depth migration Geophysics 1533-1543

Balch, A. H., and Erdemir, C. 566 Aug. 1994 Sign-change correction for prestack migration of P-S converted wave reflections Geophys. Prosp. 637-663

Bale, R., and Jakubowicz, H. 398 Oct. 1987 Post-stack prestack migration Exp. Abs., SEG Nat. Conv. 714-717

Bancroft, J. C. 91 May. 1990 Accurate Amplitude Calculations for Time Domain DMO CSEG Calgary 1990

Bancroft, J. C., and Geiger, H. D. 440 Oct. 1994 Equivalent offset CRP gathers Exp. Abs., SEG convention

Bancroft, J. C., Geiger, H. D., Foltinek, D. S., and Wang, S. 525 Dec. 1994 Prestack migration by equivalent offset and CSP gathers CREWES Research Report 27.1-27.18

Bancroft, J. C., and Wang, S. 526 Dec. 1994 Converted wave prestack migration and velocity analysis by equivalent offset and CSP gathers CREWES Research Report 28.1-28.7

Bancroft, J. C., Geiger, H. D., Foltinek, D., and Wang, S. 622 May. 1995 Prestack migration by equivalent offsets and common scatterpoint (CSP) gathers Exp. Abs. EAEG P124

Bancroft, J. C., Wang, S., Foltinek, D., and Geiger, H. D. 634 May. 1995 Common scatter point (CSP) prestack migration Exp. Abs. CSG Nat. Conv. 47-48

Bancroft, J. C. 555 Nov. 1995 3-D design, migration and aliasing CREWES Research Report Ch. 26

Bancroft, J. C., Geiger, H. D., Wang, S., and Foltinek, D. S., and 552 Nov. 1995 Prestack migration by equivalent offset and CSP gathers: An Update CREWES Research Report Ch. 23

Bancroft, J. C., and Geiger, H. D. 553 Nov. 1995 Velocity sensitivity for equivalent offsets in CSP gathers CREWES Research Report Ch. 24

Bancroft, J. C. 554 Nov. 1995 Aliasing in prestack migration CREWES Research Report Ch. 25

Bancroft, J. C., and Geiger, H. D. 677 May. 1996 Velocity sensitivity for equivalent offset migration: a contrast in robustness and fragility Exp. Abs. CSEG 149-150

Bancroft, J. C., and Geiger, H. D. 632 May. 1996 Velocity sensitivity for equivalent offset prestack migration: a contrast in robustness and fragility Exp. Abs. CSEG Nat. Conv. 149-150

Bancroft, J. C., and Wang, S. 623 Jun. 1996 Converted wave (P-S)

prestack migration and velocity analysis using equivalent offsets Exp. Abs. EAGE X023

Bancroft, J. C. 625 Nov. 1996 Natural anti-aliasing in equivalent offset prestack migration Exp. Abs. SEG 1465-1466b

Bancroft, J. C. 630 May. 1997 Fast 3-D Kirchhoff poststack migration with migration velocity analysis Exp. Abs. SEG A046

Bancroft, J. C., and Geiger, H. D. 643 May. 1997 Anatomy of CSP gathers formed during equivalent offset migration Exp. Abs. CSEG Nat. Conv.

Bancroft, J. C., Margrave, G., and Geiger, H. D. 620 Nov. 1997 A kinematic comparison of DMO-PSI and equivalent offset migration (EOM) Exp. Abs. SEG 1997 1575-1578

Bancroft, J. C., Margrave, G. F., and Geiger, H. g. 647 Dec. 1997 A kinematic comparison of conventional processing, DMO-PSI, and EOM CREWES Research Report Ch 29

Bancroft, J. C., and Geiger, H. D. 646 Dec. 1997 Anatomy of common scatter point CSP gathers formed during equivalent offset prestack migration (EOM) CREWES Research Report Ch 28

Bancroft, J. C., and Xu, L. 714 May. 1998 Equivalent offset migration for VSP's and data acquired with vertical receiver arrays Exp. Abs. CSEG 36-39

Bancroft, J. C., Margrave, G., and Geiger, H. D. 544 Jun. 1998 A comparison of prestack migration gathers formed by equivalent offset migration (EOM) and DMO-PSI Exp. Abs. EAGE 1-18

Bancroft, J. C., Geiger, H. D., and Margrave, G. F. 639 Nov. 1998 The equivalent offset method of prestack time migration Geophysics 2042-2053

Bancroft, J. C. 702 Dec. 1998 Optimum CSP gathers with a fixed equivalent offset CREWES Research report Ch 44

Bancroft, J. C. 700 Dec. 1998 Equivalent offset migration for vertical receiver arrays CREWES Research report Ch 11

Bancroft, J. C. 704 Dec. 1998 Dip limits on pre- and poststack Kirchhoff migrations CREWES Research report Ch 25

Bancroft, J. C. 701 Dec. 1998 Computational speed of EOM relative to standard Kirchhoff migration CREWES Research report Ch 43

Bancroft, J. C., and Vestrum. R. W. 710 May. 1999 Anisotropic depth migration using the equivalent offset method (EOM) EXP. Abs. CSEG 138-141

Bancroft, J. C., and Xu, L. 715 Jun. 1999 Prestack migration of vertical array data using EOM Exp. Abs. EAGE 2-47

Bancroft, J. C., and Ursenbach, C. p. 753 Sep. 2001 Prestack considerations for the migration of oblique reflectors Exp. Abs. SEG

Barnes, A. E. 499 Oct. 1994 Theory of two-dimensional complex seismic trace analysis Exp. Abs., SEG Nat. Conv. 1580-1583

Barnes, A. E. 737 Dec. 1995 Discussion: Pulse distortion in depth migration Geophysics 1942-1944

Baysal, E., Kosloff, D. D., and Sherwood, J. W. C. 374 Nov. 1983 Reverse time migration Geophysics 1514-1524

Baysal, E., Kosloff, D.D., and Sherwood, J.W.C. 304 Feb. 1984 A two-way nonreflecting wave equation Geophysics 132-141

Bazelaire, E., and Viallix, J. 434 Jul. 1994 Normal moveout in focus Geophys. Prosp. 477-499

Beasley, C., Chambers, R., and Jakubowicz, H. 413 Jun. 1985 Prestack partial migration: a comprehensive solution to problems in the processing of 3-D data Abs., EAEG Convention 80

Beasley, C., Lynn, W., Larner, K., and Nguyen, H. 99 Jul. 1988 Cascaded F-K migration: Removing the Restrictions of Depth-Varying Velocities Geophysics 881-893

Beasley, C., and Lynn, W. 93 Oct. 1989 The Zero Velocity Layer: Migration From Irregular Surfaces Exp. Abs., SEG Nat. Conv. 1179-1183

Beasley, C., and Klotz, R. 344 Nov. 1991 Migration velocity analysis by migration of velocity spectra Exp. Abs., SEG Nat. Conv. 1239-1242

Beasley, C., and Lynn, W. 175 Nov. 1992 The zero-velocity layer: Migration from irregular surfaces Geophysics 1435-1443

Beitzel, J. E., and Davis, J. M. 317 Oct. 1974 A computer oriented velocity analysis interpretation technique Geophysics 619-632

Benson, A. K. 178 Sep. 1991 An explicit, unconditionally stable, 15-degree depth migration and modeling algorithm implemented in poststack, directional, and prestack modes Geophysics 1412-1422

Berg, L. 72 Dec. 1984 Prestack Partial Migration Exp. Abs., SEG Nat. Conv. 796-799

Berkhout, A. J., and Zaanen, P. R. 685 Jan. 1976 A comparison between Wiener filtering, Kalman filtering, and deterministic least squares estimation Geophys. Prosp. 141-197

Berkhout, A. J., and Palthe, D. Van W. 36 Mar. 1979 Migration in Terms of Spatial Deconvolution Geophys. Prosp. 261-291

Berkhout, A. J., and Jong, B. 422 Oct. 1981 Recursive migration in three dimensions Geophys. Prosp. 758-781

Berkhout, A. J. 451 Dec. 1981 Wave field extrapolation techniques in seismic migration, a tutorial Geophysics 1638-1656

Berkhout, A. J. 548 Sep. 1984 Seismic Migration. (B. Practical Aspects) Imaging of Acoustic Energy by Wave Field Extrapolation Elsevier

Berkhout, A. J. 547 Jan. 1995 Seismic Migration. (A. Theoretical Aspects) Imaging of Acoustic Energy by Wave Field Extrapolation Elsevier

Berkhout, A. J. 749 Aug. 1999 Seismic inversion in steps TLE 933-939

Berryhill, J. 419 Oct. 1977 Diffraction response for nonzero separation of source and receiver Geophysics 1158-1176

Berryhill, J. 6 Aug. 1979 Wave Equation Datumming Geophysics 1329-1339

Berryhill, J. 158 Nov. 1984 Wave-equation Datuming Before Stack Geophysics 2064-2066

Berryhill, J. 212 Oct. 1991 Kinematics of crossline prestack migration Geophysics 1674-1676

Bevc, D., Black, J. L., and Palacharla, G. 582 Jul. 1995 Plumes: Response of time migration to lateral velocity variation Geophysics 1118-1127

Bevc, D. 751 Sep. 1997 Flooding the topography: Wave-equation datuming of land data with rugged acquisition topography: , 62, no. 05, . Geophysics 1558-1569

Bickel, S. H. 663 Jun. 1998 The maximum depth principle as a dual to Fermat's minimum time principle Exp. Abs. EAGE 1-55

Biondi, B. 361 Nov. 1986 Shot profile dip moveout using log-stretch transform Exp. Abs., SEG Nat. Conv. 431-434

Biondi, B., and Ronen, J. 369 Nov. 1987 Dip moveout in shot profiles Geophysics 1473-1482

Biondi, B. 353 Sep. 1990 Velocity estimation by beam stack: field-data results Exp. Abs., SEG Nat. Conv. 1271-1274

Biondi, B., and Palacharla, G. 461 Jun. 1994 3-D depth migration using rotated McClellan filter Exp. Abs. EAEG Tech. Exh. B045

Biondi, B., and Palacharla, G. 505 Oct. 1994 3-D migration using rotated McMClellan filters Exp. Abs., SEG Nat. Conv. 1278-1281

Biondi, B., and Chemingui, N. 497 Oct. 1994 Transformation of 3-D prestack data by azimuth moveout (AMO) Exp. Abs., SEG Nat. Conv. 1541-1544

Biondi, B., and Palacharla, G 597 Nov. 1995 3-D depth migration by rotated McClellan filters Geophys. Prosp. 1005-1020

Biondi, B., Fomel, S., and Chemingui, N. 669 Mar. 1998 Azimuth

References

moveout for 3-D prestack imaging Geophysics 574-588

Biondi, B. 768 May. 2002 Stable wide-angled Fourier finite-difference3 downward extrapolation of 3-D wavefields Geophysics 872-882

Black, J. L., and Leong, T. 405 Oct. 1987 A flexible, accurate approach to 1-pass 3-D migration Exp. Abs., SEG Nat. Conv. 559-560

Black, J. L., and Egan, M. 238 Nov. 1988 True-amplitude DMO in 3-D Handout, SGE convention

Black, J. L., and Brzostowski, M. A. 331 Nov. 1991 Steep-dip time migration and residual depth migration Exp. Abs., SEG Nat. Conv. 1140-1143

Black, J. L. 323 Oct. 1992 Prestack remedial migration kinematics Exp. Abs., SEG Nat. Conv. 913-916

Black, J. L., Schleicher, L., and Zhang, L. 205 Jan. 1993 True-amplitude imaging and dip moveout Geophysics 47-66

Black, J. L., and Brzostowski, M. A. 324 Sep. 1994 Systematics of time-migration errors Geophysics 1419-1434

Blacquiere, G., Debeye, W. J., Wapenaar, C. P. A., and Berkhout, A. J. 482 Nov. 1989 3-D table-driven migration Geophys. Prosp. 925-958

Blacquiere, G., Duijndam, A. J. W., and Romijn, R. 684 Jan. 1991 Efficient x-f depth migration of shot records: practical aspects First Break 9-23

Blacquiere, G. 329 Nov. 1991 Optimized McClellan transformation filters applied in one-pass 3D depth extrapolation Exp. Abs., SEG Nat. Conv. 1126-1129

Blanch, J. O., Robertsson, J. O. A., and Symes, W. W. 516 Oct. 1994 Constant Q modeling: Preliminary results for a new algorithm Exp. Abs., SEG Nat. Conv. 1414-1417

Bleistein, N 375 Jul. 1987 On the imaging of reflectors in the earth Geophysics 931-942

Bleistein, N. 748 Jan. 1984 Mathematical Methods for Wave Phenomena Academic Press

Bleistein, N., Cohen, J. K., and Hagin, F. G. 477 Aug. 1985 Computational and asymptotic aspects of velocity inversion Geophysics 1252-1265

Bleistein, N., and Gray, S. H. 198 Nov. 1985 An extension of the Born inversion method to a depth dependent reference profile Geophys. Prosp. 999-1022

Bleistein, N., Cohen, J., and Hagin, F. 81 Jan. 1987 Two and One-Half Dimensional Born Inversion with Arbitrary References Geophysics 26-36

Bleistein, N. 359 Sep. 1990 Born DMO revisited Exp. Abs., SEG Nat. Conv. 1366-1369

Bleistein, N. 744 Aug. 1999 Hagedoorn told us how to do Kirchhoff migration and inversion TLE 981-927

Bleistein, N., Cohen, J. K., and Stockwell, J. W. 745 Jan. 2001 Mathematics of Multidimensional Seismic Imaging, Migration, and Inversion Springer

Blondel, P. 495 Oct. 1994 Amplitude preservation for Kirchhoff dip moveout Exp. Abs., SEG Nat. Conv. 1529-1532

Bolondi, G., Rocca, F., and Savelli, S. 39 Dec. 1978 A Frequency Domain Approach to Two Dimensional Migration Geophys. Prosp. 750-772

Bolondi, G., Loinger, E., and Rocca, F. 450 Dec. 1982 Offset continuation of seismic sections Geophys. Prosp. 813-828

Bolondi, G., Loinger, E., Rocca, F. 82 Dec. 1984 Offset Continuation in Theory Practice Geophys. Prosp. 1045-1073

Booer, A., Chambers, J., and Mason, I. 29 Jan. 1940 Numerical Holographic Reconstruction by a Projective Transform Electron. Lett.

Bording, P. 656 Jan. 1995 Seismic Wave Modeling and Inversion www.arsc.sunyit.edu/csep/

Bortfeld, R. 534 Mar. 1989 Geometrical ray theory: Rays and traveltime in seismic systems (second order approximation of the traveltimes) Geophysics 342-349

Bortfeld, R., and Kiehn, M. 291 Nov. 1992 Reflection amplitudes and migration amplitudes (zero-offset situation) Geophys. Prosp. 873-884

Boumahdi, M., and Glangeaud, F. 500 Oct. 1994 Nonminimum phase blind deconvolution: Parametric approach Exp. Abs., SEG Nat. Conv. 1595-1598

Bracewell R. 385 Jan. 1965 The Fourier Transform and its Applications (text) McGraw-Hill

Brown, R. L. 244 Oct. 1992 Estimation of AVO in the presence of noise using prestack migration Exp. Abs., SEG Nat. Conv. 885-888

Bruin, C. G. M., Wapenaar, C. P. A., and Berkhout, A. J. 777 Sep. 1990 Angle-dependent reflectivity by means of prestack migration Geophysics 1223-1234

Brune, R. H., O'Sullivan, B., and Lu, L. 599 Jul. 1994 Comprehensive analysis of marine 3-D bin coverage The Leading Edge 757-762

Brysk, H. 100 May. 1983 Numerical Analysis of the 45-degree Finite-difference Equation for Migration Geophysics

Brysk, H. 334 Sep. 1990 Slotnick revisited: DMO with linear velocity profile Exp. Abs., SEG Nat. Conv. 1350-1353

Brysk, H., and Majesty, T. 337 Nov. 1991 3-D isochron illuminates variable-velocity DMO Exp. Abs., SEG Nat. Conv. 1197-1200

Brzostowski, M. A., Black, J. A. 179 Jul. 1990 Steep-dip time migration and residual depth migration EAEG/SEG Res. Wkp. Seismic velocities, Cambridge, England

Buchanan, D. 19 Jan. 1978 An Exact Solvable One Way Wave Equation Presented at SEG Nat. Conv.

Bunks, C. 343 Nov. 1991 Optimal velocity model inversion from analysis of Paraxial wave-equation focusing spots Exp. Abs., SEG Nat. Conv. 1226

Bunks, C. 161 Oct. 1992 Optimization of Paraxial Wave Exp. Abs., SEG Nat. Conv. 897-900

Bunks, C. 446 Oct. 1992 Paraxial wave-equation inversion with geometric constraints Exp. Abs. SEG Nat. Conv. 901-904

Bunks, C. 292 Sep. 1993 Effective filtering of artifacts for Finite-Difference implementations of paraxial wave equation migration Exp. Abs., SEG Nat. Conv. 1044-1047

Bunks, C. 536 Feb. 1995 Effective filtering of artifacts for implicit finite-difference paraxial wave equation migration Geophys. Prosp. 203-220

Butler, K. E., Russell, R.D. 232 Jun. 1993 Subtraction of powerline harmonics from geophysical records Geophysics 898-903

Cabrera, J., and Levy, S. 156 Aug. 1989 Shot Dip Moveout with Logarithmic Transformations Geophysics 1038-1041

Cambois, G. 160 Dec. 1991 A Proof for the Convergence of the 15-Degree Cascaded Migration Geophysics

Cambois, G., and Stoffa, P.L. 195 Jun. 1992 Surface-consistent deconvolution in the log/Fourier domain Geophysics 823-840

Canales, L. 429 Aug. 1976 Mixed L1 and L2 norm problems SEP 114-124

Canales, L. 432 Sep. 1976 A quantile finding program SEP 99-100

Canning, A., and Gardner, G. H. 658 Mar. 1996 A two-pass approximation to 3-D prestack migration Geophysics 409-421

Carcione, J. M., and Cavallini, F. 579 Mar. 1995 Forbidden directions for inhomogeneous pure shear waves in dissipative anisotropic media Geophysics 522-530

Carnahan, B., Luther, H.A., and Wilkes, J.O. 296 Jan. 1969 Applied Numerical Methods John Wiley and Sons

Carrion, P. M., Sato, H. K., and Buono, A.V.D. 177 Jun. 1991 Wavefront sets analysis of limited aperture migration sections Geophysics 778-784

Carter, J., and Frazer, N. 69 Jan. 1984 Accommodating Lateral Velocity changes in Kirchhoff Migration by means of Fermat's Principle Geophysics 46-53

Cary, P. W. 106 Jan. 1991 Symmetric One Pass 3-D Migration CSEG Recorder

Cary, P. W. 230 May. 1992 Applying dip moveout to land data: accounting for topography and irregular acquisition geometry Exp. Abs., CSEG Nat. Con. 25-26

Cary, P. W. 278 Sep. 1993 3-D migration with four-way splitting Exp. Abs., SEG Nat. Conv. 978-981

Castle, R. J. 285 Sep. 1993 A 2-D V(z) DMO algorithm Exp. Abs., SEG Nat. Conv. 1030-1033

Castle, R. J. 435 Jun. 1994 A theory of normal moveout Geophysics 983-999

Cerjan., C., Kosloff, D., and Kosloff, R., Reshef, M. 60 Apr. 1985 A Non-Reflecting Boundary Condition for Discrete Acoustic and Elastic Wave Equations Geophysics 705-708

Cerveny, V. 784 Jan. 1972 Seismic rays and ray intensities in inhomogeneous anisotropic media Geophysics 1-13

Cerveny, V., and Soares, J.E.P., 170 Jul. 1992 Fresnel volume ray tracing; Geophysics 902-915

Chambers, R., Jakubowicz, H., Larner, K., and Yang, M. 410 Oct. 1984 Suppression of backscattered coherent noise by prestack partial migration Exp. Abs., SEG Nat. Conv. 451-454

Chan, Wai-Kin 699 Oct. 1998 Analysing Converted-Wave Seismic Data: Statics, Interpolation, Imaging, and P-P Correlation Ph.D. Thesis, University of Calgary

Chang, H., VanDyke, J. P., Solano, M., McMechan, G.A., and Epili, D. 490 Mar. 1998 3-D prestack migration: From prototype to production in a massively parallel processor environment Geophysics 546-556

Chang, W., and McMechan, G. 103 Apr. 1989 3-D Acoustic Reverse Time Migration Geophys. Prosp.

Charles, S., Kapotas, S., and Phadke, S. 509 Oct. 1994 Absorbing boundaries with Huygens' secondary sources Exp. Abs., SEG Nat. Conv. 1363-1366

Chernis, L. 707 Sep. 1998 Depth migration by generalised equivalent migration offset mapping Exp. Abs. SEG 1550-1553

Chon, Y., and Gonzalez, A. 421 Oct. 1987 Accuracy in RMS-velocity determination using a Kirchhoff DMO algorithm Exp. Abs., SEG Nat. Conv. 722-725

Chun, J., and Jacewitz, C. 34 May. 1981 Fundamentals of Frequency Domain Migration Geophysics 717-733

Claerbout, J. F 433 Jun. 1976 Absorptive side boundaries for migration SEP 10-15

Claerbout, J. F. 9 Jun. 1970 Course Grid Calculations of Waves in Inhomogeneous Media with Application to Delineation of Complicated Seismic Structure Geophysics 407-418

Claerbout, J. F. 10 Dec. 1970 Acoustical Holography, Vol. 3. ed. A.F. Metered Plenum Press, New York 273-283

Claerbout, J. F. 7 Jun. 1971 Toward a Unified Theory of Reflector Mapping Geophysics 467-481

Claerbout, J. F., and Johnson, A., 11 Dec. 1971 Extrapolation of Time Dependent Wave Forms along their Path of Propagation Geophys. Jnl. Royal Astronomical Soc. 285-293

Claerbout, J. F., and Doherty, S.M. 293 Oct. 1972 Downward continuation of moveout corrected seismograms Geophysics 741-768

Claerbout, J. F. 430 Jul. 1976 Inverse incomplete beta functions of one half SEP 109-113

Claerbout, J. F. 431 Jul. 1976 Probability and entropy of seismic data SEP 101-108

Claerbout, J. F., and Clayton, R.W. 236 Sep. 1976 A paraxial equation for elastic waves EP Progress Report 165-170

Claerbout, J. F. 23 Dec. 1976 Fundamentals of Geophysical Data Processing McGraw-Hill, New York

Claerbout, J. F. 294 Jan. 1985 Imaging the Earth's Interior Blackwell Sci., available through Email (out of print)

Claerbout, J. F. 535 Jan. 1992 Earth Soundings Analysis: Processing versus Inversion Blackwell Scientific Publications

Claerbout, J. F. 617 May. 1996 Basic Earth Imaging (Version 2.2): A Living Document Stanford University 178

Clayton, R. 237 Sep. 1976 Programming absorptive side boundaries for migration EP Progress Report 16-29

Clayton, R., and Engquist, B. 17 Jan. 1977 Absorbing Boundary Conditions for Acoustic and Elastic Wave Equations Bull. Seis. Soc. Am. 1529-1540

Clayton, R., and Engquist, B. 114 May. 1980 Absorbing Boundary Conditions for Wave Equation Migration Geophysics 895-904

Clayton, R., and Stolt, R. 127 Nov. 1981 A Born-WKBJ Inversion Method for Acoustic Reflection Data Geophysics

Cohen, J., and Bleistein, N. 46 Jan. 1977 An Inverse Method for Determining Small Variations in Propagation Speed Soc. Ind. Appl. Math., J. Appl. Math. 784-99

Cohen, J., and Bleistein, N. 45 Jun. 1979 Velocity Inversion Procedure for Acoustic Waves Geophysics 1077-1087

Cohen, J., Hagin, F., and Bleistein, N. 126 Aug. 1986 Three-Dimensional Born Inversion with an Arbitrary Reference Geophysics

Compani-Tabrizi, B. 59 Dec. 1986 k-t Scattering Formulation of the Absorptive Acoustic Wave Equation: Wraparound and Edge Effect Elimination Geophysics 2185-2192

Coultrip, R.L. 219 Feb. 1993 High-accuracy wavefront tracing traveltime calculation Geophysics 284-292

Craig, M. S., Long, L. T., and Tie, A. 180 Dec. 1991 Modeling the seismic P coda as the response of a discrete-scatterer medium Phys. Earth Plan. Int. 20-35

Crase, E. 681 Sep. 1990 High-order (space and time) Finite-difference modeling of the elastic wave equation Exp. Abs. SEG 987-991

Cross, G. M. 260 Oct. 1992 Array responses for plane and spherical incidence Geophysics 1294-1306

Dai, N., and Kanasewich, E.R. 253 Oct. 1992 One-way wave equation: absorbing boundary and seismic migration Exp. Abs., SEG Nat. Conv. 1026-1029

Dai, N., Kanasewich, E.R., and Vafidis, A., 281 Sep. 1993 Composed absorbing boundaries for seismic modeling Exp. Abs., SEG Nat. Conv. 1052-1055

Dai, N., and West, G. F. 517 Oct. 1994 Inverse Q migration Exp. Abs., SEG Nat. Conv. 1418-1421

Dai, N., Vafidis, A., and Kanasewich, E. R. 595 Jul. 1996 Seismic migration and absorbing boundaries with one-way wave system for heterogeneous media Geophys. Prosp. 719-737

Daly, J. W. 220 Apr. 1948 An instrument for plotting reflection data on the assumption of linear increase of velocity Geophysics 153-157

Dankbaar, J. W. M. 778 Nov. 1985 Separation of P- and S-waves Geophys. Prosp. 970-986

Datta, S., and Bhowmick,A. 418 Sep. 1974 Seismic model studies on diffraction of waves by edged of varying radius of curvature and depth Geophys. Prosp. 534-545

de Bruin, C. G. M., Wapenaar, C. P. A., and Berkhout, A. J. 570 Sep. 1990 Angle-dependent reflectivity by means of prestack migration Geophysics 1223-1234

Dellinger, J., and Etgen, J. 427 Oct. 1989 Wave-type separation in

References

3-D anisotropic media Exp. Abs., SEG Nat. Conv. 977-979

Dellinger, J., Gray, S., Murphy, G., Etgen, J., and Fei, T., 717 Nov. 1999 Efficient two and one-half dimension true-amplitude migration Exp. Abs., SEG Nat. Conv. 1540-1543

Denelle, E., Dezard, Y., and Raoult, J. J. 298 Nov. 1986 2-D prestack depth migration in the (S-G-W) domain Exp. Abs., SEG Nat. Conv. 327-330

Deregowski, S. M. 14 Nov. 1978 A Finite Difference Method for CDP Stacked Section Migration Presented EAEG

Deregowski, S. M., and Rocca, F. 307 Jun. 1981 Geometrical optics and wave theory of constant offset sections in layered media Geophys. Prosp. 374-406

Deregowski, S. M. 449 Jun. 1982 Dip-moveout and reflector point dispersal Geophys. Prosp. 318-322

Deregowski, S. M. 76 Jul. 1986 What is DMO? First Break 7-24

Deregowski, S. M. 211 Jun. 1990 Common-offset migrations and velocity analysis First Break 225-234

De-Segonzac, P. D, and Laherrere, j, 733 Mar. 1959 Applications of the continuous velocity log to anisotropy measurements in northern Sahara-results and consequenses Geophys. Prosp. 202-217

Dessing, F. J., and Wapenaar, C. P. A. 465 Jun. 1994 Wave field extrapolation in the space-wavenumber domain Exp. Abs. EAEG Tech. Exh. H032

Devey, M. 42 Jan. 1979 Derivation of the Migration Integral Technical Note TN451, BP Company Ltd., Exploration and Production Department

Dickinson, J. A. 365 Jan. 1988 Evaluation of two-pass three-dimensional migration Geophysics 32-49

Diet, J. P., and Audebert, F. 318 May. 1990 A focus on focusing Handout, EAEG meeting 1-14

Diet, J. P. 319 Apr. 1991 Prestack depth migration or 'How to beat a (ray) path to your prospect' CGG handout 1-12

Diet, J. P., Audebert, F., Huard, I., Lanfranchi, P., and Zhang, X., 273 Sep. 1993 Velocity analysis with prestack time migration using the S-G method: A unified approach Exp. Abs., SEG Nat. Conv. 957-

Dietrich, M., and Cohen, J.K. 249 Oct. 1992 Three-dimensional dip moveout operator for linear velocity-depth functions Exp. Abs., SEG Nat. Conv. 958-961

Dillon, P. B. 209 Jun. 1988 Vertical seismic profile migration using the Kirchhoff integral Geophysics 786-799

Dix, C. H. 442 Jan. 1955 Seismic velocities from surface measurements Geophysics 68-86

Docherty, P. 128 Oct. 1987 Two-and-One-Half-Dimensional Poststack Inversion Exp. Abs. SEG Nat. Conv. 564

Docherty, P. 121 Aug. 1991 A Brief Comparison of some Kirchhoff Integral Formulas for Migration and Inversion Geophysics 1164-1169

Docherty, P. 259 Oct. 1992 Solving for the thickness and velocity of the weathering layer using 2-D refraction tomography Geophysics 1307-1318

Doherty, S. 22 Dec. 1975 Structure Independent Seismic Velocity Estimation Ph.D. Thesis, Stanford University, Ca.

Domenico, S. 96 Dec. 1989 Discussion on Inquest of the Flank Geophysics

Donati, M., Martin, N. B., and Bancroft, J. C. 559 Dec. 1994 Steep dip Kirchhoff migration for linear velocity gradient CREWES Research report Ch. 17

Dong, W., Emanuel, M., Bording, P., and Bleistein, N. 129 Sep. 1991 A Computer Implementation of 2.5-D Common-Shot Inversion Geophysics

Dong, Z., and McMechan, G.A. 207 Jan. 1993 3-D prestack migration in anisotropic media Geophysics 79-90

Downie, S. P., Hartley, B. M., and Uren, N. F. 589 Jun. 1995 Interactive attenuation of seismic multiples in the radial domain Exploration Geophysics 486-492

Dubrulle, A. A. 454 Sep. 1983 Numerical methods for the migration of constant-offset sections in homogeneous and horizontally layered media Geophysics 1195-1203

Dunne, J., and Beresford, G. 538 Mar. 1995 A review of the t-p transform, its implementation and its application in seismic processing Exploration Geophysics (ASEG) 19-36

Durrani, T. S., and Bisset, D. 613 Aug. 1994 The Radon transform and its properties Geophysics 1180-1187

EAEG 474 Jun. 1994 Program Listing EAEG Exp. Abs. EAEG Tech. Exh.

Eaton, D. W. S. 779 Dec. 1989 The free surface effects:implications for amplitude-verses-offset inversion Cdn. Jnl. Exp. Geo. 97-103

Eaton, D. W. S., Stewart, R., and Harrison, M. 107 Mar. 1991 The Fresnel Zone for P-SV Waves Geophysics

Eaton, D. W. S., and Stewart, R. R. 458 Jan. 1994 Migration/inversion for transversely isotropic elastic media Geophys. J. Int. 667-683

Ehinger, A., Lailly, P., and Marfurt, K. 585 Nov. 1996 Green's function implementation of common-offset, wave-equation migration Geophysics 1813-1821

Emerman, S., Schmidt, W., and Stephen R. 125 Nov. 1982 An Implicit Finite-Difference Formulation of the Elastic Wave Equation Geophysics 1521-1526

Emsley, D., Boswell, P., and Davis, P. 689 Jun. 1998 Sub-basalt imaging using long offset reflection seismic data Exp. Abs. EAGE 1-48

Engquist, B. 235 Sep. 1976 Absorbing boundary conditions for wave equations EP Progress Report 30-49

Engquist, B., and Clayton, R. 234 Sep. 1976 Difference approximations of one way elastic waves SEP Progress Report 141-164

Engquist, B. 428 Sep. 1976 One way elastic wave equations SEP 125-140

Estes, L., and Fain, G. 35 Jan. 1977 Numerical Technique for Computing the Wide Angle Acoustic Field in an Ocean with Range-Dependent Velocity Profiles J. Acoust. Soc. Am. 38-43

Estevez, R. 26 Dec. 1977 Wide Angle Diffracted Multiple Reflections Ph.D. Thesis, Stanford University, Ca.

Etgen, J. T. 321 Oct. 1988 Prestack Migration of P and SV-waves Exp. Abs., SEG Nat. Conv. 972-975

Faye, J. P., and Jeannot, J. P. 447 Nov. 1986 Prestack migration velocities from focusing depth analysis Exp. Abs. SEG Nat. Conv. 438-440

Faye, J. P., and Jeannot, J. P. 363 Nov. 1986 Prestack migration velocities from focusing depth analysis Exp. Abs., SEG Nat. Conv. 438-440

Ferber, R. G. 133 Aug. 1991 A Filter, Delay and Spread Technique for 3D DMO Geophys. Prosp.

Ferber, R. G. 352 Mar. 1994 Migration to multiple offset and velocity analysis Geophy. Prosp. 99-112

Ferguson, R. J., and Margrave, G.F. 766 Mar. 2002 Prestack depth migration by symmetric nonstationary phase shift Geophysics 594-603

Filpo, E., and Hubral, p. 533 Jan. 1995 Numerical tests of 3D true-amplitude zero-offset migration Geophys. Prosp. 119-134

Fischer, R., Lees, J.M. 217 Jul. 1993 Shortest path ray tracing with sparse graphs Geophysics 987-996

Fisher, E., McMechan, G.A., Annan, A. P., and Cosway, S. W. 488

Apr. 1992 Examples of reverse-time migration of single-channel, ground penetrating radar profiles Geophysics 577-586

Forel D., and Gardner, G. 111 May. 1988 A Three-Dimensional Perspective on Two-Dimensional Dip Moveout Geophysics 604-610

Fornberg, B. 380 May. 1988 The pseudospectral method: Accurate representation of interfaces in elastic wave calculations Geophysics 625-637

Foster, D. J., and Carrion, P. M. 780 Nov. 1985 Full wave equation downward continuation of seismic reflection data Geophys. Prosp. 929-942

Fowler, P. 680 Dec. 1984 Velocity independent imaging of seismic reflectors Exp. Abs. SEG 383-385

Fowler, P. J. 254 Oct. 1992 Improving prestack F-K migration velocity analysis using residual moveout Exp. Abs., SEG Nat. Conv. 1038-1041

Fowler, P. J. 515 Oct. 1994 Finite-difference solutions of the 3-D eikonal equation in spherical coordinates Exp. Abs., SEG Nat. Conv. 1394-1397

Fowler, P. J. 619 Nov. 1997 A comparative overview of prestack time migration methods Exp. Abs. SEG 1997 1571-1574

Fowler, P. J. 690 Jun. 1998 A comparative overview of prestack time migration Exp. Abs. EAGE 1-17

Fowler, P. J. 697 Sep. 1998 A comparative overview of dip moveout methods Exp. Abs. SEG 1744-1747

Frei, W. 539 Apr. 1995 Refined field static corrections in near-surface reflection profiling across rugged terrain The Leading Edge 259-262

Freire, S. L. M., and Ulrych, T.J. 208 Jun. 1988 Application of singular value decomposition to vertical seismic profiling Geophysics 778-785

French, W. 4 Dec. 1975 Computer Migration of Oblique Seismic Reflection Profiles Geophysics 961-980

French, W. 87 Jan. 1985 Partial Migration of True 3-D Seismic Reflection Surveys via Common Reflection-Point Stacking Pre-print, Offshore Tech. Conf..

French, W. 78 Sep. 1986 Trends in Seismic Data Processing The Leading Edge

Fricke, J 98 Sep. 1988 Reverse Time Migration in Parallel; a Tutorial Geophysics

Fruhn, J., White, R. S., Fliedner, M., Richardson, K. R., Cullen, E., Latkiewicz, C., Kirk, W., and Smallwood, J. R. 688 Jun. 1998 Two-ship large aperture seismic profiles-application to imaging through basalt Exp. Abs. EAGE 1-47

Gaiser, J. E. 504 Oct. 1994 Effective use of the Laplacian transformation for explicit finite-difference 3-D migration Exp. Abs., SEG Nat. Conv. 1274-1277

Gaiser, J. E. 660 Jul. 1996 Multicomponent Vp/Vs correlation analysis Geophysics 1137-1149

Galbraith, M. 439 Feb. 1994 3-D survey design by computer CSEG Recorder 4-8

Gardner, G. H. F., French, W., and Matzuk, T. 5 Dec. 1974 Elements of Migration and Velocity Analysis Geophysics 811-825

Gardner, G. H. F., Gardner, L. W., and Gregory, A. R. 241 Dec. 1974 Formation velocity and density - the diagnostic basics for stratigraphic traps Geophysics 770-780

Gardner, G. H. F., (Editor) 528 Jan. 1985 Migration of Seismic Data (Reprint series) SEG

Gardner, G. H. F., Wang, S. Y., Pan, N. D., and Zhang, Z. 322 May. 1986 Dip moveout and prestack imaging Exp. Abs., Offshore Technology 75-84

Gardner, G. H. F., and Canning, A. 498 Oct. 1994 Effects of irregular sampling on 3-D prestack migration Exp. Abs., SEG Nat. Conv. 1553-1556

Gardner, L. W. 443 Apr. 1947 Vertical velocities from reflection shooting Geophysics 221-228

Garibotto, G. 40 Jan. 1979 2-D Recursive Filters for the Solution of Two-Dimensional Wave Equations IEEE Trans. on ASSP 367-373

Gazdag, J. 31 Jan. 1978 Extrapolation of Seismic Waveforms by Fourier Methods IBM J. Res. Dev. 481-486

Gazdag, J. 30 Dec. 1978 Wave Equation Migration with the Phase-Shift Method Geophysics 1342-1351

Gazdag, J. 32 Feb. 1980 Wave Equation Migration with the Accurate Space Derivative Method Geophys. Prosp. 60-70

Gazdag, J. 302 Jun. 1981 Modeling of the acoustic wave equation with transform methods Geophysics 854-859

Gazdag, J., and Sguazzero, P. 303 Feb. 1984 Migration of seismic data by phase shift plus interpolation Geophysics 124-131

Gazdag, J., and Sguazzero, P. 305 Jun. 1984 Interval velocity analysis by wave extrapolation Geophys. Prosp. 454-479

Gazdag, J., and Squazzero, P. 102 Oct. 1984 Migration of Seismic Data Proceedings of IEEE 1302-1315

Geiger, H. D., Bancroft, J. C., and Foltinek, D. 635 May. 1995 Prestack migration of crustal data using common scatter point (CSP) migration technique Exp. Abs. CSG Nat. Conv. 49-50

Geiger, H. D., and Bancroft, J. C. 557 Nov. 1995 Equivalent offset prestack migration for rugged topography CREWES Research Report Ch. 28

Geiger, H. D., and Bancroft, J. C. 558 Nov. 1995 Prestack migration to an unmigrated zero offset section CREWES Research Report Ch. 29

Geiger, H. D., and Bancroft, J. C. 631 May. 1996 Forming an unmigrated zero-offset stacked section with no NMO using equivalent offset prestack migration Exp. Abs. CSEG Nat. Conv. 147-148

Geiger, H. D., and Bancroft, J. C. 624 Jun. 1996 Equivalent offset prestack migration - application to rugged topography and formation of unmigrated zero-offset section Exp. Abs. EAGE P098

Geiger, H. D., and Bancroft, J. C. 628 Nov. 1996 Equivalent offset prestack migration for rugged topography Exp. Abs. SEG 447-450

Geiger, H. D. 743 Dec. 2001 Migration Ph.D. Thesis

Geoltrain, S., and Brac, J. 148 Nov. 1991 Can We Image Complex Structures with Finite-Difference Traveltimes? Exp. Abs., SEG Nat. Conv. 1110-1113

Gesbert, S. 767 May. 2002 From acquisiton footprint to true amplitude Geophysics 830-839

Gibson, B., Larner, K., and Levin, S. 44 May. 1979 Efficient 3D migration in 2 steps Handout EAEG, Hamburg

Gibson, R. L., Lee, J. M., Dini, I., and Cameli, G. M. 507 Oct. 1994 The application of 3-D Kirchhoff migration to VSP data from complex geological settings Exp. Abs., SEG Nat. Conv. 1290-1293

Godfrey, R. J. 248 Oct. 1992 DMO and V(z). Exp. Abs., SEG Nat. Conv. 952-954

Gonzales-Serrano, and Claerbout, J. 186 Apr. 1981 Deformations of CMP gathers with v(z) to hyperbolas Stanford Exploration Project Report 137-156

Gray, S. 747 Aug. 1999 True-amplitude migration TLE 917

Gray, S. H. 457 Jan. 1984 A problem of discrete approximations to an arbitrarily varying one-dimensional seismic model Geophys. J. R. Ast. Soc. 431-437

Gray, S. H. 64 Aug. 1986 Efficient Traveltime Calculations for Kirchhoff Migration Geophysics 1685-1688

Gray, S. H., and Epton, M. A. 184 Jul. 1990 Multigrid migration:

References

Reducing the migration aperture but not the migrated dips Geophysics 856-862

Gray, S. H., Jacewitz, C. A., and Epton, M.E. 181 May. 1991 Analytic synthetic seismograms for depth migration testing Geophysics 697-700

Gray, S. H. 163 Jul. 1992 Frequency-selective design of the Kirchhoff migration operator Geophys. Prosp. 565-571

Gray, S. H., and May, W. P. 480 May. 1994 Kirchhoff migration using eikonal equation travel times Geophysics 810-817

Gray, S. H., Etgen, J., Dellinger, J. and Whitmore, D., 758 Jan. 2001 Seismic migration problems and solutions Geophysics 1622-1640

Grech, G. K. 641 May. 1998 Numerical seismic modeling of exposed structures UofC 701/CREWES/FRP 1-30

Grech, M. G.K., and Bancroft, J. C. 703 Dec. 1998 Building complex velocity models using EOM - a case study CREWES Research report Ch 27

Grechka, V., and Tsvankin, I. 771 May. 2002 NMO-velocity surfaces and Dix-type formulas in anisotropic heterogeneous media Geophysics 939-951

Gruber, T., Greenhalgh, S., and Zhou, B. 764 Mar. 2001 A later-arrival based inversion scheme to recover diffractors and reflectors Expl. Geoph. (ASEG) 1-10

Haddow, C. M. 311 Oct. 1992 Hybrid migration by Finite-Difference and Phase Shift Exp. Abs., SEG Nat. Conv. 893-896

Hagedoorn, J. 1 Jun. 1954 A Process of Seismic Reflection Interpretation Geophys. Prosp. 85-127

Hagedoorn, J. G. 731 Jan. 1954 A practical example of an anisotropic velocity-layer Geophys. Prosp. 52-60

Hagin, F., and Cohein, J. 400 Feb. 1984 Refinements to linear velocity inversion theory Geophysics 112-118

Hale, D 637 Jan. 1992 Dip Moveout Processing SEG Course notes

Hale, D. 416 Jan. 1983 Dip-Moveout by Fourier Transform Dissertation, Stanford University

Hale, D. 57 Jun. 1984 Dip Move-Out by Fourier Transform Geophysics 741-757

Hale, D. 357 Sep. 1990 3-D depth migration via McClellan transforms Exp. Abs., SEG Nat. Conv. 1325-1328

Hale, D. 101 Jun. 1991 A Nonaliased integral method for Dip Moveout Geophysics

Hale, D. 169 Nov. 1991 Stable explicit depth extrapolation of seismic wavefields Geophysics 1770-1777

Hale, D. 159 Nov. 1991 3-D Depth Migration via McClellan Transformations Geophysics

Hale, D., Hill, N. R., and Stefani, J. 174 Nov. 1992 Imaging salt with turning seismic waves Geophysics 1453-1462

Hale, D., and Artley, C. 221 Feb. 1993 Squeezing dip moveout for depth-variable velocity Geophysics 257-264

Hanitzsch, C. 473 Jun. 1994 Amplitude preserving 2.5D Kirchhoff migration for lateral inhomogeneous media Exp. Abs. EAEG Tech. Exh. P106

Hanitzsch, C. 501 Oct. 1994 Vector diffraction stack migration: A fast technique for amplitude-preserving Kirchhoff migration Exp. Abs., SEG Nat. Conv. 1183-1186

Harris, C., Marcoux, M., and Bickel, S. 563 Jun. 1998 Aperture selection to improve Kirchhoff depth imaging using the maximum depth principle Exp. Abs. EAGE 1-56

Harris, C. E., and McMachan, G. A. 489 Jun. 1992 Using downward continuation to reduce memory requirements in reverse-time-migration Geophysics 848-853

Harrison, M. P. 360 Sep. 1990 Dip moveout for converted wave (P-SV) data in constant velocity medium. Exp. Abs., SEG Nat. Conv. 1370-1373

Harrison, M. P., and Stewart, R.R. 265 Aug. 1993 Poststack migration of P-SV seismic data Geophysics 1127-1135

Hatherly, P. J., and Neville, M.J. 242 Feb. 1986 Experience with the generalized reciprocal method of seismic refraction interpretation for shallow engineering site investigation Geophysics 255-265

Hatton, L., Larner K., and Gibson, B. 12 May. 1981 Migration of Seismic Data from Inhomogeneous Media Geophysics 751-767

Hatton, L., Worthington, M. H., and Makin, J. 549 Jan. 1986 Seismic Data Processing, Theory and Practice Blackwell

Hedley, J. P. 330 Nov. 1991 3-D migration via McClellan transformations on hexagonal grids Exp. Abs., SEG Nat. Conv. 1130-1133

Helbig, K. 297 Jul. 1983 Elliptical anisotropy - Its significance and meaning Geophysics 825-832

Helbig, K. 481 Feb. 1990 Rays and wavefront charts in gradient media Geophys. Prosp. 189-220

Herbert, V. 246 Oct. 1992 F-K depth migration Exp. Abs., SEG Nat. Conv. 935-938

Herman, G. C. 164 Jan. 1992 Generalization of traveltime inversion Geophysics 9-14

Hildebrand, S. 347 Nov. 1991 Migration velocity analysis Exp. Abs., SEG Nat. Conv. 1268-1272

Hill, N. R. 134 Nov. 1990 Gausian Beam Migration Geophysics

Hill, N. R., Watson, T. H., Hassler, M. H., and Sisemore, L. K. 332 Nov. 1991 Salt-flank imaging using Gausian Beam migration Exp. Abs., SEG Nat. Conv. 1178-1180

Hilterman, F. J. 200 Dec. 1970 Three-dimensional seismic modeling Geophysics 1020-1037

Hilterman, F. J. 686 Oct. 1975 Amplitudes of seismic waves - a quick look Geophysics 745-762

Hodgkins, M. A., and O'Brien, M. J. 604 Aug. 1994 Salt sill deformation and its applications for subsalt exploration The Leading Edge 849-851

Hodgkinson, J. 735 May. 1970 The in situ determination of velocity anisotropy in sedimentary rocks bt the analysis os normal moveout data Masters Thesis, Ucalgary

Hoffmann, H. J., and Klaeschen, D. 471 Jun. 1994 Common-fault-point imaging by raytracing-based pre-stack depth migration Exp. Abs. EAEG Tech. Exh. P104

Holberg, 0. 88 Feb. 1988 Toward optimum one-way propagation Geophys. Prosp. 99-114

Hood, P. 18 Dec. 1978 Finite Difference and Wave Number Migration Geophys. Prospecting 773-789

Hood, P., (book edited by Fitch) 77 Jan. 1981 Developments in Geophysical Exploration Methods (Chapter 6 by Hood)) Applied Science Publishers Ltd. 151-230

Hornbostel, S. 424 Dec. 1991 Spatial prediction filtering in the t-x and f-x domains Geophysics 2019-2026

Hosken, J. W. J. 43 Jan. 1979 Improvements in the Practice of 2D Diffraction Stack Migration Report No. EPR/R1247, BP Company Limited, Exploration and Production Department

Hosken, J. W. J., and Deregowski, S.M. 202 Feb. 1985 Tutorial migration strategy Geophys. Prosp. 1-33

Hospers, J. 204 Feb. 1985 Sideswipe reflections and other external and internal reflections from salt plugs in the Norwegian-Danish Basin Geophys. Prosp. 52-71

House, W. M., and Pritchett, J.A. 603 Aug. 1994 Salt deformation modeling through the use of enhanced seismic imaging techniques The Leading Edge 844-848

Hovem, J. M. 776 Jul. 1995 Acoustic waves in finely layered media Geophysics 1217-1221

Howell, L. H., and Trefethen, L. N. 379 May. 1988 Ill-posedness of absorbing boundary conditions for migration Geophysics 593-603

Hubral, P. 3 Dec. 1977 Time migration-some ray theoretical aspects Geophys. Prosp. 738-745

Hubral, P., and Krey, T. 191 Apr. 1980 Interval Velocities from Seismic Reflection Time Measurements SEG

Hubral, P. 299 Aug. 1983 Computing true amplitude reflections in a laterally inhomogeneous earth Geophysics 1051-1062

Hubral, P., Tygel, M., and Zein, H. 132 Jan. 1991 Three-Dimensional True-Amplitude Zero Offset Migration Geophysics 18-26

Hubral, P., Schleicher, J., and Tygel, M. 572 May. 1996 A unified approach to 3-D seismic reflection imaging, Part1: Basic concepts Geophysics 742-758

Hunter, J. A., and Pullan, S. E. 568 Jan. 1990 A vertical array method for shallow seismic refraction of the sea floor Geophysics 92-96

Igel, H., Mora, P., and Riollet, B. 583 Jul. 1995 Anisotropic wave propagation through finite-difference grids Geophysics 1203-1216

Igel, H., Mora, P., and Riollet, B. 775 Jul. 1995 Anisotropic wave propagation through finite-difference grids Geophysics 1302-1216

Isaac, J. H., and Lawton, C. 709 Nov. 1998 Multicomponent anisotropic physical modelling FRP Research Report 9.1-9.18

Iverson, W. P. 455 Jun. 1987 Combining attenuation by Q and spherical divergence Geophysics 740-744

Jackson, G. M., Mason, I. M., and Lee, D. 167 Nov. 1991 Multicomponent common-receiver gatherer migration of single-level walk-away seismic profiles Geophys. Prosp. 1015-1029

Jackson, M. P. A., Vendeville, B. C., and Schultz-ela, D. D. 602 Aug. 1994 Salt-related structures in the Gulf of Mexico: A field guide for geophysicists The Leading Edge 837-842

Jain, S., and Wren, A. E. 476 Feb. 1980 Migration before stack - Procedure and Significance Geophysics 204-212

Jakubowicz, H., and Levin, S. 53 Feb. 1983 A Simple Exact Method of 3-D Migration-Theory Geophys. Prosp. 34-56

Jannaud, L. R. 514 Oct. 1994 Common-offset ray tracing Exp. Abs., SEG Nat. Conv. 1390-1393

Jastram, C., Behle, A. 290 May. 1993 Accurate finite-difference operators for modelling the elastic wave equation Geophys. Prosp. 453-458

Jeannot, J. P. 469 Jun. 1994 Robust Kirchhoff pre-stack depth imaging with semi-gridded rays Exp. Abs. EAEG Tech. Exh. P102

Jeannot, J. P., and Berranger, I. 508 Oct. 1994 Re-mapped focusing: A migration velocity analysis for Kirchhoff prestack imaging Exp. Abs., SEG Nat. Conv. 1326-1329

Jervis, M., Sen, M., Stoffa, P.L. 275 Sep. 1993 Optimization methods for 2-D prestack migration velocity estimation Exp. Abs., SEG Nat. Conv. 965-968

Jianlin, Z. 511 Oct. 1994 2.5-D finite element seismic modeling Exp. Abs., SEG Nat. Conv. 1375-1377

Johansen, T. A., Bruland, L.., and Lutro, J. 537 Feb. 1995 Tracking the amplitude verses offset (AVO) by using orthogonal polynomials Geophys. Prosp. 245-261

Johnson, J. 147 Apr. 1992 Structural Imaging in the Real World The Leading Edge

Jones, I. F. 277 Sep. 1993 3-D velocity model building via iterative one-pass depth migration Exp. Abs., SEG Nat. Conv. 974-977

Jones, I. F. 406 Oct. 1993 Comparative anatomy 0f 3D one-pass depth migration schemes Handout, CGG France 1-7

Jones, I. F. 460 Jun. 1994 3-D velocity-depth model building - layer-striping verses map-migration Exp. Abs. EAEG Tech. Exh. B044

Jorden, T., Bleistein, N., and Cohen, J. 401 Oct. 1987 A wave equation-based dip moveout Exp. Abs., SEG Nat. Conv. 718-721

Judson, D., Lin, J., Schultz, P., and Sherwood, J. 50 Mar. 1980 Depth Migration After Stack Geophysics 361-375

Julien, P., and Cole, S. 358 Sep. 1990 3-D prestack depth migration on real data. Exp. Abs., SEG Nat. Conv. 1329-1332

Kaila, K., and Sain, K. 438 May. 1994 Errors in RMS velocity and zero-offset two-way time as determined from wide-angle seismic reflection traveltimes using truncated series J. of Seismic Exploration 173-188

Kallweit, R., and Wood, L. 70 Jul. 1982 The Limits of Resolution of Zero-Phase Wavelets Geophysics 1035-1046

Kamel, A. H. 183 Mar. 1991 Time-domain behavior of wide-angle one-way wave equations Geophysics 382-384

Kanasewich, E. R. 551 Jan. 1985 Time Sequence Analysis in Geophysics The University of Alberta Press

Kanasewich, E. R. 366 Mar. 1988 Imaging discontinuities on seismic sections Geophysics 334-345

Karrenbach, M. 408 Jan. 1990 Three-dimensional time-slice migration Geophysics 10-19

Kaufman, H. 483 Jan. 1953 Velocity functions in seismic prospecting Geophysics 289-297

Keho, T. H., and Beydoun, W. B. 478 Dec. 1988 Paraxial ray Kirchhoff migration Geophysics 1540-1546

Kessler, D., Chan, W. 271 Oct. 1993 DMO velocity analysis with Jacubowicz's dip-decomposition method Geophysics 1517-1524

Kim, Y. C., and Gonzarlez, R. 120 Mar. 1991 Migration Velocity Analysis with the Kirchhoff Integral Geophysics

Kim, Y. C., and Krebs, J.R., 276 Sep. 1993 Pitfalls in velocity analysis using common-offset time migration Exp. Abs., SEG Nat. Conv. 969-973

Kirtland Grech, M. G., Lawton, D. C. and Spratt, D. A. 708 Nov. 1998 Numerical Seismic Modelling and Imaging of Exposed Structures FRP Research Report 1.1-1.32

Kitchenside, P. W., and Jakubowicz, H. 403 Oct. 1987 Operator design for 3-D depth migration Exp. Abs., SEG Nat. Conv. 556-558

Kitchenside, P. W., Ellis, A.N., and Bale, R.A. 231 May. 1992 Datum-consistent migration and DMO for irregular topography Exp. Abs., CSEG Nat. Conv. 27-28

Kleyn, A. H. 300 Jan. 1983 Seismic Reflection Interpretation Applied Science Publishers

Kleyn. A. H. 732 Jan. 1956 On seismic wave propagation in an anisptropic media with applications in the Betun area, South samatra Geophys. Prosp. 57-69

Knapp, R. 108 Mar. 1991 Fresnel Zones in the Light of Broad Band Data Geophysics

Koehler, F., and Taner, M. T. 668 Oct. 1977 Direct and inverse problems relating reflection coefficients and reflection response for horizontally layered media Geophysics 1199-1206

Kosloff, D. D., and Baysal, E. 65 Jun. 1982 Forward Modelling by a Fourier Method Geophysics 1402-1412

Kosloff, D. D., and Baysal, E. 295 Jun. 1983 Migration with the full acoustic wave equation Geophysics 677-687

Krail, P. M. 611 Aug. 1994 Vertical cable as a subsalt imaging tool The Leading Edge 885-887

Krebes, E. S. 794 Jul. 2004 Seismic Theory and Methods GOPH551 Course notes UC

Kuhn, M., and Alhilali, K. 38 Aug. 1977 Weighting Factors in the Construction and Reconstruction of Acoustical Wavefields Geo-

References

physics 1183-1198

Kuhn, M. 41 Mar. 1979 Acoustical Imaging of Source Receiver Coincident Profile Geophys. Prospecting 62-77

Kuo, J., and Dai, T. 67 Aug. 1984 Kirchhoff Elastic Wave Migration for the Case of Noncoincident Source and Receiver Geophysics 1223-1238

Lafond, C. F., and Levander, A.R. 206 Jan. 1993 Migration moveout analysis and depth focusing Geophysics 91-100

Lambaré, G., Lucio, P. S., and Hanyga, A. 513 Oct. 1994 2-D asymptotic Green's functions Exp. Abs., SEG Nat. Conv. 1386-1389

Landa, E., Kosloff, D., Keydar, S., Koren, Z., and Resheff, M. 316 Apr. 1988 A method for determination of velocity and depth from seismic data Geophys. Prosp. 223-243

Landa, E., Thore, P., and Reshef, M. 289 Aug. 1993 Model-based stack: A method for constructing an accurate zero-offset section for complex overburdens Geophys. Prosp. 661-670

Langan, R. T., Lerche, I., and Cutler, R. T. 785 Sep. 1985 Tracing of rays through heterogeneous media - An accurate and efficient proceedure Geophysics 1456-1465

Lapidus, L. 15 Dec. 1962 Digital Computation for Chemical Engineers McGraw-Hill, New York

Larner, K., Gibson, B., and Chambers, R. 417 Nov. 1980 Imaging beneath complex structures: a case history Handout, Western Geophysical SEG Nat. Conv. 1-26

Larner, K., Hatton., L., and Gibson, R.S., Hsu, I. 73 May. 1981 Depth Migration of Imaged Time Sections Geophysics 734-750

Larner, K., and Beasley, C. 381 Jan. 1986 DMO and steep-dip migrations Western Geophysical

Larner, K., and Beasley, C. 85 May. 1987 Cascaded Migration: Improving the Accuracy of Finite-Difference Migration Geophysics 618-643

Larner, K., Weglein, A., and Wyatt, K. 396 Nov. 1988 Migration in the real world. Research Workshop, Questions Meeting handout 1-11

Larner, K., Beasley, C., Lynn, W. 94 Jun. 1989 In quest of the flank Geophysics 701-717

Larner, K., and Hatton, L. 104 Dec. 1990 Wave Equation Migration: Two Approaches(Classic old paper from 1976) First Break

Lathi, B. P. 560 Jan. 1965 Communication Systems John Wiley and Sons, Inc.

Lawrence, C., Schneider, W., and Shurtleff, R. 411 Oct. 1985 A comparative analysis of dip moveout and prestack migration Exp. Abs., SEG Nat. Conv. 307-309

Lawton, D., and Bertram, M. 391 Nov. 1993 Azimuthal response of some three-component geophones The Leading Edge 1118-1121

Lazaratos, S., and Harris, J. M. 356 Sep. 1990 Radon transform/Gausian beam migration Exp. Abs., SEG Nat. Conv. 1314-1317

Lee, D., Mason, I.M., and Jackson, G.M., 162 Nov. 1991 Split-step Fourier shot-record migration with deconvolution imaging Geophysics 1786-1793

Lee, D., Jackson, G. M., and Mason, I.M. 270 Sep. 1993 Partially coherent migration Geophysics 1301-1313

Lee, M. W., and Suh, S. Y. 487 Oct. 1985 Optimization of one-way wave equations Geophysics 1634-1637

Lee, S., and House-Finch, N. 605 Aug. 1994 Imaging alternatives around salt bodies in the Gulf of Mexico The Leading Edge 853-857

Lee, W. B., and Zhang, L., 194 Jun. 1992 Residual shot profile migration Geophysics 815-822

Lehmann, H. J., and Houba, W. 203 Feb. 1985 Practical aspects of the determination of 3-D stacking velocities Geophys. Prosp. 34-51

Levin, F. K. 115 Jun. 1971 Apparent Velocity From Dipping Interface Reflections Geophysics

Levin, F. K. 176 Jul. 1990 Reflection from a dipping plane-Transversely isotropic solid Geophysics 851-855

Levin, S. 123 May. 1984 Principle of Reverse-Time Migration Geophysics

Levin, S. 149 Nov. 1991 Should We Ever Use Poststack Depth Migration Exp. Abs., SEG Nat. Conv. 1144-1147

Lewis, G. G., Young, K. T., Finn, C. J., and Schneider, W. A. 609 Aug. 1994 Analysis if subsalt reflections at a Gulf of Mexico salt sheet through 3-D depth migration and 3-D seismic modeling The Leading Edge 873-878

Li X., and Bancroft, J. C. 629 May. 1997 Residual statics analysis before NMO using prestack migration Exp. Abs. SEG A032

Li, X., and Bancroft, J. C. 618 Nov. 1997 Integrated residual statics analysis with prestack time migration Exp. Abs. SEG 1997 1452-1455

Li, X., and Bancroft, J. C. 621 Nov. 1997 Converted wave migration and common conversion point binning by equivalent offset Exp. Abs. SEG 1997 1587-1590

Li, X., and Bancroft, J. C. 645 Dec. 1997 A new algorithm for converted wave prestack migration CREWES Research Report Ch 26

Li, X., Xu, Y., and Bancroft, J. C. 648 Dec. 1997 Equivalent offset migration: the implementation and application update CREWES Research Report Ch 27

Li, X., and Bancroft, J. C. 644 Dec. 1997 Residual statics using CSP gathers CREWES Research Report Ch 23

Li, X. 711 Jun. 1999 Residual statics analysis using prestack equivalent offset migration M.Sc. Thesis 1-123

Li, Z. 75 Aug. 1986 Wave-Field Extrapolation by the Linearly Transformed Wave Equation Geophysics 1538-1551

Li, Z., Chambers, R., and Abma, R. 412 Oct. 1987 Suppressing spatial aliasing noise in prestack Stolt FK migration Exp. Abs., SEG Nat. Conv. 734-736

Li, Z., Lynn, W, Chambers R., Larner K., and Abma, R. 124 Jan. 1991 Enhancements to Prestack Frequency-Wavenumber (F-K) Migration Geophysics 27-40

Li, Z. 214 Oct. 1991 Compensating finite-difference errors in 3-D migration and modeling Geophysics 1650-1660

Lindseth, R. O 754 Jan. 1979 Synthetic sonic logs — a process for stratigraphic interpretation Geophysics 3-26

Lindseth, R. O. 562 Sep. 1982 Digital Processing of Geophysical Data: A Review SEG

Liner, C., and Bleistein, N. 86 Oct. 1988 Dip Moveout Processing Exp. Abs., SEG Nat. Conv.

Liner, C. L., and Bleistein, N., 239 Nov. 1988 Comparative anatomy of common offset dip moveout Exp. Abs., SEG Nat. Conv.

Liner, C. L. 89 May. 1990 General theory and comparative anatomy of Dip Moveout Geophysics 595-607

Liner, C. L. 188 Feb. 1991 Born theory of wave-equation dip moveout Geophysics 182-189

Liner, C. L. 706 Nov. 1996 Bin Size and Linear V(z) Exp. Abs. SEG 47-50

Liner, C. L. 713 Nov. 1997 3-D seismic survey design and linear v(z) Exp. Abs. SEG 43-46

Lines, L. R., Schultz, A. K., and Treitel, S. 367 Jan. 1988 Cooperative inversion of geophysical data Geophysics 8-20

Lines, R. L. Slawinski, R., and Bording, R. P. 730 May. 1999 A recipe for stability of finite-difference wave-equation computations Geophysics 967-969

Link, B., Kendrick, J., and Butler, J.P. 252 Oct. 1992 Practical approach to imaging under extreme lateral velocity variations Exp. Abs., SEG Nat. Conv. 1013-1016

Liu, N., Jin, H., and Rockwood, A. P. 596 May. 1996 Antialiasing by Gausian integration Computer Graphics 58-63

Liu, Z. 787 Sep. 1997 An analytical approach to migration velocity analysis Geophysics 1238-1249

Loewenthal, D., Lu, L., Roberson, R., and Sherwood, J. 8 Jun. 1976 The Wave Equation Applied to Migration Geophys. Prosp. 380-399

Loewenthal, D., and Mufti, I.R. 309 May. 1983 Reversed time migration in spatial frequency domain Geophysics 627-635

Loewenthal, D., and Hu, L. 110 Mar. 1991 Two Methods for Computing the Imaging Condition for Common-Shot Prestack Migration Geophysics

Lopez-Mora, S., Purnell,N., and Krigsvoll 530 Jun. 1993 Kirchhoff time migration errors for vertical gradient EAEG-1993 Exp. Abs. C-047

Loveridge, M., Parkes, G., Hatton, L. 79 Aug. 1987 3-D Modelling of Migration Velocity Fields and Velocity Error Zones First Break 281-293

Lowenthal, D., Lu, L., Robertson, R., and Sherwood, J. 327 Jun. 1976 The wave equation applied to migration Geophys. Prosp. 380-399

Lumley, D. E. 54 Oct. 1989 A generalized Kirchhoff-WKBJ depth migration theory for multi-offset seismic reflection data: reflectivity model construction by wavefield imaging and amplitude estimation. M.Sc. Thesis, U of BC

Lumley, D. E., and Claerbout, J.F. 310 Jan. 1993 Anti-aliased Kirchhoff 3-D migration: a salt intrusion example SEG 3-D Seismic Workshop 115-123

Lumley, D. E., Claerbout, J. E., and Bevc, D. 462 Jun. 1994 Anti-aliased Kirchhoff migration Exp. Abs. EAEG Tech. Exh. H027

Lumley, D. E., Claerbout, J. F., and Bevc, D. 445 Oct. 1994 Anti-aliased Kirchhoff 3-D migration Exp. Abs. SEG Nat. Conv. 1282-1285

Lynn, H. B., and Deregowski, S. 692 Oct. 1981 Dip limitations on migrated sections Geophysics 1392-1397

Lynn, W., MacKay, S., and Beasley, C.J. 189 Sep. 1990 Efficient migration through irregular water-bottom topography Exp. Abs., SEG Nat. Conv. 1297-1300

Lynn, W., Gonzarez, A., and MacKay, S. 328 Nov. 1991 Where are the fault-plane reflections? Exp. Abs., SEG Nat. Conv. 1152-1154

MacKay, S., and Abma, R. 342 Nov. 1991 Depth-focusing analysis: practical applications and potential pitfalls Exp. Abs., SEG Nat. Conv. 1222-1225

MacKay, S., and Abma, R. 166 Dec. 1992 Imaging and velocity estimation with depth-focusing analysis Geophysics 1608-1622

MacKay, S., Abma, R. 264 Aug. 1993 Depth-focusing analysis using a wavefront-curvature criterion Geophysics 1148-1156

MacKay, S., and Dragoset, B. 606 Aug. 1994 Imaging beneath a Gulf Coast salt-injection feature: A processing case history The Leading Edge 858-860

MacKay, S. 529 Oct. 1994 Efficient wavefield extrapolation to irregular surfaces using finite differences: Zero-velocity datuming Exp. Abs., SEG Nat. Conv. 1564-1567

Maeland, E. 182 May. 1991 Migration analysis with an extra time shift Geophysics 691-696

Maginness, M. 28 Jan. 1972 The Reconstruction of Elastic Wavefields from Measurements over a Transducer Array J. Sound and Vibration, 20 219-40

Mao, W., and Stuart, G. w. 651 Jan. 1997 Rapid multi-wave-type ray tracing in complex 2-D and 3-D isotropic media Geophysics 298-308

March, D., and Bailey, A. 394 Jan. 1983 A review of the two-dimensional transform and its use in seismic processing First Break 9-21

Marcoux, M. O., Harris, C., and Chernis, L. 493 Oct. 1994 Depth migration and modeling from exact time migration Exp. Abs., SEG Nat. Conv. 1647-1650

Marfurt, K. 2 Dec. 1978 Elastic Wave Equation Migration-Inversion Ph.D. Thesis, Columbia University

Marfurt, K. J. 119 May. 1984 Accuracy of Finite-Difference and Finite-Element Modeling of the Scalar and Elastic Wave Equations Geophysics

Marfurt, K. J., Scheet, R. M., Sharp, J. A., and Harper, M. G. 674 May. 1998 Suppression of the acquisition footprint for sequence attribute mapping Geophysics 1024-1035

Margrave, G. F., and Bancroft, J.C. 627 Nov. 1996 The theoretical basis for prestack migration by equivalent offset Exp. Abs. SEG 433-446

Margrave, G. F. 712 Nov. 1997 Seismic acquisition parameter considerations for a linear velocity medium Exp. Abs. SEG 47-50

Margrave, G. F. 691 Dec. 1997 Zero offset seismic resolution theory for linear V(z) CREWES Research Report Ch 1

Margrave, G. F., 755 Jan. 1998 Theory of non-stationary linear filtering in the Fourier domain with applications to time-variant filtering Geophysics 255-259

Margrave, G. F., and Ferguson, R. J. 728 Aug. 1999 Wavefield extrapolation by nonstationary phase shift Geophysics 1067-1078

Margrave, G. F., and Yao, Z. 729 Dec. 1999 Imaging from topography with Fourier methods CREWES Research report 689-698

Margrave, G. F. 781 Jan. 2001 Numerical Methods of Siesmology with Algorithms in MATLAB www.crewes.org

Margrave, G. F. 756 Sep. 2001 Direct Fourier migration for vertical velocity variations Geophysics 1504-1514

Margrave. G. F., Bancroft, J.C., and Geiger, H. D. 640 Jan. 1999 Fourier prestack migration by equivalent wavenumber Geophysics 197-207

Martin, N. W., Donati, M. S., and Bancroft, J. C. 687 Jun. 1996 Steep dip Kirchhoff migration for linear velocity gradients Exp. Abs. EAGE X024

May, B. 95 Aug. 1983 Structural Inversion of Salt Dome Flanks Geophysics

May, B. T., and Hron, F. 150 Oct. 1978 Synthetic seismic sections of typical petroleum traps Geophysics 1119-1147

McMechan, G. A. 453 Jun. 1983 Migration by extrapolation of time-dependent boundary values Geophys. Prosp. 413-420

Meinardus, H. A., and Schleicher, K. 339 Nov. 1991 3-D time-variant dip moveout by the FK method Exp. Abs., SEG Nat. Conv. 1208-1210

Meinardus, H. A., and Schleicher, K. 228 May. 1992 3-D time-variant dip-moveout by the FK method Exp. Abs., CSEG Nat. Conv. 22-23

Meinardus, H. A., Schleicher, K.L., and Sudhakar, V. 250 Oct. 1992 Processing sequence for turning wave imaging Exp. Abs., SEG Nat. Conv. 988-991

Mereu, R. F. 226 Aug. 1976 Exact wave-shaping with a time-domain digital filter of finite length Geophysics 659-672

Mereu, R. F. 225 Feb. 1978 A computer program to obtain the weights of a time-domain wave-shaping filter which is optimum in an error-distribution sense Geophysics 197-215

Meyer, H. G. 168 Nov. 1991 Some remarks on noise stability in dynamic inversion of reflection seismic data Geophys. Prosp. 1005-1014

Michelena, R. J. Ata, E., and Sierra, J. 519 Oct. 1994 Exploiting P-S converted waves: Part 1, Modeling the effects of anisotropy and heterogeneities Exp. Abs., SEG Nat. Conv. 236-239

Middleton, D. 693 Jan. 1960 An Introduction to Statistical Communication Theory McGraw-Hill

Mikulich, W., and Hale, D. 157 Jan. 1992 Steep-dip V(z) Imaging from an Ensemble of Stolt-like Migrations Geophysics

Milkereit, B., and Spencer, C. 397 Jan. 1987 A new migration method applied to the inversion of P-S converted phases Blackwell Scientific Publications, Oxford 251-266

Milkereit, B. 193 Jul. 1987 Migration of noisy crustal seismic data J. Geophys. Res. 7916-7930

Milkereit, B. 399 Nov. 1987 Decomposition and inversion of seismic data - an instantaneous slowness approach Geophys. Prosp. 875-894

Miller, D., Oristaglio, M., and Beylkin, G. 376 Jul. 1987 A new slant on seismic imaging: Migration and integral geometry Geophysics 943-964

Miller, J., and Goldman, R. 389 Mar. 1992 Using tangent balls to find plane sections of natural quadrics IEEE Computer Graphics and Applications 68-82

Moser, T. J. 139 Jan. 1991 Shortest Path Calculation of Seismic Rays Geophysics

Moser, T. J. 279 Sep. 1993 Migration using fast traveltime calculators Exp. Abs., SEG Nat. Conv. 1033-1035

Mosher, C. C., Keho, T. H., Weglein, A. B., and Foster, D. J. 584 Nov. 1996 The impact of migration on AVO Geophysics 1603-1615

Mufti, I. R., Pita, J. A., and Huntly, R. W. 506 Oct. 1994 Finite-difference depth migration of exploration scale 3-D seismic data Exp. Abs., SEG Nat. Conv. 1286-1289

Muir, F. 16 Sep. 1976 Stanford Exploration Project 54

Mukulich, W., and Hale, D. 355 Sep. 1990 Steep-dip v(z) imaging from an ensemble of constant-velocity Stolt Migrations Exp. Abs., SEG Nat. Conv. 1307-1313

Muller, G., and Temme, P. 370 Nov. 1987 Fast frequency-wavenumber migration for depth-dependent velocity Geophysics 1483-1491

Myczkowski, J., McCowan, D., and Mufti, I. 130 Jun. 1991 Seismic Modeling in Real Time The Leading Edge

Nautiyal, A., Gray, S.H., Whitmore, N.D., and Garing, J.D. 216 Feb. 1993 Stability versus accuracy for an explicit wavefield extrapolation operator Geophysics 277-283

Neidell, N. S. 395 Jul. 1994 Sampling 3-D seismic surveys: A conjecture favoring coarser but higher-fold sampling The Leading Edge 764-768

Neidell, N. S. 661 Jul. 1997 Perceptions in seismic imaging Part 1: Kirchhoff migration operators in space and offset time, an appreciation The Leading Edge 1005-1006

Newman, P. 312 Nov. 1990 Amplitude and phase properties of a digital migration process First Break 397-403

Ng, M. 243 Jun. 1994 Prestack migration of shot records using phase shift plus interpolation CSEG Journal 11-27

Ng, M. 694 Dec. 1996 3-D prestack phase shift migration of shot records Can. J. of Expl. Geophys. 130-138

Ng, M. 695 May. 1997 Compensating 3-D prestack and poststack phase shift migration errors due to X-Y splitting Exp. Abs. CSEG 195-195

Nichols, D. 463 Jun. 1994 Imaging complex structures using band-limited Green's functions Exp. Abs. EAEG Tech. Exh. H030

Nichols, D. 512 Oct. 1994 Maximum energy traveltimes calculated in the seismic frequency band Exp. Abs., SEG Nat. Conv. 1382-1385

Nichols, D., Farmer, p., and Palacharla, G. 676 Jun. 1998 Improved prestack depth migration by novel raypath selection Exp. Abs. EAGE 1-54

Nolet, G. 383 Jan. 1987 Seismic wave propagation, Chap 1 in "Seismic Tomography: with Applications in Global Seismology and Exploration Geophysics D. Reidel Publishing Co.

Notfors, C., Godfrey, R. 84 Dec. 1987 Dip moveout in the frequency-wavenumber domain Geophysics 1718-1721

Notfors, C. 141 May. 1991 Common Offset Depth Migration Using Finite Difference Raytracing Exp. Abs., EAEG Convention 28

O'Brien, M., Etgen, J., Murphy, G. and Whitmore, N. D. 759 Jan. 1999 Multicomponent imaging with reciprocal shot records 69th Ann. Internat. Mtg: Soc. of Expl. Geophys. 784-787

O'Brien, M. J., and Gray, S. 598 Jan. 1996 Can we image beneath the salt? The Leading Edge 17-22

Ohanian, V. 288 Sep. 1993 DMO by the Huygens-Fresnel diffraction integral Exp. Abs., SEG Nat. Conv. 1037-1040

Ohanian, V., Snyder, T.M., and Gunawardena, A. 287 Sep. 1993 Analytic properties of the f-k DMO operator Exp. Abs., SEG Nat. Conv. 1134-1136

Ohanian, V. 286 Sep. 1993 Virtual images and normal moveout Exp. Abs., SEG Nat. Conv. 1126-1129

Ohanian, V., Snyder, T. M., and Hampson, D. P. 496 Oct. 1994 Approximate regimes of the H-F DMO operator Exp. Abs., SEG Nat. Conv. 1533-1536

Oppenheim, A. V., and Schafer, R. W. 384 Jan. 1975 Digital Signal Processing Prentice-Hall

Oppenheim, A. V. 545 Jan. 1978 Applications of Digital Signal Processing Prentice-Hall

Osterander, W. J. 475 Oct. 1984 Plane-wave reflection coefficients for gas sands at nonnormal angles of incidence Geophysics 1637-1648

Ottolini, R. A. 373 Nov. 1982 Migration of reflection seismic data in angle-midpoint coordinates Dissertation, Stanford University

Ottolini, R. A., and Claerbout, J. F. 452 Mar. 1984 The migration of common midpoint slant stacks Geophysics 237-249

Owusu, K., Gardner, G. H. F., and Massell, W. F. 364 Nov. 1983 Velocity estimates derived from three-dimensional seismic data. Geophysics 1466-1497

Pai, D., and Chen, T. 414 Oct. 1987 A new method for wave equation migration in laterally inhomogeneous media Exp. Abs., SEG Nat. Conv. 554-555

Pai, D. 90 Dec. 1988 Generalized F-K (Frequency wavenumber) Migration in Arbitrarily Varying Media Geophysics

Palmer, D. 591 Jun. 1995 Can linear inversion achieve detailed refraction statics? Exploration Geophysics 506-511

Pan, N., and French, W. 105 May. 1989 Generalized Two Pass Three Dimensional Migration For Imaging Steep Dips in Vertically Inhomogeneous Media Geophysics 544-554

Pan, N., Perkins, W., and French, W. 122 Nov. 1991 Exact Two-Pass Full-Dip (>90 degree) 3-D Migration for Vertically Inhomogeneous Media Exp. Abs., SEG Nat. Conv. 1160-1163

Parkes, G., Hatton, L. 80 Apr. 1987 Towards a systematic understanding of the effects of velocity model errors on depth and time migration of seismic data First Break 121-132

Payne, B., and Marcoux, M. 256 Oct. 1992 Rugged topography and S-R downward continuation Exp. Abs., SEG Nat. Conv. 1124-1126

Payne, B. 468 Jun. 1994 Phase shift migration for lateral velocity variation Exp. Abs. EAEG Tech. Exh. P101

Payne, M. A., Eriksen, E. A., and Rape, T. D. 664 Mar. 1994 Considerations for high-resolution VSP imaging The Leading Edge 173-180

Peng, C., and Toksšz, M.N. 282 Sep. 1993 An optimal absorbing boundary condition for elastic wave modeling Exp. Abs., SEG Nat. Conv. 1056-1059

Perkins, W., and French, W. 153 Sep. 1990 3-D Migration to Zero Offset for a Constant Velocity Gradient Exp. Abs., SEG Nat. Conv. 1354-1357

Phinney, R., and Frazer, L. 33 Jan. 1978 On the Theory of Imaging by Fourier Transform Presented SEG Nat. Conv.

Pleshkevitch, A. L. 470 Jun. 1994 Shot-DMO by finite-difference Exp. Abs. EAEG Tech. Exh. P103

Popovici, A. M. 522 Oct. 1994 Reducing artifacts in prestack phase-shift migration of common offset gathers Exp. Abs., SEG Nat. Conv. 684-687

Popovici, A. M. 571 Sep. 1996 Prestack migration by the split-step DSR Geophysics 1412-1416

Popovici, M., and Sethian, J., 666 Jun. 1998 Three dimensional traveltimes using fast marching method Exp. Abs. EAGE 1-22

Press, W. H., Flannery, B. P., Teukolsky, S. A., and Vetterling, W. T. 187 Dec. 1986 Numerical Recipes (FORTRAN) (Text) Cambridge University Press

Press, W.H., Teukolsky, S.a., Vetterling, W.T., and Flannery, B.P., 793 Jan. 1992 Numerical Recipies in Fortran 77 (also "C") Cambridge University Press

Profeta, R. M., Moscoso, J., and Koremblit, M 523 Oct. 1994 Minimum field static corrections Exp. Abs., SEG Nat. Conv. 715-718

Purnell, G. W. 258 Nov. 1992 Imaging beneath a high-velocity layer using converted waves Geophysics 1444-1452

Qin, F., Olsen, K., and Schuster, G. 423 Sep. 1990 Solution of the Eikonal equation by a finite-difference method Exp. Abs., SEG Nat. Conv. 1004-1007

Qin, F., Luo, Y., Olsen K., Cai, W., and Schuster G. 143 Mar. 1992 Finite-Difference Solution of the Eikonal Equation along Expanding Wavefronts Geophysics

Qin, F., Schuster, G. T. 269 Sep. 1993 First-arrival traveltime calculation for anisotropic media Geophysics 1349-1358

Rajasekaran, S., and McMechan 659 Jul. 1996 Tomographic estimation of the spatial distribution of statics Geophysics 1198-1208

Randall, C. 154 May. 1988 Absorbing Boundary Condition for the Elastic Wave Equation Geophysics

Ratcliff, D. W., and Pan, N. 404 Oct. 1987 2-D and 3-D post stack dip enhancement (?) Exp. Abs., SEG Nat. Conv. 1-4

Ratcliff, D. W., Jacewitz, C. A., and Gray, S. H. 642 Mar. 1994 Subsalt imaging via target-orientated 3-D prestack depth migration The Leading Edge 163-170

Ratcliff, D. W., and Weber, D. J. 502 Oct. 1994 Subsalt imaging over the Mahogany salt sill Exp. Abs., SEG Nat. Conv. 1246-1249

Ratcliff, D.W., Grey, S., and Whitmore, N. 146 Apr. 1992 Seismic Imaging of Salt Structures in the Gulf of Mexico The Leading Edge 15-31

Rayleigh, J. 51 Jan. 1945 The Theory of Sound Dover Publications, London

Raz, S. 48 Jan. 1980 An Approximate Propagation Speed Inversion over a Prescribed Slab, Acoustic Imaging, Vol. 9 Plenum Press, New York

Raz, S. 301 Jun. 1981 Direct reconstruction of velocity and density profiles from scattered field data Geophysics 832-836

Reichenderger, H. 97 Sep. 1988 Lithotripter Systems Proceeding of IEEE

Reilly, J. M. 172 Oct. 1992 Seismic imaging adjacent to and beneath salt diapirs, UK North Sea First Break 383-397

Reiter, E. C., Toksoz, M. N., and Purdy, G.M. 224 Jan. 1993 A semblance-guided median filter Geophys. Prosp. 15-41

Reitveld, W. E. A., Berkhout, A. J., and Wapenaar, C. P. A. 348 Nov. 1991 Controlled illumination of hydrocarbon reservoirs Exp. Abs., SEG Nat. Conv. 1281-1284

Reksnes, P. A., Haugane, E., and Hegna, S. 790 Dec. 2002 How PGS created a new image for the Varg field First Break 773-777

Reksnes, P.A., and Haugane, E., 792 Dec. 2002 How PGS created a new image for the Varg field First Break 773-781

Reshef, M., and Kosloff, D. 109 Feb. 1986 Migration of Common-Shot Gathers Geophysics 324-331

Reshef, M., Kosloff, D., and Edwards, M. 393 Sep. 1988 Three-dimensional elastic modeling by the Fourier method Geophysics 1184-1193

Reshef, M. 117 Jan. 1991 Depth Migration From Irregular Surfaces with Depth Extrapolation Methods Geophysics

Reshef, M. 142 Aug. 1991 Prestack Depth Imaging of Three-Dimensional Shot Gathers Geophysics

Reshef, M. 761 Jan. 2001 Some aspects of interval velocity analysis using 3-D depth-migrated gathers Geophysica 261-266

Reshef, M. Keydar, S., and Landa, E. 791 Mar. 2003 Multiple prediction without prestackdata: an efficient tool for interpretive processing First Break 29-37

Reynolds, A. C. 325 Oct. 1978 Boundary conditions for the numerical solution of wave propagation problems Geophysics 1099-1110

Richardson, R. P. 173 Oct. 1992 Linear velocity function: Reflection times for specific ranges Geophysics 1352-1353

Rickett, J. E., and Sava, P. C. 769 May. 2002 Offet and angle-domain common image-point gathers for shot-profile migration Geophysics 883-889

Rietveld, W. E. A., and Berkhout, A.J. 245 Oct. 1992 Depth migration combined with controlled illumination Exp. Abs., SEG Nat. Conv. 931-934

Roberts, G., and Goulty, N. 387 Aug. 1988 Directional deconvolution of the seismic source signature combined with prestack migration First Break 247-253

Robinson, E. A. 71 Jan. 1982 A Historical Perspective of Spectrum Estimation Proc. IEEE 885-907

Robinson, E. A. 456 Jan. 1983 Migration of geophysical data IHRDC

Robinson, E. A. 388 Mar. 1986 Migration of seismic data by the WKBJ method Proceedings of the IEEE 428-438

Robinson, J., and Robbins, T. 61 Feb. 1978 Dip Domain Migration of Two-Dimensional Seismic Profiles Geophysics 77-93

Robinson, W. B. 788 Jan. 1957 The need for seismic dip migration Reprint, Migration of Seismic Data SEG 51-56

Rocca, F., and Ronen, J. 409 Oct. 1984 Improving resolution by dip moveout Exp. Abs., SEG Nat. Conv. 611-614

Rockwell, D. W. 789 Jan. 1971 Migration Stack aids interpretation Reprint,Migration of Seismic Data SEG, from Oil and Gas Journal 75-81

Rodriguez, S., Diet, J. P., and Paturet, D. 350 Nov. 1991 Dip moveout in cases of irregular surfaces Exp. Abs., SEG Nat. Conv. 1301-1340

Rodriguez, S., Whitsett, R., Diet, J.P., Paturet, D. 229 May. 1992 Elevation and weathering effects on dip move-out resequenced DMO (RSDMO) Exp. Abs., CSEG Nat. Conv. 24

Rodriguez, S., and Vuillermoz, C. 491 Oct. 1994 Solutions to near surface effects in mountainous thrust areas Exp. Abs., SEG Nat. Conv. 1644-1646

Rodriguez-Suarez, C., Stewart, R. R., Li, X., and Bancroft, J. C. 649 Dec. 1997 Analyzing 4-C marine data from the Valhall field, Norway CREWES Research Report Ch 42

Ronen, J. 377 Jul. 1987 Wave-equation trace interpolation Geophysics 973-984

Rosenbaum, J. H., and Boudreaux, G. F. 564 Dec. 1981 Rapid convergence of some seismic processing algorithms Geophysics

References

1667-1672

Ross, C. P. 662 Feb. 1997 AVO and nonhyperbolic moveout: a practical example First Break 43-48

Rowbotham, P. S. 614 Jul. 1997 Anisotropic migration of coincident VSP and cross-hole seismic reflection surveys Geophys. Prosp. 683-699

Rueger, A. 510 Oct. 1994 Applicability of the Gausian beam approach for modeling of seismic data in triangulated subsurface models Exp. Abs., SEG Nat. Conv. 1371-1374

Ruehl, T., Kopp, C., and Ristow, D. 467 Jun. 1994 Fourier FD migration for steeply dipping reflectors with complex overburden Exp. Abs. EAEG Tech. Exh. P100

Safar, M. 63 Jul. 1985 On the Lateral Resolution achieved by Kirchhoff Migration Geophysics 1091-1099

Sams, M., and Goldberg, D. 696 Jan. 1990 The validity of Q estimates from borehole data using spectral ratios Geophysics 97-101

Samson, J. C., and Olson, J. V. 683 Oct. 1981 Data-adaptive polarization filters for multichannel geophysical data Geophysics 1423-1431

Sattlegger, J. W., Stiller, P.K., Echterhoff, J.A., and Hentschke, M.K. 210 Dec. 1980 Common offset plane migration (COPMIG) Geophys. Prosp. 859-871

Sattlegger, J. W. 62 Feb. 1982 Migration of Seismic Interfaces Geophys. Prosp. 71-85

Sayers, C. M., and Ebrom, D. A. 752 Sep. 1997 Seismic traveltime analysis for azimuthally anisotropic media: Theory and experiment Geophysics 1570 - 1582

Scales, J. A. 716 May. 1994 Theory of Seismic Imaging Samizdat Press (CWP) Colorado School of Mines 1-204

Schleicher, J., Hubral, p., and Tygel, M. 746 Aug. 1991 Nonspecular reflections from a curved interface Geophysics 1203-1214

Schleicher, J., Tygel, M. Hubral, P. 266 Aug. 1993 3-D true-amplitude finite-offset migration Geophysics 1112-1126

Schleicher, J., Hubral, P., Tygel, M., and Jaya, M. S. 653 Jan. 1997 Minimum apertures and Fresnel zones in Migration and demigration Geophysics 183-194

Schleicher, K. L., Grygier, D. J., and Brzostowski, M. A. 326 Nov. 1991 Migration velocity analysis: a comparison of two approaches Exp. Abs., SEG Nat. Conv. 1237-1238

Schneider Jr., W. A. 774 Jul. 1995 Robust and efficient upwind finite-difference traveltime calculations in three dimensions Geophysics 1018-1117

Schneider, W. A. 313 Dec. 1971 Developments in seismic data processing and analysis (1968-1970) Geophysics 1043-1073

Schneider, W. A. 37 Feb. 1978 Integral Formulation for Migration in Two and Three Dimensions Geophysics 49-76

Schneider, W. A., Ranzinger, K., Balch, A., and Kruse, C. 145 Jan. 1992 A Dynamic Programming Approach to First Arrival Traveltime Computation in Media with Arbitrarily Distributed Velocities Geophysics

Schneider, W. A. 280 Sep. 1993 Robust, efficient upwind finite-difference traveltime calculations in 3-D Exp. Abs., SEG Nat. Conv. 1036-1039

Schneider, W. A. 581 Jul. 1995 Robust and efficient upwind finite-difference traveltime calculations in three dimensions Geophysics 1108-1117

Schoenberger, M. 577 May. 1996 Optimum weighted stack for multiple suppression Geophysics 891-901

Schultz, P. S., and Claerbout, J. 25 Jun. 1978 Velocity Estimation and Downward Continuation by Wavefront Synthesis Geophysics 691-714

Schultz, P. S., and Sherwood, J. 27 Mar. 1980 Depth Migration Before Stack Geophysics 376-393

Schultz, P. S. 308 Dec. 1982 A method for direct estimation of interval velocities Geophysics 1657-1671

Schuster, G. T., and Quintus-Bosz, A. 268 Sep. 1993 Wavepath eikonal traveltime inversion: Theory Geophysics 1314-1323

Schwab, M., and Gardner, H. F. 336 Nov. 1991 2-D DMO for a medium with constant vertical velocity gradient Exp. Abs., SEG Nat. Conv. 1194-1196

Scott, J. H., Tibbetts, B.L., and Burdick, R.G., 185 Jan. 1940 Computer analysis of seismic refraction data. ???

Selvi, O 567 Jan. 1996 Comment on 'Sign-change correction for prestack migration of P-S converted wave reflections' by Balch and Erdemir Geophys. Prosp. 175-177

Sen, M., and Stoffa, P. 390 Jan. 1992 Genetic inversion of AVO The Leading Edge 27-29

Sena, A. G., and Toksöz, M.N. 215 Feb. 1993 Kirchhoff migration and velocity analysis for converted and nonconverted waves in anisotropic media Geophysics 265-276

Shaum 382 Jan. 1968 Mathematical Handbook, Schaum's outline series McGraw-Hill

Sheriff, R. E. 55 May. 1980 Nomogram for Fresnel-Zone Calculation Geophysics 968-972

Sheriff, R. E. 543 Oct. 1990 Encyclopedic Dictionary of Exploration Geophysics (Third Edition) SEG

Sheriff, R. E., and Gelart, L. P. 550 Jan. 1995 Exploration Seismology Cambridge 592

Sherwood, J. W. C., and Poe, P.H. 315 Oct. 1972 Continuos velocity estimation and seismic wavelet processing Geophysics 769-787

Sherwood, J. W. C., Schultz, p., and Judson, D. 49 Jan. 1978 Equalizing the Stacking Velocities of Dipping Events AAPG 2362

Shih, R. C., and Levander, A. R. 368 Apr. 1994 Layer-stripping reverse-time migration Geophys. Prosp. 211-227

Shtivelman, V., and Canning, A. 201 Oct. 1988 Datum correction by wave-equation extrapolation Geophysics 1311-1322

Shuey, R. T. 485 Apr. 1985 A simplification of the Zeoppritz equations Geophysics 609-614

Silva, R., and Haseko, P. 340 Nov. 1991 Antialiasing and amplitude preserving 2-D and 3-D DMO: an integral implementation Exp. Abs., SEG Nat. Conv. 1211-1214

Silva, R. 251 Oct. 1992 Antialiasing and application of weighting factors in Kirchhoff migration Exp. Abs., SEG Nat. Conv. 995-998

Slotnick, M. M. 371 Jan. 1936 On seismic computations, with applications, I Geophysics 9-22

Slotnick, M. M. 372 Oct. 1936 On seismic computations, with applications, II Geophysics 299-305

Slotnick, M. M. 542 Jan. 1959 Lessons in Seismic Computing SEG 1-268

Sochacki, J., Kubichek, R., and George, J., Fletcher, W., and Smithson, S. 66 Jan. 1987 Absorbing Boundary Conditions and Surface Waves Geophysics 60-71

Sollie, R., Mittet, R., and Hokstad, K. 464 Jun. 1994 Pre-stack depth migration with compensation for absorption Exp. Abs. EAEG Tech. Exh. H031

Sorin, V., Keydar, S., and Landa, E. 592 Sep. 1996 Velocity analysis: Phenomenon of local homogeneous model The Leading Edge 1033-1035

Spies, B. R. 261 Oct. 1992 Information at your fingertips — using the Digital Cumulative Index The Leading Edge 65-66

Spitz, S. 116 Jun. 1991 Seismic Trace Interpolation in the F-X Domain Geophysics

Stadtlander, R., and Brown, L. 569 Jan. 1997 Turning waves and crustal reflection profiling Geophysics 335-341

Stanley, M. 314 Jun. 1998 Practical wave equation datuming Exp. Abs. EAGE 1-15

Starich, P. J., Lewis, G. G., Faulkner, J., Standley, P. G., and Setterquist, S. 610 Aug. 1994 Integrated geophysical study of an onshore salt dome The Leading Edge 880-884

Stockwell, J. W. 580 Mar. 1995 2.5-D wave equations and high-frequency asymptotic Geophysics 556-562

Stoffa, P. L., Buhl, P., and Bryan, G.M. 240 Aug. 1974 The application of homomorphic deconvolution to shallow-water marine seismology-Part I: Models Geophysics 401-416

Stolt, R. H. 21 Feb. 1978 Migration by Fourier Transform Geophysics 23-48

Stolt, R. H., and Benson, A. K. 531 Jan. 1986 Seismic Migration, Theory and Practice Geophysical Press

Stolt, R. H. 770 May. 2002 Seismic data mapping and reconstruction Geophysics 890-908

Stone, D. 402 Apr. 1988 Imaging with post-stack DMO Handout Seismograph Service Corp. 1-9

Stork, C., and Kusuma, T. 257 Oct. 1992 Hybrid genetic autostatics: New approach for large-amplitude statics with noisy data Exp. Abs., SEG Nat. Conv. 1127-1131

Sullivan, M. F., and Cohen, J. K. 444 Jun. 1987 Prestack Kirchhoff inversion of common-offset data Geophysics 745-754

Sumner, B. L. 233 Apr. 1988 Asymptotic solutions to forward and inverse problems in isotropic elastic media CSM CWP

Sun, J 593 May. 1996 The relationship between the first Fresnel zone and the normalized geometrical spreading factor Geophys. Prosp. 351-374

Sun, J., and Gajewski, D. 667 Aug. 1997 True-amplitude common-shot migration revisited Geophysics 1250-1259

Sun, J. 675 May. 1998 On the limited aperture migration in two dimensions Geophysics 984-994

Takahashi, T. 283 Sep. 1993 Multicomponent prestack depth migration Exp. Abs., SEG Nat. Conv. 1078-1081

Tallin, A. G., and Santamarina, J.C. 171 May. 1992 Digital ray tracing for geotomography IEEE Trans. 617-619

Taner, M. T., and Koehler, F. 56 Dec. 1969 Velocity Spectra-Digital Computer Derivation and Applications of Velocity Functions Geophysics 859-881

Taner, M. T. 441 Feb. 1975 Long-period multiples and their suppression Geophysics (Abstract from annual meeting) 143-144

Taner, M. T., Koehler, F., and Sheriff, R. E. 484 Jun. 1979 Complex seismic trace analysis Geophysics 1041-1063

Taner, M. T., Postman, R. W., Lu, L., and Baysal, E. 341 Nov. 1991 Depth-migration velocity analysis Exp. Abs., SEG Nat. Conv. 1218-1211

Tappert, F., and Hardin, R. 47 Jan. 1973 A Synopsis of the AESD Workshop on Acoustic Propagation Modelling by Non-Ray Tracing Techniques AD-773 741, AESD Tech. Note TN-73-05

Tarantola, A. 657 Jan. 1989 Theoretical background for the inversion of seismic waveforms, including elasticity and attenuatuin Digital Seismology..., Plenum Press 157-190

Tatalovic, R., Dillen, m. W. P., and Fokkema, J. T. 349 Nov. 1991 Prestack imaging in the double transformed Radon domain Exp. Abs., SEG Nat. Conv. 1285-1288

Tatham, R. H. 698 Jan. 1985 Vp/Vs interpretation, Chap. 5 of Developments in Geophysical exploration methods Elsevier App. Sc. Pub. 139-188

Teiman, H. J. 345 Nov. 1991 Migration velocity analysis through weighted stacking Exp. Abs., SEG Nat. Conv. 1243-1246

Thompsen, L.A., Barkved, O. I., Haggard, B., Kommedal, J. H., and Rosland, B. 650 May. 1997 Converted-wave imaging of Valhall Reservoir Exp. Abs. EAGE B048-V1

Thompsen. L. 734 Oct. 1986 Weak elastic anisotropy Geophysics 1954-1966

Thorbecke, J. W., and Reitveld, W. E. A. 472 Jun. 1994 Optimum extrapolation operators - a comparison Exp. Abs. EAEG Tech. Exh. P105

Thorbecke, J. W., and Berkhout, A. J. 503 Oct. 1994 3-D recursive extrapolation operators: An overview Exp. Abs., SEG Nat. Conv. 1262-1265

Thore, P. D., and Kelly, P. 284 Sep. 1993 Stack enhancement using the three-term equation: synthetic and real data examples Exp. Abs., SEG Nat. Conv. 1023-1025

Tjan, T., and Larner K. 494 Oct. 1994 Prestack migration for residual statics estimation in complex areas Exp. Abs., SEG Nat. Conv. 1513-1516

Tricker, R. A. R. 199 Jan. 1965 Bores, breakers, waves and wakes American Elsevier Publishing Company, Inc. 186-214

Trorey, A. 74 Oct. 1970 A Simple Theory for Seismic Diffractions Geophysics 762-784

Trorey, A. 420 Oct. 1977 Diffractions for Arbitrary source-receiver locations Geophysics 1177-1182

Tsvankin, I., and Thomsen, L. 773 Jul. 1995 Inversion of reflection traveltimes for transverse isotropy Geophysics 1095-1107

Tygel, M., Schleicher, J., Hubral, P., and Hanitzsch, C. 448 Dec. 1993 Multiple weights in diffraction stack migration Geophysics 1820-1830

Tygel, M., Schleicher, J., and Hubral. P. 736 Oct. 1994 Pulse distortion in depth migration Geophysics 1561-1569

Tygel, M., Schleicher, J., and Hubral, P. 573 May. 1996 A unified approach to 3-D seismic reflection imaging, Part II: Theory Geophysics 759-775

Tygel, M., Schleicher, J., Hubral, P., and Santos, L. T. 486 Mar. 1998 2.5-D true-amplitude Kirchhoff migration to zero offset in laterally inhomogeneous media Geophysics 557-573

Ulrych, T. J., Sacchi,M. D., and Woodbury, A. 760 Jan. 2001 A Bayes tour of inversion: A tutorial Geophysics 55-69

Uren, N. F., Gardner, G., and McDonald, J. 425 Oct. 1989 The anisotropic migrator's equation Exp. Abs., SEG Nat. Conv. 1184-1186

Uren, N. F., Gardner, G., and McDonald, J. 426 Oct. 1989 Zero-offset seismic reflection surveys using an anisotropic physical model Exp. Abs., SEG Nat. Conv. 1044-1046

Uren, N. F., Gardner, G.H.F., and McDonald, J.A. 112 Jul. 1990 Dip moveout in anisotropic media Geophysics 863-867

Uren, N. F., Gardner, G.H.F., and McDonald, J.A. 113 Nov. 1990 The migrator's equation for anisotropic media Geophysics 1429-1434

van der Schoot, A., Romijn, R., Larson, D.E., and Berkhout, A.J. 197 Jul. 1989 Prestack migration by shot record inversion and common depth point stacking: a case study First Break 293-304

Van Der Wal, L. F., and Berkhout, A.J. 306 Jun. 1984 Influence of amplitude and phase errors on migration results Geophys. Prosp. 425-453

Van Melle, F. A. 223 Apr. 1948 Wave-front circles for a linear increase of velocity with depth Geophysics 158-162

van Trier, J., and Symes, W., 140 Jun. 1991 Upwind Finite-Difference Calculation of Traveltimes Geophysics

Varela, C. L., Rosa, A. L. R., and Ulrych, T.J. 262 Aug. 1993 Modeling of attenuation and dispersion Geophysics 1617-1173

Vasco, D. W., Peterson, J. E., and Majer, E. L. 600 Jul. 1996 Nonuniqueness in traveltime tomography: Ensemble interference and cluster analysis Geophysics 1209-1227

Vermeer, G. J. O 532 Sep. 1995 Is "coarse" the right course? (Roundtable portion) The Leading Edge 989-993

References

Vermeer, G. J. O. 546 Jan. 1990 Seismic Wavefield Sampling SEG

Versteeg, R., and Grau, G. 151 May. 1990 The Marmousi Experience EAEG Workshop PASD Inversion

Versteeg, R., 152 Oct. 1991 Analyze Du Probleme De La Determination Du Modele De Vitesse Pour L'Imagerie Sismique, Velocity analysis on the Marmousi data set Doc. Thesis, text is in English.

Versteeg, R., Ehinger, A., and Geoltrain, S. 346 Nov. 1991 Sensitivity of migration coherency panels to the velocity model Exp. Abs., SEG Nat. Conv. 1251-1254

Versteeg, R. 267 Jun. 1993 Sensitivity of prestack depth migration to the velocity model Geophysics 873-8824

Versteeg, R. 437 Sep. 1994 The Marmousi experience: Velocity model determination on a synthetic complex data set The Leading Edge 927-936

Vesnaver, A. L. 587 May. 1995 The contribution of reflected, refracted and transmitted waves to seismic tomography: a tutorial First Break 159-168

Vidale, J. 136 Dec. 1988 Finite-Difference Traveltime Calculation Bull., Seis. Soc. Am. 2062-2076

Vidale, J. 137 May. 1990 Finite-Difference Calculation of Traveltimes in Three Dimensions Geophysics

Vidale, J., and Houston, H. 138 Nov. 1990 Rapid Calculation of Seismic Amplitudes Geophysics

Vinje, V., Iversen, E., Gjoystdal, H. 263 Aug. 1993 Traveltime and amplitude estimation using wavefront construction Geophysics 1157-1166

Vinje, V., Lecomte, I., Astebol, K., Iversen, E., and Gjoystdal, H. 459 Jun. 1994 Efficient Green's functions calculation for improved 3-D seismic in complex areas Exp. Abs. EAEG Tech. Exh. B043

von Seggern, D. 586 Mar. 1994 Depth-imaging resolution of 3-D seismic recording patterns Geophysics 564-576

Wang, C. 678 Oct. 1995 DMO in radon domain Exp. Abs. SEG 1441-444

Wang, C. 679 Nov. 1996 Radon DMO amplitude and frequency preservation Exp. Abs. SEG 1479-1482

Wang, C. 705 Jan. 1999 Dip Moveout in Radon Domain Geophysics 278-288

Wang, S., Bancroft, J. C., and Lawton, D. C. 527 Dec. 1994 DMO processing for mode-converted waves in a medium with linear increasing velocity with depth CREWES Research Report 26.1-26.15

Wang, S., Bancroft, J. C., Foltinek, D., and Lawton, D.C. 665 May. 1995 Converted-wave (P-SV) prestack migration and migration velocity analysis Exp. Abs. CSEG 67-68

Wang, S., Bancroft, J. C., Lawton, D. C., and Foltinek, D. S. 556 Nov. 1995 Converted wave (P-S) prestack migration and migration velocity analysis CREWES Research Report 27.1 - 27.22

Wang, S., Bancroft, J. C., and Lawton, D.C. 626 Nov. 1996 Converted-wave (P-SV) prestack migration and migration velocity analysis Exp. Abs. SEG 1575-1578

Wang, S. 633 Aug. 1997 Three-component and Three-dimensional seismic imaging M.Sc. Thesis 1-83

Wang, W., Bancroft, J. C., Foltinek, D., and Lawton, D. 636 May. 1995 Converted-wave (P-SV) prestack migration and velocity analysis. Exp. Abs. CSG Nat. Conv. 67-68

Wang, y. 765 Apr. 2002 A stable and efficient approach of inverse Q filtering Geophysics 657-663

Wapenaar, C. P. A 672 Jul. 1997 3-D migration of cross-spread data: Resolution and amplitude aspects Geophysics 1220-1225

Wapenaar, C. P. A., Blacquiere, G., Kinneging, N. A., and Berkhout, A. J. 362 Nov. 1986 Practical aspects of 3-D prestack migration Exp. Abs., SEG Nat. Conv. 436-438

Wapenaar, C. P. A., Kinneging, N., and Berkhout, A. 392 Feb. 1987 Principle of prestack migration based on the full elastic two-way wave equation Geophysics 151-173

Wapenaar, C. P. A., and Berkhout, A. 407 Jun. 1987 Three-dimensional target-oriented pre-stack migration First Break 217-227

Ward, R. W., MacKay, S., Greenlee, S. M., and Dengo, C. A. 601 Aug. 1994 Imaging sediments under salt: Where are we? The Leading Edge 834-836

Watson, K. 222 Jun. 1993 Processing remote sensing images using the 2-D FFT--Noise reduction and other applications Geophysics 835-852

Weglein, A. B., and Stolt, R. H. 750 Aug. 1999 Migration-inversion revisited TLE 950-975

Wenes, G., Kremer, S., and Shiu, J. 466 Jun. 1994 A practical implementation of a 3-D prestack depth migration algorithm Exp. Abs. EAEG Tech. Exh. H040

Wenzel, F., and Menges, D. 155 Aug. 1989 A Comparison Between Born Inversion and Frequency-Wavenumber Migration Geophysics

Wenzel, F. 213 Oct. 1991 Frequency-wavenumber migration in laterally heterogeneous media Geophysics 1671-1673

Whitmore, N. D., 757 Sep. 1983 Iterative depth migration by backward time propagation, 53rd Ann. Internat. Mtg: Soc. of Expl. Geophys., S10.1

Whitmore, N. D., Gray, S. H., and Gersztenkorn, A. 165 Jun. 1988 Two-dimensional post-stack depth migration: a survey of methods First Break 189-197

Whittlesey, J., and Quay, R. 20 Jan. 1977 Wave Equation Migration Operators Using 2-D Z-Transform Theory Presented at SEG Nat. Conv.

Wiggins, J. 68 Aug. 1984 Kirchhoff Integral Extrapolation and Migration of Nonplanar Data Geophysics 1239-1248

Williams, R. G., and Cooper, N. J. 588 Jun. 1995 Suppression of dipping noise and multiples using 3-D prestack time migration Exploration Geophysics 468-471

Winbow, G. A., and Trantham, E. C. 524 Oct. 1994 Nonaliased amplitude-preserving DMO (AVO???) Exp. Abs., SEG Nat. Conv. 719-722

Witte, D. 333 Nov. 1991 Dip moveout in vertically varying media Exp. Abs., SEG Nat. Conv. 1181-1183

Worley, S. C. 227 Feb. 1993 The geometry of reflection Geophysics 293-297

Wren, A., and Jan, S. 52 Jan. 1987 Diffractions Revisited Cdn. Jnl. of Exp. Geo. 8-13

Wu, H., and Lees, J. M. 655 Sep. 1997 Boundary conditions on a finite grid: Applications with psuedospectral wave propagation Geophysics 1544-1557

Wu, W., Lines, L. R., and Lu, H 576 May. 1996 Analysis of high-order, finite-difference schemes in 3-D reverse-time migration Geophysics 845-856

Wyatt, K. D., Towe, S.K., Layton, J.E., Wyatt, S.B., Von Seggern, D.H., and Brockmeier, C.A. 247 Oct. 1992 Ergonomics in 3-D depth migration Exp. Abs., SEG Nat. Conv. 944-947

Wyatt, K.D., Wyatt, S.B., Towe, S. K., Layton, J. E., von Seggern, D. H., and Brockmeier, C. A. 607 Aug. 1994 Building velocity-depth models for 3-D depth migration The Leading Edge 862-866

Yilmaz, (Orhan)., and Taner, M. 436 Jun. 1994 Discrete plane-wave decomposition by least-mean-squared-error method Geophysics 973-982

Yilmaz, O. 24 Dec. 1979 Pre-Stack Partial Migration Ph.D. Thesis, Stanford University, Ca.

Yilmaz, O., and Claerbout, J. 58 Dec. 1980 Prestack Partial Migration Geophysics 1753-1779

Yilmaz, O., and Chambers, R. 190 Oct. 1984 Migration velocity analysis by wave-field extrapolation Geophysics 1664-1674

Yilmaz, O. 83 Jan. 1987 Seismic Data Processing (Text) SEG

Yilmaz, O., Chambers, R., Nichols, D., and Abma, R. 131 Dec. 1987 Fundamentals of 3-D Migration, Parts 1(Nov) and 2 The Leading Edge

Yilmaz, O. 135 May. 1991 Tutorial interpretive evaluation of migrated data Exp. Abs., EAEG Convention 120

Young, R. A., Deng, Z., and Sun, J. 540 Apr. 1995 Interactive processing of GPR data The Leading Edge 275-280

Yuon, O. K., and Zhou, H. 762 Jan. 2001 Depth imaging with multiples Geophysics 246-255

Zanzi, L., and Bagaini, C. 521 Oct. 1994 The design of prestack migration-demigration operators Exp. Abs., SEG Nat. Conv. 676-679

Zelt, C., and Ellis, R. 386 Jun. 1988 Practical and efficient ray tracing in two-dimensional media for rapid traveltime and amplitude forward modelling Cdn. Jnl. Exp. Geo. 16-31

Zhang, J., and McMechan, G. A. 652 Jan. 1997 Turning wave migration by horizontal extrapolation Geophysics 291-297

Zhao, P., Uren, N. F., Wenzel, F., Hatherly, P. J., and McDonald, J. A. 786 Dec. 1998 Kirchhoff diffraction mapping in media with large veloicty contrasts Geophysics 2072-2081

Zheludev, V. A., Ragoza, E., and Kosloff, D. D. 763 Mar. 2001 Fast Kirchhoff migration in the wavelet domain Expl. Geoph. (ASEG) 23-27

Zhong, B., Zhou, X., Liu, X., and Yule, J. 578 Mar. 1995 A new strategy for CCP stacking Geophysics 517-521

Zhou, B., and Greenhalgh, A. 378 Sep. 1994 Wave-equation extrapolation-based multiple attenuation: 2-D filtering in the f-k domain Geophysics 1377-1391

Zhou, B., and Greenhalgh, S. 590 Jun. 1995 A partial; DMO operator for use with the stacking velocity Exploration Geophysics 493-496

Zhou, B., Mason I. M., and Greenhalgh S. A. 574 May. 1996 An accurate formulation of log-stretch dip moveout in the frequency-wavenumber domain Geophysics 815-821

Zhou, B., Greenhalgh, S. 594 May. 1996 Multiple suppression by 2-D filtering in the parabolic t-p domain: a wave-equation-based method Geophys. Prosp. 375-401

Zhu, J., and Lines, L. R. 782 Jul. 1998 Comparison of Kirchhoff and reverse-time migration methods with applications to prestack depth imaging of complex structures Geophysics 116-1176

Zhu, T. 196 Jun. 1988 Ray-Kirchhoff migration in inhomogeneous media Geophysics 760-768

Zhu, X., Sixta, D., and Angstman, B 255 Oct. 1992 Tomostatics: Turning-ray tomography + static correction Exp. Abs., SEG Nat. Conv. 1108-1111

Zhu, X. 612 Aug. 1994 Survey parameters for imaging salt domes The Leading Edge 888-892

References: Numerical by identification number

ID number   Author(s)   Date   Title   Journal/publisher   Volume number   Pages

1 Hagedoorn, J. Jun. 1954 A Process of Seismic Reflection Interpretation Geophys. Prosp. 2 85-127

2 Marfurt, K. Dec. 1978 Elastic Wave Equation Migration-Inversion Ph.D. Thesis, Columbia University

3 Hubral, P. Dec. 1977 Time migration-some ray theoretical aspects Geophys. Prosp. 25 738-745

4 French, W. Dec. 1975 Computer Migration of Oblique Seismic Reflection Profiles Geophysics 40 961-980

5 Gardner, G. H. F., French, W., and Matzuk, T. Dec. 1974 Elements of Migration and Velocity Analysis Geophysics 39 811-825

6 Berryhill, J. Aug. 1979 Wave Equation Datumming Geophysics 44 1329-1339

7 Claerbout, J. F. Jun. 1971 Toward a Unified Theory of Reflector Mapping Geophysics 36 467-481

8 Loewenthal, D., Lu, L., Roberson, R., and Sherwood, J. Jun. 1976 The Wave Equation Applied to Migration Geophys. Prosp. 24 380-399

9 Claerbout, J. F. Jun. 1970 Course Grid Calculations of Waves in Inhomogeneous Media with Application to Delineation of Complicated Seismic Structure Geophysics 35 407-418

10 Claerbout, J. F. Dec. 1970 Acoustical Holography, Vol. 3. ed. A.F. Metered Plenum Press, New York 273-283

11 Claerbout, J. F., and Johnson, A., Dec. 1971 Extrapolation of Time Dependent Wave Forms along their Path of Propagation Geophys. Jnl. Royal Astronomical Soc. 26 285-293

12 Hatton, L., Larner K., and Gibson, B. May. 1981 Migration of Seismic Data from Inhomogeneous Media Geophysics 46 751-767

13 Alford, R., Kelly, K., and Boore, D., Dec. 1974 Accuracy of Finite Difference Modelling of the Acoustic Wave Equation Geophysics 39 834-842

14 Deregowski, S. M. Nov. 1978 A Finite Difference Method for CDP Stacked Section Migration Presented EAEG

15 Lapidus, L. Dec. 1962 Digital Computation for Chemical Engineers McGraw-Hill, New York

16 Muir, F. Sep. 1976 Stanford Exploration Project 8 54

17 Clayton, R., and Engquist, B. Jan. 1977 Absorbing Boundary Conditions for Acoustic and Elastic Wave Equations Bull. Seis. Soc. Am. 67 1529-1540

18 Hood, P. Dec. 1978 Finite Difference and Wave Number Migration Geophys. Prospecting 26 773-789

19 Buchanan, D. Jan. 1978 An Exact Solvable One Way Wave Equation Presented at SEG Nat. Conv.

20 Whittlesey, J., and Quay, R. Jan. 1977 Wave Equation Migration Operators Using 2-D Z-Transform Theory Presented at SEG Nat. Conv.

21 Stolt, R. H. Feb. 1978 Migration by Fourier Transform Geophysics 43 23-48

22 Doherty, S. Dec. 1975 Structure Independent Seismic Velocity Estimation Ph.D. Thesis, Stanford University, Ca.

23 Claerbout, J. F. Dec. 1976 Fundamentals of Geophysical Data Processing McGraw-Hill, New York

24 Yilmaz, O. Dec. 1979 Pre-Stack Partial Migration Ph.D. Thesis, Stanford University, Ca.

25 Schultz, P. S., and Claerbout, J. Jun. 1978 Velocity Estimation and Downward Continuation by Wavefront Synthesis Geophysics 43 691-714

26 Estevez, R. Dec. 1977 Wide Angle Diffracted Multiple Reflections Ph.D. Thesis, Stanford University, Ca.

27 Schultz, P. S., and Sherwood, J. Mar. 1980 Depth Migration Before Stack Geophysics 45 376-393

28 Maginness, M. Jan. 1972 The Reconstruction of Elastic Wavefields from Measurements over a Transducer Array J. Sound and Vibration, 20 2 219-40

29 Booer, A., Chambers, J., and Mason, I. Jan. 1940 Numerical Holographic Reconstruction by a Projective Transform Electron. Lett. 13

30 Gazdag, J. Dec. 1978 Wave Equation Migration with the Phase-Shift Method Geophysics 43 1342-1351

31 Gazdag, J. Jan. 1978 Extrapolation of Seismic Waveforms by Fourier Methods IBM J. Res. Dev. 22 481-486

32 Gazdag, J. Feb. 1980 Wave Equation Migration with the Accurate Space Derivative Method Geophys. Prosp. 28 60-70

33 Phinney, R., and Frazer, L. Jan. 1978 On the Theory of Imaging by Fourier Transform Presented SEG Nat. Conv.

34 Chun, J., and Jacewitz, C. May. 1981 Fundamentals of Frequency Domain Migration Geophysics 46 717-733

35 Estes, L., and Fain, G. Jan. 1977 Numerical Technique for Computing the Wide Angle Acoustic Field in an Ocean with Range-Dependent Velocity Profiles J. Acoust. Soc. Am. 62 38-43

36 Berkhout, A. J., and Palthe, D. Van W. Mar. 1979 Migration in Terms of Spatial Deconvolution Geophys. Prosp. 27 261-291

37 Schneider, W. A. Feb. 1978 Integral Formulation for Migration in Two and Three Dimensions Geophysics 43 49-76

38 Kuhn, M., and Alhilali, K. Aug. 1977 Weighting Factors in the Construction and Reconstruction of Acoustical Wavefields Geophysics 42 1183-1198

39 Bolondi, G., Rocca, F., and Savelli, S. Dec. 1978 A Frequency Domain Approach to Two Dimensional Migration Geophys. Prosp. 26 750-772

40 Garibotto, G. Jan. 1979 2-D Recursive Filters for the Solution of Two-Dimensional Wave Equations IEEE Trans. on ASSP 27 367-373

41 Kuhn, M. Mar. 1979 Acoustical Imaging of Source Receiver Coincident Profile Geophys. Prospecting 27 62-77

42 Devey, M. Jan. 1979 Derivation of the Migration Integral Technical Note TN451, BP Company Ltd., Exploration and Production Department

43 Hosken, J. W. J. Jan. 1979 Improvements in the Practice of 2D Diffraction Stack Migration Report No. EPR/R1247, BP Company Limited, Exploration and Production Department

44 Gibson, B., Larner, K., and Levin, S. May. 1979 Efficient 3D migration in 2 steps Handout EAEG, Hamburg

45 Cohen, J., and Bleistein, N. Jun. 1979 Velocity Inversion Procedure for Acoustic Waves Geophysics 44 1077-1087

46 Cohen, J., and Bleistein, N. Jan. 1977 An Inverse Method for Determining Small Variations in Propagation Speed Soc. Ind. Appl. Math., J. Appl. Math. 32 784-99

47 Tappert, F., and Hardin, R. Jan. 1973 A Synopsis of the AESD Workshop on Acoustic Propagation Modelling by Non-Ray Tracing Techniques AD-773 741, AESD Tech. Note TN-73-05

48 Raz, S. Jan. 1980 An Approximate Propagation Speed Inversion over a Prescribed Slab, Acoustic Imaging, Vol. 9 Plenum Press, New York

49 Sherwood, J. W. C., Schultz, p., and Judson, D. Jan. 1978 Equalizing the Stacking Velocities of Dipping Events AAPG 2362

50 Judson, D., Lin, J., Schultz, P., and Sherwood, J. Mar. 1980 Depth Migration After Stack Geophysics 45 361-375

51 Rayleigh, J. Jan. 1945 The Theory of Sound Dover Publications, London

52. Wren, A., and Jan, S. Jan. 1987 Diffractions Revisited Cdn. Jnl. of Exp. Geo. 12 8-13
53. Jakubowicz, H., and Levin, S. Feb. 1983 A Simple Exact Method of 3-D Migration-Theory Geophys. Prosp. 31 34-56
54. Lumley, D. E. Oct. 1989 A generalized Kirchhoff-WKBJ depth migration theory for multi-offset seismic reflection data: reflectivity model construction by wavefield imaging and amplitude estimation. M.Sc. Thesis, U of BC
55. Sheriff, R. E. May. 1980 Nomogram for Fresnel-Zone Calculation Geophysics 45 968-972
56. Taner, M. T., and Koehler, F. Dec. 1969 Velocity Spectra-Digital Computer Derivation and Applications of Velocity Functions Geophysics 34 859-881
57. Hale, D. Jun. 1984 Dip Move-Out by Fourier Transform Geophysics 49 741-757
58. Yilmaz, O., and Claerbout, J. Dec. 1980 Prestack Partial Migration Geophysics 45 1753-1779
59. Compani-Tabrizi, B. Dec. 1986 k-t Scattering Formulation of the Absorptive Acoustic Wave Equation: Wraparound and Edge Effect Elimination Geophysics 51 2185-2192
60. Cerjan., C., Kosloff, D., and Kosloff, R., Reshef, M. Apr. 1985 A Non-Reflecting Boundary Condition for Discrete Acoustic and Elastic Wave Equations Geophysics 50 705-708
61. Robinson, J., and Robbins, T. Feb. 1978 Dip Domain Migration of Two-Dimensional Seismic Profiles Geophysics 43 77-93
62. Sattlegger, J. W. Feb. 1982 Migration of Seismic Interfaces Geophys. Prosp. 30 71-85
63. Safar, M. Jul. 1985 On the Lateral Resolution achieved by Kirchhoff Migration Geophysics 50 1091-1099
64. Gray, S. H. Aug. 1986 Efficient Traveltime Calculations for Kirchhoff Migration Geophysics 51 1685-1688
65. Kosloff, D. D., and Baysal, E. Jun. 1982 Forward Modelling by a Fourier Method Geophysics 42 1402-1412
66. Sochacki, J., Kubichek, R., and George, J., Fletcher, W., and Smithson, S. Jan. 1987 Absorbing Boundary Conditions and Surface Waves Geophysics 52 60-71
67. Kuo, J., and Dai, T. Aug. 1984 Kirchhoff Elastic Wave Migration for the Case of Noncoincident Source and Receiver Geophysics 49 1223-1238
68. Wiggins, J. Aug. 1984 Kirchhoff Integral Extrapolation and Migration of Nonplanar Data Geophysics 49 1239-1248
69. Carter, J., and Frazer, N. Jan. 1984 Accommodating Lateral Velocity changes in Kirchhoff Migration by means of Fermat's Principle Geophysics 49 46-53
70. Kallweit, R., and Wood, L. Jul. 1982 The Limits of Resolution of Zero-Phase Wavelets Geophysics 47 1035-1046
71. Robinson, E. A. Jan. 1982 A Historical Perspective of Spectrum Estimation Proc. IEEE 70 885-907
72. Berg, L. Dec. 1984 Prestack Partial Migration Exp. Abs., SEG Nat. Conv. 796-799
73. Larner, K., Hatton., L., and Gibson, R.S., Hsu, I. May. 1981 Depth Migration of Imaged Time Sections Geophysics 46 734-750
74. Trorey, A. Oct. 1970 A Simple Theory for Seismic Diffractions Geophysics 35 762-784
75. Li, Z. Aug. 1986 Wave-Field Extrapolation by the Linearly Transformed Wave Equation Geophysics 51 1538-1551
76. Deregowski, S. M. Jul. 1986 What is DMO? First Break 4 7-24
77. Hood, P., (book edited by Fitch) Jan. 1981 Developments in Geophysical Exploration Methods (Chapter 6 by Hood)) Applied Science Publishers Ltd. 2 151-230
78. French, W. Sep. 1986 Trends in Seismic Data Processing The Leading Edge
79. Loveridge, M., Parkes, G., Hatton, L. Aug. 1987 3-D Modelling of Migration Velocity Fields and Velocity Error Zones First Break 5 281-293
80. Parkes, G., Hatton, L. Apr. 1987 Towards a systematic understanding of the effects of velocity model errors on depth and time migration of seismic data First Break 5 121-132
81. Bleistein, N., Cohen, J., and Hagin, F. Jan. 1987 Two and One-Half Dimensional Born Inversion with Arbitrary References Geophysics 52 26-36
82. Bolondi, G., Loinger, E., Rocca, F. Dec. 1984 Offset Continuation in Theory Practice Geophys. Prosp. 32 1045-1073
83. Yilmaz, O. Jan. 1987 Seismic Data Processing (Text) SEG
84. Notfors, C., Godfrey, R. Dec. 1987 Dip moveout in the frequency-wavenumber domain Geophysics 52 1718-1721
85. Larner, K., and Beasley, C. May. 1987 Cascaded Migration: Improving the Accuracy of Finite-Difference Migration Geophysics 52 618-643
86. Liner, C., and Bleistein, N. Oct. 1988 Dip Moveout Processing Exp. Abs., SEG Nat. Conv.
87. French, W. Jan. 1985 Partial Migration of True 3-D Seismic Reflection Surveys via Common Reflection-Point Stacking Pre-print, Offshore Tech. Conf..
88. Holberg, 0. Feb. 1988 Toward optimum one-way propagation Geophys. Prosp. 36 99-114
89. Liner, C. L. May. 1990 General theory and comparative anatomy of Dip Moveout Geophysics 55 595-607
90. Pai, D. Dec. 1988 Generalized F-K (Frequency wavenumber) Migration in Arbitrarily Varying Media Geophysics
91. Bancroft, J. C. May. 1990 Accurate Amplitude Calculations for Time Domain DMO CSEG Calgary 1990
92. Andrews, H. C., and Hunt, B. R. Jan. 1977 Digital Image Restoration (Text) Prentice-Hall Inc.
93. Beasley, C., and Lynn, W. Oct. 1989 The Zero Velocity Layer: Migration From Irregular Surfaces Exp. Abs., SEG Nat. Conv. 1179 - 1183
94. Larner, K., Beasley, C., Lynn, W. Jun. 1989 In quest of the flank Geophysics 54 701-717
95. May, B. Aug. 1983 Structural Inversion of Salt Dome Flanks Geophysics
96. Domenico, S. Dec. 1989 Discussion on Inquest of the Flank Geophysics
97. Reichenderger, H. Sep. 1988 Lithotripter Systems Proceeding of IEEE
98. Fricke, J Sep. 1988 Reverse Time Migration in Parallel; a Tutorial Geophysics
99. Beasley, C., Lynn, W., Larner, K., and Nguyen, H. Jul. 1988 Cascaded F-K migration: Removing the Restrictions of Depth-Varying Velocities Geophysics 53 881-893
100. Brysk, H. May. 1983 Numerical Analysis of the 45-degree Finite-difference Equation for Migration Geophysics
101. Hale, D. Jun. 1991 A Nonaliased integral method for Dip Moveout Geophysics
102. Gazdag, J., and Squazzero, P. Oct. 1984 Migration of Seismic Data Proceedings of IEEE 72 1302-1315
103. Chang, W., and McMechan, G. Apr. 1989 3-D Acoustic Reverse Time Migration Geophys. Prosp.
104. Larner, K., and Hatton, L. Dec. 1990 Wave Equation Migration: Two Approaches(Classic old paper from 1976) First Break
105. Pan, N., and French, W. May. 1989 Generalized Two Pass Three Dimensional Migration For Imaging Steep Dips in Vertically Inhomogeneous Media Geophysics 54 544-554
106. Cary, P. W. Jan. 1991 Symmetric One Pass 3-D Migration CSEG Recorder
107. Eaton, D. W. S., Stewart, R., and Harrison, M. Mar. 1991 The

References

Fresnel Zone for P-SV Waves Geophysics

108 Knapp, R. Mar. 1991 Fresnel Zones in the Light of Broad Band Data Geophysics

109 Reshef, M., and Kosloff, D. Feb. 1986 Migration of Common-Shot Gathers Geophysics 51 324-331

110 Loewenthal, D., and Hu, L. Mar. 1991 Two Methods for Computing the Imaging Condition for Common-Shot Prestack Migration Geophysics

111 Forel D., and Gardner, G. May. 1988 A Three-Dimensional Perspective on Two-Dimensional Dip Moveout Geophysics 53 604-610

112 Uren, N. F., Gardner, G.H.F., and McDonald, J.A. Jul. 1990 Dip moveout in anisotropic media Geophysics 55 863-867

113 Uren, N. F., Gardner, G.H.F., and McDonald, J.A. Nov. 1990 The migrator's equation for anisotropic media Geophysics 55 1429-1434

114 Clayton, R., and Engquist, B. May. 1980 Absorbing Boundary Conditions for Wave Equation Migration Geophysics 45 895-904

115 Levin, F. K. Jun. 1971 Apparent Velocity From Dipping Interface Reflections Geophysics

116 Spitz, S. Jun. 1991 Seismic Trace Interpolation in the F-X Domain Geophysics

117 Reshef, M. Jan. 1991 Depth Migration From Irregular Surfaces with Depth Extrapolation Methods Geophysics

118 Ameely, L., Krey, T., Muhtadie, F., Rau, H., and Wrist, H. Aug. 1983 Migration in the Presence of a Rugged Interface with High Velocity Contrast Geophys. Prosp. 31 561-573

119 Marfurt, K. J. May. 1984 Accuracy of Finite-Difference and Finite-Element Modeling of the Scalar and Elastic Wave Equations Geophysics

120 Kim, Y. C., and Gonzarlez, R. Mar. 1991 Migration Velocity Analysis with the Kirchhoff Integral Geophysics

121 Docherty, P. Aug. 1991 A Brief Comparison of some Kirchhoff Integral Formulas for Migration and Inversion Geophysics 56 1164-1169

122 Pan, N., Perkins, W., and French, W. Nov. 1991 Exact Two-Pass Full-Dip (>90 degree) 3-D Migration for Vertically Inhomogeneous Media Exp. Abs., SEG Nat. Conv. 1160-1163

123 Levin, S. May. 1984 Principle of Reverse-Time Migration Geophysics

124 Li, Z., Lynn, W, Chambers R., Larner K., and Abma, R. Jan. 1991 Enhancements to Prestack Frequency-Wavenumber (F-K) Migration Geophysics 56 27-40

125 Emerman, S., Schmidt, W., and Stephen R. Nov. 1982 An Implicit Finite-Difference Formulation of the Elastic Wave Equation Geophysics 47 1521-1526

126 Cohen, J., Hagin, F., and Bleistein, N. Aug. 1986 Three-Dimensional Born Inversion with an Arbitrary Reference Geophysics

127 Clayton, R., and Stolt, R. Nov. 1981 A Born-WKBJ Inversion Method for Acoustic Reflection Data Geophysics

128 Docherty, P. Oct. 1987 Two-and-One-Half-Dimensional Poststack Inversion Exp. Abs. SEG Nat. Conv. 564

129 Dong, W., Emanuel, M., Bording, P., and Bleistein, N. Sep. 1991 A Computer Implementation of 2.5-D Common-Shot Inversion Geophysics

130 Myczkowski, J., McCowan, D., and Mufti, I. Jun. 1991 Seismic Modeling in Real Time The Leading Edge

131 Yilmaz, O., Chambers, R., Nichols, D., and Abma, R. Dec. 1987 Fundamentals of 3-D Migration, Parts 1(Nov) and 2 The Leading Edge

132 Hubral, P., Tygel, M., and Zein, H. Jan. 1991 Three-Dimensional True-Amplitude Zero Offset Migration Geophysics 56 18-26

133 Ferber, R. G. Aug. 1991 A Filter, Delay and Spread Technique for 3D DMO Geophys. Prosp.

134 Hill, N. R. Nov. 1990 Gausian Beam Migration Geophysics

135 Yilmaz, O. May. 1991 Tutorial interpretive evaluation of migrated data Exp. Abs., EAEG Convention 120

136 Vidale, J. Dec. 1988 Finite-Difference Traveltime Calculation Bull., Seis. Soc. Am. 78 2062-2076

137 Vidale, J. May. 1990 Finite-Difference Calculation of Traveltimes in Three Dimensions Geophysics

138 Vidale, J., and Houston, H. Nov. 1990 Rapid Calculation of Seismic Amplitudes Geophysics

139 Moser, T. J. Jan. 1991 Shortest Path Calculation of Seismic Rays Geophysics

140 van Trier, J., and Symes, W., Jun. 1991 Upwind Finite-Difference Calculation of Traveltimes Geophysics

141 Notfors, C. May. 1991 Common Offset Depth Migration Using Finite Difference Raytracing Exp. Abs., EAEG Convention 28

142 Reshef, M. Aug. 1991 Prestack Depth Imaging of Three-Dimensional Shot Gathers Geophysics

143 Qin, F., Luo, Y., Olsen K., Cai, W., and Schuster G. Mar. 1992 Finite-Difference Solution of the Eikonal Equation along Expanding Wavefronts Geophysics

144 Aldridge, D., and Oldenburg D. Mar. 1992 Refractor Imaging Using an Automated Wavefront Reconstruction Method Geophysics 57 378-385

145 Schneider, W. A., Ranzinger, K., Balch, A., and Kruse, C. Jan. 1992 A Dynamic Programming Approach to First Arrival Traveltime Computation in Media with Arbitrarily Distributed Velocities Geophysics

146 Ratcliff, D.W., Grey, S., and Whitmore, N. Apr. 1992 Seismic Imaging of Salt Structures in the Gulf of Mexico The Leading Edge 11 15-31

147 Johnson, J. Apr. 1992 Structural Imaging in the Real World The Leading Edge 11

148 Geoltrain, S., and Brac, J. Nov. 1991 Can We Image Complex Structures with Finite-Difference Traveltimes? Exp. Abs., SEG Nat. Conv. 2 1110-1113

149 Levin, S. Nov. 1991 Should We Ever Use Poststack Depth Migration Exp. Abs., SEG Nat. Conv. 2 1144-1147

150 May, B. T., and Hron, F. Oct. 1978 Synthetic seismic sections of typical petroleum traps Geophysics 43 1119-1147

151 Versteeg, R., and Grau, G. May. 1990 The Marmousi Experience EAEG Workshop PASD Inversion

152 Versteeg, R., Oct. 1991 Analyze Du Probleme De La Determination Du Modele De Vitesse Pour L'Imagerie Sismique, Velocity analysis on the Marmousi data set Doc. Thesis, text is in English.

153 Perkins, W., and French, W. Sep. 1990 3-D Migration to Zero Offset for a Constant Velocity Gradient Exp. Abs., SEG Nat. Conv. 2 1354-1357

154 Randall, C. May. 1988 Absorbing Boundary Condition for the Elastic Wave Equation Geophysics

155 Wenzel, F., and Menges, D. Aug. 1989 A Comparison Between Born Inversion and Frequency-Wavenumber Migration Geophysics

156 Cabrera, J., and Levy, S. Aug. 1989 Shot Dip Moveout with Logarithmic Transformations Geophysics 54 1038-1041

157 Mikulich, W., and Hale, D. Jan. 1992 Steep-dip V(z) Imaging from an Ensemble of Stolt-like Migrations Geophysics

158 Berryhill, J. Nov. 1984 Wave-equation Datuming Before Stack Geophysics 49 2064-2066

159 Hale, D. Nov. 1991 3-D Depth Migration via McClellan Transformations Geophysics

160 Cambois, G. Dec. 1991 A Proof for the Convergence of the 15-Degree Cascaded Migration Geophysics

161 Bunks, C. Oct. 1992 Optimization of Paraxial Wave Exp. Abs., SEG Nat. Conv. 897-900

162 Lee, D., Mason, I.M., and Jackson, G.M., Nov. 1991 Split-step Fourier shot-record migration with deconvolution imaging Geophysics 6 1786-1793

163 Gray, S. H. Jul. 1992 Frequency-selective design of the Kirchhoff migration operator Geophys. Prosp. 40 565-571

164 Herman, G. C. Jan. 1992 Generalization of traveltime inversion Geophysics 57 9-14

165 Whitmore, N. D., Gray, S. H., and Gersztenkorn, A. Jun. 1988 Two-dimensional post-stack depth migration: a survey of methods First Break 6 189-197

166 MacKay, S., and Abma, R. Dec. 1992 Imaging and velocity estimation with depth-focusing analysis Geophysics 57 1608-1622

167 Jackson, G. M., Mason, I. M., and Lee, D. Nov. 1991 Multicomponent common-receiver gatherer migration of single-level walkaway seismic profiles Geophys. Prosp. 39 1015-1029

168 Meyer, H. G. Nov. 1991 Some remarks on noise stability in dynamic inversion of reflection seismic data Geophys. Prosp. 39 1005-1014

169 Hale, D. Nov. 1991 Stable explicit depth extrapolation of seismic wavefields Geophysics 56 1770-1777

170 Cerveny, V., and Soares, J.E.P., Jul. 1992 Fresnel volume ray tracing; Geophysics 57 902-915

171 Tallin, A. G., and Santamarina, J.C. May. 1992 Digital ray tracing for geotomography IEEE Trans. 30 617-619

172 Reilly, J. M. Oct. 1992 Seismic imaging adjacent to and beneath salt diapirs, UK North Sea First Break 10 383-397

173 Richardson, R. P. Oct. 1992 Linear velocity function: Reflection times for specific ranges Geophysics 57 1352-1353

174 Hale, D., Hill, N. R., and Stefani, J. Nov. 1992 Imaging salt with turning seismic waves Geophysics 57 1453-1462

175 Beasley, C., and Lynn, W. Nov. 1992 The zero-velocity layer: Migration from irregular surfaces Geophysics 57 1435-1443

176 Levin, F. K. Jul. 1990 Reflection from a dipping plane-Transversely isotropic solid Geophysics 55 851-855

177 Carrion, P. M., Sato, H. K., and Buono, A.V.D. Jun. 1991 Wavefront sets analysis of limited aperture migration sections Geophysics 56 778-784

178 Benson, A. K. Sep. 1991 An explicit, unconditionally stable, 15-degree depth migration and modeling algorithm implemented in poststack, directional, and prestack modes Geophysics 56 1412-1422

179 Brzostowski, M. A., Black, J. A. Jul. 1990 Steep-dip time migration and residual depth migration EAEG/SEG Res. Wkp. Seismic velocities, Cambridge, England

180 Craig, M. S., Long, L. T., and Tie, A. Dec. 1991 Modeling the seismic P coda as the response of a discrete-scatterer medium Phys. Earth Plan. Int. 67 20-35

181 Gray, S. H., Jacewitz, C. A., and Epton, M.E. May. 1991 Analytic synthetic seismograms for depth migration testing Geophysics 56 697-700

182 Maeland, E. May. 1991 Migration analysis with an extra time shift Geophysics 56 691-696

183 Kamel, A. H. Mar. 1991 Time-domain behavior of wide-angle one-way wave equations Geophysics 56 382-384

184 Gray, S. H., and Epton, M. A. Jul. 1990 Multigrid migration: Reducing the migration aperture but not the migrated dips Geophysics 55 856-862

185 Scott, J. H., Tibbetts, B.L., and Burdick, R.G., Jan. 1940 Computer analysis of seismic refraction data. ???

186 Gonzales-Serrano, and Claerbout, J. Apr. 1981 Deformations of CMP gathers with v(z) to hyperbolas Stanford Exploration Project Report 26 137-156

187 Press, W. H., Flannery, B. P., Teukolsky, S. A., and Vetterling, W. T. Dec. 1986 Numerical Recipes (FORTRAN) (Text) Cambridge University Press

188 Liner, C. L. Feb. 1991 Born theory of wave-equation dip moveout Geophysics 56 182-189

189 Lynn, W., MacKay, S., and Beasley, C.J. Sep. 1990 Efficient migration through irregular water-bottom topography Exp. Abs., SEG Nat. Conv. 1297-1300

190 Yilmaz, O., and Chambers, R. Oct. 1984 Migration velocity analysis by wave-field extrapolation Geophysics 49 1664-1674

191 Hubral, P., and Krey, T. Apr. 1980 Interval Velocities from Seismic Reflection Time Measurements SEG

192 Al-Yahya, K. Jun. 1989 Velocity analysis by iterative profile migration Geophysics 54 718-729

193 Milkereit, B. Jul. 1987 Migration of noisy crustal seismic data J. Geophys. Res. 92 7916-7930

194 Lee, W. B., and Zhang, L., Jun. 1992 Residual shot profile migration Geophysics 57 815-822

195 Cambois, G., and Stoffa, P.L. Jun. 1992 Surface-consistent deconvolution in the log/Fourier domain Geophysics 57 823-840

196 Zhu, T. Jun. 1988 Ray-Kirchhoff migration in inhomogeneous media Geophysics 53 760-768

197 van der Schoot, A., Romijn, R., Larson, D.E., and Berkhout, A.J. Jul. 1989 Prestack migration by shot record inversion and common depth point stacking: a case study First Break 7 293-304

198 Bleistein, N., and Gray, S. H. Nov. 1985 An extension of the Born inversion method to a depth dependent reference profile Geophys. Prosp. 33 999-1022

199 Tricker, R. A. R. Jan. 1965 Bores, breakers, waves and wakes American Elsevier Publishing Company, Inc. 186-214

200 Hilterman, F. J. Dec. 1970 Three-dimensional seismic modeling Geophysics 35 1020-1037

201 Shtivelman, V., and Canning, A. Oct. 1988 Datum correction by wave-equation extrapolation Geophysics 53 1311-1322

202 Hosken, J. W. J., and Deregowski, S.M. Feb. 1985 Tutorial migration strategy Geophys. Prosp. 33 1-33

203 Lehmann, H. J., and Houba, W. Feb. 1985 Practical aspects of the determination of 3-D stacking velocities Geophys. Prosp. 33 34-51

204 Hospers, J. Feb. 1985 Sideswipe reflections and other external and internal reflections from salt plugs in the Norwegian-Danish Basin Geophys. Prosp. 33 52-71

205 Black, J. L., Schleicher, L., and Zhang, L. Jan. 1993 True-amplitude imaging and dip moveout Geophysics 58 47-66

206 Lafond, C. F., and Levander, A.R. Jan. 1993 Migration moveout analysis and depth focusing Geophysics 58 91-100

207 Dong, Z., and McMechan, G.A. Jan. 1993 3-D prestack migration in anisotropic media Geophysics 58 79-90

208 Freire, S. L. M., and Ulrych, T.J. Jun. 1988 Application of singular value decomposition to vertical seismic profiling Geophysics 53 778-785

209 Dillon, P. B. Jun. 1988 Vertical seismic profile migration using the Kirchhoff integral Geophysics 53 786-799

210 Sattlegger, J. W., Stiller, P.K., Echterhoff, J.A., and Hentschke, M.K. Dec. 1980 Common offset plane migration (COPMIG) Geophys. Prosp. 28 859-871

211 Deregowski, S. M. Jun. 1990 Common-offset migrations and velocity analysis First Break 8 225-234

References

212 Berryhill, J. Oct. 1991 Kinematics of crossline prestack migration Geophysics 56 1674-1676

213 Wenzel, F. Oct. 1991 Frequency-wavenumber migration in laterally heterogeneous media Geophysics 56 1671-1673

214 Li, Z. Oct. 1991 Compensating finite-difference errors in 3-D migration and modeling Geophysics 56 1650-1660

215 Sena, A. G., and Toksöz, M.N. Feb. 1993 Kirchhoff migration and velocity analysis for converted and nonconverted waves in anisotropic media Geophysics 58 265-276

216 Nautiyal, A., Gray, S.H., Whitmore, N.D., and Garing, J.D. Feb. 1993 Stability versus accuracy for an explicit wavefield extrapolation operator Geophysics 58 277-283

217 Fischer, R., Lees, J.M. Jul. 1993 Shortest path ray tracing with sparse graphs Geophysics 58 987-996

218 Asakawa, E., and Kawanaka, T. Jan. 1993 Seismic ray tracing using linear traveltime interpolation Geophys. Prosp. 41 99-111

219 Coultrip, R.L. Feb. 1993 High-accuracy wavefront tracing traveltime calculation Geophysics 58 284-292

220 Daly, J. W. Apr. 1948 An instrument for plotting reflection data on the assumption of linear increase of velocity Geophysics 13 153-157

221 Hale, D., and Artley, C. Feb. 1993 Squeezing dip moveout for depth-variable velocity Geophysics 58 257-264

222 Watson, K. Jun. 1993 Processing remote sensing images using the 2-D FFT--Noise reduction and other applications Geophysics 58 835-852

223 Van Melle, F. A. Apr. 1948 Wave-front circles for a linear increase of velocity with depth Geophysics 13 158-162

224 Reiter, E. C., Toksoz, M. N., and Purdy, G.M. Jan. 1993 A semblance-guided median filter Geophys. Prosp. 41 15-41

225 Mereu, R. F. Feb. 1978 A computer program to obtain the weights of a time-domain wave-shaping filter which is optimum in an error-distribution sense Geophysics 43 197-215

226 Mereu, R. F. Aug. 1976 Exact wave-shaping with a time-domain digital filter of finite length Geophysics 41 659-672

227 Worley, S. C. Feb. 1993 The geometry of reflection Geophysics 58 293-297

228 Meinardus, H. A., and Schleicher, K. May. 1992 3-D time-variant dip-moveout by the FK method Exp. Abs., CSEG Nat. Conv. 22-23

229 Rodriguez, S., Whitsett, R., Diet, J.P., Paturet, D. May. 1992 Elevation and weathering effects on dip move-out resequenced DMO (RSDMO) Exp. Abs., CSEG Nat. Conv. 24

230 Cary, P. W. May. 1992 Applying dip moveout to land data: accounting for topography and irregular acquisition geometry Exp. Abs., CSEG Nat. Con. 25-26

231 Kitchenside, P. W., Ellis, A.N., and Bale, R.A. May. 1992 Datum-consistent migration and DMO for irregular topography Exp. Abs., CSEG Nat. Conv. 27-28

232 Butler, K. E., Russell, R.D. Jun. 1993 Subtraction of powerline harmonics from geophysical records Geophysics 58 898-903

233 Sumner, B. L. Apr. 1988 Asymptotic solutions to forward and inverse problems in isotropic elastic media CSM CWP

234 Engquist, B., and Clayton, R. Sep. 1976 Difference approximations of one way elastic waves SEP Progress Report 141-164

235 Engquist, B. Sep. 1976 Absorbing boundary conditions for wave equations EP Progress Report 30-49

236 Claerbout, J. F., and Clayton, R.W. Sep. 1976 A paraxial equation for elastic waves EP Progress Report 165-170

237 Clayton, R. Sep. 1976 Programming absorptive side boundaries for migration EP Progress Report 10 16-29

238 Black, J. L., and Egan, M. Nov. 1988 True-amplitude DMO in 3-D Handout, SGE convention

239 Liner, C. L., and Bleistein, N., Nov. 1988 Comparative anatomy of common offset dip moveout Exp. Abs., SEG Nat. Conv.

240 Stoffa, P. L., Buhl, P., and Bryan, G.M. Aug. 1974 The application of homomorphic deconvolution to shallow-water marine seismology-Part I: Models Geophysics 39 401-416

241 Gardner, G. H. F., Gardner, L. W., and Gregory, A. R. Dec. 1974 Formation velocity and density - the diagnostic basics for stratagraphic traps Geophysics 39 770-780

242 Hatherly, P. J., and Neville, M.J. Feb. 1986 Experience with the generalized reciprocal method of seismic refraction interpretation for shallow engineering site investigation Geophysics 51 255-265

243 Ng, M. Jun. 1994 Prestack migration of shot records using phase shift plus interpolation CSEG Journal 30 11-27

244 Brown, R. L. Oct. 1992 Estimation of AVO in the presence of noise using prestack migration Exp. Abs., SEG Nat. Conv. 885-888

245 Rietveld, W. E. A., and Berkhout, A.J. Oct. 1992 Depth migration combined with controlled illumination Exp. Abs., SEG Nat. Conv. 931-934

246 Herbert, V. Oct. 1992 F-K depth migration Exp. Abs., SEG Nat. Conv. 935-938

247 Wyatt, K. D., Towe, S.K., Layton, J.E., Wyatt, S.B., Von Seggern, D.H., and Brockmeier, C.A. Oct. 1992 Ergonomics in 3-D depth migration Exp. Abs., SEG Nat. Conv. 944-947

248 Godfrey, R. J. Oct. 1992 DMO and V(z). Exp. Abs., SEG Nat. Conv. 952-954

249 Dietrich, M., and Cohen, J.K. Oct. 1992 Three-dimensional dip moveout operator for linear velocity-depth functions Exp. Abs., SEG Nat. Conv. 958-961

250 Meinardus, H. A., Schleicher, K.L., and Sudhakar, V. Oct. 1992 Processing sequence for turning wave imaging Exp. Abs., SEG Nat. Conv. 988-991

251 Silva, R. Oct. 1992 Antialiasing and application of weighting factors in Kirchhoff migration Exp. Abs., SEG Nat. Conv. 995-998

252 Link, B., Kendrick, J., and Butler, J.P. Oct. 1992 Practical approach to imaging under extreme lateral velocity variations Exp. Abs., SEG Nat. Conv. 1013-1016

253 Dai, N., and Kanasewich, E.R. Oct. 1992 One-way wave equation: absorbing boundary and seismic migration Exp. Abs., SEG Nat. Conv. 1026-1029

254 Fowler, P. J. Oct. 1992 Improving prestack F-K migration velocity analysis using residual moveout Exp. Abs., SEG Nat. Conv. 1038-1041

255 Zhu, X., Sixta, D., and Angstman, B Oct. 1992 Tomostatics: Turning-ray tomography + static correction Exp. Abs., SEG Nat. Conv. 1108-1111

256 Payne, B., and Marcoux, M. Oct. 1992 Rugged topography and S-R downward continuation Exp. Abs., SEG Nat. Conv. 1124-1126

257 Stork, C., and Kusuma, T. Oct. 1992 Hybrid genetic autostatics: New approach for large-amplitude statics with noisy data Exp. Abs., SEG Nat. Conv. 1127-1131

258 Purnell, G. W. Nov. 1992 Imaging beneath a high-velocity layer using converted waves Geophysics 57 1444-1452

259 Docherty, P. Oct. 1992 Solving for the thickness and velocity of the weathering layer using 2-D refraction tomography Geophysics 57 1307-1318

260 Cross, G. M. Oct. 1992 Array responses for plane and spherical incidence Geophysics 57 1294-1306

261 Spies, B. R. Oct. 1992 Information at your fingertips — using the

Digital Cumulative Index The Leading Edge 11 65-66
262 Varela, C. L., Rosa, A. L. R., and Ulrych, T.J. Aug. 1993 Modeling of attenuation and dispersion Geophysics 58 1617-1173
263 Vinje, V., Iversen, E., Gjoystdal, H. Aug. 1993 Traveltime and amplitude estimation using wavefront construction Geophysics 58 1157-1166
264 MacKay, S., Abma, R. Aug. 1993 Depth-focusing analysis using a wavefront-curvature criterion Geophysics 58 1148-1156
265 Harrison, M. P., and Stewart, R.R. Aug. 1993 Poststack migration of P-SV seismic data Geophysics 58 1127-1135
266 Schleicher, J., Tygel, M. Hubral, P. Aug. 1993 3-D true-amplitude finite-offset migration Geophysics 58 1112-1126
267 Versteeg, R. Jun. 1993 Sensitivity of prestack depth migration to the velocity model Geophysics 58 873-8824
268 Schuster, G. T., and Quintus-Bosz, A. Sep. 1993 Wavepath eikonal traveltime inversion: Theory Geophysics 58 1314-1323
269 Qin, F., Schuster, G. T. Sep. 1993 First-arrival traveltime calculation for anisotropic media Geophysics 58 1349-1358
270 Lee, D., Jackson, G. M., and Mason, I.M. Sep. 1993 Partially coherent migration Geophysics 58 1301-1313
271 Kessler, D., Chan, W. Oct. 1993 DMO velocity analysis with Jacubowicz's dip-decomposition method Geophysics 58 1517-1524
272 Alkhalifah, T. Sep. 1993 Gaussian Beam Migration for Anisotropic Media Exp. Abs., SEG Nat. Conv. 847-851
273 Diet, J. P., Audebert, F., Huard, I., Lanfranchi, P., and Zhang, X., Sep. 1993 Velocity analysis with prestack time migration using the S-G method: A unified approach Exp. Abs., SEG Nat. Conv. 957-
274 Audebert, F., and Diet, J. P., Sep. 1993 Migrated focus panels: focusing analysis reconciled with prestack depth migration Exp. Abs., SEG Nat. Conv. 961-964
275 Jervis, M., Sen, M., Stoffa, P.L. Sep. 1993 Optimization methods for 2-D prestack migration velocity estimation Exp. Abs., SEG Nat. Conv. 965-968
276 Kim, Y. C., and Krebs, J.R., Sep. 1993 Pitfalls in velocity analysis using common-offset time migration Exp. Abs., SEG Nat. Conv. 969-973
277 Jones, I. F. Sep. 1993 3-D velocity model building via iterative one-pass depth migration Exp. Abs., SEG Nat. Conv. 974-977
278 Cary, P. W. Sep. 1993 3-D migration with four-way splitting Exp. Abs., SEG Nat. Conv. 978-981
279 Moser, T. J. Sep. 1993 Migration using fast traveltime calculators Exp. Abs., SEG Nat. Conv. 1033-1035
280 Schneider, W. A. Sep. 1993 Robust, efficient upwind finite-difference traveltime calculations in 3-D Exp. Abs., SEG Nat. Conv. 1036-1039
281 Dai, N., Kanasewich, E.R., and Vafidis, A., Sep. 1993 Composed absorbing boundaries for seismic modeling Exp. Abs., SEG Nat. Conv. 1052-1055
282 Peng, C., and Toksšz, M.N. Sep. 1993 An optimal absorbing boundary condition for elastic wave modeling Exp. Abs., SEG Nat. Conv. 1056-1059
283 Takahashi, T. Sep. 1993 Multicomponent prestack depth migration Exp. Abs., SEG Nat. Conv. 1078-1081
284 Thore, P. D., and Kelly, P. Sep. 1993 Stack enhancement using the three-term equation: synthetic and real data examples Exp. Abs., SEG Nat. Conv. 1023-1025
285 Castle, R. J. Sep. 1993 A 2-D V(z) DMO algorithm Exp. Abs., SEG Nat. Conv. 1030-1033
286 Ohanian, V. Sep. 1993 Virtual images and normal moveout Exp. Abs., SEG Nat. Conv. 1126-1129
287 Ohanian, V., Snyder, T.M., and Gunawardena, A. Sep. 1993 Analytic properties of the f-k DMO operator Exp. Abs., SEG Nat. Conv. 1134-1136
288 Ohanian, V. Sep. 1993 DMO by the Huygens-Fresnel diffraction integral Exp. Abs., SEG Nat. Conv. 1037-1040
289 Landa, E., Thore, P., and Reshef, M. Aug. 1993 Model-based stack: A method for constructing an accurate zero-offset section for complex overburdens Geophys. Prosp. 41 661-670
290 Jastram, C., Behle, A. May. 1993 Accurate finite-difference operators for modelling the elastic wave equation Geophys. Prosp. 41 453-458
291 Bortfeld, R., and Kiehn, M. Nov. 1992 Reflection amplitudes and migration amplitudes (zero-offset situation) Geophys. Prosp. 40 873-884
292 Bunks, C. Sep. 1993 Effective filtering of artifacts for Finite-Difference implementations of paraxial wave equation migration Exp. Abs., SEG Nat. Conv. 1044-1047
293 Claerbout, J. F., and Doherty, S.M. Oct. 1972 Downward continuation of moveout corrected seismograms Geophysics 37 741-768
294 Claerbout, J. F. Jan. 1985 Imaging the Earth's Interior Blackwell Sci., available through Email (out of print)
295 Kosloff, D. D., and Baysal, E. Jun. 1983 Migration with the full acoustic wave equation Geophysics 48 677-687
296 Carnahan, B., Luther, H.A., and Wilkes, J.O. Jan. 1969 Applied Numerical Methods John Wiley and Sons
297 Helbig, K. Jul. 1983 Elliptical anisotropy - Its significance and meaning Geophysics 48 825-832
298 Denelle, E., Dezard, Y., and Raoult, J. J. Nov. 1986 2-D prestack depth migration in the (S-G-W) domain Exp. Abs., SEG Nat. Conv. 327-330
299 Hubral, P. Aug. 1983 Computing true amplitude reflections in a laterally inhomogeneous earth Geophysics 48 1051-1062
300 Kleyn, A. H. Jan. 1983 Seismic Reflection Interpretation Applied Science Publishers
301 Raz, S. Jun. 1981 Direct reconstruction of velocity and density profiles from scattered field data Geophysics 46 832-836
302 Gazdag, J. Jun. 1981 Modeling of the acoustic wave equation with transform methods Geophysics 46 854-859
303 Gazdag, J., and Sguazzero, P. Feb. 1984 Migration of seismic data by phase shift plus interpolation Geophysics 49 124-131
304 Baysal, E., Kosloff, D.D., and Sherwood, J.W.C. Feb. 1984 A two-way nonreflecting wave equation Geophysics 49 132-141
305 Gazdag, J., and Sguazzero, P. Jun. 1984 Interval velocity analysis by wave extrapolation Geophys. Prosp. 32 454-479
306 Van Der Wal, L. F., and Berkhout, A.J. Jun. 1984 Influence of amplitude and phase errors on migration results Geophys. Prosp. 32 425-453
307 Deregowski, S. M., and Rocca, F. Jun. 1981 Geometrical optics and wave theory of constant offset sections in layered media Geophys. Prosp. 29 374-406
308 Schultz, P. S. Dec. 1982 A method for direct estimation of interval velocities Geophysics 47 1657-1671
309 Loewenthal, D., and Mufti, I.R. May. 1983 Reversed time migration in spatial frequency domain Geophysics 48 627-635
310 Lumley, D. E., and Claerbout, J.F. Jan. 1993 Anti-aliased Kirchhoff 3-D migration: a salt intrusion example SEG 3-D Seismic Workshop 115-123
311 Haddow, C. M. Oct. 1992 Hybrid migration by Finite-Difference and Phase Shift Exp. Abs., SEG Nat. Conv. 893-896
312 Newman, P. Nov. 1990 Amplitude and phase properties of a digital migration process First Break 8 397-403
313 Schneider, W. A. Dec. 1971 Developments in seismic data pro-

cessing and analysis (1968-1970) Geophysics 36 1043-1073
314 Stanley, M. Jun. 1998 Practical wave equation datuming Exp. Abs. EAGE 1-15
315 Sherwood, J. W. C., and Poe, P.H. Oct. 1972 Continuos velocity estimation and seismic wavelet processing Geophysics 37 769-787
316 Landa, E., Kosloff, D., Keydar, S., Koren, Z., and Resheff, M. Apr. 1988 A method for determination of velocity and depth from seismic data Geophys. Prosp. 36 223-243
317 Beitzel, J. E., and Davis, J. M. Oct. 1974 A computer oriented velocity analysis interpretation technique Geophysics 39 619-632
318 Diet, J. P., and Audebert, F. May. 1990 A focus on focusing Handout, EAEG meeting 1-14
319 Diet, J. P. Apr. 1991 Prestack depth migration or 'How to beat a (ray) path to your prospect' CGG handout 1-12
320 Agarwal, B. N. P., and Lal, T. Jun. 1971 Application of rational approximation in calculation of the second derivative of the gravity field Geophysics 36 571-581
321 Etgen, J. T. Oct. 1988 Prestack Migration of P and SV - waves Exp. Abs., SEG Nat. Conv. 972-975
322 Gardner, G. H. F., Wang, S. Y., Pan, N. D., and Zhang, Z. May. 1986 Dip moveout and prestack imaging Exp. Abs., Offshore Technology 75-84
323 Black, J. L. Oct. 1992 Prestack remedial migration kinematics Exp. Abs., SEG Nat. Conv. 913-916
324 Black, J. L., and Brzostowski, M. A. Sep. 1994 Systematics of time-migration errors Geophysics 59 1419-1434
325 Reynolds, A. C. Oct. 1978 Boundary conditions for the numerical solution of wave propagation problems Geophysics 43 1099-1110
326 Schleicher, K. L., Grygier, D. J., and Brzostowski, M. A. Nov. 1991 Migration velocity analysis: a comparison of two approaches Exp. Abs., SEG Nat. Conv. 1237-1238
327 Lowenthal, D., Lu, L., Robertson, R., and Sherwood, J. Jun. 1976 The wave equation applied to migration Geophys. Prosp. 24 380-399
328 Lynn, W., Gonzarez, A., and MacKay, S. Nov. 1991 Where are the fault-plane reflections? Exp. Abs., SEG Nat. Conv. 1152-1154
329 Blacquiere, G. Nov. 1991 Optimized McClellan transformation filters applied in one-pass 3D depth extrapolation Exp. Abs., SEG Nat. Conv. 1126-1129
330 Hedley, J. P. Nov. 1991 3-D migration via McClellan transformations on hexagonal grids Exp. Abs., SEG Nat. Conv. 1130-1133
331 Black, J. L., and Brzostowski, M. A. Nov. 1991 Steep-dip time migration and residual depth migration Exp. Abs., SEG Nat. Conv. 1140-1143
332 Hill, N. R., Watson, T. H., Hassler, M. H., and Sisemore, L. K. Nov. 1991 Salt-flank imaging using Gausian Beam migration Exp. Abs., SEG Nat. Conv. 1178-1180
333 Witte, D. Nov. 1991 Dip moveout in vertically varying media Exp. Abs., SEG Nat. Conv. 1181-1183
334 Brysk, H. Sep. 1990 Slotnick revisited: DMO with linear velocity profile Exp. Abs., SEG Nat. Conv. 1350-1353
335 Alfaraj, M., and Larner, K. Nov. 1991 Dip moveout for mode-converted waves Exp. Abs., SEG Nat. Conv. 1191-1193
336 Schwab, M., and Gardner, H. F. Nov. 1991 2-D DMO for a medium with constant vertical velocity gradient Exp. Abs., SEG Nat. Conv. 1194-1196
337 Brysk, H., and Majesty, T. Nov. 1991 3-D isochron illuminates variable-velocity DMO Exp. Abs., SEG Nat. Conv. 1197-1200
338 Artley, C. T. Nov. 1991 Dip-moveout processing for depth-variable velocity Exp. Abs., SEG Nat. Conv. 1204-1207
339 Meinardus, H. A., and Schleicher, K. Nov. 1991 3-D time-variant dip moveout by the FK method Exp. Abs., SEG Nat. Conv. 1208-1210
340 Silva, R., and Haseko, P. Nov. 1991 Antialiasing and amplitude preserving 2-D and 3-D DMO: an integral implementation Exp. Abs., SEG Nat. Conv. 1211-1214
341 Taner, M. T., Postman, R. W., Lu, L., and Baysal, E. Nov. 1991 Depth-migration velocity analysis Exp. Abs., SEG Nat. Conv. 1218-1211
342 MacKay, S., and Abma, R. Nov. 1991 Depth-focusing analysis: practical applications and potential pitfalls Exp. Abs., SEG Nat. Conv. 1222-1225
343 Bunks, C. Nov. 1991 Optimal velocity model inversion from analysis of Paraxial wave-equation focusing spots Exp. Abs., SEG Nat. Conv. 1226
344 Beasley, C., and Klotz, R. Nov. 1991 Migration velocity analysis by migration of velocity spectra Exp. Abs., SEG Nat. Conv. 1239-1242
345 Teiman, H. J. Nov. 1991 Migration velocity analysis through weighted stacking Exp. Abs., SEG Nat. Conv. 1243-1246
346 Versteeg, R., Ehinger, A., and Geoltrain, S. Nov. 1991 Sensitivity of migration coherency panels to the velocity model Exp. Abs., SEG Nat. Conv. 1251-1254
347 Hildebrand, S. Nov. 1991 Migration velocity analysis Exp. Abs., SEG Nat. Conv. 1268-1272
348 Reitveld, W. E. A., Berkhout, A. J., and Wapenaar, C. P. A. Nov. 1991 Controlled illumination of hydrocarbon reservoirs Exp. Abs., SEG Nat. Conv. 1281-1284
349 Tatalovic, R., Dillen, m. W. P., and Fokkema, J. T. Nov. 1991 Prestack imaging in the double transformed Radon domain Exp. Abs., SEG Nat. Conv. 1285-1288
350 Rodriguez, S., Diet, J. P., and Paturet, D. Nov. 1991 Dip moveout in cases of irregular surfaces Exp. Abs., SEG Nat. Conv. 1301-1340
351 Apostoiu-Marin, I., and Ehinger, A. Sep. 1993 Kinematics of prestack depth migration Exp. Abs., SEG Nat. Conv. 873-875
352 Ferber, R. G. Mar. 1994 Migration to multiple offset and velocity analysis Geophy. Prosp. 42 99-112
353 Biondi, B. Sep. 1990 Velocity estimation by beam stack: field-data results Exp. Abs., SEG Nat. Conv. 1271-1274
354 Alkhalifah, T., and Larner, K. Sep. 1994 Migration error in transversely isotropic media Geophysics 59 1405-1418
355 Mukulich, W., and Hale, D. Sep. 1990 Steep-dip v(z) imaging from an ensemble of constant-velocity Stolt Migrations Exp. Abs., SEG Nat. Conv. 1307-1313
356 Lazaratos, S., and Harris, J. M. Sep. 1990 Radon transform/ Gausian beam migration Exp. Abs., SEG Nat. Conv. 1314-1317
357 Hale, D. Sep. 1990 3-D depth migration via McClellan transforms Exp. Abs., SEG Nat. Conv. 1325-1328
358 Julien, P., and Cole, S. Sep. 1990 3-D prestack depth migration on real data. Exp. Abs., SEG Nat. Conv. 1329-1332
359 Bleistein, N. Sep. 1990 Born DMO revisited Exp. Abs., SEG Nat. Conv. 1366-1369
360 Harrison, M. P. Sep. 1990 Dip moveout for converted wave (P-SV) data in constant velocity medium. Exp. Abs., SEG Nat. Conv. 1370-1373
361 Biondi, B. Nov. 1986 Shot profile dip moveout using log-stretch transform Exp. Abs., SEG Nat. Conv. 431-434
362 Wapenaar, C. P. A., Blacquiere, G., Kinneging, N. A., and Berkhout, A. J. Nov. 1986 Practical aspects of 3-D prestack migration Exp. Abs., SEG Nat. Conv. 436-438

363 Faye, J. P., and Jeannot, J. P. Nov. 1986 Prestack migration velocities from focusing depth analysis Exp. Abs., SEG Nat. Conv. 438-440

364 Owusu, K., Gardner, G. H. F., and Massell, W. F. Nov. 1983 Velocity estimates derived from three-dimensional seismic data. Geophysics 48 1466-1497

365 Dickinson, J. A. Jan. 1988 Evaluation of two-pass three-dimensional migration Geophysics 53 32-49

366 Kanasewich, E. R. Mar. 1988 Imaging discontinuities on seismic sections Geophysics 53 334-345

367 Lines, L. R., Schultz, A. K., and Treitel, S. Jan. 1988 Cooperative inversion of geophysical data Geophysics 53 8-20

368 Shih, R. C., and Levander, A. R. Apr. 1994 Layer-stripping reverse-time migration Geophys. Prosp. 41 211-227

369 Biondi, B., and Ronen, J. Nov. 1987 Dip moveout in shot profiles Geophysics 52 1473-1482

370 Muller, G., and Temme, P. Nov. 1987 Fast frequency-wavenumber migration for depth-dependent velocity Geophysics 52 1483-1491

371 Slotnick, M. M. Jan. 1936 On seismic computations, with applications, I Geophysics 1 9-22

372 Slotnick, M. M. Oct. 1936 On seismic computations, with applications, II Geophysics 1 299-305

373 Ottolini, R. A. Nov. 1982 Migration of reflection seismic data in angle-midpoint coordinates Dissertation, Stanford University

374 Baysal, E., Kosloff, D. D., and Sherwood, J. W. C. Nov. 1983 Reverse time migration Geophysics 48 1514-1524

375 Bleistein, N Jul. 1987 On the imaging of reflectors in the earth Geophysics 52 931-942

376 Miller, D., Oristaglio, M., and Beylkin, G. Jul. 1987 A new slant on seismic imaging: Migration and integral geometry Geophysics 52 943-964

377 Ronen, J. Jul. 1987 Wave-equation trace interpolation Geophysics 52 973-984

378 Zhou, B., and Greenhalgh, A. Sep. 1994 Wave-equation extrapolation-based multiple attenuation: 2-D filtering in the f-k domain Geophysics 59 1377-1391

379 Howell, L. H., and Trefethen, L. N. May. 1988 Ill-posedness of absorbing boundary conditions for migration Geophysics 53 593-603

380 Fornberg, B. May. 1988 The pseudospectral method: Accurate representation of interfaces in elastic wave calculations Geophysics 53 625-637

381 Larner, K., and Beasley, C. Jan. 1986 DMO and steep-dip migrations Western Geophysical

382 Shaum Jan. 1968 Mathematical Handbook, Schaum's outline series McGraw-Hill

383 Nolet, G. Jan. 1987 Seismic wave propagation, Chap 1 in "Seismic Tomography: with Applications in Global Seismology and Exploration Geophysics D. Reidel Publishing Co.

384 Oppenheim, A. V., and Schafer, R. W. Jan. 1975 Digital Signal Processing Prentice-Hall

385 Bracewell R. Jan. 1965 The Fourier Transform and its Applications (text) McGraw-Hill

386 Zelt, C., and Ellis, R. Jun. 1988 Practical and efficient ray tracing in two-dimensional media for rapid traveltime and amplitude forward modelling Cdn. Jnl. Exp. Geo. 24 16-31

387 Roberts, G., and Goulty, N. Aug. 1988 Directional deconvolution of the seismic source signature combined with prestack migration First Break 6 247-253

388 Robinson, E. A. Mar. 1986 Migration of seismic data by the WKBJ method Proceedings of the IEEE 74 428-438

389 Miller, J., and Goldman, R. Mar. 1992 Using tangent balls to find plane sections of natural quadrics IEEE Computer Graphics and Applications 68-82

390 Sen, M., and Stoffa, P. Jan. 1992 Genetic inversion of AVO The Leading Edge 11 27-29

391 Lawton, D., and Bertram, M. Nov. 1993 Azimuthal response of some three-component geophones The Leading Edge 12 1118-1121

392 Wapenaar, C. P. A., Kinneging, N., and Berkhout, A. Feb. 1987 Principle of prestack migration based on the full elastic two-way wave equation Geophysics 52 151-173

393 Reshef, M., Kosloff, D., and Edwards, M. Sep. 1988 Three-dimensional elastic modeling by the Fourier method Geophysics 53 1184-1193

394 March, D., and Bailey, A. Jan. 1983 A review of the two-dimensional transform and its use in seismic processing First Break 1 9-21

395 Neidell, N. S. Jul. 1994 Sampling 3-D seismic surveys: A conjecture favoring coarser but higher-fold sampling The Leading Edge 764-768

396 Larner, K., Weglein, A., and Wyatt, K. Nov. 1988 Migration in the real world. Research Workshop, Questions Meeting handout 1-11

397 Milkereit, B., and Spencer, C. Jan. 1987 A new migration method applied to the inversion of P-S converted phases Blackwell Scientific Publications, Oxford 251-266

398 Bale, R., and Jakubowicz, H. Oct. 1987 Post-stack prestack migration Exp. Abs., SEG Nat. Conv. 714-717

399 Milkereit, B. Nov. 1987 Decomposition and inversion of seismic data - an instantaneous slowness approach Geophys. Prosp. 35 875-894

400 Hagin, F., and Cohein, J. Feb. 1984 Refinements to linear velocity inversion theory Geophysics 49 112-118

401 Jorden, T., Bleistein, N., and Cohen, J. Oct. 1987 A wave equation-based dip moveout Exp. Abs., SEG Nat. Conv. 718-721

402 Stone, D. Apr. 1988 Imaging with post-stack DMO Handout Seismograph Service Corp. 1-9

403 Kitchenside, P. W., and Jakubowicz, H. Oct. 1987 Operator design for 3-D depth migration Exp. Abs., SEG Nat. Conv. 556-558

404 Ratcliff, D. W., and Pan, N. Oct. 1987 2-D and 3-D post stack dip enhancement (?) Exp. Abs., SEG Nat. Conv. 1-4

405 Black, J. L., and Leong, T. Oct. 1987 A flexible, accurate approach to 1-pass 3-D migration Exp. Abs., SEG Nat. Conv. 559-560

406 Jones, I. F. Oct. 1993 Comparative anatomy 0f 3D one-pass depth migration schemes Handout, CGG France 1-7

407 Wapenaar, C. P. A., and Berkhout, A. Jun. 1987 Three-dimensional target-oriented pre-stack migration First Break 5 217-227

408 Karrenbach, M. Jan. 1990 Three-dimensional time-slice migration Geophysics 55 10-19

409 Rocca, F., and Ronen, J. Oct. 1984 Improving resolution by dip moveout Exp. Abs., SEG Nat. Conv. 611-614

410 Chambers, R., Jakubowicz, H., Larner, K., and Yang, M. Oct. 1984 Suppression of backscattered coherent noise by prestack partial migration Exp. Abs., SEG Nat. Conv. 451-454

411 Lawrence, C., Schneider, W., and Shurtleff, R. Oct. 1985 A comparative analysis of dip moveout and prestack migration Exp. Abs., SEG Nat. Conv. 307-309

412 Li, Z., Chambers, R., and Abma, R. Oct. 1987 Suppressing spatial aliasing noise in prestack Stolt FK migration Exp. Abs., SEG Nat. Conv. 734-736

413 Beasley, C., Chambers, R., and Jakubowicz, H. Jun. 1985 Prestack partial migration: a comprehensive solution to problems in the processing of 3-D data Abs., EAEG Convention

References

414 Pai, D., and Chen, T. Oct. 1987 A new method for wave equation migration in laterally inhomogeneous media Exp. Abs., SEG Nat. Conv. 554-555

415 Amestoy, P., Gardner, G., Leiss, E. Oct. 1987 Phase shift - Based prestack depth migration for laterally varying velocities Exp. Abs., SEG Nat. Conv. 737-740

416 Hale, D. Jan. 1983 Dip-Moveout by Fourier Transform Dissertation, Stanford University

417 Larner, K., Gibson, B., and Chambers, R. Nov. 1980 Imaging beneath complex structures: a case history Handout, Western Geophysical SEG Nat. Conv. 1-26

418 Datta, S., and Bhowmick,A. Sep. 1974 Seismic model studies on diffraction of waves by edged of varying radius of curvature and depth Geophys. Prosp. 22 534-545

419 Berryhill, J. Oct. 1977 Diffraction response for nonzero separation of source and receiver Geophysics 42 1158-1176

420 Trorey, A. Oct. 1977 Diffractions for Arbitrary source-receiver locations Geophysics 42 1177-1182

421 Chon, Y., and Gonzalez, A. Oct. 1987 Accuracy in RMS-velocity determination using a Kirchhoff DMO algorithm Exp. Abs., SEG Nat. Conv. 722-725

422 Berkhout, A. J., and Jong, B. Oct. 1981 Recursive migration in three dimensions Geophys. Prosp. 29 758-781

423 Qin, F., Olsen, K., and Schuster, G. Sep. 1990 Solution of the Eikonal equation by a finite-difference method Exp. Abs., SEG Nat. Conv. 1004-1007

424 Hornbostel, S. Dec. 1991 Spatial prediction filtering in the t-x and f-x domains Geophysics 56 2019-2026

425 Uren, N. F., Gardner, G., and McDonald, J. Oct. 1989 The anisotropic migrator's equation Exp. Abs., SEG Nat. Conv. 1184-1186

426 Uren, N. F., Gardner, G., and McDonald, J. Oct. 1989 Zero-offset seismic reflection surveys using an anisotropic physical model Exp. Abs., SEG Nat. Conv. 1044-1046

427 Dellinger, J., and Etgen, J. Oct. 1989 Wave-type separation in 3-D anisotropic media Exp. Abs., SEG Nat. Conv. 977-979

428 Engquist, B. Sep. 1976 One way elastic wave equations SEP 125-140

429 Canales, L. Aug. 1976 Mixed L1 and L2 norm problems SEP 114-124

430 Claerbout, J. F. Jul. 1976 Inverse incomplete beta functions of one half SEP 109-113

431 Claerbout, J. F. Jul. 1976 Probability and entropy of seismic data SEP 101-108

432 Canales, L. Sep. 1976 A quantile finding program SEP 99-100

433 Claerbout, J. F Jun. 1976 Absorptive side boundaries for migration SEP 10-15

434 Bazelaire, E., and Viallix, J. Jul. 1994 Normal moveout in focus Geophys. Prosp. 42 477-499

435 Castle, R. J. Jun. 1994 A theory of normal moveout Geophysics 59 983-999

436 Yilmaz, (Orhan)., and Taner, M. Jun. 1994 Discrete plane-wave decomposition by least-mean-squared-error method Geophysics 59 973-982

437 Versteeg, R. Sep. 1994 The Marmousi experience: Velocity model determination on a synthetic complex data set The Leading Edge 927-936

438 Kaila, K., and Sain, K. May. 1994 Errors in RMS velocity and zero-offset two-way time as determined from wide-angle seismic reflection traveltimes using truncated series J. of Seismic Exploration 3 173-188

439 Galbraith, M. Feb. 1994 3-D survey design by computer CSEG Recorder 4-8

440 Bancroft, J. C., and Geiger, H. Oct. 1994 Equivalent offset CRP gathers Exp. Abs., SEG convention

441 Taner, M. T. Feb. 1975 Long-period multiples and their suppression Geophysics (Abstract from annual meeting) 40 143-144

442 Dix, C. H. Jan. 1955 Seismic velocities from surface measurements Geophysics 20 68-86

443 Gardner, L. W. Apr. 1947 Vertical velocities from reflection shooting Geophysics 12 221-228

444 Sullivan, M. F., and Cohen, J. K. Jun. 1987 Prestack Kirchhoff inversion of common-offset data Geophysics 52 745-754

445 Lumley, D. E., Claerbout, J. F., and Bevc, D. Oct. 1994 Antialiased Kirchhoff 3-D migration Exp. Abs. SEG Nat. Conv. 1282-1285

446 Bunks, C. Oct. 1992 Paraxial wave-equation inversion with geometric constraints Exp. Abs. SEG Nat. Conv. 901-904

447 Faye, J. P., and Jeannot, J. P. Nov. 1986 Prestack migration velocities from focusing depth analysis Exp. Abs. SEG Nat. Conv. 438-440

448 Tygel, M., Schleicher, J., Hubral, P., and Hanitzsch, C. Dec. 1993 Multiple weights in diffraction stack migration Geophysics 59 1820-1830

449 Deregowski, S. M. Jun. 1982 Dip-moveout and reflector point dispersal Geophys. Prosp. 30 318-322

450 Bolondi, G., Loinger, E., and Rocca, F. Dec. 1982 Offset continuation of seismic sections Geophys. Prosp. 30 813-828

451 Berkhout, A. J. Dec. 1981 Wave field extrapolation techniques in seismic migration, a tutorial Geophysics 46 1638-1656

452 Ottolini, R. A., and Claerbout, J. F. Mar. 1984 The migration of common midpoint slant stacks Geophysics 49 237-249

453 McMechan, G. A. Jun. 1983 Migration by extrapolation of time-dependent boundary values Geophys. Prosp. 31 413-420

454 Dubrulle, A. A. Sep. 1983 Numerical methods for the migration of constant-offset sections in homogeneous and horizontally layered media Geophysics 48 1195-1203

455 Iverson, W. P. Jun. 1987 Combining attenuation by Q and spherical divergence Geophysics 52 740-744

456 Robinson, E. A. Jan. 1983 Migration of geophysical data IHRDC

457 Gray, S. H. Jan. 1984 A problem of discrete approximations to an arbitrarily varying one-dimensional seismic model Geophys. J. R. Ast. Soc. 78 431-437

458 Eaton, D. W. S., and Stewart, R. R. Jan. 1994 Migration/inversion for transversely isotropic elastic media Geophys. J. Int. 119 667-683

459 Vinje, V., Lecomte, I., Astebol, K., Iversen, E., and Gjoystdal, H. Jun. 1994 Efficient Green's functions calculation for improved 3-D seismic in complex areas Exp. Abs. EAEG Tech. Exh. B043

460 Jones, I. F. Jun. 1994 3-D velocity-depth model building - layer-striping verses map-migration Exp. Abs. EAEG Tech. Exh. B044

461 Biondi, B., and Palacharla, G. Jun. 1994 3-D depth migration using rotated McClellan filter Exp. Abs. EAEG Tech. Exh. B045

462 Lumley, D. E., Claerbout, J. E., and Bevc, D. Jun. 1994 Antialiased Kirchhoff migration Exp. Abs. EAEG Tech. Exh. H027

463 Nichols, D. Jun. 1994 Imaging complex structures using band-limited Green's functions Exp. Abs. EAEG Tech. Exh. H030

464 Sollie, R., Mittet, R., and Hokstad, K. Jun. 1994 Pre-stack depth migration with compensation for absorption Exp. Abs. EAEG Tech. Exh. H031

465 Dessing, F. J., and Wapenaar, C. P. A. Jun. 1994 Wave field extrapolation in the space-wavenumber domain Exp. Abs. EAEG

466 Wenes, G., Kremer, S., and Shiu, J. Jun. 1994 A practical implementation of a 3-D prestack depth migration algorithm Exp. Abs. EAEG Tech. Exh. H040

467 Ruehl, T., Kopp, C., and Ristow, D. Jun. 1994 Fourier FD migration for steeply dipping reflectors with complex overburden Exp. Abs. EAEG Tech. Exh. P100

468 Payne, B. Jun. 1994 Phase shift migration for lateral velocity variation Exp. Abs. EAEG Tech. Exh. P101

469 Jeannot, J. P. Jun. 1994 Robust Kirchhoff pre-stack depth imaging with semi-gridded rays Exp. Abs. EAEG Tech. Exh. P102

470 Pleshkevitch, A. L. Jun. 1994 Shot-DMO by finite-difference Exp. Abs. EAEG Tech. Exh. P103

471 Hoffmann, H. J., and Klaeschen, D. Jun. 1994 Common-fault-point imaging by raytracing-based pre-stack depth migration Exp. Abs. EAEG Tech. Exh. P104

472 Thorbecke, J. W., and Reitveld, W. E. A. Jun. 1994 Optimum extrapolation operators - a comparison Exp. Abs. EAEG Tech. Exh. P105

473 Hanitzsch, C. Jun. 1994 Amplitude preserving 2.5D Kirchhoff migration for lateral inhomogeneous media Exp. Abs. EAEG Tech. Exh. P106

474 EAEG Jun. 1994 Program Listing EAEG Exp. Abs. EAEG Tech. Exh.

475 Osterander, W. J. Oct. 1984 Plane-wave reflection coefficients for gas sands at nonnormal angles of incidence Geophysics 49 1637-1648

476 Jain, S., and Wren, A. E. Feb. 1980 Migration before stack - Procedure and Significance Geophysics 45 204-212

477 Bleistein, N., Cohen, J. K., and Hagin, F. G. Aug. 1985 Computational and asymptotic aspects of velocity inversion Geophysics 50 1252-1265

478 Keho, T. H., and Beydoun, W. B. Dec. 1988 Paraxial ray Kirchhoff migration Geophysics 53 1540-1546

479 Alfaraj, M., and Larner, K. Oct. 1994 Anti-aliased Kirchhoff 3-D migration Exp. Abs. SEG Nat. Conv. 1282-1285

480 Gray, S. H., and May, W. P. May. 1994 Kirchhoff migration using eikonal equation travel times Geophysics 59 810-817

481 Helbig, K. Feb. 1990 Rays and wavefront charts in gradient media Geophys. Prosp. 38 189-220

482 Blacquiere, G., Debeye, W. J., Wapenaar, C. P. A., and Berkhout, A. J. Nov. 1989 3-D table-driven migration Geophys. Prosp. 37 925-958

483 Kaufman, H. Jan. 1953 Velocity functions in seismic prospecting Geophysics 23 289-297

484 Taner, M. T., Koehler, F., and Sheriff, R. E. Jun. 1979 Complex seismic trace analysis Geophysics 44 1041-1063

485 Shuey, R. T. Apr. 1985 A simplification of the Zeoppritz equations Geophysics 50 609-614

486 Tygel, M., Schleicher, J., Hubral, P., and Santos, L. T. Mar. 1998 2.5-D true-amplitude Kirchhoff migration to zero offset in laterally inhomogeneous media Geophysics 63 557-573

487 Lee, M. W., and Suh, S. Y. Oct. 1985 Optimization of one-way wave equations Geophysics 50 1634-1637

488 Fisher, E., McMechan, G.A., Annan, A. P., and Cosway, S. W. Apr. 1992 Examples of reverse-time migration of single-channel, ground penetrating radar profiles Geophysics 57 577-586

489 Harris, C. E., and McMachan, G. A. Jun. 1992 Using downward continuation to reduce memory requirements in reverse-time-migration Geophysics 57 848-853

490 Chang, H., VanDyke, J. P., Solano, M., McMechan, G.A., and Epili, D. Mar. 1998 3-D prestack migration: From prototype to production in a massively parallel processor environment Geophysics 63 546-556

491 Rodriguez, S., and Vuillermoz, C. Oct. 1994 Solutions to near surface effects in mountainous thrust areas Exp. Abs., SEG Nat. Conv. 1644-1646

492 Amundsen, L., and Reitan, A. Oct. 1994 The relationship between 2-D and 3-D wave propagation Exp. Abs., SEG Nat. Conv. 1387-1381

493 Marcoux, M. O., Harris, C., and Chernis, L. Oct. 1994 Depth migration and modeling from exact time migration Exp. Abs., SEG Nat. Conv. 1647-1650

494 Tjan, T., and Larner K. Oct. 1994 Prestack migration for residual statics estimation in complex areas Exp. Abs., SEG Nat. Conv. 1513-1516

495 Blondel, P. Oct. 1994 Amplitude preservation for Kirchhoff dip moveout Exp. Abs., SEG Nat. Conv. 1529-1532

496 Ohanian, V., Snyder, T. M., and Hampson, D. P. Oct. 1994 Approximate regimes of the H-F DMO operator Exp. Abs., SEG Nat. Conv. 1533-1536

497 Biondi, B., and Chemingui, N. Oct. 1994 Transformation of 3-D prestack data by azimuth moveout (AMO) Exp. Abs., SEG Nat. Conv. 1541-1544

498 Gardner, G. H. F., and Canning, A. Oct. 1994 Effects of irregular sampling on 3-D prestack migration Exp. Abs., SEG Nat. Conv. 1553-1556

499 Barnes, A. E. Oct. 1994 Theory of two-dimensional complex seismic trace analysis Exp. Abs., SEG Nat. Conv. 1580-1583

500 Boumahdi, M., and Glangeaud, F. Oct. 1994 Nonminimum phase blind deconvolution: Parametric approach Exp. Abs., SEG Nat. Conv. 1595-1598

501 Hanitzsch, C. Oct. 1994 Vector diffraction stack migration: A fast technique for amplitude-preserving Kirchhoff migration Exp. Abs., SEG Nat. Conv. 1183-1186

502 Ratcliff, D. W., and Weber, D. J. Oct. 1994 Subsalt imaging over the Mahogany salt sill Exp. Abs., SEG Nat. Conv. 1246-1249

503 Thorbecke, J. W., and Berkhout, A. J. Oct. 1994 3-D recursive extrapolation operators: An overview Exp. Abs., SEG Nat. Conv. 1262-1265

504 Gaiser, J. E. Oct. 1994 Effective use of the Laplacian transformation for explicit finite-difference 3-D migration Exp. Abs., SEG Nat. Conv. 1274-1277

505 Biondi, B., and Palacharla, G. Oct. 1994 3-D migration using rotated McMClellan filters Exp. Abs., SEG Nat. Conv. 1278-1281

506 Mufti, I. R., Pita, J. A., and Huntly, R. W. Oct. 1994 Finite-difference depth migration of exploration scale 3-D seismic data Exp. Abs., SEG Nat. Conv. 1286-1289

507 Gibson, R. L., Lee, J. M., Dini, I., and Cameli, G. M. Oct. 1994 The application of 3-D Kirchhoff migration to VSP data from complex geological settings Exp. Abs., SEG Nat. Conv. 1290-1293

508 Jeannot, J. P., and Berranger, I. Oct. 1994 Re-mapped focusing: A migration velocity analysis for Kirchhoff prestack imaging Exp. Abs., SEG Nat. Conv. 1326-1329

509 Charles, S., Kapotas, S., and Phadke, S. Oct. 1994 Absorbing boundaries with Huygens' secondary sources Exp. Abs., SEG Nat. Conv. 1363-1366

510 Rueger, A. Oct. 1994 Applicability of the Gausian beam approach for modeling of seismic data in triangulated subsurface models Exp. Abs., SEG Nat. Conv. 1371-1374

511 Jianlin, Z. Oct. 1994 2.5-D finite element seismic modeling Exp. Abs., SEG Nat. Conv. 1375-1377

512 Nichols, D. Oct. 1994 Maximum energy traveltimes calculated in the seismic frequency band Exp. Abs., SEG Nat. Conv. 1382-1385

References

513 Lambaré, G., Lucio, P. S., and Hanyga, A. Oct. 1994 2-D asymptotic Green's functions Exp. Abs., SEG Nat. Conv. 1386-1389

514 Jannaud, L. R. Oct. 1994 Common-offset ray tracing Exp. Abs., SEG Nat. Conv. 1390-1393

515 Fowler, P. J. Oct. 1994 Finite-difference solutions of the 3-D eikonal equation in spherical coordinates Exp. Abs., SEG Nat. Conv. 1394-1397

516 Blanch, J. O., Robertsson, J. O. A., and Symes, W. W. Oct. 1994 Constant Q modeling: Preliminary results for a new algorithm Exp. Abs., SEG Nat. Conv. 1414-1417

517 Dai, N., and West, G. F. Oct. 1994 Inverse Q migration Exp. Abs., SEG Nat. Conv. 1418-1421

518 Akbar, F. E., Sen, M. K., and Stoffa, P. L. Oct. 1994 Prestack plane-wave Kirchhoff depth migration using a cluster of workstations Exp. Abs., SEG Nat. Conv. 225-228

519 Michelena, R. J. Ata, E., and Sierra, J. Oct. 1994 Exploiting P-S converted waves: Part 1, Modeling the effects of anisotropy and heterogeneities Exp. Abs., SEG Nat. Conv. 236-239

520 Ata, E., Michelena, R. J., Gonzales, M., Cerquone, H., and Carry, M Oct. 1994 Exploiting P-S converted waves: Part 2, Application to a fracture reservoir Exp. Abs., SEG Nat. Conv. 240-243

521 Zanzi, L., and Bagaini, C. Oct. 1994 The design of prestack migration-demigration operators Exp. Abs., SEG Nat. Conv. 676-679

522 Popovici, A. M. Oct. 1994 Reducing artifacts in prestack phase-shift migration of common offset gathers Exp. Abs., SEG Nat. Conv. 684-687

523 Profeta, R. M., Moscoso, J., and Koremblit, M Oct. 1994 Minimum field static corrections Exp. Abs., SEG Nat. Conv. 715-718

524 Winbow, G. A., and Trantham, E. C. Oct. 1994 Nonaliased amplitude-preserving DMO (AVO???) Exp. Abs., SEG Nat. Conv. 719-722

525 Bancroft, J. C., Geiger, H. D., Foltinek, D. S., and Wang, S. Dec. 1994 Prestack migration by equivalent offset and CSP gathers CREWES Research Report 6 27.1-27.18

526 Bancroft, J. C., and Wang, S. Dec. 1994 Converted wave prestack migration and velocity analysis by equivalent offset and CSP gathers CREWES Research Report 6 28.1-28.7

527 Wang, S., Bancroft, J. C., and Lawton, D. C. Dec. 1994 DMO processing for mode-converted waves in a medium with linear increasing velocity with depth CREWES Research Report 6 26.1-26.15

528 Gardner, G. H. F., (Editor) Jan. 1985 Migration of Seismic Data (Reprint series) SEG

529 MacKay, S. Oct. 1994 Efficient wavefield extrapolation to irregular surfaces using finite differences: Zero-velocity datuming Exp. Abs., SEG Nat. Conv. 1564-1567

530 Lopez-Mora, S., Purnell,N., and Krigsvoll Jun. 1993 Kirchhoff time migration errors for vertical gradient EAEG-1993 Exp. Abs. C-047

531 Stolt, R. H., and Benson, A. K. Jan. 1986 Seismic Migration, Theory and Practice Geophysical Press

532 Vermeer, G. J. O Sep. 1995 Is "coarse" the right course? (Roundtable portion) The Leading Edge 989-993

533 Filpo, E., and Hubral, p. Jan. 1995 Numerical tests of 3D true-amplitude zero-offset migration Geophys. Prosp. 43 119-134

534 Bortfeld, R. Mar. 1989 Geometrical ray theory: Rays and traveltime in seismic systems (second order approximation of the traveltimes) Geophysics 54 342-349

535 Claerbout, J. F. Jan. 1992 Earth Soundings Analysis: Processing versus Inversion Blackwell Scientific Publications

536 Bunks, C. Feb. 1995 Effective filtering of artifacts for implicit finite-difference paraxial wave equation migration Geophys. Prosp. 43 203-220

537 Johansen, T. A., Bruland, L.., and Lutro, J. Feb. 1995 Tracking the amplitude verses offset (AVO) by using orthogonal polynomials Geophys. Prosp. 43 245-261

538 Dunne, J., and Beresford, G. Mar. 1995 A review of the t-p transform, its implementation and its application in seismic processing Exploration Geophysics (ASEG) 26 19-36

539 Frei, W. Apr. 1995 Refined field static corrections in near-surface reflection profiling across rugged terrain The Leading Edge 14 259-262

540 Young, R. A., Deng, Z., and Sun, J. Apr. 1995 Interactive processing of GPR data The Leading Edge 14 275-280

541 Armstrong, P. N., Chmela, W., and Leaney, W. S. Aug. 1995 AVO calibration using borehole data First Break 13 319-328

542 Slotnick, M. M. Jan. 1959 Lessons in Seismic Computing SEG 1-268

543 Sheriff, R. E. Oct. 1990 Encyclopedic Dictionary of Exploration Geophysics (Third Edition) SEG

544 Bancroft, J. C., Margrave, G., and Geiger, H. D. Jun. 1998 A comparison of prestack migration gathers formed by equivalent offset migration (EOM) and DMO-PSI Exp. Abs. EAGE 1-18

545 Oppenheim, A. V. Jan. 1978 Applications of Digital Signal Processing Prentice-Hall

546 Vermeer, G. J. O. Jan. 1990 Seismic Wavefield Sampling SEG

547 Berkhout, A. J. Jan. 1995 Seismic Migration. (A. Theoretical Aspects) Imaging of Acoustic Energy by Wave Field Extrapolation Elsevier

548 Berkhout, A. J. Sep. 1984 Seismic Migration. (B. Practical Aspects) Imaging of Acoustic Energy by Wave Field Extrapolation Elsevier

549 Hatton, L., Worthington, M. H., and Makin, J. Jan. 1986 Seismic Data Processing, Theory and Practice Blackwell

550 Sheriff, R. E., and Gelart, L. P. Jan. 1995 Exploration Seismology Cambridge 2 592

551 Kanasewich, E. R. Jan. 1985 Time Sequence Analysis in Geophysics The University of Alberta Press

552 Bancroft, J. C., Geiger, H. D., Wang, S., and Foltinek, D. S., and Nov. 1995 Prestack migration by equivalent offset and CSP gathers: An Update CREWES Research Report 7 Ch. 23

553 Bancroft, J. C., and Geiger, H. D. Nov. 1995 Velocity sensitivity for equivalent offsets in CSP gathers CREWES Research Report 7 Ch. 24

554 Bancroft, J. C. Nov. 1995 Aliasing in prestack migration CREWES Research Report 7 Ch. 25

555 Bancroft, J. C. Nov. 1995 3-D design, migration and aliasing CREWES Research Report 7 Ch. 26

556 Wang, S., Bancroft, J. C., Lawton, D. C., and Foltinek, D. S. Nov. 1995 Converted wave (P-S) prestack migration and migration velocity analysis CREWES Research Report 7 27.1 - 27.22

557 Geiger, H. D., and Bancroft, J. C. Nov. 1995 Equivalent offset prestack migration for rugged topography CREWES Research Report 7 Ch. 28

558 Geiger, H. D., and Bancroft, J. C. Nov. 1995 Prestack migration to an unmigrated zero offset section CREWES Research Report 7 Ch. 29

559 Donati, M., Martin, N. B., and Bancroft, J. C. Dec. 1994 Steep dip Kirchhoff migration for linear velocity gradient CREWES Research report 6 Ch. 17

560 Lathi, B. P. Jan. 1965 Communication Systems John Wiley and Sons, Inc.

561 Allan D., and Bale, R. Jun. 1998 Components of 3-D converted wave imaging Exp. Abs. EAGE 1-46

562 Lindseth, R. O. Sep. 1982 Digital Processing of Geophysical Data: A Review SEG

563 Harris, C., Marcoux, M., and Bickel, S. Jun. 1998 Aperture selection to improve Kirchhoff depth imaging using the maximum depth principle Exp. Abs. EAGE 1-56

564 Rosenbaum, J. H., and Boudreaux, G. F. Dec. 1981 Rapid convergence of some seismic processing algorithms Geophysics 46 1667-1672

565 Alkhalifah, T., Tsvankin, I., Larner, K., and Toldi, J May. 1996 Velocity analysis and imaging in transverse isotropic media: Methology and a case study. The Leading Edge 15 371-378

566 Balch, A. H., and Erdemir, C. Aug. 1994 Sign-change correction for prestack migration of P-S converted wave reflections Geophys. Prosp. 42 637-663

567 Selvi, O Jan. 1996 Comment on 'Sign-change correction for prestack migration of P-S converted wave reflections' by Balch and Erdemir Geophys. Prosp. 44 175-177

568 Hunter, J. A., and Pullan, S. E. Jan. 1990 A vertical array method for shallow seismic refraction of the sea floor Geophysics 55 92-96

569 Stadtlander, R., and Brown, L. Jan. 1997 Turning waves and crustal reflection profiling Geophysics 62 335-341

570 de Bruin, C. G. M., Wapenaar, C. P. A., and Berkhout, A. J. Sep. 1990 Angle-dependent reflectivity by means of prestack migration Geophysics 55 1223-1234

571 Popovici, A. M. Sep. 1996 Prestack migration by the split-step DSR Geophysics 61 1412-1416

572 Hubral, P., Schleicher, J., and Tygel, M. May. 1996 A unified approach to 3-D seismic reflection imaging, Part1: Basic concepts Geophysics 61 742-758

573 Tygel, M., Schleicher, J., and Hubral, P. May. 1996 A unified approach to 3-D seismic reflection imaging, Part II: Theory Geophysics 61 759-775

574 Zhou, B., Mason I. M., and Greenhalgh S. A. May. 1996 An accurate formulation of log-stretch dip moveout in the frequency-wavenumber domain Geophysics 61 815-821

575 Anderson, J., Alkhalifah, T., and Tsvankin, I May. 1996 Fowler DMO and time migration for transversely isotropic media Geophysics 61 835-856

576 Wu, W., Lines, L. R., and Lu, H May. 1996 Analysis of high-order, finite-difference schemes in 3-D reverse-time migration Geophysics 61 845-856

577 Schoenberger, M. May. 1996 Optimum weighted stack for multiple suppression Geophysics 61 891-901

578 Zhong, B., Zhou, X., Liu, X., and Yule, J. Mar. 1995 A new strategy for CCP stacking Geophysics 60 517-521

579 Carcione, J. M., and Cavallini, F. Mar. 1995 Forbidden directions for inhomogeneous pure shear waves in dissipative anisotropic media Geophysics 60 522-530

580 Stockwell, J. W. Mar. 1995 2.5-D wave equations and high-frequency asymptotic Geophysics 60 556-562

581 Schneider, W. A. Jul. 1995 Robust and efficient upwind finite-difference traveltime calculations in three dimensions Geophysics 60 1108-1117

582 Bevc, D., Black, J. L., and Palacharla, G. Jul. 1995 Plumes: Response of time migration to lateral velocity variation Geophysics 60 1118-1127

583 Igel, H., Mora, P., and Riollet, B. Jul. 1995 Anisotropic wave propagation through finite-difference grids Geophysics 60 1203-1216

584 Mosher, C. C., Keho, T. H., Weglein, A. B., and Foster, D. J. Nov. 1996 The impact of migration on AVO Geophysics 61 1603-1615

585 Ehinger, A., Lailly, P., and Marfurt, K. Nov. 1996 Green's function implementation of common-offset, wave-equation migration Geophysics 61 1813-1821

586 von Seggern, D. Mar. 1994 Depth-imaging resolution of 3-D seismic recording patterns Geophysics 59 564-576

587 Vesnaver, A. L. May. 1995 The contribution of reflected, refracted and transmitted waves to seismic tomography: a tutorial First Break 14 159-168

588 Williams, R. G., and Cooper, N. J. Jun. 1995 Suppression of dipping noise and multiples using 3-D prestack time migration Exploration Geophysics 26 468-471

589 Downie, S. P., Hartley, B. M., and Uren, N. F. Jun. 1995 Interactive attenuation of seismic multiples in the radial domain Exploration Geophysics 26 486-492

590 Zhou, B., and Greenhalgh, S. Jun. 1995 A partial; DMO operator for use with the stacking velocity Exploration Geophysics 26 493-496

591 Palmer, D. Jun. 1995 Can linear inversion achieve detailed refraction statics? Exploration Geophysics 26 506-511

592 Sorin, V., Keydar, S., and Landa, E. Sep. 1996 Velocity analysis: Phenomenon of local homogeneous model The Leading Edge 15 1033-1035

593 Sun, J May. 1996 The relationship between the first Fresnel zone and the normalized geometrical spreading factor Geophys. Prosp. 44 351-374

594 Zhou, B., Greenhalgh, S. May. 1996 Multiple suppression by 2-D filtering in the parabolic t-p domain: a wave-equation-based method Geophys. Prosp. 44 375-401

595 Dai, N., Vafidis, A., and Kanasewich, E. R. Jul. 1996 Seismic migration and absorbing boundaries with one-way wave system for heterogeneous media Geophys. Prosp. 44 719-737

596 Liu, N., Jin, H., and Rockwood, A. P. May. 1996 Antialiasing by Gausian integration Computer Graphics 0 58-63

597 Biondi, B., and Palacharla, G Nov. 1995 3-D depth migration by rotated McClellan filters Geophys. Prosp. 43 1005-1020

598 O'Brien, M. J., and Gray, S. Jan. 1996 Can we image beneath the salt? The Leading Edge 15 17-22

599 Brune, R. H., O'Sullivan, B., and Lu, L. Jul. 1994 Comprehensive analysis of marine 3-D bin coverage The Leading Edge 13 757-762

600 Vasco, D. W., Peterson, J. E., and Majer, E. L. Jul. 1996 Nonuniqueness in traveltime tomography: Ensemble interference and cluster analysis Geophysics 61 1209-1227

601 Ward, R. W., MacKay, S., Greenlee, S. M., and Dengo, C. A. Aug. 1994 Imaging sediments under salt: Where are we? The Leading Edge 13 834-836

602 Jackson, M. P. A., Vendeville, B. C., and Schultz-ela, D. D. Aug. 1994 Salt-related structures in the Gulf of Mexico: A field guide for geophysicists The Leading Edge 13 837-842

603 House, W. M., and Pritchett, J.A. Aug. 1994 Salt deformation modeling through the use of enhanced seismic imaging techniques The Leading Edge 13 844-848

604 Hodgkins, M. A., and O'Brien, M. J. Aug. 1994 Salt sill deformation and its applications for subsalt exploration The Leading Edge 13 849-851

605 Lee, S., and House-Finch, N. Aug. 1994 Imaging alternatives around salt bodies in the Gulf of Mexico The Leading Edge 13 853-857

606 MacKay, S., and Dragoset, B. Aug. 1994 Imaging beneath a Gulf Coast salt-injection feature: A processing case history The Leading Edge 13 858-860

607 Wyatt, K.D., Wyatt, S.B., Towe, S. K., Layton, J. E., von Seggern, D. H., and Brockmeier, C. A. Aug. 1994 Building veloc-

References

ity-depth models for 3-D depth migration The Leading Edge 13 862-866
608 Abrial, W. L., and Wright, R. M. Aug. 1994 The shape of Gulf Coast salt intrusions related to seismic imaging The Leading Edge 13 868-872
609 Lewis, G. G., Young, K. T., Finn, C. J., and Schneider, W. A. Aug. 1994 Analysis if subsalt reflections at a Gulf of Mexico salt sheet through 3-D depth migration and 3-D seismic modeling The Leading Edge 13 873-878
610 Starich, P. J., Lewis, G. G., Faulkner, J., Standley, P. G., and Setterquist, S. Aug. 1994 Integrated geophysical study of an onshore salt dome The Leading Edge 13 880-884
611 Krail, P. M. Aug. 1994 Vertical cable as a subsalt imaging tool The Leading Edge 13 885-887
612 Zhu, X. Aug. 1994 Survey parameters for imaging salt domes The Leading Edge 13 888-892
613 Durrani, T. S., and Bisset, D. Aug. 1994 The Radon transform and its properties Geophysics 49 1180-1187
614 Rowbotham, P. S. Jul. 1997 Anisotropic migration of coincident VSP and cross-hole seismic reflection surveys Geophys. Prosp. 45 683-699
615 Al-Chalabi, M. Jul. 1997 Time-depth relationships for multilayer depth conversions Geophys. Prosp. 45 715-720
616 Akbar, F. E., Sen, M., and Stoffa, P. L. Sep. 1996 Prestack plane-wave Kirchhoff migration in laterally varying media Geophysics 61 1068-1079
617 Claerbout, J. F. May. 1996 Basic Earth Imaging (Version 2.2): A Living Document Stanford University 2 178
618 Li, X., and Bancroft, J. C. Nov. 1997 Integrated residual statics analysis with prestack time migration Exp. Abs. SEG 1997 1452-1455
619 Fowler, P. J. Nov. 1997 A comparative overview of prestack time migration methods Exp. Abs. SEG 1997 1571-1574
620 Bancroft, J. C., Margrave, G., and Geiger, H. D. Nov. 1997 A kinematic comparison of DMO-PSI and equivalent offset migration (EOM) Exp. Abs. SEG 1997 1575-1578
621 Li, X., and Bancroft, J. C. Nov. 1997 Converted wave migration and common conversion point binning by equivalent offset Exp. Abs. SEG 1997 1587-1590
622 Bancroft, J. C., Geiger, H. D., Foltinek, D., and Wang, S. May. 1995 Prestack migration by equivalent offsets and common scatterpoint (CSP) gathers Exp. Abs. EAEG P124
623 Bancroft, J. C., and Wang, S. Jun. 1996 Converted wave (P-S) prestack migration and velocity analysis using equivalent offsets Exp. Abs. EAGE X023
624 Geiger, H. D., and Bancroft, J. C. Jun. 1996 Equivalent offset prestack migration - application to rugged topography and formation of unmigrated zero-offset section Exp. Abs. EAGE P098
625 Bancroft, J. C. Nov. 1996 Natural anti-aliasing in equivalent offset prestack migration Exp. Abs. SEG 1465-1466b
626 Wang, S., Bancroft, J. C., and Lawton, D.C. Nov. 1996 Converted-wave (P-SV) prestack migration and migration velocity analysis Exp. Abs. SEG 1575-1578
627 Margrave, G. F., and Bancroft, J.C. Nov. 1996 The theoretical basis for prestack migration by equivalent offset Exp. Abs. SEG 433-446
628 Geiger, H. D., and Bancroft, J. C. Nov. 1996 Equivalent offset prestack migration for rugged topography Exp. Abs. SEG 447-450
629 Li X., and Bancroft, J. C. May. 1997 Residual statics analysis before NMO using prestack migration Exp. Abs. SEG A032
630 Bancroft, J. C. May. 1997 Fast 3-D Kirchhoff poststack migration with migration velocity analysis Exp. Abs. SEG A046
631 Geiger, H. D., and Bancroft, J. C. May. 1996 Forming an unmigrated zero-offset stacked section with no NMO using equivalent offset prestack migration Exp. Abs. CSEG Nat. Conv. 147-148
632 Bancroft, J. C., and Geiger, H. D. May. 1996 Velocity sensitivity for equivalent offset prestack migration: a contrast in robustness and fragility Exp. Abs. CSEG Nat. Conv. 149-150
633 Wang, S. Aug. 1997 Three-component and Three-dimensional seismic imaging M.Sc. Thesis 1-83
634 Bancroft, J. C., Wang, S., Foltinek, D., and Geiger, H. D. May. 1995 Common scatter point (CSP) prestack migration Exp. Abs. CSG Nat. Conv. 47-48
635 Geiger, H. D., Bancroft, J. C., and Foltinek, D. May. 1995 Prestack migration of crustal data using common scatter point (CSP) migration technique Exp. Abs. CSG Nat. Conv. 49-50
636 Wang, W., Bancroft, J. C., Foltinek, D., and Lawton, D. May. 1995 Converted-wave (P-SV) prestack migration and velocity analysis. Exp. Abs. CSG Nat. Conv. 67-68
637 Hale, D Jan. 1992 Dip Moveout Processing SEG Course notes
638 Aminzadeh, F., Brac, J., and Kunz, T. Jan. 1997 SEG/EAGE 3-D Salt and Overthrust Models SEG/EAGE 2-CD's
639 Bancroft, J. C., Geiger, H. D., and Margrave, G. F. Nov. 1998 The equivalent offset method of prestack time migration Geophysics 2042-2053
640 Margrave. G. F., Bancroft, J.C., and Geiger, H. D. Jan. 1999 Fourier prestack migration by equivalent wavenumber Geophysics 197-207
641 Grech, G. K. May. 1998 Numerical seismic modeling of exposed structures UofC 701/CREWES/FRP 1-30
642 Ratcliff, D. W., Jacewitz, C. A., and Gray, S. H. Mar. 1994 Subsalt imaging via target-orientated 3-D prestack depth migration The Leading Edge 13 163-170
643 Bancroft, J. C., and Geiger, H. D. May. 1997 Anatomy of CSP gathers formed during equivalent offset migration Exp. Abs. CSEG Nat. Conv.
644 Li, X., and Bancroft, J. C. Dec. 1997 Residual statics using CSP gathers CREWES Research Report 9 Ch 23
645 Li, X., and Bancroft, J. C. Dec. 1997 A new algorithm for converted wave prestack migration CREWES Research Report 9 Ch 26
646 Bancroft, J. C., and Geiger, H. D. Dec. 1997 Anatomy of common scatter point CSP gathers formed during equivalent offset prestack migration (EOM) CREWES Research Report 9 Ch 28
647 Bancroft, J. C., Margrave, G. F., and Geiger, H. g. Dec. 1997 A kinematic comparison of conventional processing, DMO-PSI, and EOM CREWES Research Report 9 Ch 29
648 Li, X., Xu, Y., and Bancroft, J. C. Dec. 1997 Equivalent offset migration: the implementation and application update CREWES Research Report 9 Ch 27
649 Rodriguez-Suarez, C., Stewart, R. R., Li, X., and Bancroft, J. C. Dec. 1997 Analyzing 4-C marine data from the Valhall field, Norway CREWES Research Report 9 Ch 42
650 Thompsen, L. A., Barkved, O. I., Haggard, B., Kommedal, J. H., and Rosland, B. May. 1997 Converted-wave imaging of Valhall Reservoir Exp. Abs. EAGE B048-V1
651 Mao, W., and Stuart, G. w. Jan. 1997 Rapid multi-wave-type ray tracing in complex 2-D and 3-D isotropic media Geophysics 62 298-308
652 Zhang, J., and McMechan, G. A. Jan. 1997 Turning wave migration by horizontal extrapolation Geophysics 62 291-297

653 Schleicher, J., Hubral, P., Tygel, M., and Jaya, M. S. Jan. 1997 Minimum apertures and Fresnel zones in Migration and demigration Geophysics 62 183-194

654 Audebert, F., Nichols, D., Rekdal, T., Biondi, B., Lumley, D. E., and Urdaneta, H. Sep. 1997 Imaging complex geologic structure with single-arrival Kirchhoff prestack depth migration Geophysics 62 1533-1543

655 Wu, H., and Lees, J. M. Sep. 1997 Boundary conditions on a finite grid: Applications with psuedospectral wave propagation Geophysics 62 1544-1557

656 Bording, P. Jan. 1995 Seismic Wave Modeling and Inversion www.arsc.sunyit.edu/csep/

657 Tarantola, A. Jan. 1989 Theoretical background for the inversion of seismic waveforms, including elasticity and attenuatuin Digital Seismology..., Plenum Press 157-190

658 Canning, A., and Gardner, G. H. Mar. 1996 A two-pass approximation to 3-D prestack migration Geophysics 61 409-421

659 Rajasekaran, S., and McMechan Jul. 1996 Tomographic estimation of the spatial distribution of statics Geophysics 61 1198-1208

660 Gaiser, J. E. Jul. 1996 Multicomponent Vp/Vs correlation analysis Geophysics 61 1137-1149

661 Neidell, N. S. Jul. 1997 Perceptions in seismic imaging Part 1: Kirchhoff migration operators in space and offset time, an appreciation The Leading Edge 16 1005-1006

662 Ross, C. P. Feb. 1997 AVO and nonhyperbolic moveout: a practical example First Break 15 43-48

663 Bickel, S. H. Jun. 1998 The maximum depth principle as a dual to Fermat's minimum time principle Exp. Abs. EAGE 1-55

664 Payne, M. A., Eriksen, E. A., and Rape, T. D. Mar. 1994 Considerations for high-resolution VSP imaging The Leading Edge 13 173-180

665 Wang, S., Bancroft, J. C., Foltinek, D., and Lawton, D.C. May. 1995 Converted-wave (P-SV) prestack migration and migration velocity analysis Exp. Abs. CSEG 67-68

666 Popovici, M., and Sethian, J., Jun. 1998 Three dimensional traveltimes using fast marching method Exp. Abs. EAGE 1-22

667 Sun, J., and Gajewski, D. Aug. 1997 True-amplitude common-shot migration revisited Geophysics 62 1250-1259

668 Koehler, F., and Taner, M. T. Oct. 1977 Direct and inverse problems relating reflection coefficients and reflection response for horizontally layered media Geophysics 42 1199-1206

669 Biondi, B., Fomel, S., and Chemingui, N. Mar. 1998 Azimuth moveout for 3-D prestack imaging Geophysics 63 574-588

670 Anderson, J. E., and Tsvankin, I. Jul. 1997 Dip-moveout processing by Fourier transform in anisotropic media Geophysics 62 1260-1269

671 Apostoiu-Marin, I., and Ehinger, A. Jul. 1997 Kinematic interpretation in the prestack depth-migrated domain Geophysics 62 1226-1237

672 Wapenaar, C. P. A Jul. 1997 3-D migration of cross-spread data: Resolution and amplitude aspects Geophysics 62 1220-1225

673 Alkhalifah, T. Jul. 1997 Kinematics of 3-D DMO operators in transversely isotropic media Geophysics 62 1214-1219

674 Marfurt, K. J., Scheet, R. M., Sharp, J. A., and Harper, M. G. May. 1998 Suppression of the acquisition footprint for sequence attribute mapping Geophysics 63 1024-1035

675 Sun, J. May. 1998 On the limited aperture migration in two dimensions Geophysics 63 984-994

676 Nichols, D., Farmer, p., and Palacharla, G. Jun. 1998 Improved prestack depth migration by novel raypath selection Exp. Abs. EAGE 1-54

677 Bancroft, J. C., and Geiger, H. D. May. 1996 Velocity sensitivity for equivalent offset migration: a contrast in robustness and fragility Exp. Abs. CSEG 149-150

678 Wang, C. Oct. 1995 DMO in radon domain Exp. Abs. SEG 1441-444

679 Wang, C. Nov. 1996 Radon DMO amplitude and frequency preservation Exp. Abs. SEG 1479-1482

680 Fowler, P. Dec. 1984 Velocity independent imaging of seismic reflectors Exp. Abs. SEG 383-385

681 Crase, E. Sep. 1990 High-order (space and time) Finite-difference modeling of the elastic wave equation Exp. Abs. SEG 987-991

682 Alfaraj, M. Oct. 1992 Transformation to zero offset for mode-converted waves by Fourier transform Exp. Abs. SEG 974-978

683 Samson, J. C., and Olson, J. V. Oct. 1981 Data-adaptive polarization filters for multichannel geophysical data Geophysics 46 1423-1431

684 Blacquiere, G., Duijndam, A. J. W., and Romijn, R. Jan. 1991 Efficient x-f depth migration of shot records: practical aspects First Break 9 9-23

685 Berkhout, A. J., and Zaanen, P. R. Jan. 1976 A comparison between Wiener filtering, Kalman filtering, and deterministic least squares estimation Geophys. Prosp. 24 141-197

686 Hilterman, F. J. Oct. 1975 Amplitudes of seismic waves - a quick look Geophysics 40 745-762

687 Martin, N. W., Donati, M. S., and Bancroft, J. C. Jun. 1996 Steep dip Kirchhoff migration for linear velocity gradients Exp. Abs. EAGE X024

688 Fruhn, J., White, R. S., Fliedner, M., Richardson, K. R., Cullen, E., Latkiewicz, C., Kirk, W., and Smallwood, J. R. Jun. 1998 Two-ship large aperture seismic profiles - application to imaging through basalt Exp. Abs. EAGE 1-47

689 Emsley, D., Boswell, P., and Davis, P. Jun. 1998 Sub-basalt imaging using long offset reflection seismic data Exp. Abs. EAGE 1-48

690 Fowler, P. J. Jun. 1998 A comparative overview of prestack time migration Exp. Abs. EAGE 1-17

691 Margrave, G. F. Dec. 1997 Zero offset seismic resolution theory for linear V(z) CREWES Research Report Ch 1

692 Lynn, H. B., and Deregowski, S. Oct. 1981 Dip limitations on migrated sections Geophysics 46 1392-1397

693 Middleton, D. Jan. 1960 An Introduction to Statistical Communication Theory McGraw-Hill

694 Ng, M. Dec. 1996 3-D prestack phase shift migration of shot records Can. J. of Expl. Geophys. 32 130-138

695 Ng, M. May. 1997 Compensating 3-D prestack and poststack phase shift migration errors due to X-Y splitting Exp. Abs. CSEG 195-195

696 Sams, M., and Goldberg, D. Jan. 1990 The validity of Q estimates from borehole data using spectral ratios Geophysics 55 97-101

697 Fowler, P. J. Sep. 1998 A comparative overview of dip moveout methods Exp. Abs. SEG 1744-1747

698 Tatham, R. H. Jan. 1985 Vp/Vs interpretation, Chap. 5 of Developments in Geophysical exploration methods Elsevier App. Sc. Pub. 139-188

699 Chan, Wai-Kin Oct. 1998 Analysing Converted-Wave Seismic Data: Statics, Interpolation, Imaging, and P-P Correlation Ph.D. Thesis, University of Calgary

700 Bancroft, J. C. Dec. 1998 Equivalent offset migration for vertical receiver arrays CREWES Research report 10 Ch 11

701 Bancroft, J. C. Dec. 1998 Computational speed of EOM relative to standard Kirchhoff migration CREWES Research report 10 Ch 43

References

702 Bancroft, J. C. Dec. 1998 Optimum CSP gathers with a fixed equivalent offset CREWES Research report 10 Ch 44

703 Grech, M. G.K., and Bancroft, J. C. Dec. 1998 Building complex velocity models using EOM - a case study CREWES Research report 10 Ch 27

704 Bancroft, J. C. Dec. 1998 Dip limits on pre- and poststack Kirchhoff migrations CREWES Research report 10 Ch 25

705 Wang, C. Jan. 1999 Dip Moveout in Radon Domain Geophysics 64 278-288

706 Liner, C. L. Nov. 1996 Bin Size and Linear V(z) Exp. Abs. SEG 66 47-50

707 Chernis, L. Sep. 1998 Depth migration by generalised equivalent migration offset mapping Exp. Abs. SEG 1550-1553

708 Kirtland Grech, M. G., Lawton, D. C. and Spratt, D. A. Nov. 1998 Numerical Seismic Modelling and Imaging of Exposed Structures FRP Research Report 4 1.1-1.32

709 Isaac, J. H., and Lawton, C. Nov. 1998 Multicomponent anisotropic physical modelling FRP Research Report 4 9.1-9.18

710 Bancroft, J. C., and Vestrum. R. W. May. 1999 Anisotropic depth migration using the equivalent offset method (EOM) EXP. Abs. CSEG 138-141

711 Li, X. Jun. 1999 Residual statics analysis using prestack equivalent offset migration M.Sc. Thesis 1-123

712 Margrave, G. F. Nov. 1997 Seismic acquisition parameter considerations for a linear velocity medium Exp. Abs. SEG 67 47-50

713 Liner, C. L. Nov. 1997 3-D seismic survey design and linear v(z) Exp. Abs. SEG 67 43-46

714 Bancroft, J. C., and Xu, L. May. 1998 Equivalent offset migration for VSP's and data acquired with vertical receiver arrays Exp. Abs. CSEG 36-39

715 Bancroft, J. C., and Xu, L. Jun. 1999 Prestack migration of vertical array data using EOM Exp. Abs. EAGE 2-47

716 Scales, J. A. May. 1994 Theory of Seismic Imaging Samizdat Press (CWP) Colorado School of Mines 1-204

717 Dellinger, J., Gray, S., Murphy, G., Etgen, J., and Fei, T., Nov. 1999 Efficient two and one-half dimension true-amplitude migration Exp. Abs., SEG Nat. Conv. 69 1540-1543

728 Margrave, G. F., and Ferguson, R. J. Aug. 1999 Wavefield extrapolation by nonstationary phase shift Geophysics 64 1067-1078

729 Margrave, G. F., and Yao, Z. Dec. 1999 Imaging from topography with Fourier methods CREWES Research report 11 689-698

730 Lines, R. L. Slawinski, R., and Bording, R. P. May. 1999 A recipe for stability of finite-difference wave-equation computations Geophysics 64 967-969

731 Hagedoorn, J. G. Jan. 1954 A practical example of an anisotropic velocity-layer Geophys. Prosp. 2 52-60

732 Kleyn. A. H. Jan. 1956 On seismic wave propagation in an anisptropic media with applications in the Betun area, South samatra Geophys. Prosp. 4 57-69

733 De-Segonzac, P. D, and Laherrere, j, Mar. 1959 Applications of the continuous velocity log to anisotropy measurements in northern Sahara-results and consequenses Geophys. Prosp. 7 202-217

734 Thompsen. L. Oct. 1986 Weak elastic anisotropy Geophysics 51 1954-1966

735 Hodgkinson, J. May. 1970 The in situ determination of velocity anisotropy in sedimentary rocks bt the analysis os normal moveout data Masters Thesis, Ucalgary

736 Tygel, M., Schleicher, J., and Hubral. P. Oct. 1994 Pulse distortion in depth migration Geophysics 59 1561-1569

737 Barnes, A. E. Dec. 1995 Discussion: Pulse distortion in depth migration Geophysics 60 1942-1944

743 Geiger, H. D. Dec. 2001 Migration Ph.D. Thesis

744 Bleistein, N. Aug. 1999 Hagedoorn told us how to do Kirchhoff migration and inversion TLE 981-927

745 Bleistein, N., Cohen, J. K., and Stockwell, J. W. Jan. 2001 Mathematics of Multidimensional Seismic Imaging, Migration, and Inversion Springer

746 Schleicher, J., Hubral, p., and Tygel, M. Aug. 1991 Nonspecular reflections from a curved interface Geophysics 56 1203-1214

747 Gray, S. Aug. 1999 True-amplitude migration TLE 917

748 Bleistein, N. Jan. 1984 Mathematical Methods for Wave Phenomena Academic Press

749 Berkhout, A. J. Aug. 1999 Seismic inversion in steps TLE 933-939

750 Weglein, A. B., and Stolt, R. H. Aug. 1999 Migration-inversion revisited TLE 950-975

751 Bevc, D. Sep. 1997 Flooding the topography: Wave-equation datuming of land data with rugged acquisition topography:, 62, no. 05, Geophysics 62 1558-1569

752 Sayers, C. M., and Ebrom, D. A. Sep. 1997 Seismic traveltime analysis for azimuthally anisotropic media: Theory and experiment Geophysics 62 1570-1582

753 Bancroft, J. C., and Ursenbach, C. p. Sep. 2001 Prestack considerations for the migration of oblique reflectors Exp. Abs. SEG

754 Lindseth, R. O Jan. 1979 Synthetic sonic logs — a process for stratigraphic interpretation Geophysics 44 3-26

755 Margrave, G. F., Jan. 1998 Theory of non-stationary linear filtering in the Fourier domain with applications to time-variant filtering Geophysics 63 255-259

756 Margrave, G. F., Sep. 2001 Direct Fourier migration for vertical velocity variations Geophysics 66 1504-1514

757 Whitmore, N. D., Sep. 1983 Iterative depth migration by backward time propagation, 53rd Ann. Internat. Mtg: Soc. of Expl. Geophys., S10.1

758 Gray, S. H., Etgen, J., Dellinger, J. and Whitmore, D., Jan. 2001 Seismic migration problems and solutions Geophysics 66 1622-1640

759 O'Brien, M., Etgen, J., Murphy, G. and Whitmore, N. D. Jan. 1999 Multicomponent imaging with reciprocal shot records 69th Ann. Internat. Mtg: Soc. of Expl. Geophys. 784-787

760 Ulrych, T. J., Sacchi, M. D., and Woodbury, A. Jan. 2001 A Bayes tour of inversion: A tutorial Geophysics 66 55-69

761 Reshef, M. Jan. 2001 Some aspects of interval velocity analysis using 3-D depth-migrated gathers Geophysica 66 261-266

762 Yuon, O. K., and Zhou, H. Jan. 2001 Depth imaging with multiples Geophysics 66 246-255

763 Zheludev, V. A., Ragoza, E., and Kosloff, D. D. Mar. 2001 Fast Kirchhoff migration in the wavelet domain Expl. Geoph. (ASEG) 33 23-27

764 Gruber, T., Greenhalgh, S., and Zhou, B. Mar. 2001 A later-arrival based inversion scheme to recover diffractors and reflectors Expl. Geoph. (ASEG) 33 1-10

765 Wang, y. Apr. 2002 A stable and efficient approach of inverse Q filtering Geophysics 67 657-663

766 Ferguson, R. J., and Margrave, G.F. Mar. 2002 Prestack depth migration by symmetric nonstationary phase shift Geophysics 67 594-603

767 Gesbert, S. May. 2002 From acquisiton footprint to true amplitude Geophysics 67 830-839

768 Biondi, B. May. 2002 Stable wide-angled Fourier finite-difference3 downward extrapolation of 3-D wavefields Geophysics 67 872-882

769 Rickett, J. E., and Sava, P. C. May. 2002 Offet and angle-domain common image-point gathers for shot-profile migration

770 Stolt, R. H. May. 2002 Seismic data mapping and reconstruction Geophysics 67 883-889
770 Stolt, R. H. May. 2002 Seismic data mapping and reconstruction Geophysics 67 890-908
771 Grechka, V., and Tsvankin, I. May. 2002 NMO-velocity surfaces and Dix-type formulas in anisotropic heterogeneous media Geophysics 67 939-951
772 Alkhalifah, T., and Tsvankin, I. Sep. 1995 Velocity analysis for transversly isotropic media Geophysics 60 1550-1566
773 Tsvankin, I., and Thomsen, L. Jul. 1995 Inversion of reflection traveltimes for transverse isotropy Geophysics 60 1095-1107
774 Schneider Jr., W. A. Jul. 1995 Robust and efficient upwind finite-difference traveltime calculations in three dimensions Geophysics 60 1018-1117
775 Igel,H., Mora, P., and Riollet, B. Jul. 1995 Anisotropic wave propagation through finite-difference grids Geophysics 60 1302-1216
776 Hovem, J. M. Jul. 1995 Acoustic waves in finely layered media Geophysics 60 1217-1221
777 Bruin, C. G. M., Wapenaar, C. P. A., and Berkhout, A. J. Sep. 1990 Angle-dependent reflectivity by means of prestack migration Geophysics 55 1223-1234
778 Dankbaar, J. W. M. Nov. 1985 Separation of P- and S-waves Geophys. Prosp. 33 970-986
779 Eaton, D. W. S. Dec. 1989 The free surface effects:implications for amplitude-verses-offset inversion Cdn. Jnl. Exp. Geo. 25 97-103
780 Foster, D. J., and Carrion, P. M. Nov. 1985 Full wave equation downward continuation of seismic reflection data Geophys. Prosp. 33 929-942
781 Margrave, G. F. Jan. 2001 Numerical Methods of Siesmology with Algorithms in MATLAB www.crewes.org
782 Zhu, J., and Lines, L. R. Jul. 1998 Comparison of Kirchhoff and reverse-time migration methods with applications to prestack depth imaging of complex structures Geophysics 63 116-1176
784 Cerveny, V. Jan. 1972 Seismic rays and ray intensities in inhomogeneous anisotropic media Geophysics 28 1-13
785 Langan, R. T., Lerche, I., and Cutler, R. T. Sep. 1985 Tracing of rays through heterogeneous media — An accurate and efficient proceedure Geophysics 50 1456-1465
786 Zhao, P., Uren, N. F., Wenzel, F., Hatherly, P. J., and McDonald, J. A. Dec. 1998 Kirchhoff diffraction mapping in media with large veloicty contrasts Geophysics 63 2072-2081
787 Liu, Z. Sep. 1997 An analytical approach to migration velocity analysis Geophysics 62 1238-1249
788 Robinson, W. B. Jan. 1957 The need for seismic dip migration Reprint, Migration of Seismic Data SEG 4 51-56
789 Rockwell, D. W. Jan. 1971 Migration Stack aids interpretation Reprint,Migration of Seismic Data SEG, from Oil and Gas Journal 75-81
790 Reksnes, P. A., Haugane, E., and Hegna, S. Dec. 2002 How PGS created a new image for the Varg field First Break 20 773-777
791 Reshef, M. Keydar, S., and Landa, E. Mar. 2003 Multiple prediction without prestackdata: an efficient tool for interpretive processing First Break 21 29-37
792 Reksnes, P.A., and Haugane, E., Dec. 2002 How PGS created a new image for the Varg field First Break 20 773-781
793 Press, W.H., Teukolsky, S.a., Vetterling, W.T., and Flannery, B.P., Jan. 1992 Numerical Recipies in Fortran 77 (also "C") Cambridge University Press 1
794 Krebes, E. S. Jul. 2004 Seismic Theory and Methods GOPH551 Course notes UC